QUANTUM MECHANICAL ELECTRONIC STRUCTURE CALCULATIONS WITH CHEMICAL ACCURACY

Understanding Chemical Reactivity

Volume 13

The titles published in this series are listed at the end of this volume.

Quantum Mechanical Electronic Structure Calculations with Chemical Accuracy

edited by

Stephen R. Langhoff

NASA Ames Research Center,
Moffett Field, California, U.S.A.

SPRINGER-SCIENCE+BUSINESS MEDIA, B.V.

Library of Congress Cataloging-in-Publication Data

Quantum mechanical electronic structure calculations with chemical accuracy / edited by Stephen R. Langhoff.
p. cm. -- (Understanding chemical reactivity ; v. 13)

1. Quantum chemistry. I. Langhoff, Stephen R. II. Series.
QD462.Q347 1995
541.2'8--dc20 94-39289

DOI 10.1007/978-94-011-0193-6

Printed on acid-free paper

Contents

Exact Quantum Chemistry by Monte Carlo Methods

James B. Anderson
Department of Chemistry, The Pennsylvania State University, 152 Davey Laboratory, University Park, Pennsylvania 16802

June-1994

Abstract

With Monte Carlo methods it is now possible to solve the Schrödinger equation for systems of a few electrons without systematic error. The first quantum calculation to achieve an absolute accuracy of 1.0 microhartree for a polyatomic system was a quantum Monte Carlo calculation for the H_3^+ molecular ion. The first to achieve an absolute accuracy of 0.01 kcal/mole for the H-H-H system was a Monte Carlo calculation for the barrier for the reaction $H + H_2 \rightarrow H_2 + H$. The first to achieve an absolute accuracy of 0.1 K for the stable dimer He-He was a quantum Monte Carlo calculation. In this chapter we review the exact quantum Monte Carlo method, its successful applications thus far to systems of a few electrons, and its prospects for successful applications to larger systems.

S. R. Langhoff (Ed.), Quantum Mechanical Electronic Structure Calculations with Chemical Accuracy, 1–45.

Contents

1 Introduction

The theoretical chemist is accustomed to judging the success of a theoretical prediction according to how well it agrees with an experimental measurement. Since the object of theory is the prediction of the results of experiment, that would appear to be an entirely satisfactory state of affairs. However, if it is true that "the underlying physical laws ... for the whole of chemistry are ... completely known" (1), then it should be possible, at least in principle, to predict the results of experiment more accurately than they can be measured. If the theoretical chemist could obtain exact solutions of the Schrödinger equation for many-body systems, then the experimental chemist would soon become accustomed to judging the success of an experimental measurement by how well it agrees with a theoretical prediction.

In fact, it is now possible to obtain exact solutions of the Schrödinger equation for systems of a few electrons(2-8). These systems include the molecular ion H_3^+, the molecule H_2, the reaction intermediate H-H-H, the unstable pair H-He, the stable dimer He_2, and the trimer He_3. The quantum Monte Carlo method used in solving the time-independent Schrödinger equation for these systems is exact in that it requires no physical or mathematical assumptions beyond those of the Schrödinger equation. As in most Monte Carlo methods there is a statistical or sampling error which is readily estimated.

We use the term 'exact' as it has been described by Zabolitzky(9) in the sense that $2\pi r$ is the 'exact' circumference of a circle of radius r. The numerical value of π is not known exactly but may be computed to any precision desired and the same is true for the circumference of a circle of a given r. A calculation method is 'exact' if there exist finite amounts of computer time such that a result may be determined with any arbitrary precision.

The exact solution of the Schrödinger equation for many-electron systems is one of the Grand Challenges to Computational Science (10). More specifically, the challenge is that of overcoming 'the node problem' or 'the sign problem in quantum Monte Carlo' (11,12). This challenge has now been met for systems of a few electrons.

In this chapter we review the exact quantum Monte Carlo method, its successful applications thus far to systems of a few electrons, and the prospects for successful applications to larger systems. Approximate methods applicable to larger systems are mentioned, but the emphasis is on 'exact' quantum chemistry.

2 Overview of Quantum Monte Carlo

There are at least a dozen different Monte Carlo methods which might be used for solving the Schrödinger equation, but only three of these are currently popular in calculations of electronic structure. The three are collectively referred to as quantum Monte Carlo methods (QMC). They are the variational quantum Monte

Carlo method (VQMC), the diffusion quantum Monte Carlo method (DQMC), and the Green's function quantum Monte Carlo method (GFQMC).

The variational method is the same as conventional analytic variational methods except that the integrals required are evaluated by Monte Carlo procedures. The first applications in quantum chemistry were those by Conroy(13) in the 1960's. The expectation value of the energy is determined for a trial function Ψ_T using Metropolis sampling based on Ψ_T^2. The expectation value $< E >$ is given by

$$< E >= \frac{\int \frac{H\Psi_0}{\Psi_0} {\Psi_0}^2 d\vec{X}}{\int {\Psi_0}^2 d\vec{X}} \cong \frac{\sum_{i=1}^{N} \frac{H\Psi_0}{\Psi_0}}{\sum_{i=1}^{N} 1} \quad , \tag{1}$$

where the summations are for samples of equal weights selected with probabilities proportional to Ψ_T^2. As in analytic variational calculations the expectation value $< E >$ is an upper limit to the true value of the energy E,

$$< E > \;\geq\; E \;. \tag{2}$$

The term $\frac{H\Psi_T}{\Psi_T}$ is a local energy E_{loc}. In determining $< E >$ it is not necessary to carry out analytic integrations; and, since only differentiation of the trial wave function is required to evaluate the local energy, the trial wave function may take any desired functional form. It may even include inter-electron distances r_{ij} explicitly. Thus, relatively simple trial functions may incorporate electron correlation effects rather accurately and produce expectation values of the energy well below those of the Hartree-Fock limit. Except in the limit of a large number of terms the VQMC method is not an exact method.

The diffusion and Green's function methods are based on the similarity of the Schrödinger equation and the diffusion equation and the use of a random walk process to simulate diffusion. The relation of the Schrödinger equation to the diffusion equation and to diffusion was noted as early as 1932 by Wigner(14), and a random walk simulation of the Schrödinger equation was discussed by Metropolis and Ulam(15) and by King(16) in 1949.

The time-independent Schrödinger equation, $H\Psi = E\Psi$, for a single particle is given by

$$\frac{\hbar^2}{2m}\nabla^2\Psi(\vec{X}) - V(\vec{X})\Psi(\vec{X}) = -E\Psi(\vec{X}) \quad , \tag{3}$$

where the potential energy V is a function of position. The equation has as solutions the eigenfunctions $\Psi(\vec{X})$ together with their eigenvalues E_i.

In the solution of partial differential equations it is in some cases easier to solve the corresponding transient equation than to solve a steady-state equation directly. If we introduce an imaginary time τ and define the transient function

$$\Psi(\vec{X},\tau) = \Psi(\vec{X})e^{-E\tau} \quad , \tag{4}$$

$\Psi(\vec{X},\tau)$ obeys the relations

$$\frac{\partial\Psi(\vec{X},\tau)}{\partial\tau} = -E\Psi(\vec{X},\tau) \tag{5}$$

and

$$\frac{\partial\Psi}{\partial\tau} = \frac{\hbar^2}{2m}\nabla^2\Psi - V\Psi \ . \tag{6}$$

The wave function $\Psi(\vec{X},\tau)$ consists of a spatial function $\Psi(\vec{X})$ changing exponentially with time. The function $\Psi(\vec{X})$ is a solution to the time-independent equation. There is an infinite number of solutions having energies higher than the lowest-energy or ground-state solution. If an initial function contains components of the ground-state wave function and higher-state wave functions the more rapid exponential decay of the higher-energy wave functions, as indicated by Eq. (4), ensures that the solution at large values of τ is the lowest-energy solution.

Thus, we have

$$\Psi(\vec{X},0) = \sum_i c_i\Psi_i(\vec{X}) \ , \tag{7}$$

$$\Psi(\vec{X},\tau) = \sum_i c_i\Psi_i(\vec{X})e^{-E_i\tau} \ , \tag{8}$$

$$\Psi(\vec{X},\tau\to\infty) = c_0\Psi_0(\vec{X})e^{-E_0\tau} \ , \tag{9}$$

where $\Psi_0(\vec{X})$ is the ground-state wave function.

The Schrödinger equation in imaginary time, Eq. (6), is identical in form to Fick's diffusion equation to which a first-order reaction rate term is added,

$$\frac{\partial C(\vec{X},t)}{\partial t} = D\nabla^2 C(\vec{X},t) - kC(\vec{X},t) \ . \tag{10}$$

The concentration C is equivalent to the wave function Ψ, the diffusion coefficient D is equivalent to the group $\frac{\hbar^2}{2m}$, and the rate constant k is equivalent to the potential energy V. In the language of mathematics the two equations are 'isomorphic'.

The diffusion of molecules occurs physically by a random walk process. The molecules are subject to random motion, moving in a time interval Δt a distance Δx in one direction or the other in each of the three dimensions. The time and distance steps are related to the diffusion coefficient by the Einstein equation(17),

$$D = \frac{\overline{(\Delta x)^2}}{2\Delta\tau} \ . \tag{11}$$

Usually, the diffusion equation is used to simulate a random walk process (namely diffusion) in a physical system. In quantum Monte Carlo the procedure is reversed: the random walk process is used to simulate the differential equation.

We describe first the diffusion quantum Monte Carlo method. The DQMC procedure for solving the Schrödinger equation in one dimension is a repeated sequence of operations. An initial collection of particles (hereafter, psi particles or psips) is distributed on a line. Time is advanced one step Δt and each psip is moved right or left one step Δx. Then each psip gives birth to a new psip with a probability $P_b = -V\Delta t$ if V is negative or disappears with a probability $P_d = V\Delta t$ if V is positive. Time is advanced another step and the process is repeated. With increasing time and number of iterations the normalized distribution of psips approaches a 'steady-state' distribution fluctuating about an average steady-state distribution, which corresponds to the lowest-energy wave function satisfying the time-independent Schrödinger equation.

The Green's function quantum Monte Carlo method is similar to the DQMC method, but takes advantage of the properties of Green's functions useful in solving the steady-state diffusion equation. The GFQMC method was proposed first by Kalos(18) for nodeless systems and we describe it here for nodeless systems. The challenge of systems with nodes is discussed in a later section.

The time-independent Schrödinger equation, Eq. (6), may be written in the form

$$-\nabla^2\Psi(\vec{X}) + k^2\Psi(\vec{X}) = k^2\frac{V(\vec{X})}{E}\Psi(\vec{X}) \ , \tag{12}$$

where

$$k^2 = -\frac{2mE}{\hbar^2} \ . \tag{13}$$

Since k^2 must be positive the energy must be negative, but this can be accomplished by adjusting the reference or zero of the potential energy.

The Green's function for Eq. (12) which satisfies the boundary conditions for a problem in electronic structure (i.e., $\Psi \to 0$ as $X \to \infty$) is known and is given by

$$G(\vec{X},\vec{X}') = \frac{1}{(2\pi)^{\frac{3N}{2}}} K_{\frac{3N}{2}-1}(k|\vec{X}-\vec{X}'|)/(k|\vec{X}-\vec{X}'|)^{\frac{3N}{2}-1}, \tag{14}$$

where K_ν is the modified Bessel function of the second kind. Sampling is based on that property of the Green's function which reproduces the wave function from itself,

$$\Psi(\vec{X}) = \int d\vec{X}' G_o(\vec{X},\vec{X}')\frac{V(\vec{X}')}{E}\Psi(\vec{X}') \ . \tag{15}$$

Repeated application of Eq. (15) to an initially arbitrary wave function $\Psi(\vec{X}')$ produces a wave function $\Psi(\vec{X})$ which is the lowest-energy solution to the Schrödinger equation for the boundary conditions specified. A psip in the distribution $\Psi(\vec{X}')$ may be transferred to the distribution $\Psi(\vec{X})$ by multiplying its weight by $\frac{V(\vec{X}')}{E}$, sampling the Green's function distribution $G_o(\vec{X}, \vec{X}')$, and moving the psip to its new position $\vec{X}$. Repetition for an initially arbitrary collection of psips leads to a set of psips which is a sample of points from the lowest-energy wave function for the boundary conditions and any other constraints imposed. Further repetition continues the sampling.

The DQMC and GFQMC methods yield samples of the wave function from which, in principle, all the properties of a system may be determined. The energy E may be determined directly from the growth rate of a psip population according to

$$\frac{\partial \Psi}{\partial \tau} = -E\Psi \tag{16}$$

or

$$E = -\frac{1}{\Psi}\frac{\partial \Psi}{\partial \tau} = -\frac{1}{N}\frac{\partial N}{\partial \tau} \quad , \tag{17}$$

where N is the population (or weight) of psips(11). An alternative method, an importance sampling scheme proposed by Grimm and Storer (19), usually gives a higher accuracy in the energy.

Importance sampling can be used in a way which takes advantage of any prior knowledge of the properties of the wave function. This prior knowledge is utilized in the form of a trial wave function Ψ_T, which might be simply a good guess at the true wave function or the result of an earlier calculation of any type. Multiplying each side of the Schrödinger equation, $H\Psi = E\Psi$, by the trial function and integrating over all space we obtain

$$\int \Psi H \Psi_T d\vec{X} = \int \Psi_T E \Psi d\vec{X} \quad . \tag{18}$$

Taking advantage of the Hermitian properties of wave functions we have

$$\int \Psi_T H \Psi d\vec{X} = \int \Psi H \Psi_T d\vec{X} \tag{19}$$

and, noting that E is everywhere the same, we obtain an expression for the energy,

$$E = \frac{\int \Psi H \Psi_T d\vec{X}}{\int \Psi \Psi_T d\vec{X}} \quad . \tag{20}$$

This may be rewritten as

$$E = \frac{\int \Psi \Psi_T \frac{H\Psi_T}{\Psi_T} d\vec{X}}{\int \Psi \Psi_T d\vec{X}} \quad , \tag{21}$$

in which $\frac{H\Psi_T}{\Psi_T}$ is the local energy for the trial wave function.

For samples of the true wave function Ψ the integrals may be determined in Monte Carlo fashion according to

$$E = \frac{\sum \Psi_T \frac{H\Psi_T}{\Psi_T}}{\sum \Psi_T} \ . \tag{22}$$

It is somewhat more convenient and more efficient to arrange for samples based on $\Psi\Psi_T$ and in this case the energy is given by

$$E = \frac{\sum \frac{H\Psi_T}{\Psi_T}}{\sum 1} = \frac{\sum E_{loc}}{\sum 1} \ . \tag{23}$$

The uncertainty or sampling error σ_E in the estimate of the energy depends on the variance or the standard deviation $\sigma_{E_{loc}}$ in single values of E_{loc} according to

$$\sigma_E = \frac{1}{\sqrt{N-1}}\sigma_{E_{loc}} \tag{24}$$

where N is the number of independent samples.

If the trial function is exactly correct the local energy is everywhere exactly the value of the true energy, the standard deviation in E_{loc} is zero, and the error in E is zero. If the trial wave function is approximately correct the standard deviation in E_{loc} is small and the error in E is small. Thus, for high accuracy in evaluating the energy one desires an accurate trial wave function and a large number of statistically independent samples N. Combining Eqs. (23) and (24) we have

$$E = \overline{E_{loc}} \pm \frac{1}{\sqrt{N-1}}\sigma_{E_{loc}} \tag{25}$$

In practice, it is more convenient and probably safer to compute energies E for blocks of iterations large enough to ensure that successive block energies are uncorrelated. The uncertainty in the overall energy for equal block weights may then be estimated from the standard deviation in block energies σ_b

$$\sigma_E = \sigma_b \frac{1}{\sqrt{N_b - 1}} \tag{26}$$

where N_b is the number of blocks.

The variances in the local energy and in the estimate of the true energy are greatly reduced by a good trial wave function. An ideal trial wave function is not only accurate, but simple and compact to facilitate computations of the local energy which must be made repeatedly. The most successful thus far have been based on LCAO functions multiplied by electron-electron correlation functions of the 'Jastrow' type, originally due to Bijl (20),

$$J = \exp[\frac{(br_{ij})}{(1 + cr_{ij})}] \tag{27}$$

where b and c are constants and r_{ij} is an electron-electron distance. The cusp condition at $r_{ij} = 0$ is treated exactly if $b = \frac{1}{4}$ for electrons of the same spin or $\frac{1}{2}$ for electrons of opposite spin. Several more complex functions have been proposed and used successfully for a number of different molecular systems (21-25).

For a system of either bosons or fermions the wave function must have the correct properties of symmetry and antisymmetry. Except in the simplest cases the wave function for a system of n fermions is positive and negative in different regions of the $3n$-dimensional space of the fermions. The regions are separated by one or more $(3n-1)$-dimensional hypersurfaces which cannot be specified except by solution of the Schrödinger equation.

The procedures described above for DQMC and GFQMC lead to the lowest-energy solutions for boson systems which are nodeless ground-state wave functions. They also lead to the ground-state in the case of two electrons (fermions) of opposite spin for which the wave function is symmetric to the exchange of the two electrons. For a system of two or more electrons of the same spin the wave function must be antisymmetric to the exchange of electrons of the same spin and must contain one or more nodal hypersurfaces. The treatment of systems with nodes requires that the solutions be constrained to the appropriate antisymmetry.

3 Node Structure

In the first forty years of quantum chemistry, the question of the structure and properties of nodal hypersurfaces received very little attention. In analytic variational calculations the wave functions obtained are seldom examined and although electron densities are often examined, these reveal little or nothing about the node structure. Examination of the basis set of a determinantal wave function also reveals little or nothing, because the many operations of the determinant scramble the properties of the basis functions. Only recently, with a knowledge of node structure required for developing Monte Carlo methods, have the structure and properties of nodal hypersurfaces been examined in detail.

We are particularly concerned with the nodes in systems of electrons. Let us first simplify to a system of two electrons (12), without an interaction potential, restricted to a line, i.e., two fermions in a one-dimensional box as illustrated in Fig. 1. The ground state is that for two electrons of opposite spin and the wave function is simply the product

$$\Psi = \Psi_1\Psi_2 = sin(\pi\frac{x_1}{L})sin(\pi\frac{x_2}{L}) \quad , \tag{28}$$

which is everywhere positive as shown in Fig. 1a. The first excited state for two electrons of opposite spin is the symmetric combination of independent excited states

$$\Psi = sin(\pi\frac{x_1}{L})sin(2\pi\frac{x_2}{L}) + sin(\pi\frac{x_2}{L})sin(2\pi\frac{x_1}{L}) \quad , \tag{29}$$

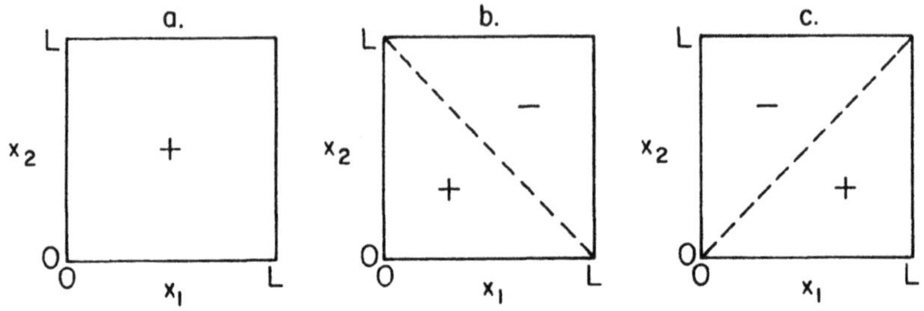

Fig. 1. Node structure for two electrons on a line. a) Ground state for two electrons of opposite spin. b) First excited state for two electrons of opposite spin. c) Ground state for two electrons of the same spin.

as illustrated in Fig. 1b. The lowest state for two electrons of the same spin is the antisymmetric combination,

$$\Psi = sin(\pi\frac{x_1}{L})sin(2\pi\frac{x_2}{L}) - sin(\pi\frac{x_2}{L})sin(2\pi\frac{x_1}{L}) \quad , \tag{30}$$

as shown in Fig. 1c.

Several properties should be noted for the system of two electrons of the same spin, Fig. 1c. The configuration space of the electrons is divided in half by the nodal line. The two halves are similar in shape and are nested together face-to-face. The positions of the two electrons are represented by a single point in configuration space and interchange of the two electrons moves the point across the diagonal to a similar position in the other half of configuration space.

Another simple case is that of the $1s2s$ ^{3}S helium atom for which the wave function may be regarded as a function of the electron-nucleus radii r_1, r_2 and the angle θ_{12} between them. In this case the r_1, r_2, θ_{12} configuration space is divided into two equivalent sections, one positive and one negative, by the nodal surface having the equation $r_1 = r_2$. This is the nodal structure given by the simplest single-determinant wave function

$$\Psi = \begin{vmatrix} 1s(1)\alpha(1) & 1s(2)\alpha(2) \\ 2s(1)\alpha(1) & 2s(2)\alpha(2) \end{vmatrix} \quad . \tag{31}$$

When the electron spins are specified as α spins, such that $\alpha(1)$ and $\alpha(2)$ are unity, Ψ becomes

$$\Psi = 1s(1)2s(2) - 1s(2)2s(1) \tag{32}$$

or

$$\Psi = e^{-ar_1}e^{-br_2} - e^{-ar_2}e^{-br_1} \quad . \tag{33}$$

For $r_1 = r_2$ the wave function is zero and the nodal surface occurs for $r_1 = r_2$ regardless of the functions $1s$ and $2s$ provided they are functions of r_1 only and r_2 only.

In the case of $1s2p\ ^3P$ helium the situation is not so simple. The node structure is not determined by geometric symmetry alone because there are many possible wave functions which have the required anti-symmetry on reflection in the $z = 0$ plane and on exchange of the two electrons. A single-determinant function,

$$\Psi = \begin{vmatrix} 1s(1)\alpha(1) & 1s(2)\alpha(2) \\ 2p(1)\alpha(1) & 2p(2)\alpha(2) \end{vmatrix} \tag{34}$$

or, for $\alpha(1)$ and $\alpha(2)$ equal to unity,

$$\Psi = 1s(1)2p(2) - 1s(2)2p(1) \quad , \tag{35}$$

meets the symmetry requirements. But, there is an infinite number of different $1s$ and $2p$ functions which may be used and the node structures of the resulting wave functions are different. Thus, the symmetry properties alone are insufficient to specify the node structure for this case.

The node structure of $1s2p\ ^3P$ helium is shown in Fig. 2. A detailed examination(26) has revealed that the nodal surface is symmetric to rotation about the z-axis, so that the structure may be illustrated in two dimensions. The position of the nucleus is shown in Fig. 2 along with several different positions for electron 1, indicated by filled circles. For each position of electron 1 the positions of electron 2 for which the wave function is zero (i.e., the nodal surface) are indicated by dashed lines passing through electron 1.

It should be noted that the nodal surface is not a simple plane passing through the origin in the three-dimensional space of one or the other of the electrons. The wave function is not the product function $\Psi = 1s(1)2p(2)$ and its node structure is not that of the product function. The node structure is similar to that of the determinantal function and entirely different from that of the product function.

DQMC calculations for atoms and molecules such as H_2, H_4, Be, H_2O, HF using fixed node structures (see below) obtained from optimized single-determinant SCF calculations typically recover more than 90 precent of the correlation energies of these species and yield total electronic energies lower than the lowest-energy analytic variational calculations. These results suggest that the optimized single-determinant wave functions have node structures that are reasonably correct.

Investigation of the node structure of the wave function in the 30-dimensional configuration space of a 10-electron molecule is not an easy task, but it has been

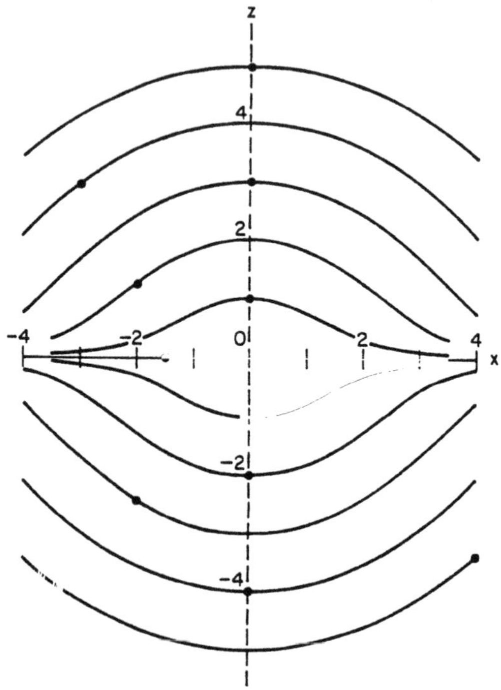

Fig. 2. Node structure for $1s2p\ ^3P$ helium. The structure is symmetric to rotation about the z-axis. For one electron located at any filled circle the wave function is zero if the second electron is on the line passing through that filled circle.

carried out for the water molecule (27). The node structure for the ground state is that suggested by earlier investigations for smaller systems. For a molecule with m spin-up electrons and n spin-down electrons the node structure is approximately that of the product of two functions, one for spin-up and the other for spin-down electrons. The $3m$-dimensional configuration space of spin-up electrons is divided by a $(3m-1)$-dimensional nodal hypersurface into two geometrically similar regions nested together face-to-face. The $(3n-1)$-dimensional nodal hypersurface for the spin-down electrons has equivalent characteristics.

4 Solutions to the Node Problem

Unless the psip distribution in a DQMC or GFQMC calculation is restricted in some way it will proceed to the nodeless ground-state wave function of a boson system. The problem of restricting the distribution in order to obtain the desired solutions for fermion systems - the node problem - has been the subject of a number of investigations and these have led to a variety of methods for restricting the solutions. Two of these are especially useful: the fixed-node method and the exact cancellation method.

The fixed-node method, proposed by Anderson(11) in 1975 for use with diffusion Monte Carlo calculations, allows the treatment of ground and excited states with use of approximate nodal surfaces that act as sinks for diffusing psips. The nodal hypersurfaces are boundaries on which the wave function is fixed at zero and the Schrödinger equation is, or may be, solved exactly within these boundaries. The fixed-node method has been found to be remarkably successful in many cases, for both small and large molecules (12,28-34). For large systems it remains the only demonstrated successful method for handling the node problem, and it yields total energies well below those from analytic variational calculations (31,32). Nevertheless, unless the nodal surfaces used are exactly correct, the fixed-node method yields only an upper bound to the true energy and is not an exact method.

In another method, devised by Ceperley and Alder(35) and applied in a number of cases (35,36), the nodes are released to obtain a 'transient estimate' of the wave function. In this method the wave function is obtained as the difference between positive and negative populations which, in effect, are released from an approximate distribution obtained from a fixed-node calculation. The initial distribution contains contributions from all states but is primarily that of the fermion ground state. The higher states decay away to produce a transient approximation to the fermion ground state before both positive and negative distributions approach the boson ground state and the difference between them disappears. The computational effort for high accuracy is large, and use of the transient-estimate method is restricted to systems with characteristics especially favorable to it.

In another method cancellation of positive and negative psips in close proximity is used to stabilize the positive and negative populations. Arnow, Kalos, Lee, and Schmidt(37) devised an exact cancellation method for GFQMC in which partial cancellations are made on the basis of overlapping distributions of positions for pairs of positive and negative psips. Coker and Watts(38) used cancellation of psips in close proximity in DQMC to obtain stable populations. Anderson, Traynor, and Boghosian(39) reported an improved exact cancellation method for GFQMC, described in detail below, which has produced excellent results for a number of small systems. Zhang and Kalos(40) have reported a somewhat different type of cancellation scheme.

Four other methods have been found useful in several applications. A combination of the ground-state method with a local-mode approximation by Coker, Miller, and Watts(41) has been used for molecular vibrations. The transient-estimate method along with subspace projection methods by Ceperley and Bernu(42) has also been applied to molecular vibrations. A mirror potential method by Carlson and Kalos(43) may have applications to electronic systems. A completely different scheme has been proposed recently by Bianchi *et al.* (44) for use with DQMC

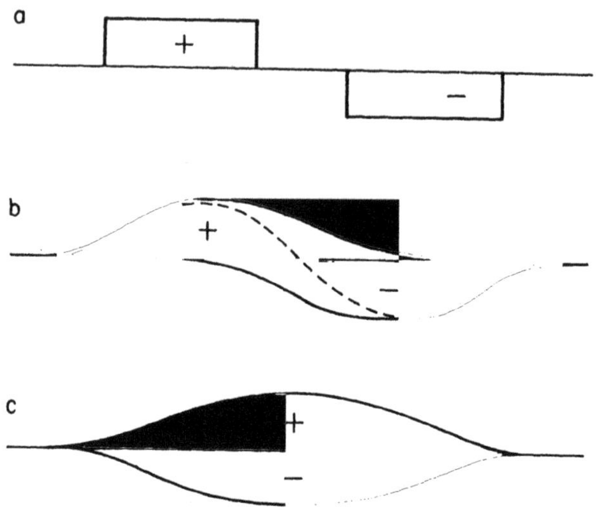

Fig. 3. Inter-penetrating distributions of positive and negative psips. With cancellation the net distribution approaches the antisymmetric distribution shown by the dotted line of the center panel.

calculations.

None of these methods provides the solution to the node problem for systems with large numbers of highly correlated electrons. The choice of methods depends on the problem. Although we must admit to a little bias, we have found the improved method developed in our own laboratory to be the most useful for few-electron molecular problems.

5 Exact Cancellation Schemes: General

We describe here an exact cancellation scheme which is based on the earlier exact cancellation scheme of Arnow *et al.* (37) and incorporates some of the best features of fixed-node, released-node, and other cancellation methods. The method takes full advantage of the symmetric and antisymmetric properties of wave functions and it offers pairwise cancellations of psips as well as self-cancellations and multiple collective cancellations. The method is 'exact' as defined above and it has been applied successfully to a number of few-electron systems.

The basic idea of cancellation is most easily illustrated with the case of the first excited state of a particle in a two-dimensional box. A quantum Monte Carlo calculation for the excited state can be carried out with positive and negative psips,

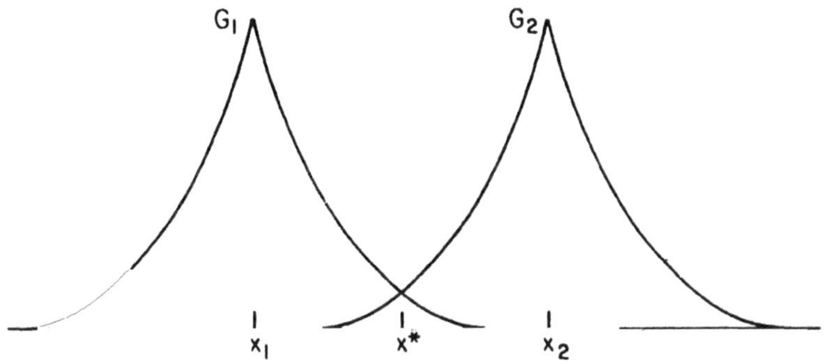

Fig. 4. Schematic diagram of overlapping Green's functions for two psips in close proximity.

initially separated left and right as shown in Fig. 3a. In the absence of cancellation the two populations spread throughout the available configuration space, penetrate each other, and independently approach the symmetric distribution for the ground state. If positive and negative psips in close proximity are allowed to cancel each other, the two populations tend to cancel each other and produce separated distributions as shown in Fig. 3b, in which the net population on the left of center is positive and that on the right of center is negative. Without any control a fluctuation in populations will eventually lead to the dominance of either positive or negative psips and one of the ground-state distributions of Fig 3c. But, if the two populations are controlled to maintain equal numbers of positive and negative psips and if cancellations are properly executed, the net distribution evolves to that of the dotted line of Fig. 3b.

There are several ways of cancelling positive and negative particles. Some of these are rigorously correct but not efficient and some are efficient but not rigorously correct. One might cancel positive and negative particles occupying the same position, but the probability of two psips occupying the same position is vanishingly small. One might cancel psips within an arbitrary distance of each other, but that would lead to a bias in the distributions. Fortunately, there is one way which is rigorously correct and reasonably efficient for systems of a few electrons: cancellation on the basis of the overlap of the distributions to which the psips are moved.

The distributions of moves for positive and negative psips are illustrated schematically in Fig. 4. It is convenient to assign separate weights W_i (positive only, $W_i > 0$) and signs s_i ($s_i = +1$ or -1) for each of the psips. The distribution of weight for a psip at the end of its next move is given by the products $W_1 G_o(\vec{X}, \vec{X}_1{}')$ or $W_2 G_o(\vec{X}, \vec{X}_2{}')$ as indicated by the solid lines. The Green's function $G_o(\vec{X}, \vec{X}_i{}')$

is normalized so that its integral over all space is unity.

The two distributions overlap by the amount O_{lap} ,

$$O_{lap} = \int Min(W_1G_1, W_2G_2)\delta\vec{X} \quad , \tag{36}$$

where $Min(W_1G_1, W_2G_2)$ is the smaller of $W_1G_o(\vec{X},\vec{X_1}')$ and $W_2G_o(\vec{X},\vec{X_2}')$. The overlap is the intersection of the two regions under the curves of Fig. 4. If the distance R separating the two psips at positions $\vec{X_1}'$ and $\vec{X_2}'$ is zero, the overlap is equal to the lesser of the two weights. For large separations the overlap approaches zero.

In the procedure proposed by Arnow *et al.* (37) two psips of opposite sign are partially cancelled to the extent of their overlap by reducing the weights of each prior to their moves. Each weight is altered to become $W_i - O_{lap}$. The distribution of new positions for each psip must be altered to account for the effects of the partial cancellation, i.e., the removal of the overlapping portions of the distributions. This may be accomplished by a rejection procedure. The new position for the first psip is selected from the distribution $G_o(\vec{X},\vec{X_1}')$ in the normal way, but the move is accepted with the probability given by

$$P_a = \frac{W_1G_0(\vec{X},\vec{X_1}') - W_2G_0(\vec{X},\vec{X_2}')}{W_1G_0(\vec{X},\vec{X_1}')} \quad . \tag{37}$$

If P_a is less than zero P_a is replaced by zero. The rejection process is repeated until a new move is accepted. The move for the second psip is selected in a similar fashion.

In our own work(39) we have used an improved procedure, which eliminates the overlap integral, facilitates multiple cancellations, and eliminates the computationally expensive conditional acceptance of moves. The improved procedure may be considered a Monte Carlo version of the earlier procedure. We describe first the partial cancellation of a pair of psips as illustrated in Fig. 4 and then generalize to multiple or collective cancellations.

The move for the first psip of Fig. 4 is selected unconditionally from the distribution $G_o(\vec{X},\vec{X_1}')$ and its weight at the new position $\vec{X_1}$ becomes

$$W_1(new) = \frac{Max([s_1W_1G_1 - s_2W_2G_2], 0)}{s_1G_1} \quad , \tag{38}$$

where, as before, $G_1 = G_0(\vec{X},\vec{X_1}')$ and $G_2 = G_0(\vec{X},\vec{X_2}')$. The move for the second psip is treated similarly and its new weight is given by

$$W_2(new) = \frac{Max([s_2W_2G_2 - s_1W_1G_1], 0)}{s_2G_2} \quad . \tag{39}$$

Two psips of the same weight and opposite sign at the same position cancel completely. In the limit of large separation the Green's function for the partner's move falls to zero and each psip keeps its original weight.

For the two psips of Fig. 4 the position for equal values of the distributions, $W_1G_1 = W_2G_2$, is indicated as X*. The procedure samples moves of the first psip to the region to the left of X* and adjusts its weight according to the effect of the second psip. It also samples moves for the second psip to the right of X* and adjusts its weight according to the effect of the first psip. The point X* divides the configuration space into two Voronoy cells, the left with $W_1G_1 > W_2G_2$ and the right with $W_2G_2 > W_1G_1$.

Multiple cancellations may be carried out with a similar procedure. For n psips chosen to participate, the configuration space is divided into n or fewer Voronoy cells, each one corresponding to a region in which W_iG_i is largest for one of the psips. In some cases there may be fewer than n cells. The sampling of moves for each psip is carried out unconditionally and its new weight becomes

$$
\begin{aligned}
W_j(new) &= \sum_i \frac{s_iW_iG_i}{s_jG_j} \quad , W_jG_j > all \;\; W_iG_i(i \neq j) \;\; , \\
&= \qquad 0 \;\; , W_jG_j < any \;\; W_iG_i(i \neq j) \;\; .
\end{aligned}
\tag{40}
$$

For the interaction of only two psips the procedure reduces to that described above for a single pair.

For the exact cancellation on the basis of overlapping Green's functions to be useful, cancellations must occur often enough to maintain an adequate ratio of positive to negative psips in regions where the wave function is positive and a similar ratio of negative to positive psips in regions where the wave function is negative. Since there are multiple steady-state solutions for the ground state, fluctuations can shift the system from one solution to the other. For the case of a particle in a two-dimensional box, illustrated in Fig. 5, one solution is left-positive/right-negative and the other is left-negative/right-positive. To prevent shifts from one to the other and the resulting loss of information there must be an adequate number of psips as well as an adequate cancellation rate. The required number of psips and the required cancellation rate depend on the system investigated. Some systems are inherently more stable than others.

There are many ways in which cancellations can be carried out and, if stability is not a problem, most of them would probably serve quite well. The best that one can do is to use as many psips as possible and to allow each to participate in multiple cancellations with every psip of opposite sign. We have found that pairwise cancellations with the nearest psip of opposite sign are nearly as effective and are much easier to carry out. If stability is a problem the density of psips is usually low and, as in a dilute gas, the number of significant three-body interactions is negligible in comparison to the number of two-body interactions.

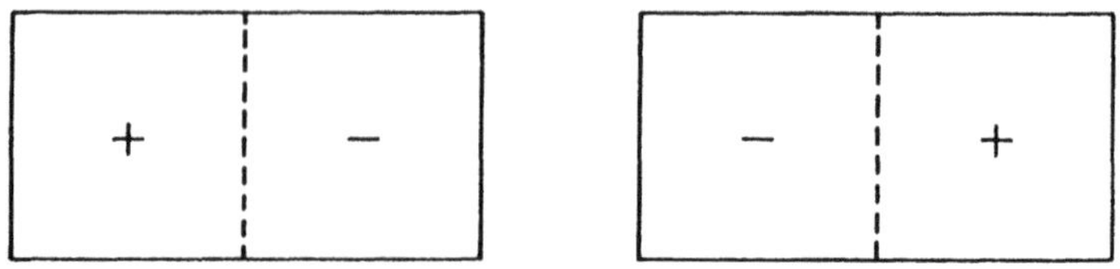

Fig. 5. **Two solutions for the first excited state of a particle in a two-dimensional rectangular box.**

For fermions it is in general necessary to restrict the evolving populations to equal numbers of positive and negative psips or to effect an equivalent restriction in some way. As in cancellation there are many ways of accomplishing this. One might simply adjust the weights of negative psips at the end of each step to obtain a total weight of negatives equal to that of positives, but to avoid any bias the adjustment should be delayed by at least several relaxation times for the system. One might occasionally permute two electrons of the same spin and change the sign to move a psip from a position in one region to an equivalent position in a region of opposite sign. The resulting exchanges of positive and negative psips tend to keep the populations approximately equal in number. But, one may make use of the knowledge that the configuration space is divided into geometrically similar regions by permuting electrons for psips leaving a single chosen region to restore them to that region. The effect is the same as that of maintaining equal populations of positive and negative psips for the entire space. The procedure is illustrated below for the case of a particle in a two-dimensional box.

The energy for collections of positive and negative psips may be determined with the aid of an importance sampling trial function having the same symmetry properties imposed on the collection. Equation (23) may be applied directly and, with the use of signs, becomes

$$E = \frac{\sum_i s_i W_i \Psi_{Ti} \left(\frac{H\Psi_T}{\Psi_T}\right)_i}{\sum_i s_i W_i \Psi_{Ti}}, \tag{41}$$

where the summation is over all psips.

The most efficient calculations are those for which the sum of positive products $\Psi\Psi_T$ (or $s_i W_i \Psi_{Ti}$) is large compared to the sum of negative products $\Psi\Psi_T$ (or $s_i W_i \Psi_{Ti}$). This gives the highest signal-to-noise ratio in computing the energy.

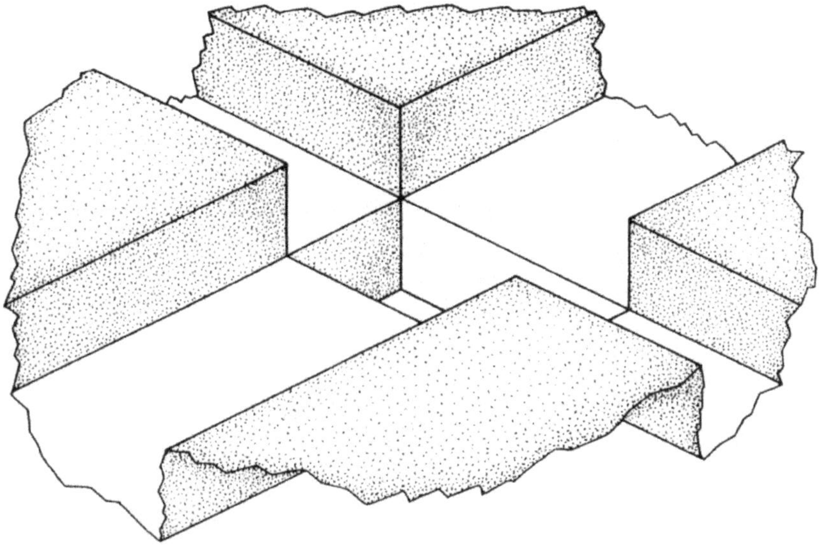

Fig. 6. Intersecting channels of the sample problem.

6 Exact Cancellation: A Simple Example

The problem of the first excited state of a particle in a two-dimensional box exhibits most of the features of a many-electron system, but it can be illustrated in two-dimensional drawings and is therefore an excellent example problem. To avoid the complications of infinities in the potential over finite regions we consider the case of intersecting channels of finite depth as illustrated in Fig. 6.

The potential energies are given by (in units of hartrees and bohrs)

$$\begin{aligned} V &= V_x + V_y \ , \\ V_x &= 0, \quad -1 < x < +1 \ , \\ V_x &= 20, \quad x < -1 \quad or \quad +1 < x \ , \\ V_y &= 0, \quad -0.5 < y < +0.5 \ , \\ V_y &= 20, \quad y < -0.5 \quad or \quad +0.5 < y \ . \end{aligned} \tag{42}$$

The mass of the particle is specified as that of the electron. The exact wave function Ψ of the first excited state is separable and is the product of functions for the x and y directions,

$$\begin{aligned}
\Psi &= \Psi_x \Psi_y \ , \\
\Psi_x &= C_A sin(Ax) \ , \quad -1 < x < +1 \ , \\
\Psi_x &= C_B e^{-Bx} \ , \quad +1 < x \ , \\
\Psi_x &= -C_B e^{Bx} \ , \quad x < -1 \ , \\
\\
\Psi_y &= C_C cos(Cy) \ , \quad -0.5 < y < 0.5 \ , \\
\Psi_y &= C_D e^{-Dy} \ , \quad 0.5 < y \ , \\
\Psi_y &= -C_D e^{Dy} \ , \quad y < -0.5 \ .
\end{aligned} \tag{43}$$

where

$$\begin{aligned}
C_A &= 1.0 \\
A &= 2.700 \\
C_B &= 130.65 \\
B &= 5.719 \\
C_C &= 1.0 \\
C &= 2.373 \\
C_D &= 7.035 \\
D &= 5.862
\end{aligned}$$

The wave function, which has a node at $x = 0$, is plotted in Fig. 7. The exact energy for the first excited state is 6.4606 hartrees.

The exact wave function is positive to the left and negative to the right of the exact node or *vice versa* and it is antisymmetric to reflection in the nodal line. It is also antisymmetric to inversion through the origin at the center of the box. The inversion antisymmetry is similar in character to the antisymmetry for permutation of two electrons of the same spin in molecular problems. The space can be divided into two regions overall symmetric to inversion using a dividing line which itself has inversion symmetry. Any one of the dividing lines of Fig. 8 - the correct node i, the straight line ii, or the yin-yang line iii - has the specified symmetry. If we require the desired wave function to have antisymmetry to inversion, then any of these lines is a possible nodal line. In a molecular problem, if we require the desired wave function to have antisymmetry to permutation of two electrons of like spin, then any of a large number of hypersurfaces is similarly a possible nodal hypersurface.

Since the left and right regions are equivalent one need not calculate the wave function in both. We make use of the dividing line to specify the boundary of a

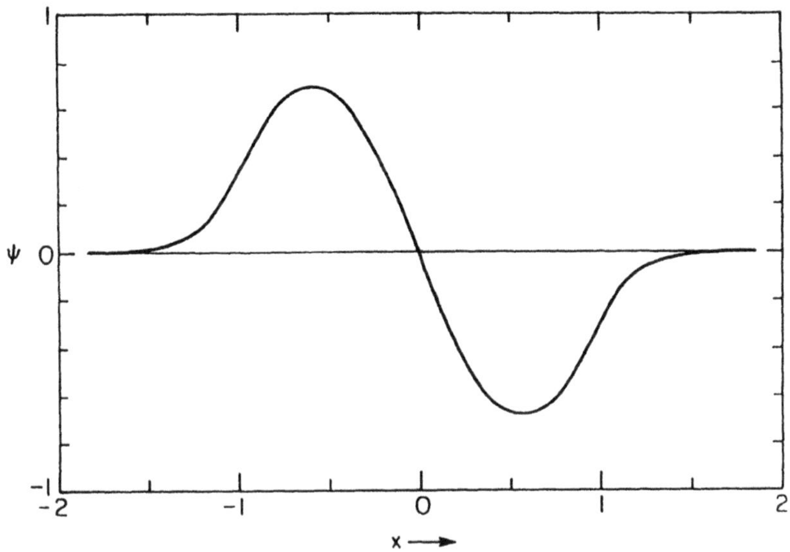

Fig. 7. wave function for sample problem of intersecting channels. The wave function is that along a line parallel to one of the channels.

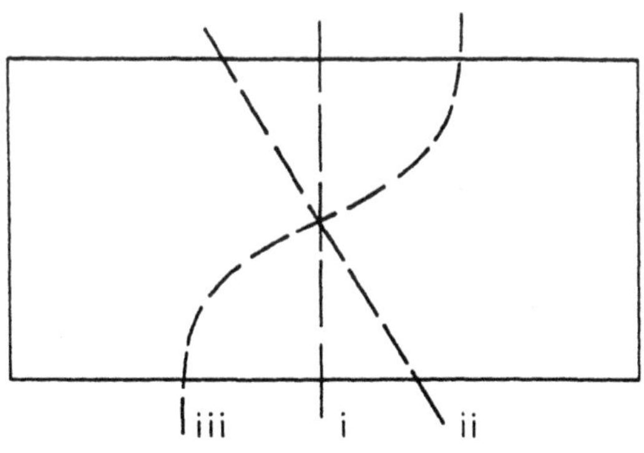

Fig. 8. Illustration of possible nodal surfaces satisfying the requirement of inversion symmetry imposed on solutions of the problem of intersecting channels.

single region and carry out a GFQMC calculation for positive and negative psips in that region. The calculation is started with positive psips on the left of the dividing line and any psip crossing the line is returned to the region by inverting its position through the origin and changing its sign. Cancellations are carried out among positive and negative psips in the occupied region.

To avoid the possibility of making use of the known solution we specify the dividing line as a straight line passing through the origin twisted by 30 degrees from the node line.

An importance sampling trial function Ψ_T having inversion antisymmetry and an incorrect node line matching the dividing line was used in energy determinations. The function is that for a harmonic oscillator in two dimensions twisted by the angle $\theta = 30^0$ about the origin,

$$\begin{aligned} \Psi_T &= e^{-au^2} e^{-bv_{BB}^2} \quad , \\ u &= x\cos\theta + y\sin\theta \quad , \\ v &= -x\sin\theta + y\cos\theta \quad . \end{aligned} \tag{44}$$

The zero of potential energy was shifted by 50 hartrees so that the potential at the bottom of the box was -50 hartrees and the maximum of the potential energy was -10 hartrees. The input energy E required for the scaling parameter k in the Green's function expression and the multiplication factor $(\frac{V}{E})$ of Eqs. (13) and (15) was specified as the shifted exact value, -43.5394 hartrees. If the energy were not known, an iteration would be required to determine E for adjusting the potential energy. It may seem odd to need the value of E to calculate E, but the calculated value of E has been found to be insensitive to small errors in the input value of E. Thus, the choice of the input E is not a problem, although one should test the sensitivity in every case.

A typical calculation was begun with 5000 positive psips of equal weight distributed at random on the left side of the dividing line. An iteration was carried out by first multiplying the weight of each psip by $\frac{V(\vec{X}')}{E}$ and dividing each into psips of unit weight with any fraction remaining retained at unit weight with a probability equal to the fraction. Any psips on the right side of the dividing line were moved to the left by inversion of their coordinates and reversal of their signs.

Cancellations were carried out in a pairwise fashion. The available configuration space was divided into cells, 0.2 bohrs on a side, and those psips within each cell were identified. For cells with more positive psips than negative, each negative psip was paired for cancellation with the nearest unpaired positive psip. For cells with more negative psips than positive, each positive psip was paired with the nearest unpaired negative psip. The pairwise cancellations produced new weights for the affected psips as given by Eq. (38). The required values of the Green's functions were determined with the aid of look-up tables for the coefficients of polynomials fit for small intervals of the argument of the required Bessel function. For an odd number of dimensions these may be calculated directly. The selection of psip moves, by sampling the Green's function $G_o(\vec{X}, \vec{X}')$, was carried out as described by Kalos(18) for a system without cancellations.

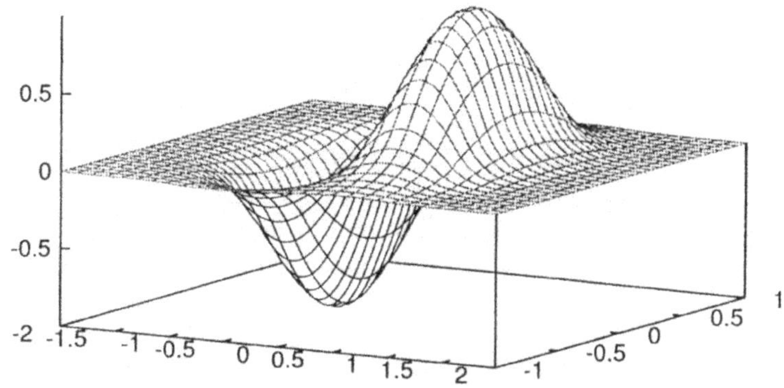

Fig. 9. Calculated wave function for intersecting channels.

The distribution of psips evolved in fewer than 50 iterations to a distribution approximating the true wave function on the left of the dividing line. Continuing the calculation for 10,000 iterations allowed the determination of the energy to high accuracy and the accumulation of weights within cells of dimension 0.02 bohrs on a side for a histogram of the psip distribution. The node of the calculated distribution appeared along the line $x = 0$ to duplicate that of the known wave function. The energy determined using the trial function for importance sampling was 6.4603 ± 0.0005 hartrees, nicely overlapping the known energy of 6.4606 hartrees.

The ratio of positive to negative psips at steady state was approximately 100 to 1. In the energy determination the ratio of the sum of positive products $\Psi\Psi_T$ to the sum of negative products $\Psi\Psi_T$ was 150 to 1.

The histogram of the psip distribution, which represents the calculated wave function, is shown in Fig. 9. In an extended calculation the density of psips found for each cell was within 1 unit of the density computed for the exact wave function when the peak density was specified as 1000 units.

7 Exact Cancellation: Stability

The essential requirement for a successful calculation is a high ratio of positive to negative psips or, to be precise, a high ratio of positive to negative elements in the summations of Eq. (23). In the absence of cancellation equal numbers of positive and negative psips in all regions will lead to an indeterminate value for the energy. In the words of the well known song, "You've got to ac-cent-tchu-ate the positive, E-lim-my-nate the negative, Latch on to the affirmative, Don't mess with Mister

Inbetween." (45).

Of course, a high ratio of negative to positive would serve just as well. It is the in-between that one must avoid. The way to accentuate the positive is to promote cancellation in every way possible. The number of ways is at present limited to increasing the density of psips and optimizing the step sizes and energy, which are not independent. A variety of other possibilities remains unexplored.

The obvious way of increasing psip density is to increase the number of psips up to the limit of available computer memory. Beyond that one can make full use of symmetry to concentrate psips in a single region of configuration space. In the case of the particle in a two-dimensional box described above the psips were limited to one-half of the available configuration space and their density was effectively doubled.

For a system of electrons the configuration space can be divided into sections designated positive and negative according to the sign of a trial wave function of appropriate symmetry. If a psip crosses into a region of negative trial wave function it can be returned to the positive region by permuting any two electrons of the same spin and changing its sign.

Psips can be further concentrated in configuration space by a process of 'ordering'. Electrons of the same spin can be ordered spatially by even numbers of permutations without changing the sign of their psip. Thus, a positive psip with spin-up electrons 1,2,3 having $x(3) < x(1) < x(2)$ can be changed to $x(1) < x(2) < x(3)$ by exchanging electrons 1,3 followed by 2,3 to be closer in configuration space to a negative psip having $x(1) < x(2) < x(3)$. In practice, one may simply order the electrons of all the psips as required to place them in a specified region.

If the nuclear framework has symmetry or a particular electronic state has appropriate symmetry, the electrons can be subjected to symmetry operations to further concentrate psips in a region of configuration space. Thus, in the case of He $1s2s\ {}^3S$, which has spherical symmetry about the nucleus, all systems can be rotated to place electron 1 on the x-axis in the positive direction and further rotated to place electron 2 in the xy-plane in the direction of positive y. For $H_2({}^3\Sigma_u{}^+)$, which has rotational symmetry about the internuclear axis (the z-axis), all configurations can be rotated to place electron 1 in the xz-plane with $y(1) = 0$ and $x(1) > 0$, followed by reflection in the xz-plane to place electron 2 in the region of $y(2) > 0$.

A self-cancellation procedure may be carried out in systems with a configuration space divided into equivalent regions. Even if the psips are confined by permutations to a single region, a psip in one region may be considered a duplicate of itself with the same or the opposite sign at a corresponding position in another region. With the proper consideration of signs one may consider the psips in one region to be duplicated in every other equivalent region. Thus, in the case of a reflection a psip may be partially cancelled by its image. The psip is completely cancelled in any move across the reflection plane, and as the psip approaches the reflection plane the fraction of its weight which is cancelled approaches unity. This suggests a convenient way of handling the boundary conditions for the problem of a particle in a box with infinitely high walls.

For certain many-electron systems self-cancellation can be an effective aid to cancellation. In place of large numbers of psips only one is required for self-cancellation. The permutation of any two electrons of the same spin leaves the others in place so that the distance in configuration space between the original psip and its permuted partner is only about the square-root-of-two larger than the distance between the permuted electrons. That distance can be small enough for an appreciable cancellation effect.

The choice of E affects the ratio of positive to negative psips in several ways. Since E can be adjusted by arbitrary shifts in the zero of potential energy it can be chosen to optimize the ratio. The Green's function $G_o(\vec{X}, \vec{X}')$ is, in fact, a function $G_o(\frac{\vec{X}}{\sqrt{-2mE}}, \frac{\vec{X}'}{\sqrt{-2mE}})$ which depends on E and the average step size is inversely proportional to $\sqrt{-E}$. If E is more negative the steps are smaller and two psips 'diffusing' past each other have an increased likelihood of landing in close proximity at least once as they pass, but their Green's functions are narrower and their ranges for effective cancellation are shorter. Nevertheless, the net effect is a somewhat greater cancellation for smaller steps.

The multiplication term $(\frac{V}{E})$ can switch the sign of a psip when V is positive and E is negative. Since E must be negative, the switching of signs can be reduced by shifting the zero of potential energy to make V negative in most regions of configuration space. It is essential that V be negative or near zero for configurations in which an electron is separated from the remainder of the system. If not, then the low values of the wave function for these regions will be obtained from large populations of psips, which alternate in sign in successive steps rather than from low populations of psips. Since the cancellation of psips is promoted by their division into more psips with smaller weights which, in effect, mix more intimately, it may be useful to divide psips in the regions in which cancellation is likely. These regions are the ones that already have a high concentration of psips and they correspond to regions of high values for good trial wave functions.

It is advantageous, in reducing the statistical error associated with the energies computed using Eq. (23) or Eq. (41), to have equal absolute values of the quantity $W_i\Psi_{T_i}$ rather than equal weights. The division of psips in regions of high absolute values of Ψ_T has the additional advantage of a tendency to equalize these quantities. Thus, we have found it useful in promoting cancellation and in reducing statistical error to 'split and kill' psips to produce approximately equal absolute values of $W_i\Psi_{T_i}$.

An additional effect of small step sizes produced by more negative values of E is a greater serial correlation of psip positions and local energies. The advantage of a higher cancellation rate for smaller steps may be offset by a greater statistical error resulting from the greater serial correlation.

In calculations to obtain high accuracies the computation time invested in experimentation to select optimum trial functions, cancellation procedures, and choice of E is likely to be regained many times over in a reduced computation effort required for a given level of accuracy.

One may, of course, introduce various approximations and use inexact devices to promote cancellation or to ensure a stable net positive population of psips, but such actions would introduce error. For example, one may limit the fluctuations of the negative population (40) or the positive to negative ratio, perform calculations with various total numbers of psips, and extrapolate the results to the stable limit of a large number of psips. One may fix the number of psips, artificially increase the range of cancellation by artificially varying the Green's function $G_0(\vec{X}, \vec{X}_1{}')$ used in the cancellation formula, and extrapolate the results to those for the correct value of the Green's function. These devices and extrapolations may possibly lead to exact results; but, since the functional form of the variation of the quantity extrapolated is not easily established, the results must always be questioned. However, such devices and extrapolations may be extremely useful in obtaining approximate or estimated values for energies and other properties that, while inexact, may be more accurate than those available from any other method.

8 Exact Cancellation: Computational Details

Computations using the exact cancellation scheme outlined here are relatively easy to carry out. For systems of 2-6 electrons the programs used consist of 200-800 Fortran statements and the bulk of the statements are those for evaluating trial functions and their derivatives. The memory requirements vary with the number of electrons and the number of psips required for stability, but work stations such as the IBM RISC 6000-580 have been found to be adequate for the few-electron problems we have treated.

Our calculations have been performed on various IBM RISC 6000 machines, the IBM 3090-600S, the Cray C90, and the Thinking Machines CM-2 and CM-5 Connection Machines. Each of these machines, as well as an individual processor on the massively parallel CM-5, is easily capable of running programs with 20,000 psips and with a little care as many as 200,000 psips.

The calculations are by their nature easily carried out on vector and parallel machines. For very high accuracies one needs an accumulation of results and it is unimportant whether one has a few long runs or many short runs. The runs can be carried out entirely independently on separate processors of a massively parallel machine.

In using IBM RISC work stations we found it convenient, especially when using machines not our own, to set up these machines for runs at low priority of about 24 hours duration, which on completion automatically E-mail the results to us, select a new random number seed, and begin another run. In this way we could keep dozens of low cost computers fully and usefully occupied with very little human effort.

A full two-electron, two-proton calculation (2) for H_2 to determine its exact energy without the Born-Oppenheimer approximation, performed on a 32,768-processor Thinking Machines CM-2, was the first quantum calculation to be reported for a massively parallel machine.

9 The Molecular Ion $H_3{}^+$

The ground state of the molecular ion $H_3{}^+$ has two electrons of opposite spin and thus no nodes in its wave function. Exact Green's function cancellation is not required in QMC calculations for this system; but, the results are, of course, exact for such calculations and the use of cancellation permits large step sizes and leads to very high accuracies. The most recent exact QMC calculation, which is now a few years old, yielded an energy accurate to within ±0.000 001 hartrees (i.e., within 0.0016 kcal/mole).

The molecular ion $H_3{}^+$ has a long history in quantum chemistry. It is the simplest polyatomic molecule (or molecular ion) and for that reason has served as a test species for developing theoretical methods in quantum mechanics. The potential energy surface of ground state of $H_3{}^+$ is required for the prediction of the rates and molecular dynamics of the reaction $H^+ + H_2 \rightarrow H_2 + H^+$ and for the prediction of the rotation-vibration states of $H_3{}^+$. The infrared spectra of transitions among these states for H_3^+ and its isotopic equivalents were not (knowingly) observed until Carney and Porter(46) made fully *ab initio* calculations of the potential energy surface, the rotation-vibration energy levels for that surface, and the expected wavelengths of lines in the spectra. With that information available Oka(47) and Shy *et al.* (48) were able to observe the spectra of H_3^+ and D_3^+.

The history of calculations for H_3^+ at or near its equilibrium nuclear configuration, a triangular configuration with protons separated by 1.65 bohrs, is summarized in Fig. 10. The first calculations were analytic variational calculations by Hirschfelder(49) in 1938 yielding an energy of −1.2929 hartrees. The most recent Monte Carlo calculations are our own (3) of a few years ago yielding −1.343 835 ±0.000 001 hartrees. In the figure it may be seen that with time the lowest-energy analytic variational calculations approach asymptotically the exact value for the energy. The several perturbation and other unbounded calculations approach the exact value with increasing time. The error bars for Monte Carlo calculations, including DQMC and GFQMC calculations, gradually narrow around the exact value of the energy. The error bars in later years are too narrow to be seen and are about $\frac{1}{200}$ the width of the lines.

Our own experience with H_3^+ reflects the improvement in methods and computers since 1975. The results are listed in Table I and they show an improvement in accuracy from ±0.013 to ±0.000 001 hartrees, a factor of 13,000.

The most recent Monte Carlo calculations (3) were carried out using the GFQMC method with cancellation. The trial function for importance sampling in determining the energy was a simple product of one-electron LCAO functions, multiplied

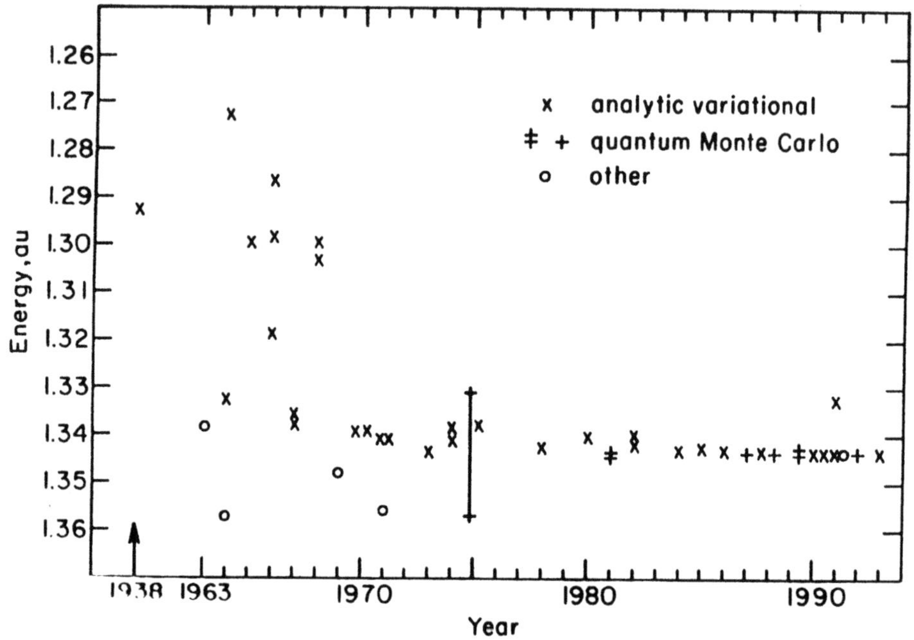

Fig. 10. A history of quantum chemistry as shown by energies determined in calculations for the molecular ion H_3^+. X's: analytic variational calculations. Circles: unbounded calculations of various types. Error bars: quantum Monte Carlo calculations

TABLE I
Some QMC results for the ground-state energy of $H_3{}^+$ in its equilateral-triangle nuclear configuration with R = 1.6500 to 1.6600 bohrs.

Reference	Year	Method	Separation bohrs	Energy hartrees
Anderson (11)	1975	DQMC	1.66	−1.344 ± 0.013
Anderson (12)	1976	DQMC	1.66	−1.341 4 ± 0.004 3
Mentch & Anderson (50)	1981	DQMC	1.65	−1.343 9 ± 0.000 2
Anderson (51)	1987	GFQMC	1.65	−1.343 76 ± 0.000 03
Traynor & Anderson (52)	1988	GFQMC	1.65	−1.343 877 ± 0.000 054
Anderson (3)	1992	GFQMC + cancellation	1.65	−1.343 835 ± 0.000 001

by exponential functions to account for electron-electron correlation effects,

$$\Psi_T = f_1 f_2 J \tag{45}$$

where

$$\begin{aligned}
f_1 &= \exp(-c_a r_{1A}) + \exp(-c_a r_{1B}) + \exp(-c_a r_{1C}) \\
f_2 &= \exp(-c_a r_{2A}) + \exp(-c_a r_{2B}) + \exp(-c_a r_{2C}) \\
J &= \exp(U) \\
U &= \sum_{k=1,20} \frac{b_k w_k}{1 + c_k w_k} \\
w_k &= r_{1A}, r_{12}, r_{1A}^2, r_{12}^2, r_{1A} r_{12} \quad etc.
\end{aligned}$$

The calculations were carried out with approximately 15,000 psips after 'steady state' was reached. The population was controlled by random duplication or elimination of psips at intervals of 55 iterations, and data from five iterations after each renormalization was discarded to avoid any bias which might be caused by artificially decreasing a growing population or artificially increasing a shrinking population.

The equilateral-triangle symmetry allowed the concentration of psips by transformations of their coordinates. With the nuclear framework placed in the xy-plane and symmetric to reflection in the xz-plane, the electrons were reflected in either the xy-plane or the xz-plane or both to place electron 1 in the region of positive x and positive z.

The six-dimensional electron configuration space in the region of the nuclei was divided into 6125 cells of side length 0.5 to 1.0 bohr in each of the five dimensions x_1, y_1, z_1, y_2, z_2. Pairs of positive and negative psips in close proximity within each box were chosen for cancellation.

Preliminary runs gave an estimate of $-1.343\ 836$ hartrees for the energy and this energy was used for the input energy in the final runs for the determination of the energy. The sensitivity of the calculated output energy to the input energy was checked in separate calculations. It was found, for example, that an input error of -0.004 000 hartrees produced an output error of about $-0.000\ 070$ hartrees, and it was concluded that the actual input error of $-0.000\ 001$ hartrees was very much smaller than the statistical uncertainty of $\pm 0.000\ 001$ hartrees in the final result. The potential energy was shifted by $+1.0$ hartree which, together with removal of the inter-nucleus potential energies, gave an input energy of $-2.162\ 018$ hartrees used in the final calculation.

The exact quantum Monte Carlo result of $-1.343\ 835 \pm 0.000\ 001$ hartrees was 0.000 007 hartrees below the lowest-energy analytic variational result at the time of the calculation, $-1.343\ 828$ hartrees by Frye, Preiskorn, Lie, and Clementi(53).

Since then the lowest-energy analytic variational energy has been dropped to −1.343 835 hartrees in a calculation by Rohse, Klopper, and Kutzelnigg(54), which "is supposed to be accurate to all indicated figures"(54).

As may be seen in Fig. 10 there has been considerable progress in quantum chemistry in the past few years, not only in quantum Monte Carlo calculations, but also in analytic variational calculations.

10 The Molecule H_2

The ground state energy of the molecule H_2 has been determined in GFQMC calculations by Traynor, Anderson, and Boghosian(2) without the use of the Born-Oppenheimer or any other adiabatic approximations. The complete four-body problem of two electrons and two protons was treated as inseparable. The calculations were carried out on a massively parallel computer and were the first reported quantum calculations of any type to make use of a massively parallel computer.

The system has no nodes in the wave function and the calculations of Traynor, Anderson, and Boghosian (2) were made without use of cancellation. The psips were maintained positive with a shift in the potential energy to prevent switches of sign for close encounters of electrons, which otherwise would give positive values of V in Eq. (15). A new set of calculations currently in progress makes use of cancellation and has achieved a higher accuracy, but this newer work is not yet complete. It will be reported in due course.

The history of accurate calculations for H_2 begins with the work of James and Coolidge (55) in 1933, which represented one of the first successes in solving the Schrödinger equation for molecules. More accurate results have since been obtained by many others, including in the 1960's Kolos and Roothaan (56) and Kolos and Wolniewicz (57) who used a variational approach together with a method by Born (58) for separating the complete Hamiltonian into two parts and neglecting off-diagonal terms in what is called an adiabatic approximation. This is an improvement over the Born-Oppenheimer approximation in which additional terms are neglected.

The calculated energies have been used to predict the dissociation energy of H_2 and that in turn has been compared with the results of experimental measurements. As the accuracies of experiments and theory both improved the results shifted back and forth a bit, sometimes in agreement, sometimes not. Wolniewicz(59) made improvements to include off-diagonal contributions to the Hamiltonian and incorporated relativistic and radiative corrections. Very recently, Kolos, Szalewicz, and Monkhorst(60) incorporated similar corrections with more accurate variational calculations. Analytic variational calculations for the full four-body problem had been made earlier by Bishop and Cheung(61). The results of these calculations and others are listed in Table II along with some experimental values for the dissociation energy.

The description of GFQMC given above for electrons must be modified to allow

TABLE II
Dissociation energy of H_2 (D_0).

Reference		Energy[a] (cm^{-1})
Expt.	Herzberg (62)	$36116.3 < D_0 < 36118.3$
Expt., analysis	Stwalley (63)	36118.6 ± 0.5
Expt.	McCormack *et al.* (64)	36118.1 ± 0.2
Nonadiabatic	Wolniewicz (59)	36118.01
Nonadiabatic	Kolos *et al.* (60)	36116.088
Nonadiabatic	Bishop, Cheung (61)	36117.92
GFQMC	Anderson *et al.* (2)	36117.9 ± 2.0
Nonadiabatic	Wolniewicz (65)	36118.060
Expt.	Balakrishnan *et al.* (66)	36118.11 ± 0.08

[a]Including relativistic and radiative corrections as listed by the authors. Typical values for these corrections are -0.54 cm^{-1} and -0.22 cm^{-1}, respectively.

for the different masses of the protons. This is most easily accomplished by using scale factors for the distances and step sizes associated with the Green's function. The constant k remains as it was, in terms of the electron mass m_e,

$$k^2 = \frac{-2m_e E}{\hbar^2} \quad , \tag{46}$$

and scaled coordinates, defined by

$$Q_j = \sqrt{\frac{m_j}{m_e}} X_j \quad , \tag{47}$$

replace the coordinates X of Eqs. (12) to (15). The resulting step size for a proton is approximately 1/45 that of the electron and the range of the Green's function is similarly shorter.

The trial wave function for importance sampling was the product of four terms,

$$\Psi_T = \phi_1\phi_2\phi_3\phi_4 \quad , \tag{48}$$

$$\phi_1 = \exp(-ar_{13}) + \exp(-ar_{14}) \quad , \tag{49}$$

$$\phi_2 = \exp(-ar_{23}) + \exp(-ar_{24}) \quad , \tag{50}$$

$$\phi_3 = \exp(\frac{br_{12}}{1 + br_{12}}) \quad , \tag{51}$$

$$\phi_3 = \exp[-d(r_{34} - c)^2] \quad , \tag{52}$$

where the labels 1 and 2 refer to the electrons and 3 and 4 refer to the protons. The function includes a Jastrow term for the inter-electron effects and a harmonic oscillator term for inter-nucleus effects. The trial function could be improved substantially with more elaborate terms and a full optimization.

The calculations were performed on a 32,768-processor Thinking Machines CM-2 Connection Machine operated with 20-30 psips per processor. The processors were operated independently for blocks of iterations, in which the number of psips in a processor but not their total weight was fixed, interrupted at intervals of 100 iterations for redistributions among the processors to obtain approximately equal weights. The calculation time was about 60 hours. On a full-size 65,536-processor CM-2 the time would be about 30 hours and the average speed would be 1500 megaflops or $1.5\text{x}10^9$ floating point operations per second.

The energy obtained, which includes the vibrational zero-point energy, was $-1.164024 \pm 0.000\ 009$ hartrees. This corresponds to a dissociation energy of $36\ 118.6 \pm 2.0\ \text{cm}^{-1}$, without taking into account any relativistic effects. With relativistic and radiative corrections the dissociation energy of H_2 becomes $36\ 117.9 \pm 2.0\ \text{cm}^{-1}$. This value overlaps the best prior theoretical values as well as the best experimental values listed in Table II, but it does not challenge the accuracy of either of these.

For the more accurate quantum Monte Carlo calculations currently underway we expect the higher accuracy afforded by the exact cancellation algorithm, the better trial function, and a still faster computer will allow us to challenge the accuracies of the best existing calculations and experiments.

11 Potential Energy Surface for the Reaction $H + H_2 \rightarrow H_2 + H$

In 1956 Boys, Cook, Reeves, and Shavitt(67) published their pioneering paper entitled "Automatic Fundamental Calculations of Molecular Structure". In this paper they described the first 'computer' calculations of electronic structure and reported results for several systems including the system of three H atoms. They found the minimum in the classical barrier for the reaction $H + H_2 \rightarrow H_2 + H$ to occur for the collinear configuration H-H-H with separations of 1.757 bohrs having a height of 9.7 kcal/mol relative to reactants $H + H_2$ calculated with the same basis set. They noted that the "advent of automatic machines has simplified and made practical such calculations" and they expressed the hope that their prototype calculations would "help to demonstrate the significance and future possibilities of such calculations".

Some 38 years later we find their calculations to be remarkably accurate and their predictions remarkably correct. An exact quantum Monte Carlo calculation(4) shows the barrier height to be 9.61 ± 0.01 kcal/mol relative to $H + H_2$ at the minimum which occurs for a collinear configuration with separations of 1.757 bohrs.

The exact quantum Monte Carlo calculations described here have produced a

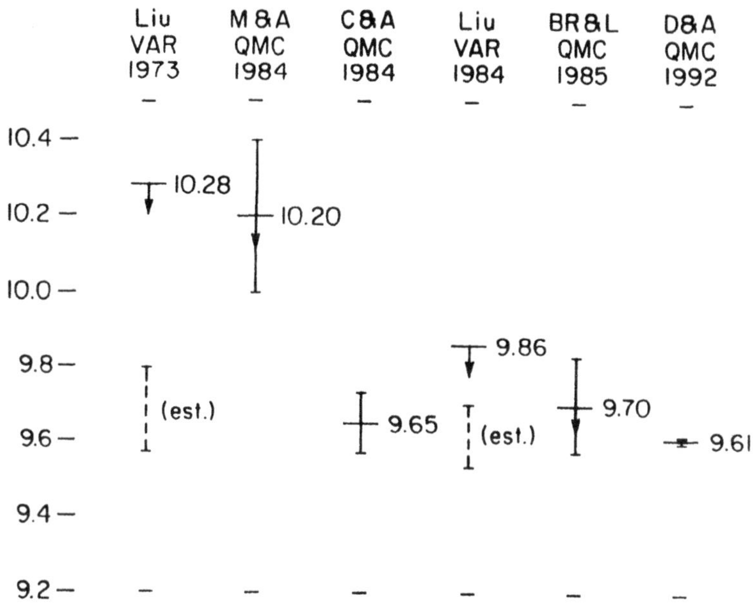

Fig. 11. Some recent calculations of the barrier height for the reaction $H + H_2 \rightarrow H_2 + H$. A description is given in the text.

potential energy surface accurate to within ±0.01 kcal/mol at the saddle point and within ±0.10 kcal/mol or better elsewhere on the surface. With this surface, essentially exact full-quantum calculations of reactive scattering, now carried out in a number of laboratories, can predict all the major observables for this reaction more accurately than they can be measured at the present time.

It is likely that, having moved ahead of experiment for the reaction $H + H_2 \rightarrow H_2 + H$, theory will remain ahead forever. Both theory and experiment for this reaction will certainly be improved, but theory is likely to maintain its lead.

Recent calculations of the barrier height for the reaction $H + H_2 \rightarrow H_2 + H$ are summarized in Fig. 11. The energies determined for the collinear saddle point configuration with separations of 1.757 bohrs are shown. The calculated energies in kcal/mol relative to the exact energy for separated $H + H_2$ are indicated by solid lines. The estimated energies, also relative to exact $H + H_2$, as given by several authors are indicated by dashed lines. Energies that are upper bounds in variational calculations are indicated by the use of arrows.

The first result shown is that by Liu(68) for analytic variational calculations in 1972 with an upper bound of 10.28 kcal/mol and an estimated range of 9.56

to 9.80 kcal/mol. Second is a fixed-node DQMC calculation, variational with respect to node locations, by Mentch and Anderson(30) giving an upper bound of 10.2±0.2 kcal/mol. Third is a released node calculation by Ceperley and Alder(22) giving a barrier height of 9.65 ± 0.08 kcal/mol. Fourth is an analytic variational calculation by Liu(68) in 1984 with an improved basis set giving an upper bound of 9.86 kcal/mol and an estimated range of 9.53 to 9.65 kcal/mol. Fifth is a fixed-node DQMC calculation by Barnett, Reynolds, and Lester(22), giving as the lowest upper bound for several different node structures an energy of 9.70 ± 0.13 kcal/mol. Sixth is the exact GFQMC calculation by Diedrich and Anderson(4) of our laboratory giving a barrier height of 9.61 ± 0.01 kcal/mol.

The H-H-H system is ideally suited to exact quantum Monte Carlo calculations. There are only three electrons, the nuclear configuration of most interest is symmetric to reflection and rotation, and the two electrons of opposite spin tend to be well separated. All of these factors tend to favor a high ratio of positive to negative psips and a stable calculation.

The exact QMC calculations(4) were carried out using excellent trial wave functions for specifying dividing surfaces and for determining the energies. Each function was the product of LCAO's ϕ_i for each electron, an antisymmetric term $(x_1 - x_2)$, electron-electron Jastrow terms J_{ij}, and electron-nucleus Jastrow terms J_{iN},

$$\Psi_T = \phi_1\phi_2\phi_3(x_2 - x_1)J_{12}J_{13}J_{23} \times J_{1A}...J_{3C} \quad , \tag{53}$$

$$\phi_i = c_{iA}e^{-ar_{iA}} + c_{iB}e^{-ar_{iB}} + c_{iC}e^{-ar_{iC}} \quad , \tag{54}$$

$$J_{ij} = \exp[\frac{(b_{ij})}{(1 + c_{ij}r_{ij})}] \quad , \tag{55}$$

$$J_{iN} = \exp[\frac{(b_{iN})}{(1 + c_{iN}r_{iN})}] \quad . \tag{56}$$

The calculations were performed on IBM RISC 6000-550 and -320H work stations with typically 30,000 psips, but as many as 300,000 psips for some nuclear configurations. The total calculation time for the most accurate calculation, that of the saddle point, was about 80 machine days. For less accurate calculations at some of the easier nuclear configurations the times were as little as a few hours.

Some of the results are listed in Table III and plotted in Fig. 12. The minimum energy calculated for collinear nuclear configurations with equal nearest-neighbor separations R_1 and R_2 was that for $R_1 = R_2 = 1.757$ bohrs with a value of $-1.659\,154 \pm 0.000\,014$ hartrees, which corresponds to a barrier height of 9.61 ± 0.01 kcal/mol relative to the minimum for separated H + H_2.

The calculated potential energy surface for collinear configurations in the region of the saddle point is shown in Fig. 13. A comparison with the lowest-energy

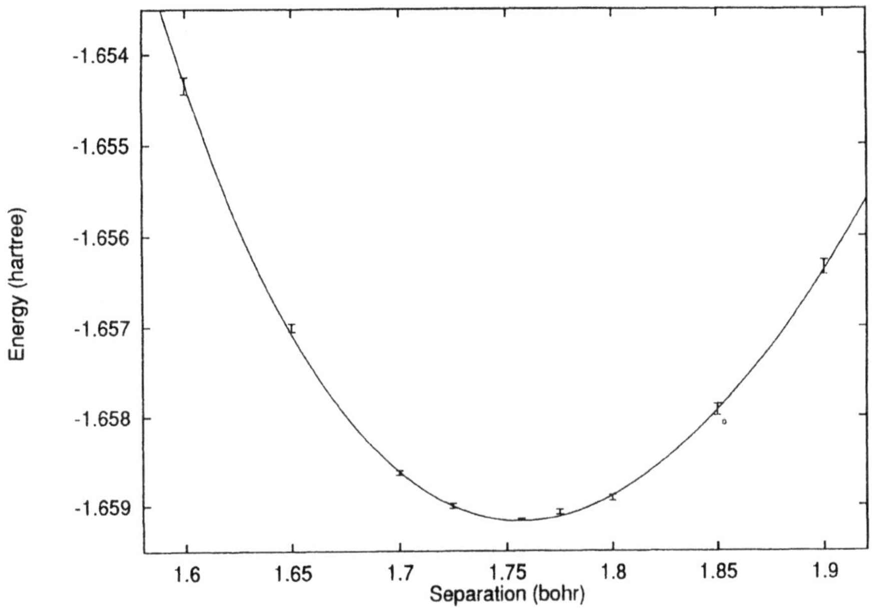

Fig. 12. Electronic energies calculated for collinear symmetric H-H-H. From Diedrich and Anderson (4).

TABLE III

Energies for the reaction intermediate H-H-H as determined in exact quantum Monte Carlo calculations (4). Collinear symmetric configurations near the saddle point.

Internuclear Separation bohrs	Total Energy hartrees	Elevation[a] kcal/mol
1.600	-1.65442 ± 0.00012	12.59 ± 0.07
1.650	-1.65701 ± 0.00005	10.96 ± 0.03
1.700	-1.658636 ± 0.000024	9.94 ± 0.02
1.725	-1.659004 ± 0.000026	9.71 ± 0.02
1.757	-1.659154 ± 0.000014	9.61 ± 0.01
1.775	-1.659072 ± 0.000027	9.67 ± 0.02
1.800	-1.65891 ± 0.00003	9.77 ± 0.02
1.850	-1.65793 ± 0.00006	10.38 ± 0.04
1.900	-1.65634 ± 0.00008	11.38 ± 0.05

[a] Elevation above that for separated H + H_2, −1.674 476 hartrees.

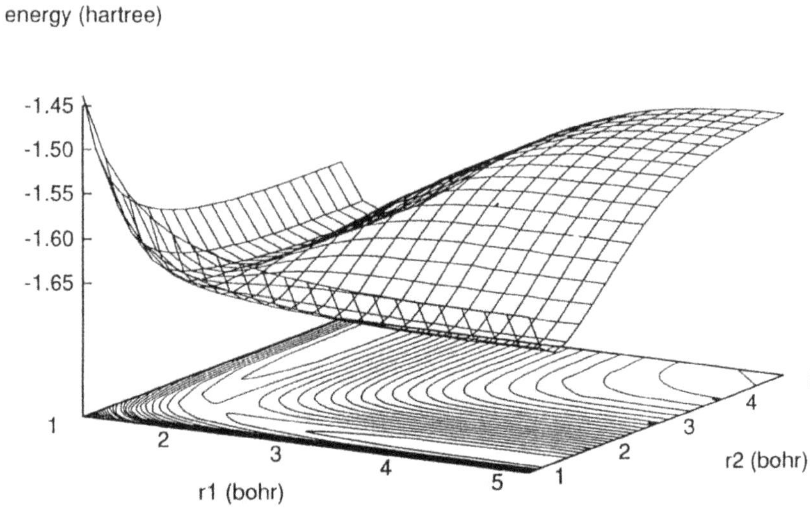

Fig. 13. Potential energy surface for collinear H-H-H. From Diedrich and Anderson (5).

variational calculations by Liu(69) for collinear configurations shows the Monte Carlo energies to be 0.25 to 0.28 kcal/mol lower. Liu's surface and the Monte Carlo surface are identical in shape within the Monte Carlo error bars in this region. Liu's estimate of 9.59 ± 0.06 kcal/mol for the minimum in the barrier overlaps the exact quantum Monte Carlo barrier height of 9.61 kcal/mol. In fact, Liu's earlier estimate(68) and that of Ceperley and Alder(36) also straddle the 'exact' result.

The region of the Jahn-Teller cusp occurring for equilateral- triangle nuclear configurations was investigated in detail. The cusp is the result of a conical intersection of the ground state with an excited state. For D_{3h} configurations there are two wave functions with the same energy but with different symmetries for reflection in a plane passing through the axis of rotation and one nucleus. One of the states is symmetric to reflection in the plane and the other is antisymmetric. When the D_{3h} symmetry is broken and reduced to C_{2v} symmetry, as by varying one of the angles of the triangle, one of the energies rises and the other falls. Thus, the ground state wave function switches symmetry as the nuclear configuration passes through D_{3h} symmetry.

To distinguish the degenerate wave functions, which may tend to contaminate each other, it was necessary to impose the appropriate symmetry on the psip population at each step of the calculations and to use trial functions of the appropriate

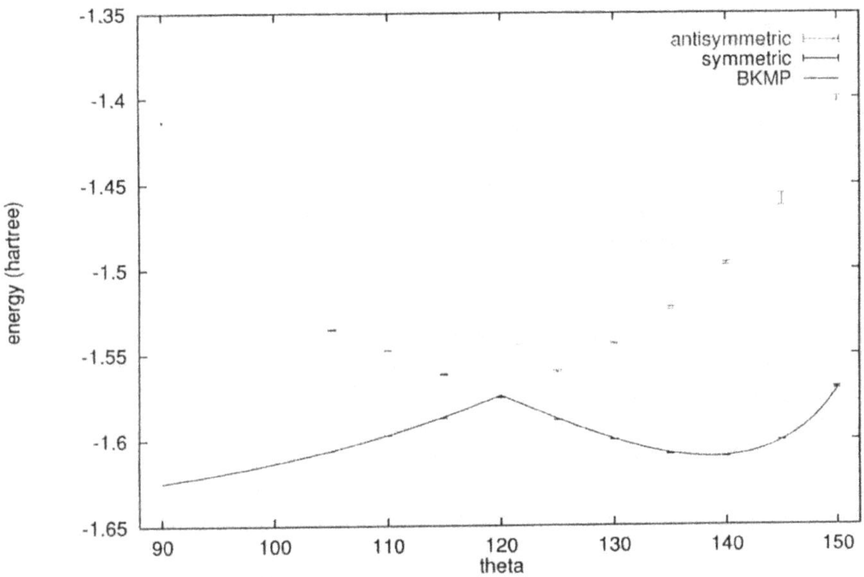

Fig. 14. The Jahn-Teller cusp for H-H-H. With H-H-H identified as A-B-C, the distances A-B and B-C are fixed at 2.0 bohrs and θ is the angle A-B-C. From Diedrich and Anderson (5).

symmetry. The calculated energies for $r_{AB} = r_{BC} = 2.0$ bohrs are shown in Fig. 14.

The accuracy of the Monte Carlo calculation for the H-H-H saddle point configuration is $\pm 0.000\,014$ hartrees or 3 cm^{-1}. Clementi *et al.* (70) recently speculated that an analytic variational calculation for H-H-H accurate to within 10 cm^{-1} of the exact value "would call for somewhere around 3500 years" on a computer similar to the one used for the Monte Carlo calculation. As Clementi *et al.* pointed out, there are many opportunities for significant reductions in the computational effort for analytic variational calculations and it may be possible to "reduce the task to 10 to 20 years" or even less (70). Other workers suggest 10 to 20 hours is a possibility. There are also many opportunities for improving the Monte Carlo calculations. In any case three months is a lot less time than three millenia.

12 The Helium Dimer He_2

The helium dimer presents another one of those problems which has been waiting a long time for an exact solution. The exact cancellation scheme has now provided a solution in the form of exact values of the potential energy for points on the Born-Oppenheimer potential energy curve. These points provide an excellent confirmation of experimental measurements: their error bars straddle the experimental-theoretical compromise potential energy curve of Aziz and Slaman(71) which successfully predicts a variety of experimental measurements.

The history of theoretical predictions of the attractive van der Waals potential for two helium atoms begins with an estimate by Slater(72) in 1928, which was followed in 1930 by the work of Eisenshitz and London (73) and London(74) who first explained the nature of the fluctuating moment/induced moment attraction and provided the simple approximate double perturbation formula for estimating what are now called London dispersion forces.

Very simple SCF calculations in the 1930's were unsuccessful in predicting the attractive potential well, but SCF calculations in the 1950's with somewhat larger basis sets gave rather good predictions. However, SCF calculations in the 1960's with still larger basis sets predicted no attractive well even for calculations approaching the Hartree-Fock limit. The wells of the 1950's were the fortunate or unfortunate result of the what is now called 'basis set superposition error'.

The term 'basis set superposition error' was introduced to the literature of quantum mechanics by Liu and McLean(75) in their 1973 paper describing their first calculation of the helium-helium interaction. In that work and in two subsequent investigations Liu and McLean (76,77) made repeated improvements in eliminating basis set superposition error in extensive CI calculations. Their sequence of results for the minimum in the potential curve relative to separated atoms is −9.5 K in 1973, −10.76 K in 1975, and −10.94 in 1989. Recent extensive CI calculations by Vos *et al.* (78) reported in 1991, also designed to eliminate basis set superposition error, give a minimum of −10.92 K.

The accurate prediction of the helium-helium potential is an extremely difficult problem. The well depth is about 11 K (about 0.000 035 hartrees) and one would like an accuracy of at least one percent of the depth or 0.1 K (0.000 000 3 hartrees). The total electronic energy is about that of two separated helium atoms or −5.807 448 hartrees, so that the desired accuracy is about one part in 2×10^7 of the total energy and one part in 3×10^5 of the correlation energy.

The recent calculations of Liu and McLean(77) yield total electronic energies 1200 K above the exact values for the total energies. Yet, the relative values of the energies, obtained as the differences in total energies of interacting atoms and separated atoms calculated with the same basis set, are in excellent agreement with the exact Monte Carlo results and the potential energy curve derived from experiments. Nevertheless, one should be cautious about relative errors of less than 0.1 K in calculations with total errors of 1200 K.

The helium dimer is a good candidate for an exact Monte Carlo calculation. An excellent trial wave function with constants optimized for several separations is available from earlier variational Monte Carlo calculations. For the nuclear separations of interest the electrons of the same spin are well separated and tend to be held at different nuclei, so that a high ratio of positive to negative psips is easily maintained.

The exact quantum Monte Carlo calculations of Anderson *et al.* (7) gave very high accuracies for the helium-helium interaction energies. The trial function used was that of Lowther and Coldwell(21) consisting of the antisymmetrized product of the wave functions for two helium atoms together with exponential cross terms accounting for the interactions of electrons of one atom with the electrons and nucleus of the other atom. The function is

$$\Psi_T = \phi_{A13}\phi_{B24}\exp(-\frac{1}{2}U_{1234}) - \phi_{A14}\phi_{B23}\exp(-\frac{1}{2}U_{1423}) \quad (57)$$
$$\phi_{A23}\phi_{B14}\exp(-\frac{1}{2}U_{2314}) + \phi_{A24}\phi_{B13}\exp(-\frac{1}{2}U_{2413})$$

where electrons 1 and 2 are spin up, electrons 3 and 4 are spin down, ϕ_{A13} is an atomic wave function for electrons 1 and 3 interacting with nucleus A, ϕ_{B24} is an atomic wave function for electrons 2 and 4 interacting with nucleus B, U_{1234} is a cross term accounting for interactions of 1 and 3 near A with 2 and 4 near B, and so forth.

The function ϕ_{A13} (and others like it) is the 189-term Hylleraas wave function for helium due to Schwartz(79) which yields in variational calculations an energy within 10^{-9} of the exact value for helium. The function U_{1324} (and others like it) is a sum of five complex terms terms containing 29 adjustable parameters devised by Lowther and Coldwell(21) and optimized for a number of internuclear separations.

Cancellations were carried out only for psips with all electrons within 1.0 bohr of one or the other nucleus. With the nuclei placed on the z-axis psips eligible for cancellation were concentrated in configuration space by rotation of all electrons about the z-axis to place electron 1 in the $y = 0$ plane and by reflection in the $y = 0$ plane to place electron 2 in the region $y > 0$. Exchange symmetry for electrons of the same spin was imposed by permuting electrons 1 and 2 if $z_2 < z_1$ and permuting electrons 3 and 4 if $z_4 < z_3$. This also concentrated psips into the regions corresponding to electrons 1 and 3 near nucleus A and 2 and 4 near nucleus B.

The zero of potential energy was fixed at -6.5 hartrees to obtain reasonably large step sizes and to limit the frequency of sign changes due to positive values for the potential for close electron-electron encounters. For a total population of 20,000 psips the positive to negative ratio was typically 50 to 1.

The calculations were performed on Thinking Machines CM-5 Connection Machine computers with 32 to 256 processors. The speed and memory of each processor were sufficiently high that completely independent calculations with 20,000 psips could be performed on each processor. Thus, the calculations were fully parallel.

Energies were determined for eight internuclear distances in the range of 3.0 to 15.0 bohrs as well as for infinite separation. Results are listed in Tables IV and V

TABLE IV
Total electronic energies for the helium-helium system calculated by exact quantum Monte Carlo method (7).

Internuclear distance R, bohrs	Total electronic energy V, hartrees
3.0	$-5.79537 \pm 0.000\,30$
4.5	$-5.807259 \pm 0.000\,006$
5.0	$-5.807450\,0 \pm 0.000\,002\,5$
5.6	$-5.807483\,41 \pm 0.000\,000\,53$
6.6	$-5.807468\,55 \pm 0.000\,000\,32$
7.5	$-5.807458\,42 \pm 0.000\,000\,32$
9.0	$-5.807451\,81 \pm 0.000\,000\,32$
15.0	$-5.807448\,87 \pm 0.000\,000\,16$
∞	$-5.807448\,71 \pm 0.000\,000\,06$

TABLE V
Intermolecular potential for the helium-helium system calculated by exact quantum Monte Carlo method (7). Other values are listed for comparison.

Internuclear distance	Potential Energy, K		
R bohrs	Expt./Theory Exact QMC[a]	Theory LM-2 Aziz-Slaman[b]	Liu-McLean[c]
3.0	$3812. \pm 96.$	3792.	3767.
4.5	60.0 ± 1.9	58.5	
5.0	-0.41 ± 0.78	−0.39	−0.33
5.6	-10.94 ± 0.17	−10.97	−10.94
6.6	-6.25 ± 0.10	−6.36	
7.5	-3.05 ± 0.10	−3.00	−3.007
9.0	-0.96 ± 0.10	−0.99	−0.985
15.0	-0.04 ± 0.05	−0.042	
	$+0.01 \pm 0.02$	(0.0)	(0.0)

[a]Ref. (7). [b]Ref. (71). [c]Ref. (77).

and plotted in Fig. 15. The error bars of the calculated interaction energies straddle the analytic variational results of Liu and McLean(77), the analytic variational results of Vos *et al.* (78), and the experimental-theoretical compromise potential of Aziz and Slaman(71). The exact quantum Monte Carlo interaction energies are not 'better' but they are known to be free of basis set superposition error, any other systematic error, and any inputs beyond the basic constants. In the exact quantum Monte Carlo case one may judge the experimental curve by how well it

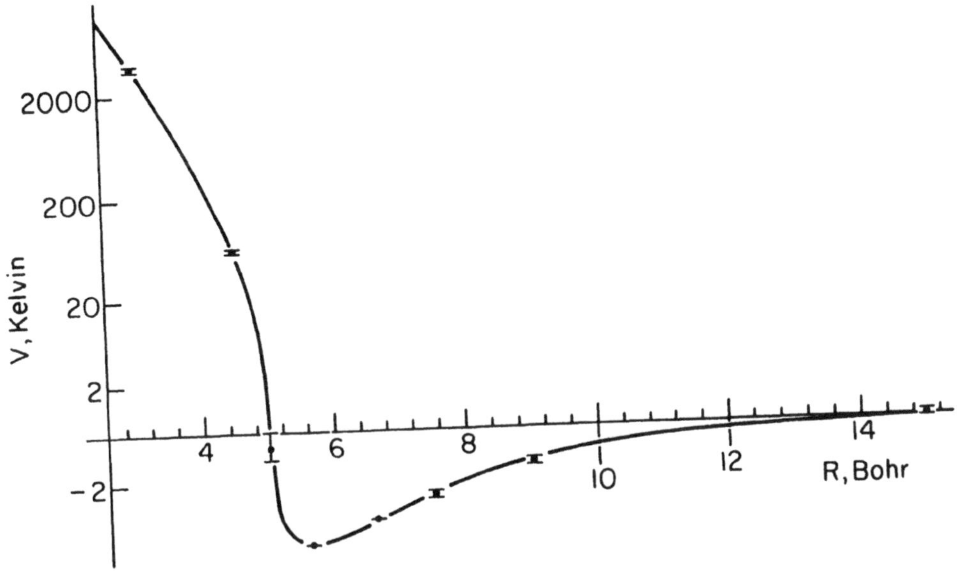

Fig. 15. The He-He interatomic potential. Potential energy in K vs internuclear distance in bohrs. The solid curve is the Aziz-Slaman(71) experimental/theoretical compromise potential. The points with error bars are the results of exact quantum Monte Carlo calculations by Anderson, Traynor, and Boghosian (7). The ordinate is scaled with the function $f(V) = sign(V)ln(1+|V|)$.

agrees with the theoretical curve rather than the reverse.

13 The H-He Interaction

The van der Waals attraction in the H-He system is one of the weakest for pairs of neutral atoms. The attractive well depth lies between that of the ${}^3\Sigma_u{}^+$ H-H system (6.46 K) and that of the He-He system (11.0 K). Exact quantum Monte Carlo calculations (8) performed for H-He in our laboratory have yielded an interaction potential with a depth of 6.8 ± 0.5 K at the equilibrium distance of 7.0 bohrs.

Since experimental measurements are much more difficult for the H-He system than for the He-He system there is much less experimental information for the H-He system and there have been correspondingly fewer calculations of the H-He interaction. Nevertheless, high-energy total scattering experiments (80) have given information on the repulsive part of the potential, low-energy scattering experiments (81) have given information on the attractive part, and these are supplemented with measurements (82) of the diffusion of H in He at low temperatures. The experimental results are consistent with several similar potential energy curves (83) having well depths of 6.8 to 7.2 K.

TABLE VI
The interaction potential for H-He determined in exact quantum Monte Carlo calculations (8). Partial list of points.

Internuclear Separation bohrs	Total Energy hartrees	Relative Energy K
5.0	$-3.403506 \pm 0.000\,014$	$69. \pm 4.$
6.0	$-3.403739 \pm 0.000\,003$	-4.5 ± 0.9
7.0	$-3.4037459 \pm 0.000\,0016$	-6.8 ± 0.5
8.0	$-3.4037382 \pm 0.000\,0006$	-4.4 ± 0.2
9.0	$-3.4037306 \pm 0.000\,0006$	-1.9 ± 0.2
10.0	$-3.4037276 \pm 0.000\,0007$	-1.0 ± 0.2
11.0	$-3.4037250 \pm 0.000\,0006$	-0.2 ± 0.2
15.0	$-3.4037246 \pm 0.000\,0003$	-0.06 ± 0.10
∞	$-3.4037243 \pm 0.000\,0001$	-0.02 ± 0.02

Previous theoretical work follows the pattern of that for the helium-helium interaction with about a dozen calculations from the 1930's to the present. The lowest-energy analytic variational calculation is that by Das, Wagner, and Wahl (84) yielding a well depth of 6.5 K at a distance of 6.8 bohrs. The total energy in that calculation was 12,000 K above the exact total energy.

In the exact quantum Monte Carlo calculation the trial function used was the antisymmetrized product of a Hylleraas function for the helium atom (A) and a $1s$ function for the hydrogen atom (B) with Jastrow functions. The Hylleraas function was the 189-term series due to Schwartz(78), the $1s$ function was the exact function for H, and the Jastrow function included cross terms for the electron-electron and electron-nucleus interactions. The calculations were run on a Cray C90 using a code adapted to take advantage of the vector capabilities of that machine to obtain sustained speeds of about 350 megaflops. The total population of psips was 30,000 and the positive-to-negative ratio was about 100 to 1.

At its minimum at a separation of 7.0 bohrs the calculated energy was $-3.403\,746 \pm 0.000\,002$ hartrees. Potential energies for nuclear separations of 5.0 to 15.0 bohrs and for infinite separation are listed in Table VI. The calculated potentials and those derived from experiment were found consistent with each other.

14 Prognostication

It is clear that the Green's function cancellation method described in these pages can provide 'exact' results for molecular systems of a few electrons. In the past half century available computer speeds have increased by a factor of ten about every four years. Speeds have increased from about 10 floating point operations per second in 1945 to about 10^{12} floating point operations per second in 1994 without a significant change in cost. And there is no end in sight. Thus, one can expect calculations for few-electron systems to become relatively cheaper and correspondingly more accurate.

For not-so-few-electron systems and larger systems predictions are much more difficult. Adding an electron increases the dimensionality of a problem by three and increases the degree of difficulty of cancellation by a factor of ten at the very least. Without any improvements in technique, one might then expect to progress to larger problems at a rate no faster than an additional electron every four years. That would seem discouraging except for the fact that improvements in quantum Monte Carlo methods continue to be made at a fairly rapid pace. There is a backlog of good ideas in the quantum Monte Carlo area waiting for its practitioners to find the time to develop them. Some of these are sure to lead to better 'exact' quantum Monte Carlo calculations. Others are sure to lead to better approximate quantum Monte Carlo calculations - not exact, but very, very accurate - perhaps more accurate than those of any other method for larger systems.

15 Acknowledgments

Support of this work by the National Science Foundation and the Office of Naval Research is gratefully acknowledged. The work was facilitated by an award from the Humboldt Foundation.

References

1. P. A. M. Dirac, Proc. Roy. Soc. (London) **A123**, 714 (1929).
2. C. A. Traynor, J. B. Anderson, and B. M. Boghosian, J. Chem. Phys. **94**, 3657 (1991).
3. J. B. Anderson, J. Chem. Phys. **96**, 3702 (1992).
4. D. L. Diedrich and J. B. Anderson, Science **258**, 786 (1992).
5. D. L. Diedrich and J. B. Anderson, J. Chem. Phys. **100**, 8089 (1994).
6. A. Bhattacharya and J. B. Anderson, Phys. Rev. A **49**, 2441 (1994).
7. J. B. Anderson, C. A. Traynor, and B. M. Boghosian, J. Chem. Phys. **99**, 345 (1993).
8. A. Bhattacharya and J. B. Anderson, J. Chem. Phys. **100**, 8999 (1994).
9. J. G. Zabolitzky, Prog. Part. Nucl. Phys. **16**, 103 (1985).
10. B. J. Alder, Fut. Gen. Comp. Sys. **5**, 191 (1989).
11. J. B. Anderson, J. Chem. Phys. **63**, 1499 (1975).
12. J. B. Anderson, J. Chem. Phys. **65**, 4121 (1976).
13. H. Conroy, J. Chem. Phys. **51**, 3979 (1969).
14. E. P. Wigner, Phys. Rev. **40**, 749 (1932).
15. N. Metropolis and S. Ulam, J. Am. Stat. Assoc. **44**, 335 (1949).
16. G. W. King, in Proceedings IBM Computation Seminar, Endicott, New York, 1949 (International Business Machines Corp., New York, 1951), p. 92.
17. A. Einstein, Ann. Phys. 17, 549 (1905); **19**, 371 (1906).
18. M. H. Kalos, Phys. Rev. A **128**, 1791 (1962).
19. R. C. Grimm and R. G. Storer, J. Comput. Phys. **7**, 134 (1971).
20. A. Bijl, Physica 7, 869 (1940); R. B. Dingle, Phil. Mag. **40**, 573 (1949); R. Jastrow, Phys. Rev. **98**, 1479 (1955).
21. R. E. Lowther and R. L. Coldwell, Phys. Rev. A **22**, 14 (1980).
22. R. N. Barnett, P. J. Reynolds, and W. A. Lester, J. Chem. Phys. **84**, 4922 (1986).
23. N. C. Handy and S. F. Boys, Proc. Roy. Soc. (London) **A310**, 63 (1969).
24. K. E. Schmidt and J. W. Moskowitz, J. Chem. Phys. **93**, 4172 (1990).
25. C. J. Umrigar, K. G. Wilson, and J. W. Wilkins, Phys. Rev. Lett. **60**, 1719 (1988).
26. J. B. Anderson, Phys. Rev. A **35**, 3550 (1987).
27. D. M. Ceperley, J. Stat. Phys. **63**, 1237 (1991).
28. J. B. Anderson, Int. J. Quant. Chem. **15**, 109 (1979).
29. P. J. Reynolds, D. M. Ceperley, B. J. Alder, and W. A. Lester, J. Chem. Phys. **77**, 5593 (1982).
30. F. Mentch and J. B. Anderson, J. Chem. Phys. **80**, 2675 (1984).
31. D. R. Garmer and J. B. Anderson, J. Chem. Phys. **86**, 4025 (1987).
32. D. R. Garmer and J. B. Anderson, J. Chem. Phys. **86**, 7237 (1987).
33. L. Mitas, E. L. Shirley, and D. M. Ceperley, J. Chem. Phys. **95**, 3467 (1991).
34. D. M. Schrader and K. Iguchi, J. Chem. Phys. **98**, 7185 (1993).
35. D. M. Ceperley and B. J. Alder, Phys. Rev. Lett. **45**, 566 (1980).
36. D. M. Ceperley and B. J. Alder, J. Chem. Phys. **81**, 5833 (1984).
37. D. M. Arnow, M. H. Kalos, M. A. Lee, and K. E. Schmidt, J. Chem. Phys. **77**, 5562 (1982).
38. D. F. Coker and R. O. Watts, Mol. Phys. **58**, 1113 (1986).
39. J. B. Anderson, C. A. Traynor, and B. M. Boghosian, J. Chem. Phys. **95**, 7418 (1991).
40. S. Zhang and M. H. Kalos, Phys. Rev. Lett. **67**, 3074 (1991).
41. D. F. Coker, R. E. Miller, and R. O. Watts, J. Chem. Phys. **82**, 3554 (1985).
42. D. M. Ceperley and B. Bernu, J. Chem. Phys. **93**, 552 (1990).
43. J. Carlson and M. H. Kalos, Phys. Rev. C **32**, 1735 (1985).
44. R. Bianchi, D. Bressanini, P. Cremaschi, and G. Morosi, Comp. Phys. Comm. **74**, 153 (1993); J. Chem. Phys. **98**, 7204 (1993).
45. J. Mercer and H. Allen, Ac-cent-tchu-ate The Positive, Harwin Music Corp., New York, 1944.
46. G. Carney and R. N. Porter, J. Chem. Phys. **60**, 4251 (1974); **65**, 3547 (1976).
47. T. Oka, Phys. Rev. Lett. **45**, 531 (1980).
48. J.-T. Shy, J. W. Farley, W. E. Lamb, and W. H. Wing, Phys. Rev. Lett. **45**, 535 (1980).
49. J. O. Hirschfelder, J. Chem. Phys. **6**, 795 (1938).

50. F. Mentch and J. B. Anderson, J. Chem. Phys. **74**, 6307 (1981).
51. J. B. Anderson, J. Chem. Phys. **86**, 2839 (1987).
52. C. A. Traynor and J. B. Anderson, Chem. Phys. Lett. **147**, 389 (1988).
53. D. Frye, A. Preiskorn, G. C. Lie, and E. Clementi, J. Chem. Phys. **92**, 4948 (1990).
54. R. Rohse, W. Klopper, and W. Kutzelnigg, J. Chem. Phys. **99**, 8830 (1993).
55. H. M. James and A. S. Coolidge, J. Chem. Phys. **1**, 825 (1933).
56. W. Kolos and C. C. J. Roothaan, Rev. Mod. Phys. **32**, 219 (1960).
57. W. Kolos and L. Wolniewicz, Rev. Mod. Phys. **35**, 473 (1963); J. Chem. Phys. **43**, 2429 (1965); **49**, 404 (1968).
58. M. Born, Nachr. Akad. Wiss. Göttingen **1**(1951).
59. L. Wolniewicz, J. Chem. Phys. **78**, 6173 (1983).
60. W. Kolos, K. Szalewicz, and H. Monkhorst, J. Chem. Phys. **84**, 3278 (1986).
61. D. M. Bishop and Cheung, Phys. Rev. A **18**, 1846 (1978).
62. G. Herzberg, J. Mol. Spectrosc. **33**, 147 (1970).
63. W. C. Stwalley, Chem. Phys. Lett. **6**, 241 (1970).
64. E. McCormack, J. M. Milligan, C. Cornaggia, and E. E. Eyler Phys. Rev. A **39**, 2260 (1990).
65. L. Wolniewicz, J. Chem. Phys. **99**, 1851 (1993).
66. A. Balakrishnan, V. Smith, and B. Stoicheff, Phys. Rev. Lett. **68**, 2149 (1992).
67. S. F. Boys, G. B. Cook, C. M. Reeves, and I. Shavitt, Nature **178**, 1207 (1956).
68. B. Liu, J. Chem. Phys. **58**, 1925 (1973).
69. B. Liu, J. Chem. Phys. **80**, 581 (1984).
70. E. Clementi, G. Corongiu, D. Bahattacharya, B. Feuston, D. Frye, A. Preiskorn, A. Rizzo, and W. Xue, Chem. Rev. **91**, 679 (1991).
71. R. A. Aziz and M. J. Slaman, J. Chem. Phys. **94**, 8047 (1991).
72. J. C. Slater, Phys. Rev. **32**, 349 (1928).
73. R. Eisenshitz and F. London, Zeits. f. Physik **60**, 491 (1930).
74. F. London, Zeits. f. Physik **63**, 245 (1930).
75. B. Liu and A. D. McLean, J. Chem. Phys. **59**, 4557 (1973).
76. B. Liu and A. D. McLean, unpublished results (1975). Cited by B. Liu and A. D. McLean, J. Chem. Phys. **72**, 3418 (1980).
77. B. Liu and A. D. McLean, J. Chem. Phys. **92**, 2348 (1989).
78. R. J. Vos, T. van Mourik, J. H. van Lenthe, and F. B. van Duijneveldt, to be published. Cited by Aziz and Slaman, Ref. 71.
79. C. Schwartz, Phys. Rev. **128**, 1146 (1962).
80. R. Gengenbach, C. Hahn, and J. P. Toennies, Phys. Rev. A **7**, 98 (1973).
81. J. P. Toennies, W. Welz, and G. Wolf, Chem. Phys. Lett. **44**, 5 (1976).
82. W. N. Hardy, M. Morrow, R. Jochemson, B. W. Statt, P. R. Kubik, R. M. Marsolais, A. J. Berlinsky, and A. Landesman, Phys. Rev. Lett. **45**, 453 (1980).
83. R. Jochemson, A. J. Berlinsky, and W. N. Hardy, Can. J. Phys. **62**, 751 (1984).
84. G. Das, A. F. Wagner, and A. C. Wahl, J. Chem. Phys. **68**, 4917 (1978).

Achieving Chemical Accuracy with Coupled-Cluster Theory

Timothy J. Lee
NASA Ames Research Center, Moffett Field, California 94035-1000

and

Gustavo E. Scuseria
Department of Chemistry and Rice Quantum Institute, Rice University, Houston, Texas 77251-1892

August-1994

Abstract

Due to formal and computational advances in coupled-cluster theory over the past few years, it is now possible to obtain very accurate molecular geometries, vibrational frequencies, heats of formation, binding energies, and vertical electronic excitation energies. For example, based on statistical analyses of a large number of calculations, it is shown that the CCSD(T)/*spdfg* level of theory gives r_{XH}, r_{XY} (double bonds), and r_{XY} (triple bonds) with an average error of 0.0010, 0.0020, and 0.0026 Å, respectively, with the theoretical bond distances usually too long relative to experiment. This level of theory yields bond angle predictions that are too small by 0.21 degrees on average. Fundamental vibrational frequencies predicted at the CCSD(T)/*spdfg* level of theory are accurate to better than 8.0 cm^{-1} on average, but the remaining errors are less systematic than those found for the geometrical parameters, except for X-Y stretches which are usually underestimated relative to experiment. For molecules described reasonably well by a single determinant reference function, single- and multiple-bond energies are given to within 1.0 and 2.0 kcal/mol, respectively, at the CCSD(T)/*spdfg* level of theory. The present monograph reviews the advances that have lead to the current state-of-the art, and also summarizes selected examples from the published literature.

S. R. Langhoff (Ed.), Quantum Mechanical Electronic Structure Calculations with Chemical Accuracy, 47–108.

Contents

1 Introduction

Coupled-cluster (CC) theory, in its various forms, has become one of the most widely used electron correlation procedures over the last several years, especially for calculations involving a single-reference wave function. It is now generally believed that, as far as single-reference electron correlation methods are concerned, coupled-cluster theory provides the most accurate description of electronic structure, and this is the primary reason for its popularity. It should be appreciated, however, that the ultimate accuracy of any *ab initio* calculation is firstly limited by the choice of one-particle basis set (an excellent review of this topic has recently been given [1] by Helgaker and Taylor). It is then important to use an electron correlation procedure that: a) recovers a large fraction of the dynamical correlation energy; b) includes important non-dynamical electron correlation effects; and c) gives correlation energies that scale linearly with respect to the number of electrons included in the procedure, or in other words, is size-extensive [2]. Much research over the past twenty years has shown that CC theory provides several approximate methods that satisfy these conditions. There have been several excellent reviews of the theoretical basis and mathematical foundations of CC methodology [3-11], and for these aspects the interested reader is referred to those. Some of these reviews also present discussion of selected chemical applications of coupled-cluster theory. The present article is concerned primarily with recent applications of CC theory in which the prediction of molecular equilibrium structures, vibrational spectra, heats of formation, binding energies, and vertical electronic excitation energies provides accuracy rivaling the best experiments, and also with the methodological advances that have made this accuracy possible.

There are several different types of coupled-cluster methods that are in common use today, and various forms and implementations are readily available in several quantum chemistry program packages such as ACES II [12], CADPAC [13], GAUSSIAN [14], MOLCAS [15], MOLPRO [16], PSI [17], TITAN [18], and others. It would be impossible to review or summarize recent applications from all of the different CC methods. Thus some restrictions in this review article are necessary, and the first shall be to discuss only single-reference based coupled-cluster techniques. The theoretical foundations of single-reference based methods are generally more complete with the one exception of open-shell CC theories that are based on a restricted Hartree-Fock (RHF) reference function (see later discussion), whereas the best approach for a multireference based coupled-cluster procedure is not generally agreed upon at this juncture (for some recent discussion of multireference coupled-cluster methods see Refs. [11, 19, 20]). However, even limiting the present review to single-reference based CC methods leads to an enormous literature. For example, it is widely known that the quadratic configuration-interaction (QCI) method [21] may be viewed as an approximate coupled-cluster technique [22-24], and indeed it has been shown explicitly that results from CC and analogous QCI methods are generally very similar [10, 25]. In addition, there has been some recent interest in so-called Brueckner [26-36] CC methods in which a specific type of orbital optimization is performed simultaneously. The literature involving the QCI method is particularly extensive, in large part due to its availability in the GAUSSIAN [14] series of electronic structure programs. Raghavachari [10] and Curtiss and Raghavachari [37] have recently presented very interesting reviews primarily aimed at the QCI method and many of its applications, and therefore we shall mainly concentrate on the singles and doubles coupled-cluster (CCSD) [38] and CCSD(T) [39] methods (CCSD plus a perturbational estimate of the effects of connected triple excitations). However, we stress that for many of the applications reviewed herein, the analogous QCI or Brueckner CC methods would yield similar accuracy to the CCSD or CCSD(T) approaches provided that the same one-particle basis sets were used.

In order to set the stage for the current chapter, it is instructive to review some of the key studies in which it was demonstrated that very accurate molecular geometries, vibrational frequencies, binding energies, and other molecular properties could be obtained at modest computational cost using coupled-cluster methods. There are two basic procedures by which a given *ab initio* electron correlation method may be benchmarked or judged. That is, results from the method may be compared to full configuration-interaction (FCI) calculations for a given one-particle basis set, or if large enough basis sets are used, results from the method may be compared to accurate experimental values. In the mid 1980's the computational chemistry group at NASA Ames provided many such FCI benchmark studies [40] in realistic basis sets, and among other aspects they demonstrated that specific types of multireference CI (MRCI) methods reproduce FCI total energies very well, often to microhartree accuracy. Unfortunately, as the number of electrons in the molecule increases, these MRCI approaches become computationally expensive very quickly. Nonetheless, they are still less costly than FCI methodology, and thus they may be used to calibrate other correlation methods when FCI is not feasible. Bartlett and co-workers [41, 42] were the first to compare CC methods with the new FCI results and they showed that it was necessary to include some estimate of connected triple excitations in the coupled-cluster wave function in order to reduce errors in the correlation energy to the 10–500 microhartree range. Since the CCSDT method [42, 43] is iterative and the computational cost scales as the eighth power of the basis in each iteration, several less costly approaches to approximating the contribution of connected triple excitations have been proposed and these are discussed later. The CCSD(T) approach [39] has become the method of choice as it gives very accurate molecular properties, and yet is economical allowing the use of large one-particle basis sets on medium sized molecules. Soon after Raghavachari *et al.* proposed the CCSD(T) method, it was shown to yield binding energies for small Be_n and Mg_n (n=3-5) clusters that are in excellent agreement with high-level MRCI values [25, 44]. This is despite the fact that the CCSD binding energies were much too small. Several studies soon followed in which the accuracy and usefulness of the CCSD(T) method used in conjunction with basis sets that include f and g-type functions was demonstrated. For example, it was shown to yield equilibrium geometries, and harmonic and fundamental vibrational frequencies for small Be_n, Mg_n, and Ca_n clusters that agree very well with high-level MRCI quantities [45-47]; it was shown to predict the equilibrium geometries of the notorious O_3 molecule [48] and the triply-bonded N_2 diatomic [49] to within 0.001 Å and 0.3° relative to experiment, and the fundamental vibrational frequencies to within ± 2 cm^{-1} excluding the very difficult O-O antisymmetric stretch, although the harmonic frequency of even this mode agreed with experiment to within 30 cm^{-1}. In a comparison of CCSDT to the various approximate methods [50], CCSD(T) was shown on average to perform much better than any of the others, and the direct comparison of CCSD(T) with FCI for the binding energy of Be_3 demonstrated that this agreement was not fortuitous [51]. The equilibrium structure of the difficult FOOF molecule was shown to be in good agreement with experiment [52], the equilibrium structures and harmonic frequencies of the HCN and HNC molecules were shown to be in excellent agreement with experiment [53], and the hyperpolarizability γ of Ne was shown [54, 55] to be in excellent agreement with CCSDT and experiment. The energy differences between the classical and non-classical forms of protonated acetylene [56] and between the linear and bent isomers of C_3^+ [57], and the classical barrier height of the prototypical $F + H_2 \rightarrow FH + H$ reaction [58] were shown to be in excellent agreement with high-level MRCI calculations. Last but not least, CCSD(T) was shown to give a correct qualitative picture for Cr_2, a molecule that has defied *ab initio* quantum chemistry methods due to its short metallic hextuple chemical bond [52, 59]. By this point, the potential of the CCSD(T) method when used in conjunction with basis sets that include at least f-type functions, had been established. Accordingly, there has been a literal explosion of the number of published studies using the CCSD(T) method since 1990, and there would appear to be no end in sight. With regards to applications, the present chapter is aimed simply at summarizing those CCSD(T) results in which chemical accuracy has been unambiguously obtained. This is not to say that CCSD(T) used in conjunction with smaller basis sets, such as triple-zeta double polarized (TZ2P), are not useful. It is just that in order to attain chemical accuracy for molecular geometries, fundamental frequencies, binding energies and other molecular properties, it is necessary to incorporate at least f-type functions in the one-particle basis set.

The present article is organized in the following way: section II, in several subsections, contains an overview of CC theory including the CCSD(T) method. Particular attention is given to extrapolation techniques and the computational cost of the various methods together with discussion of efficient and highly vectorized implementations. Some recent advances in open-shell CC theory are discussed and contrasted, together with some discussion of open-shell CCSD(T) methods. Finally, analytical gradient and response theory as applied to the CC method are summarized. Section III contains discussion regarding the type of accuracy that is currently achievable with state-of-the art CC methods, together with explicit examples that demonstrate the target accuracy. This section is split into several subsections based on the type of molecular property being considered. Section IV contains a summary and a discussion concerning very recent and possible future advances in CC methodology.

2 Theory

2.1 Coupled-Cluster Theory and Extrapolation Techniques

The foundation of coupled-cluster electronic structure theory is based on representing the electronic wave function as

$$|\Psi_e\rangle = e^T|\psi_0\rangle \quad , \tag{1}$$

where ψ_0 is a single-determinant reference function, usually taken as the Hartree-Fock determinant, and T is the cluster operator defined according to

$$T = T_1 + T_2 + T_3 + \dots \quad , \tag{2}$$

$$T_1 = \sum_{ia} t_i^a X_a^\dagger X_i \quad , \tag{3}$$

$$T_2 = \frac{1}{4}\sum_{iajb} t_{ij}^{ab} X_a^\dagger X_i X_b^\dagger X_j \quad , \tag{4}$$

where T_3 and higher cluster operators are defined similarly to T_1 and T_2. In equations (3) and (4) and throughout, we have adopted the notation that indices i, j, k, and l shall refer to occupied spin-orbitals in the reference function while a, b, c, and d shall refer to virtual spin-orbitals. In addition, $X_a^\dagger$ is a creation operator, X_i is an annihilation operator, and t_i^a and t_{ij}^{ab} are the unknown cluster amplitudes (or coefficients). Coupled sets of non-linear equations, which define the unknown amplitudes, may be derived in several ways. The most common approach, following Čižek [3], is to substitute $e^T|\psi_0\rangle$ into the electronic Schrödinger equation,

$$He^T|\psi_0\rangle = Ee^T|\psi_0\rangle \quad , \tag{5}$$

left-multiply by e^{-T}, and subtract the reference energy, $\langle\psi_0|H|\psi_0\rangle$, to yield

$$e^{-T}H'e^T|\psi_0\rangle = \Delta E|\psi_0\rangle \quad . \tag{6}$$

The correlation energy is given by $\Delta E \equiv E - \langle\psi_0|H|\psi_0\rangle$, and is obtained by projecting Eq. (6) onto the reference function,

$$\langle\psi_0|e^{-T}H'e^T|\psi_0\rangle = \Delta E \quad . \tag{7}$$

H' is given by $H - \langle\psi_0|H|\psi_0\rangle$. Coupled sets of non-linear equations defining the t_1 and t_2 amplitudes are then obtained, respectively, by projecting equation (6) onto the sets of singly excited determinants $\{\langle{}_i^a|\}$ and doubly excited determinants $\{\langle{}_{ij}^{ab}|\}$,

$$\langle{}_i^a|e^{-T}H'e^T|\psi_0\rangle = 0 \quad , \tag{8}$$

$$\langle {}^{ab}_{ij}|e^{-T}H'e^{T}|\psi_0\rangle = 0 \quad . \tag{9}$$

The CCSD model [38,60-66] is obtained by setting $T = T_1 + T_2$, whereas the CCSDT method [42, 43] is obtained by setting $T = T_1 + T_2 + T_3$ and by additionally left-projecting equation (6) onto the set of triply excited determinants $\{\langle {}^{abc}_{ijk}|\}$, in order to obtain a coupled set of non-linear equations defining the t_3 amplitudes. Higher-order methods are obtained in similar fashion and up to CCSDTQ ($T = T_1 + T_2 + T_3 + T_4$) has been implemented [67].

There are other approaches that may be used to derive the coupled set of non-linear equations defining the unknown amplitudes (for example, see Ref. [68]). In particular, diagrammatic approaches to the CC equations based on many-body perturbation theory usually defined a "connected" form of the effective Hamiltonian, which leads to a direct cancellation of low-order terms in higher-order CC equations. In this formalism, equation (6) is written as

$$\{H_N e^{T}\}_c|\psi_0\rangle = \Delta E|\psi_0\rangle \quad , \tag{10}$$

where H_N indicates that the normal product form of the Hamiltonian is used and $\{\}_c$ indicates that only the connected terms are included [3-9]. In principle, these different approaches should yield the same equations. However, it has been pointed out by Purvis and Bartlett [38] (and more recently by Hampel *et al.* [36]) that some approaches to deriving the t_1 and t_2 equations lead to the situation where equations defining the t_2 amplitudes may contain the t_1 equations. In other words, equation (9) may contain $\langle {}^{a}_{i}|e^{-T}H'e^{T}|\psi_0\rangle$, equation (8). While these additional terms will not affect the rate limiting computational steps and also will not affect the final results, they certainly unnecessarily complicate the t_2 equations and could conceivably affect the rate of convergence.

The existence and reality of solutions to the coupled, non-linear, and non-Hermitian equations that occur in coupled-cluster theory has been studied [69, 70], but it should be realized that due to the non-linearity, non-physical solutions (i.e., solutions that do not correspond to a root of the Hamiltonian) also exist. In general, it is not possible to guarantee that an obtained solution is physical (e.g., see Ref. [71]), but experience suggests that as long as the reference function is a reasonable approximation, the obtained solution will be the desired one. Of course, accuracy of the obtained solution is an entirely different matter. The coupled non-linear equations (8) and (9) for the CCSD model are solved by using an initial guess for the unknown amplitudes and then iterating the equations to self-consistency. It is usual, although by no means required, to partition the equations such that the initial guess for the t_2 amplitudes will be the coefficients of the first-order wave function in Møller-Plesset (MP) perturbation theory [72], which also requires that canonical molecular orbitals (MOs) be used. If canonical MOs are not used, this same partitioning would correspond to many body, but not MP perturbation theory. For the MP partitioning, the CCSD correlation energy from the 0th iteration corresponds to the second-order (MP2) perturbation theory energy, and the third-order (MP3) and fourth-order [MP4(SDQ), i.e., triples not included] energies can be obtained in subsequent iterations. Due to the rather slow rate of convergence for this iterative procedure, it is common to use some sort of extrapolation technique. Purvis and Bartlett [73] developed a reduced linear equation (RLE) method for this purpose, and found that atoms or molecules exhibiting large near-degeneracies (or non-dynamical correlation effects) showed better convergence with the RLE approach relative to a perturbation theory iterative method or to one that used Padé approximants. Oliphant and Adamowicz [74] have recently presented an improved RLE method specifically aimed at very difficult cases.

In 1986, Scuseria *et al.* [75] applied the DIIS (direct inversion in the iterative subspace) method of Pulay [76-78] to the problem of converging the CCSD equations and found markedly better convergence relative to a non-extrapolated method (at the time, RLE results specifically for the CCSD procedure were not available in the literature). In this study, several different factors were investigated, such as a) how small or large the norm of the error vector should be before an extrapolation was performed; b) how many iterations the system should be allowed to "relax" before performing another extrapolation; and c) how many error vectors and parameter vectors should be used in an extrapolation. However, not long thereafter, we discovered a better approach to solving the CCSD equations based on DIIS, and this method is still used in our programs [61, 62]. Since the details of this DIIS approach have never been published, we summarize them here.

The DIIS equations have been given in detail by Pulay and we shall not repeat them here. The main problems encountered when applying this approach to the CCSD equations are defining the error vector, and determining what parameter vector should be associated with a given error vector. One of the great benefits of applying DIIS to the self-consistent field (SCF) equations [76-78] is that the extrapolation can be applied in every iteration, thereby greatly increasing the rate of convergence. Hence our goal was to develop a procedure in which the DIIS extrapolation could be used every iteration. Following our earlier work [75], we let X be the array that contains the t_1 and t_2 amplitudes, $X = (t_1, t_2)$. Using the partitioning of the CCSD equations mentioned earlier and using no extrapolation procedure, we have $X_{i+1} = f(X_i)$, or in other words, the X_{i+1} vector is obtained by inserting the X_i vector into the partitioned CCSD equations. Now let Y designate an array of extrapolated t_1 and t_2 amplitudes. Using a DIIS extrapolation every iteration, the working equation becomes $X_{i+1} = f(Y_i)$, where X represents an array obtained from the partitioned CCSD equations and *not* from a DIIS extrapolation. Note that it is important not to include Y as a parameter vector in the DIIS procedure as this leads to a linear dependence in the DIIS equations (this is obvious since Y is simply a linear combination of the previous parameter vectors). Thus the parameter vectors are given by X_i and the associated error vector is $\delta_i = X_i - Y_{i-1}$. The procedure is started by defining Y_0 to be the coefficients of the first-order wave function (for closed-shell cases the initial t_1 amplitudes are zero), and then $X_1 = f(Y_0)$ and $\delta_1 = X_1 - Y_0$. Since the DIIS procedure requires at least two sets of parameter and error vectors to make an extrapolation, Y_1 is set equal to X_1, and then $X_2 = f(Y_1)$ and $\delta_2 = X_2 - Y_1$. In this case, it is valid to set $X_1 = Y_1$ since X_1 is not a linear combination of previous parameter vectors. Y_2 is obtained from the first DIIS extrapolation and the procedure is self sustaining. The possibility of improved

TABLE I
Comparison of the DIIS, RLE and modified RLE extrapolation procedures.[a]

Iteration	DIIS	RLE[b]	MRLE[c]	MRLE[d]
0	−0.057 4435	−0.057 4435	−0.057 4435	−0.057 4435
1	−0.054 4042	−0.045 5130	−0.048 9001	−0.048 9001
2	−0.112 8283	−0.059 1427	−0.064 9951	−0.065 5902
3	−0.095 6628	−0.065 7546	−0.077 1893	−0.083 8579
4	−0.136 8019	−0.058 8695	−0.093 1733	−0.105 3388
5	−0.098 6249	−0.058 1183	−0.100 6861	−0.115 9605
6	−0.106 4858	−0.055 0702	−0.099 9628	−0.110 5552
7	−0.106 9585	−0.009 8079	−0.100 5889	−0.108 4581
8	−0.108 7359	+0.084 2952	−0.102 8159	−0.108 2772
9	−0.108 6952	+0.143 0188	−0.105 9927	−0.108 6502
10	−0.108 8481	+0.018 8158	−0.106 6463	−0.108 8358
11	−0.108 8741	+0.015 5439	−0.108 0918	−0.108 8854
12	−0.108 8660	+0.055 2485	−0.107 9650	−0.108 8903
13	–	+0.042 6726	−0.108 0633	−0.108 7717
14	–	+0.030 4231	−0.108 2260	−0.108 8605
15	–	+0.034 9017	−0.108 3443	−0.108 8679
16	–	+0.051 1223	−0.108 5167	−0.108 8680
17	–	+0.043 1904	−0.108 5440	−0.108 8652
18	–	+0.048 5439	−0.108 5418	−0.108 8658
19	–	+0.053 9394	−0.108 5784	–
20	–	+0.062 8030	−0.108 6359	–

[a] Correlation energies given in E_h. Iteration 0 corresponds to the MP2 correlation energy. Test case is LiH at $\approx 3 * r_e$; see text for details. RLE and MRLE results taken from Ref. [74].
[b] Eventually diverges.
[c] Modified RLE; case 2 in Ref. [74]; converges in 30 iterations.
[d] Modified RLE; case 4 in Ref. [74].

performance by not extrapolating until the norm of the error vector was below some threshold was again investigated, but the best convergence behavior was obtained by extrapolating as soon as possible. The only remaining question is the maximum number of parameter and error vectors that should be used. As before [75], we have found eight to be a good compromise. Note that in order to avoid numerical instabilities, we scale the matrix containing the product of error vectors using the maximum absolute value. The above procedure has also been found to be effective in solving coupled sets of linear equations, although detailed comparisons with the many other approaches available for solving coupled sets of linear equations have not been performed.

As indicated above, Oliphant and Adamowicz [74] have published an example of the convergence of the CCSD equations using the RLE method and a modified RLE approach. The molecule was LiH with a bond distance of 9.045 a_0 (which

TABLE II
Two examples of the DIIS extrapolation procedure in CCSD calculations.[a]

Iteration	H_2O		O_3	
	Energy	RMS	Energy	RMS
0	−0.261 461 7994	–	−0.778 190 0271	–
1	−0.261 844 8077	0.20E-1	−0.696 906 2007	0.47E+0
2	−0.266 561 7777	0.47E-1	−0.753 303 6772	0.26E+0
3	−0.266 936 5056	0.52E-2	−0.740 956 7821	0.18E-1
4	−0.267 351 5853	0.10E-2	−0.743 171 8666	0.17E-1
5	−0.267 376 2270	0.34E-3	−0.744 850 8858	0.17E-2
6	−0.267 377 8152	0.57E-4	−0.745 214 3453	0.14E-2
7	−0.267 378 1460	0.15E-4	−0.745 134 0749	0.54E-3
8	−0.267 378 0610	0.36E-5	−0.745 129 9786	0.36E-3
9	−0.267 378 0029	0.85E-6	−0.745 141 3966	0.63E-4
10	−0.267 377 9833	0.12E-6	−0.745 142 9320	0.53E-4
11	−0.267 377 9807	0.73E-7	−0.745 143 7627	0.12E-4
12	−0.267 377 9816	0.28E-7	−0.745 144 2191	0.26E-5
13	−0.267 377 9819	0.28E-8	−0.745 144 0220	0.59E-6
14	−0.267 377 9820	0.32E-9	−0.745 144 1955	0.20E-6
15	–	–	−0.745 144 1608	0.14E-6
16	–	–	−0.745 144 1680	0.57E-7
17	–	–	−0.745 144 1693	0.23E-7
18	–	–	−0.745 144 1691	0.11E-7
19	–	–	−0.745 144 1701	0.50E-8
20	–	–	−0.745 144 1704	0.21E-8
21	–	–	−0.745 144 1706	0.49E-9

[a] Correlation energies given in E_h. Iteration 0 corresponds to the MP2 correlation energy. RMS is the root mean square of the DIIS error vector.

is about $3^* r_e$), using a small *sp* basis set. They investigated several modified RLE approaches. Results from the best two of these methods, together with the unmodified RLE approach, and the DIIS procedure described above are given in Table I. For this example, it is apparent that the DIIS method performs much better than the RLE or modified RLE approaches. Indeed, the unmodified RLE method diverges and DIIS takes only 12 iterations to converge the correlation energy to 10^{-6} E_h, compared to 18 iterations for the best modified RLE approach. Better extrapolation techniques may very well exist, but the DIIS procedure outlined above seems to yield the best convergence behavior of currently available methods.

To give a better idea of the convergence behavior for CCSD calculations employing the DIIS extrapolation technique, two additional examples are given in Table II. The CCSD correlation energy is given in E_h, and RMS is the root mean square of the DIIS error vector. The O_3 molecule provides an example where non-dynamical electron correlation is important, whereas the H_2O molecule is an example where non-dynamical correlation is relatively unimportant. Dunning's [79] cc-pVTZ basis set was used in both cases, and the heavy atom 1s-like MOs were constrained to be doubly occupied in all configurations (i.e., the frozen core approximation). Experimental geometries [80, 81] were used. It is evident that the CCSD equations converge more slowly for systems exhibiting large non-dynamical correlation effects. However, even for the worst case (O_3), the CCSD correlation energy is converged to 10^{-6} E_h in 11 iterations, and to 10^{-10} E_h in 21 iterations. In contrast, the CCSD correlation energy is converged to 10^{-6} E_h in 6 iterations and to 10^{-10} E_h in 14 iterations, for H_2O. As can be seen from Table II, the RMS value, which is a measure of the convergence of the CCSD wave function, is usually one or two orders of magnitude behind the convergence of the correlation energy.

In order to obtain quantitative or chemical accuracy, it is generally necessary to go beyond the CCSD approximation *and* it is necessary to use large one-particle basis sets. The CCSDT method is computationally very expensive, being an iterative n^8 process where n is the number of MOs, and thus it is difficult if not impossible to use large one-particle basis sets except on very small atoms or molecules. For comparison, CCSD is an iterative n^6 procedure. In order to avoid the n^8 computational cost, but still include the effects of higher connected excitations, several different approaches have been proposed. One approach that Bartlett and co-workers [42, 82, 83] have pioneered, is to write out the connected t_3 equations and then to eliminate the less important expensive terms. This series of approximations are denoted CCSDT-1a, CCSDT-1b, CCSDT-2, CCSDT-3, and CCSDT-4. The CCSDT-1 approximations have the advantage that the t_3 amplitudes need not actually be stored. However, these methods range in computational cost from iterative n^7 to iterative n^8, and so they are still too expensive to be generally applicable.

Due to the close connection between CC theory and many body perturbation theory (MBPT) [8, 39, 84], several approaches have been proposed in which the effects of higher connected excitations are included through perturbation theory. The most commonly used methods are CCSD+T(CCSD) [83] and CCSD(T) [39], where the effects of connected triple excitations are included perturbationally. The latter method is CCSD+T(CCSD) plus an additional fifth-order singles-triples interaction term. This additional fifth-order term has been shown to be very important in preventing an overestimation of the contribution from connected triple excitations when the effects of non-dynamical electron correlation are large [44-48]. In cases where non-dynamical correlation is unimportant, the CCSD+T(CCSD) and CCSD(T) methods give similar results. The computational cost of both of these methods is iterative n^6 (for the CCSD) plus two n^7 steps, and are thus much less expensive than the series of CCSDT-n approximations. More importantly, however, the CCSD(T) method has been shown to reproduce full CCSDT energies to better accuracy than any of the CCSDT-n methods [50], especially as non-dynamical electron correlation becomes important. Thus the CCSD(T) method provides the best compromise between accuracy and computational cost, allowing the use of large one-particle basis sets in order to achieve chemical accuracy. Applications of the CCSD(T) method presented later indeed show that very accurate molecular properties may be obtained with this approach.

2.2 Non-dynamical Correlation and the $\mathcal{T}_1$ Diagnostic

One of the difficulties in applying *ab initio* quantum mechanical methodology to study the properties of atoms and molecules is that it is all too easy to run one calculation and then have no idea concerning the accuracy of the obtained result. The accuracy of a specific level of theory is determined by restrictions in the one- and n-particle basis sets. One way to address this problem is to design a series of one-particle basis sets [79, 85] that may be used to extrapolate to the infinite basis set limit (see Ref. [86], for example). Similarly, a series of electron correlation techniques may be used to extrapolate to the full configuration-interaction (FCI) limit. While this approach is useful and can lead to very accurate predictions, it is limited to small to medium sized molecules due to the cost of high-level calculations in which large basis sets are used. Another approach is to perform a series of calculations at a given level of theory (including definition of the basis set) on molecules for which accurate experimental data exist. Experiment and theory are then compared and a statistical analysis yields a standard deviation σ that may be applied to the results of calculations (at the identical level of theory) on similar molecules. (As an aside, we note that the uncertainty estimate associated with the former approach is usually 3σ, i.e., 99.9% confidence limit, whereas the uncertainty associated with the latter approach is 1σ, or the 66.7% confidence limit.) Pople and co-workers [87-89] have adopted the latter approach in their G1 and G2 models. This method has the advantage that in principle it can be applied to a much broader range of molecules, although it has the disadvantage that the uncertainty estimate could be unrealistically small for particularly "difficult" molecules that exhibit an unusual electronic structure and more importantly, there is no way of knowing when an unusual molecule is encountered. For example, it is unwise to draw any conclusions about the uncertainty in the MP2 F-O bond distance in FOOF based on comparison of experimental and calculated MP2 geometries of F_2O, HOF, and HOOF; Ref. [90] shows that the MP2 F-O bond distance in FOOF can vary by ≈ 0.2 Å depending on the basis set.

Another approach that has been used to examine and estimate the quality of results obtained from a single calculation is to perform some analysis of the electronic wave function Ψ_e that results from the approximate electron correlation procedure. One method, commonly used in CI studies, is based on determining the weight of the reference function Ψ_0 in Ψ_e. However, this procedure is flawed from the beginning since truncated CI methods are not size-extensive and therefore as the number of electrons included in the approximate CI procedure increases, the weight of the reference function does not decrease rapidly enough [91]. Hence this procedure always overestimates the importance of the reference configuration (provided that the CI is not a FCI). In addition, due to the non size-extensivity, it is not meaningful to compare reference weights for molecules with different numbers of electrons. A related method commonly used in perturbation theory calculations is based on determining the norm of the first-order wave function and then using this value to determine the weight of the reference function Ψ_0 in $\Psi_e = \Psi_0 + \Psi_1$ (note that Ψ_0 is orthogonal to Ψ_n, $n \neq 0$). It is easily shown, however, that since intermediate normalization is used (i.e., $\langle\Psi_0|\Psi_e\rangle = 1$), this approach is not size-extensive. For example, the weight of the reference function obtained for one Ne atom will be considerably larger than the value obtained for 100 non-interacting Ne atoms, which is clearly not desirable. In this simple example, one could renormalize based on the number of non-interacting Ne atoms. However, it is uncertain how to proceed for a molecule with many more electrons — clearly, with this approach it is not meaningful to compare reference weights for systems with different numbers of electrons.

For the CC method, another approach that has been advocated [90, 92, 93] is based on examination of the Euclidian norm of the vector of t_1 amplitudes normalized by the number of electrons correlated, and is denoted the $\mathcal{T}_1$ diagnostic.

$$\mathcal{T}_1 = \frac{||t_1||}{N_{elec}^{1/2}} \quad . \tag{11}$$

The definition given in equation (11) is for closed-shell CC methods based on an RHF reference function. The $\mathcal{T}_1$ diagnostic was defined specifically to be size-extensive, to have the same invariance properties that the CC energy possesses, and to be a measure of the importance of non-dynamical electron correlation effects. $\mathcal{T}_1$ is also defined only for CC procedures that use SCF MOs in the reference function (although these need not be canonical), and only for the situation where valence electrons are included in the CC procedure; i.e., the frozen core approximation must be used. The first detailed investigation of the $\mathcal{T}_1$ diagnostic for closed-shell systems was presented in 1989 [92]. In this study, it was pointed out that several other contributions [28, 94, 95] had shown that use of different types of MOs in CCSD procedures, such as SCF MOs or complete active space SCF (CASSCF) MOs, resulted in very little difference in total correlation energies. Indeed, subsequent studies [32, 33] comparing CCSD with SCF MOs to the CC Brueckner doubles method have shown that there is usually very little difference in molecular properties from the two methods (the two exceptions being infrared intensities [32] and symmetry breaking phenomena [34, 35]). The earlier studies had thus concluded that, in agreement with Thouless' theorem [96], the presence of e^{T_1} in the CC wave function essentially "optimized" the starting MOs thereby overcoming orbital deficiencies in the reference function. Thus, CC theory is much more effective in accounting for orbital relaxation than analogous CI procedures due to the exponential character of the wave operator. In other words, when including the effects of electron correlation, truncated CI methods have much greater difficulty in overcoming a bias in the orbitals of the reference wave function.

The use of $\mathcal{T}_1$ as a diagnostic is based on an observed empirical correlation: the larger the $\mathcal{T}_1$ value, the less accurate were molecular structures, binding energies, and vibrational frequencies obtained with the CCSD or MP2 methods [90, 92]. This can be understood by recognizing that larger $\mathcal{T}_1$ values indicate that the SCF reference function is a poorer approximation to the true electronic wave function. Except for simple two electron systems, it is reasonable to conclude, therefore, that higher than connected double excitations will be more important for systems that exhibit a larger $\mathcal{T}_1$ value relative to those with small $\mathcal{T}_1$ values. Based on this analysis and interpretation of $\mathcal{T}_1$, it was recommended [92] that $\mathcal{T}_1$ values larger than 0.02 indicate that results from single-reference electron correlation methods limited to single and double excitations should be viewed with caution. In other words, it may be necessary to use higher levels of theory just to obtain qualitatively correct results, such as for the equilibrium structure of the FOOF molecule [52], the ordering of the vibrational frequencies for the O_3 molecule [48], or the binding energies of small Be and Mg clusters [25, 45, 46]. At the time of this recommendation, it was the opinion of Lee and Taylor [92] that the only practical (i.e., generally applicable) higher level of theory was some sort of multireference CI (MRCI), and therefore they recommended using an MRCI approach if $\mathcal{T}_1$ was larger than 0.02. However, since that time the single reference CCSD(T) method [39] has been shown to describe the effects of connected triple excitations very ably, and hence in many cases this method may be used rather than an MRCI approach. Indeed, CCSD(T) has been shown [97] to perform reasonably well in situations where $\mathcal{T}_1$ is as large as 0.08. It should be noted, that some researchers have confused Lee and Taylor's recommendation with the interpretation of $\mathcal{T}_1$, suggesting that $\mathcal{T}_1$ is a measure of multireference effects. The discussion presented above and in Refs. [90, 92, 93] clearly suggest that $\mathcal{T}_1$ is a measure of non-dynamical electron correlation, which is not necessarily the same as multireference character. For example, consider the two molecules O_3 and FOOF. Both of these molecules exhibit large $\mathcal{T}_1$ diagnostics [48, 90] ($\approx$0.028 for O_3 and $\approx$0.033 for FOOF), but the spectrum of excited state configurations (i.e., those other than the reference) is very different for these two molecules. It is well known that O_3 has about a dozen low-lying excited state configurations with one being very important, whereas it has been established [98] that FOOF possesses many configurations that are moderately important (i.e., a cumulative effect). The $\mathcal{T}_1$ diagnostic does not distinguish between these two situations, and thus it yields no information concerning which specific configurations should be used in a multireference treatment of electron correlation. This information would need to be obtained by examination of the individual t_1 and t_2 amplitudes; no statement to the contrary was ever made or implied. In fact, in the conclusions section to their paper, Lee and Taylor suggest that "specific important configurations may be identified by analysis of the CCSD t_1 and t_2 amplitudes." However, for both O_3 and FOOF, $\mathcal{T}_1$ does indicate that the SCF reference function is a poor starting point for single-reference based correlation techniques limited to single and double excitations. It should be noted that Jayatilaka and Lee [93] have pointed out that the proper definition of $\mathcal{T}_1$ for open-shell CCSD wave functions based on an RHF reference function is somewhat more complicated than a simple generalization of the closed-shell definition (see Ref. [99] for some additional numerical examples). Taylor also has presented [68, 100] very useful discussion of the $\mathcal{T}_1$ diagnostic.

It should be realized that the usefulness of the $\mathcal{T}_1$ diagnostic is based on empirical observation, and that clearly there are situations where its value may not be indicative of the true situation. For example, it was shown [92] that for the Be atom $\mathcal{T}_1$ is 0.021, which indicates that a CCSD treatment of electron correlation may not be adequate — but in this case CCSD is equivalent to a FCI. Another example where the $\mathcal{T}_1$ diagnostic may fail is for a very large molecule which contains one localized feature with large non-dynamical correlation effects, but where the vast majority of the molecule is well represented by an SCF reference function. Nevertheless, the $\mathcal{T}_1$ diagnostic has proven to be a very useful measure of the importance of non-dynamical correlation in many small to medium sized molecules. As indicated above, using $\mathcal{T}_1$ together with other wave function analyses, such as explicitly examining the largest t_1 and t_2 amplitudes, is probably the best approach.

2.3 Closed-Shell and Molecular Symmetry, and Vectorization

It was pointed out in 1987 by Scuseria *et al.* [60] that a considerable computational savings could be obtained for closed-shell atoms and molecules by spin adapting the CCSD equations for use with an RHF reference wave function. Several reformulations of the spin-adapted closed-shell coupled-cluster equations have been published subsequently [61-65], and some of these improvements are based on the work of Pulay *et al.* [101] concerning the closed-shell self-consistent electron pair (SCEP) theory. It is generally agreed that the minimum n^6 cost [62] is proportional to $\frac{1}{4}n_o^2n_v^4 + 4n_o^3n_v^3 + \frac{1}{2}n_o^4n_v^2$, where n_o is the number of doubly occupied MOs and n_v is the number of virtual MOs (note that $n = n_o + n_v$). Spin adaptations for closed-shell (T) [25, 102] [the perturbational triples contribution to CCSD(T)], and the closed-shell CCSDT procedure [42, 43] have led to similar computational savings. The computational cost for the closed-shell (T) step scales as $1n_o^3n_v^4 + 1n_o^4n_v^3$. In addition, Scuseria *et al.* [60] and independently Čársky *et al.* [103] showed how molecular point group symmetry could be used to reduce the computational work in CC procedures. More recently, Stanton *et al.* [64] have presented an elegant and thorough treatment concerning the use of molecular point group symmetry in CC procedures, although they missed the fact that earlier workers [60-62] had employed symmetry in the way outlined. While Stanton *et al.* discuss all possible molecular point groups, their implementation and the earlier CC codes that employ symmetry [60-62,103] have all been limited to D_{2h} and subgroups.

Another important computational advance has been the design and implementation of highly vectorized CC procedures. Highly vectorized CI procedures have been in existence for some time, but Lee and Rice [61] reported the first implementation of a highly vectorized closed-shell CCSD procedure in 1988, calling their program VCCSD (for vectorized CCSD). Their implementation also incorporated molecular point group symmetry in a direct product decomposition scheme, showing that the efficient use of both vectorization and symmetry are not incompatible. The CCSD equations were formulated using intermediate matrices (similar to Scuseria *et al.* [60]) thereby making it very easy to use standardized matrix multiply routines. This approach has the advantage of allowing the CC code to run efficiently on different computers since the matrix multiply routines are optimized for the individual computers. An example of this approach is given by the following,

$$\sigma_{ij}^{ab} \leftarrow \sum_{kl} t_{kl}^{ab} \sum_{cd} t_{ij}^{cd}(kc|ld) \quad , \tag{12}$$

where σ_{ij}^{ab} represents the "new" t_2 being constructed from the "old" t_1 and t_2 amplitudes and $(kc|ld)$ is a two-electron repulsion integral in Mulliken notation.

For this term the sum over cd is performed first by collecting the t_2 amplitudes and the (ov|ov) integrals (o is generic for an occupied index and v is generic for a virtual index) into two dimensional arrays. In other words, the ij, cd, and kl pairs of indices are treated as single compound indices. A matrix multiply routine is then called to perform the sum over cd thus forming the intermediate matrix

$$D_{ij}^{kl} \equiv \sum_{cd} t_{ij}^{cd}(kc|ld) \quad . \tag{13}$$

The sum over kl is performed similarly by a matrix multiply of the two dimensional arrays D_{ij}^{kl} and t_{kl}^{ab} (i.e., ij, kl, and ab are now the compound indices). For n^5 steps that involve either t_2 amplitudes or two-electron integrals, it is necessary to incorporate three indices into one compound index, while the other will consist of a single index. The simultaneous use of molecular symmetry complicates this procedure somewhat, but can be straightforwardly accomplished by defining several small indexing arrays. It should be noted that in practice, the matrix multiply has a much higher peak performance on vector computers, such as the Cray C90, than other vector operations, such as saxpy or dot product, and thus implementations that rely on the matrix multiply will generally perform better. Combining efficient use of matrix multiply routines with minimizing disk usage, it is necessary to perform several in-core sorts [61], but the arrays to be sorted are never larger than $n_o^2 n_v^2$. In order to avoid input/output bottlenecks, it is also necessary that out-of-core sorts are *never performed* during the iterative procedure.

Rendell *et al.* [102] later extended VCCSD to compute CCSD(T) correlation energies using a highly vectorized formulation and implementation of the (T) step. The high degree of vectorization in the VCCSD program may be demonstrated by a recent set of CCSD(T) calculations on different C_{20} isomers performed by Taylor *et al.* [104, 105]. The largest calculations were performed in low symmetry (C_s and C_2), and consisted of approximately 19.5 million $t_1 + t_2$ amplitudes, and about 29 billion t_3 amplitudes were estimated by perturbation theory. One of these calculations used approximately 67 hours of CPU time (for the CCSD plus (T) steps) on a Cray C90 computer (one processor), with the (T) step taking about 30% longer than the CCSD iterative procedure. For the 67 hours, a sustained megaflop rate of 857 was achieved, which compares very well with the peak performance of a matrix multiply on the Cray C90 (about 990 megaflops). Stanton *et al.* [64], Hampel *et al.* [36], and Koch *et al.* [106] have also recently implemented highly vectorized CCSD procedures.

2.4 Open-Shell Theories Based on an RHF Reference Function

Until recently, most open-shell CC calculations based on a specific state, single-reference formalism used an unrestricted Hartree-Fock (UHF) reference function. However, there are two disadvantages associated with the use of a UHF reference function. Firstly, and most importantly, open-shell UHF-CC calculations are considerably more costly than RHF-CC closed-shell calculations, and yet there is no inherent reason that open-shell CC calculations should be that much more expensive than a corresponding closed-shell calculation (i.e., similar numbers of electrons and basis functions). This increased computational cost of UHF-CC methods compared to closed-shell RHF-CC is a major concern. The second undesirable feature concerns spin contamination. There have been several studies over the past twenty years that have shown when performing approximate electron correlation calculations, it is better to use a symmetry constrained reference function rather than a symmetry broken reference function (for example, see Refs. [57,107-110]). This is in spite of the fact that the symmetry broken reference function possesses a lower energy, and these conclusions appear to be valid whether discussing spin symmetry or spatial symmetry. Therefore, it is not surprising that there has been considerable interest over the last several years in developing an open-shell CC theory that is based on an RHF reference function. However, it should be noted that at least for CC calculations, the concern about spin contamination is perhaps more aesthetic than based in fact, since there is a growing body of evidence [111-114], which suggest that the spin contamination in UHF-CCSD wave functions is generally modest except in severe cases. In fact, Schlegel [114] has shown that the CCSD procedure exactly cancels out the first contaminant in the UHF-CCSD wave function. In addition, Stanton [115] has recently compared S^2 values computed from a UHF-CCSD wave function and from the wave function of one of the recently developed open-shell RHF-CCSD methods. Only small differences were found, even when spin contamination in the UHF reference function was reasonably large. More importantly, the computed S^2 values were all very close to the appropriate "exact" values, with the largest difference being only 0.012 (for systems with no spatial symmetry breaking). Stanton also noted that the total energies from the two methods were very close, with the largest difference being about 4 millihartrees.

The open-shell RHF-CCSD method investigated by Stanton is that proposed by Rittby and Bartlett [116] in 1989, who suggested simply using an RHF reference function in the spin-orbital formulation of CC theory. This leads to correlation energies that are spin projected, but the CC wave function is spin contaminated. Thus it may be argued that while there is no direct contribution from a spin contaminant to the correlation energy, since the spin contaminants make non-zero contributions to the t_1 and t_2 equations there exists an indirect contribution. As indicated above, however, this spin contamination is generally very minor. Although spin contamination is only minor in the open-shell CC method of Rittby and Bartlett, the computational cost of this method is of much greater concern. Scuseria recently extended the Rittby/Bartlett open-shell RHF CCSD formalism to include a (T) correction for connected triple excitations [57]. In his paper, spin-integrated equations for this open-shell CCSD method were presented for the first time (some minor typographical errors exist in the CCSD equations of Ref. [57]), and examination of Scuseria's equations readily shows that open-shell CCSD calculations using this formalism are about five times more expensive than spin-adapted closed-shell CCSD calculations. The much greater expense of this open-shell RHF-CCSD approach, relative to closed-shell theory, has led to several other proposed formalisms. In 1992, Jayatilaka and Lee [66, 93] proposed an approach that is based on using symmetric spin-orbitals. Symmetric spin-orbitals consist of the standard RHF spatial MOs, with the standard α and β spin functions assigned to the doubly occupied and unoccupied MOs. The singly occupied MOs, however, are assigned the σ^+ and σ^- spin functions, where

$$\sigma^+ = \frac{1}{\sqrt{2}}(\alpha + \beta) \quad , \tag{14}$$

and

$$\sigma^- = \frac{1}{\sqrt{2}}(\alpha - \beta) \quad . \tag{15}$$

It has been shown [93] that n-particle functions constructed with symmetric spin-orbitals are S_x eigenfunctions rather than S_z eigenfunctions. It was also shown that use of symmetric spin-orbitals leads to a definition of an interacting space [117], where spin flips are counted as an excitation as well as spatial or orbital excitations. The interacting space is a direct result of mixing the α and β spin spaces that is inherent in the definitions of σ^+ and σ^-. The use of symmetric spin-orbitals does not of itself lead to an RHF-CCSD method that is different to the method proposed by Rittby and Bartlett. The symmetry of the RHF-CCSD wave function is certainly more apparent when using symmetric spin-orbitals, leading to a reduced computational cost, but the correlation energies obtained when using S_x or S_z eigenfunctions are identical (provided that the reference function is an S^2 eigenfunction). However, Jayatilaka and Lee noted that based on the interacting space definition, certain classes of "t_2" amplitudes were more properly labelled triple excitations, and furthermore they showed via explicit numerical comparisons that exclusion of these triples had only a small effect on the correlation energy, as would be expected from a triple excitation. Hence they proposed an RHF-CCSD method in which these triple excitations are eliminated. The computational cost of this open-shell RHF-CCSD approach is somewhat less than twice the cost of closed-shell theory, and therefore it is significantly less expensive than the method proposed by Rittby and Bartlett as implemented by Scuseria.

For convenience, we shall denote Jayatilaka and Lee's RHF-CCSD method by JL-CCSD while the Rittby and Bartlett approach shall be denoted RB-CCSD. It is also interesting to note that the wave function in the JL-CCSD approach will possess less spin contamination than the wave function in the RB-CCSD method. This becomes evident by writing the T_2 cluster operator in terms of the unitary group generators. Following Jayatilaka and Lee [93], the T_2 operator has the form

$$\begin{aligned} T_2 = \; & \tfrac{1}{2}(d_{ij}^{ab} E_{ai} E_{bj} + \ddot{d}_{ij}^{ab} \dot{E}_{ai} \dot{E}_{bj} + d_{ij}^{st} E_{si}^{-} E_{tj}^{-} + d_{wx}^{ab} E_{aw}^{+} E_{bx}^{+}) + \dot{d}_{ij}^{ab} E_{ai} \dot{E}_{bj} \\ & + d_{ij}^{as} E_{ai} E_{sj}^{-} + \ddot{d}_{ij}^{as} \dot{E}_{ai} E_{sj}^{-} + d_{wj}^{ab} E_{aw}^{+} E_{bj} + \ddot{d}_{wj}^{ab} E_{aw}^{+} \dot{E}_{bj} + d_{wj}^{as} E_{aw}^{+} E_{sj} \quad , \end{aligned} \tag{16}$$

where

$$E_{pq} = (p\alpha)^{\dagger}(q\alpha) + (p\beta)^{\dagger}(q\beta) \quad , \tag{17}$$

is the usual unitary group generator,

$$\dot{E}_{pq} = (p\alpha)^{\dagger}(q\beta) + (p\beta)^{\dagger}(q\alpha) \quad , \tag{18}$$

is a generalized unitary group generator defined previously [93], and E^{+}_{pq}, E^{-}_{pq} are defined according to

$$E^{+}_{pq} = \frac{1}{\sqrt{2}}(E_{pq} + \dot{E}_{pq}) \quad , \tag{19}$$

$$E^{-}_{pq} = \frac{1}{\sqrt{2}}(E_{pq} - \dot{E}_{pq}) \quad . \tag{20}$$

The d_2 coefficients in equation (16) are unknown amplitudes — relationships between these and the usual t_2 amplitudes are given in Ref. [93]. In the above, the $\{i,j\}$, $\{a,b\}$, $\{w,x,s,t\}$, and $\{p,q\}$ sets of indices represent doubly occupied, unoccupied, singly occupied, and all orbitals in the reference determinant, respectively. The T_2 operator given by equation (16) is appropriate for the RB-CCSD method, whereas the T_2 operator for the JL-CCSD approach is obtained by eliminating the fifth term (i.e., $\dot{d}^{ab}_{ij}E_{ai}\dot{E}_{bj}$). The T_1 operators for the two methods are the same, so the fifth term of equation (16) represents the only formal difference between the two approaches. Jayatilaka and Lee [93] showed that the $\dot{E}_{pq}$ operators do not commute with S^2, whereas it is well known that the E_{pq} generators do commute with S^2. It is therefore apparent that the RB-CCSD wave function will contain more terms that do not commute with S^2 than does the JL-CCSD wave function, and thus, discounting a fortuitous cancellation of errors, the RB-CCSD wave function will be more spin contaminated than the JL-CCSD wave function. It must be emphasized, however, that while this reduced spin contamination is a desirable feature, the main advantage of the JL-CCSD approach is the significantly reduced computational cost relative to the RB-CCSD method.

The Rittby/Bartlett and Jayatilaka/Lee versions of open-shell CC theory are basically spin-orbital formulations that utilize an RHF reference function (with the Jayatilaka/Lee method having the additional difference of using symmetric spin-orbitals and the interacting space). Spin-orbital formulations have the advantage that it is relatively straightforward to devise a many body perturbation theory (MBPT) that is entirely consistent with the CC theory. For example, it is possible to obtain the second-order MBPT energy from a CCSD program, although due to the appearance of "singles" in the first-order wave function this is more complicated than for the closed-shell case; that is, the non-linear part of the coupled-cluster energy expression must be excluded, see Ref. [93]. More importantly, it is possible to write down the MBPT terms and determine unambiguously what fraction of a given term is included in the CCSD approximation, for example. The perturbation theory that is related to the RB-CC theory [118, 119] is necessarily different to that related to the JL-CC theory [120, 121], but it is important that a consistent perturbation theory may be defined for both approaches so that a (T) contribution for connected triple excitations may be unambiguously defined. As noted earlier, Scuseria [57] first defined a (T) correction for the RB-CCSD method. Recent studies [113, 122] that have proposed a general definition for the (T) contribution to be used with spin-orbital open-shell RHF-CCSD methods, have shown that a term arising from the presence of single excitations in the first-order wave function was not included in Scuseria's formulation. Although this term contributes to the fourth-order perturbation theory correction, its numerical importance is apparently insignificant (see later discussion).

Another approach that has been used in formulating an open-shell RHF-CCSD method is to spin adapt the T cluster operator. In 1991 Janssen and Schaefer [123] investigated a fully spin-adapted open-shell CC theory, and showed that because of the nature of the exponential operator in the CC wave function, and the requirements of full spin-adaptation, hundreds of additional terms appear in the CCSD equations, some as high as the eighth power of T. Although this approach leads to a small number of independent amplitudes, the hundreds of terms make the method impractical for efficient implementation. Janssen and Schaefer also investigated a partially spin-adapted approach in which the coupled-cluster wave function would retain some spin contamination (due to products of the cluster operator T; the linear terms are not contaminated), but the set of projecting functions were fully spin-adapted. This approach contains essentially the same number of terms as the spin-orbital RHF-CCSD methods (i.e., only through the fourth power of T appears), but the number of independent parameters is identical to the fully spin-adapted theory. Only pilot programs of these two methods have been implemented. More recently, Knowles *et al.* [124] and Neogrády *et al.* [125] have independently proposed partially spin adapted open-shell CCSD procedures in which 1) the number of independent amplitudes is equivalent to the fully spin-adapted theory; 2) the computational cost is essentially equivalent to a similar closed-shell calculation; and 3) the defining equations do not contain many additional terms compared to the determinant based spin-orbital equations, and therefore an efficient implementation is practical. The method proposed by Knowles *et al.* is closely related to the second approach investigated by Janssen and Schaefer, although they do not use fully spin adapted projecting functions, but rather use the functions generated by applying T (i.e., linear) to the reference function (they also note that the T operator used by Janssen and Schaefer did not actually span the first-order interacting space — this was corrected).

In the scheme of Neogrády, Urban and Hubač [125], the spin-adaptation scheme previously proposed by the same authors [126] was employed to yield a single-reference model which removes the spin-contamination in both the CC wave function and energy. The number of CC amplitudes in this model is similar to that in a comparable closed-shell calculation. However, one complication in this model is that individual spin states require specific spin-adaptation schemes, thus requiring a different computational implementation for each spin state. This has been avoided by approximating the spin-adaptation (limiting this to the "closed-shell" amplitudes), which also increases the computational cost somewhat and allows some spin contamination into the CCSD wave function.

Other computational schemes for spin-adapting coupled-cluster (or similar) wave functions have also been available for quite some time. In particular, the scheme developed by Nakatsuji, and denoted as symmetry-adapted-cluster (SAC) theory, has been known and employed in quantum chemistry applications since the late 1970's (e.g., see Ref. [127], and references therein).

It would seem that the major drawback for these non-determinant based open-shell methods concerns a rigorous relationship to a well defined open-shell RHF perturbation theory that also exhibits good convergence properties. In other words, for closed-shell theory the CCSD(T) method [39] has proven to yield very accurate molecular properties, and thus it is desirable that any open-shell CCSD theory be easily extended to an analogous open-shell CCSD(T) method. As indicated above, the RB-CCSD and JL-CCSD methods both have a very close connection to a well defined open-shell RHF perturbation theory (that has good convergence characteristics), and thus it is possible to obtain a second-order perturbation theory energy on the first iteration of the CCSD procedure. However, the connection between the partially spin adapted CCSD methods and the above referenced open-shell RHF perturbation theories or other proposed open-shell RHF perturbation theories [128-130] is either not precise [124, 127], or the connection is to an open-shell perturbation theory [128] that has been shown [129] to possess poor convergence properties [125]. For those methods where the connection is unclear, this means that if one writes out a perturbation theory term by term, it is not possible to conclude whether some terms are included in the CCSD procedure completely, or in some unambiguous fraction. In other words, it is not possible to obtain the second-order energy of a well defined RHF open-shell perturbation theory from the first iteration of these partially spin-adapted CCSD procedures.

Initially, it may seem that this lack of straightforward extension to a well defined open-shell CCSD(T) method is a severe set-back for the partially spin adapted CCSD approaches. However, it may be that any reasonable definition of the (T) contribution will suffice. This assertion is based on the observation that UHF-CCSD(T), Scuseria's [57] open-shell CCSD(T), and Bartlett and co-workers [113] open-shell RHF CCSD(T) methods all seem to yield similar results. In any case, development of an open-shell RHF CCSD procedure remains an active area of research. The goal, of course, is to develop a method that has a computational cost essentially identical to the closed-shell RHF CCSD method, can be easily extended to a well defined CCSD(T) method, and provides the same degree of accuracy for molecular energies and properties that the closed-shell theory yields.

Very little has been said about the usefulness of the Brueckner CC method [29-36]. It has been noted [32, 33], that generally there is very little difference between molecular geometries and properties obtained with Hartree-Fock based CC approaches and the Brueckner CC method. However, one notable exception, which primarily concerns open-shell species, is for molecules that exhibit spatial symmetry breaking in the Hartree-Fock reference function [34, 35]. While the Brueckner CC method has been applied to only a small number of symmetry breaking problems, it has shown great promise in that the method has always been able to fully correct the symmetry breaking. The ability of Brueckner CC theory to correct for symmetry breaking is probably related to the observation of Raghavachari *et al.* [30] that the Brueckner CC method "extends" the point at which unrestricted and restricted theories diverge.

2.5 ANALYTICAL GRADIENT AND RESPONSE THEORY

The development of analytical gradient formalisms for CC theory over approximately the last 12 years has been an active area of research, and some of this development has been reviewed previously (e.g., see Ref. [131]). Analytical gradient methods [132] are very important for the efficient characterization of many dimensional potential energy surfaces, as well as the efficient determination of molecular properties, such as dipole moments, vibrational frequencies, infrared intensities, polarizabilities, and others.

Before reviewing the specific details of analytical gradient developments in CC theory, it is important to consider progress in analytical gradient theory that had a major impact on CC theory, but that was not specific to CC theory. In 1984 Handy and Schaefer [133] introduced an exchange theorem, which demonstrated that it was unnecessary to solve the first-order coupled-perturbed Hartree-Fock (CPHF) equations [134] in order to evaluate the gradient of a CI energy. The power and usefulness of this exchange theorem — often called the Z-vector method — has been extended to virtually every *ab initio* electron correlation method. Adamowicz *et al.* [135] showed that this exchange theorem could be used to avoid solving the coupled-perturbed CC (CPCC) equations in coupled-cluster theory gradients, and Rice and Amos [136] showed that in general the Handy-Schaefer exchange theorem could be used for any *ab initio* electron correlation method to incorporate all orbital and coefficient contributions into effective (also denoted "relaxed") density matrices that would then be used in a dot product with the atomic orbital derivative integrals as they are evaluated. For example, the energy expression for any correlation method may be written as

$$E = \sum_{pq} \gamma_{pq} h_{pq} + \sum_{pqrs} \Gamma_{pqrs} (pq|rs) \quad , \tag{21}$$

where p, q, r, and s label MOs, γ and Γ are one- and two-particle density matrices, respectively, and the notation for the one- and two-electron integrals is obvious. Differentiating this expression with respect to a perturbation λ, and with some manipulation, the energy gradient may be written as

$$\frac{\partial E}{\partial \lambda} = \sum_{pq} \gamma'_{pq} h^{\lambda}_{pq} + \sum_{pqrs} \Gamma'_{pqrs} (pq|rs)^{\lambda} + \sum_{pq} P_{pq} S^{\lambda}_{pq} \quad , \tag{22}$$

where

$$h^{\lambda}_{pq} = \sum_{\mu\nu} c_{p\mu} c_{q\nu} \frac{\partial h_{\mu\nu}}{\partial \lambda} \quad , \tag{23}$$

$$(pq|rs)^{\lambda} = \sum_{\mu\nu\rho\sigma} c_{p\mu} c_{q\nu} c_{r\rho} c_{s\sigma} \frac{\partial (\mu\nu|\rho\sigma)}{\partial \lambda} \quad , \tag{24}$$

$$S^{\lambda}_{pq} = \sum_{\mu\nu} c_{p\mu} c_{q\nu} \frac{\partial S_{\mu\nu}}{\partial \lambda} \quad , \tag{25}$$

and **S** represents the overlap matrix. In Eqs.(23)–(25), the μ, ν, ρ, and σ indices represent atomic orbitals (AOs) and **c** is the matrix of MO coefficients. The primed matrices in Eq. (22) will be different than the unprimed matrices, and these together with the **P** matrix are referred to as effective density matrices. Following Rice and Amos, the final gradient expression is then obtained by back transforming the effective density matrices into the AO basis

$$\frac{\partial E}{\partial \lambda} = \sum_{\mu\nu} \gamma'_{\mu\nu} h^{\lambda}_{\mu\nu} + \sum_{\mu\nu\rho\sigma} \Gamma'_{\mu\nu\rho\sigma} (\mu\nu|\rho\sigma)^{\lambda} + \sum_{\mu\nu} P_{\mu\nu} S^{\lambda}_{\mu\nu} \quad . \tag{26}$$

Hence, Rice and Amos showed that for any *ab initio* electron correlation method it was not necessary to transform derivative integrals into the MO basis, and that the Handy-Schaefer exchange theorem could be used to define effective density matrices that, once back transformed into the AO basis, would be dotted against the AO derivative integrals as they were constructed in order to obtain the analytical energy gradient. The only change for different correlation techniques was the specific formula for the effective density matrices. As well as CC methods (discussed below), these ideas have been used to formulate and sometimes implement efficient gradient methods for the CI [137, 138], perturbation theory [139, 140], coupled-pair functional [141], and multireference CI [142-144] correlation procedures. A few years ago, Helgaker and Jørgensen devised an elegant scheme whereby the features of the Handy-Schaefer exchange theorem are built into the equations for a correlation technique from the outset, and the interested reader is referred to Refs. [145-147] for a detailed account of the Lagrangian formalism.

The first derivation of the time-independent analytical gradient for CC theory was presented in 1983 by Jørgensen and Simons [148]. As indicated previously, shortly thereafter Adamowicz *et al.* [135] showed that by using the Handy-Schaefer [133] Z-vector technique, it was not necessary to solve the CPCC equations explicitly, but rather only one set of coupled, linear, perturbation independent equations need be solved. For convenience, this set of coupled equations is denoted the ZCC equations or the ZCCSD equations when applied to the CCSD approximation. In general form, the ZCC equations may be written as

$$A^T Z = I \quad , \tag{27}$$

where the **A** matrix is defined by the CPCC equations, **Z** is the vector of unknown coefficients, and the vector **I**, the term that multiplies the derivative amplitudes T^{λ}, is determined from differentiation of the coupled-cluster energy expression with respect to any perturbation λ. Fitzgerald *et al.* [149] first implemented analytical energy gradients for the CCD method — approximating T by T_2 (see Eqs. (2)–(4)), although they did not make use of the Handy-Schaefer exchange theorem, but instead solved for the perturbation dependent T_2^{λ} amplitudes. Later, Scheiner *et al.* [150] first implemented analytical energy gradients for the closed-shell CCSD method where they did make use of the Handy-Schaefer Z-vector technique. Moreover, they showed that it was necessary to solve the ZCCSD equations in an iterative fashion in order to reduce the computational cost. That is, the ZCCSD equations scale as $a * n^6$ when solving them iteratively, where a is the number of iterations, which is similar to the number of iterations necessary to solve the CCSD equations, and n is the number of MOs, but the equations scale as n^7

should the A^T matrix be explicitly constructed. In 1990 Koch *et al.* [151] presented a reformulation of the closed-shell ZCCSD equations that was considerably less costly than that presented by Scheiner *et al.*, although they did not implement the new formulation at the time. Scuseria has since incorporated this formulation into his CC gradient codes [17]. In 1991 Rendell and Lee [152] reported a highly vectorized implementation of an even more efficient formulation of the closed-shell ZCCSD equations. The rate limiting steps in the iterative part of the closed-shell ZCCSD equations are generally agreed to scale as $\frac{1}{4}n_o^2n_v^4 + 2n_o^3n_v^3 + \frac{1}{4}n_o^4n_v^2$, which is about half that of the closed-shell CCSD amplitude equations (see earlier discussion). The latter paper also presented a highly vectorized implementation of an efficient formulation of the equations defining the CCSD effective one- and two-particle density matrices, and showed that the rate limiting computational cost was identical to one iteration of the closed-shell CCSD equations. In addition, the development by Rendell and Lee incorporated point group symmetry in the same way as the VCCSD energy program [61] mentioned earlier, and this represents the first highly vectorized CCSD analytical gradient procedure. Gauss *et al.* [111, 153, 154] have since reported efficient implementations of analytical gradient methods for several CCSD procedures including open-shell techniques. Finally, we note that Kobayashi *et al.* [155] have reported a formulation and implementation of an analytical gradient procedure for the Brueckner CC doubles (i.e., BD) method.

Since it is necessary to use coupled-cluster methods that include some estimate from the contribution of connected triple excitations in order to approach chemical accuracy, there has also been much effort to devise and implement efficient analytical gradient methods for these techniques. The first development in this area was the implementation of the CCSDT-1 gradient by Scuseria [156]. Gauss and Cremer [157] implemented a spin-orbital formulation of the QCISD(T) analytical gradient, which followed MP4 analytical gradient formalisms [158-160] for the (T) part of the energy derivative. Scuseria [161] first implemented the closed-shell CCSD(T) energy gradient in a formulation that closely followed that of Gauss and Cremer for QCISD(T). Shortly thereafter, Lee and Rendell [162] reported a more efficient formulation and implementation of the CCSD(T) analytical gradient. Specifically, they showed that the computational cost of the (T) part of the gradient should scale as only $1n_o^4n_v^3 + 1n_o^3n_v^4$ rather than $2n_o^4n_v^3 + 2n_o^3n_v^4$, which is the cost that Gauss and Cremer [157] and Scuseria [161] had obtained. The reduced computational cost results from including the orbital contribution via a formulation that contains potential singularities. These singularities are avoided by threshold testing of numerators. Lee *et al.* [163] have recently devised a hybrid scheme that is numerically stable, does not increase the computational cost, and is applicable to the analytical gradient of any perturbation theory energy. The implementation by Lee and Rendell again incorporated molecular point group symmetry and was highly vectorized. Following on from the BD analytical gradient, Kobayashi *et al.* [164] also reported the implementation of an analytical gradient procedure for the BD(T) method. Finally, Watts *et al.* [113, 165] have implemented analytical gradient procedures for the general (open-shell) CCSD(T) methods.

To conclude this section, we briefly summarize recent developments in time-dependent response theory. This is by no means an exhaustive survey of time-dependent CC methods, and the interested reader is referred to Ref. [166] for a more detailed discussion. Our interest in these methods is stimulated by their relation to time-independent response or gradient theory and the fact that linear response CCSD (LRCCSD), in particular, has been shown [167-171] to yield vertical electronic excitation energies that are accurate to about ± 0.1 eV provided that the excited electronic state is primarily described as a single excitation away from the ground or reference electronic state. It has been shown [167, 170] that singlet excited state energies could be obtained by diagonalization of the **A** matrix that appears in the closed-shell CPCC (or ZCCSD) equations, and thus straightforward modification of existing computer programs provides for investigation of electronic spectra and searches of excited state potential energy surfaces. The procedure is complicated by the fact that the **A** matrix is non-Hermitian, and thus it possesses different right and left eigenvectors. This leads to an additional complication in computing oscillator strengths, although Koch *et al.* [172] have recently addressed this issue. It has been noted by several researchers that the LRCCSD approach yields identical excitation energies to the equations of motion CC (EOMCC) method [168], although Koch *et al.* [172] have recently shown that oscillator strengths from the two approaches are different, and moreover, that the EOMCC oscillator strengths are not "size-intensive" or size-extensive. It may be that some modification of the EOMCC oscillator strength formalism [168] will correct this obvious deficiency. Head-Gordon and co-workers [170, 171] using the time-dependent linear response approach, and Comeau and Bartlett [169] using the EOM methodology, have recently investigated electronic excitation energies from QCISD theory. Head-Gordon and co-workers have shown that in general excitation energies from LRQCISD are not as accurate as those from LRCCSD and that the majority of this difference is due to an important singles-doubles interaction term in the T_2 equations, which is present in CCSD but not in QCISD. Finally, Stanton and Gauss [173, 174] have recently formulated and implemented analytical energy gradients for the EOMCCSD method, which should facilitate searches and characterization of excited state potential energy surfaces.

3 Applications Attaining Chemical Accuracy

3.1 Equilibrium Geometries

As indicated in the Introduction, there has been a plethora of studies in the last few years that utilize the CCSD(T) method. While there are many fewer that use one-particle basis sets that are capable of achieving chemical accuracy, they are still numerous, and it is therefore not practical to summarize them all here. In the following, we review many such studies, concentrating mainly on those with which we are most familiar. In order to approach chemical accuracy, it is necessary to include at least f-type functions in the basis set (for heavy atoms) and usually d-type functions for H, and therefore we do not consider results (for comparison to experiment) in which smaller basis sets were used. The use of smaller basis

TABLE III
Comparison of CCSD(T) and experimental X-H bond distances (Å), for basis sets containing f functions on the heavy atom.[a]

Molecule/ Parameter	Basis set[b]	CCSD(T)	Expt	Δ	Ref[c]
HF/r_{FH}	cc-pVTZ	0.9172	0.9168	0.0004	[176, 186]
HCN/r_{CH}	TZ2Pf	1.0684	1.0650	0.0034	[53, 187]
HCN/r_{CH}	[4321]	1.0668	1.0650	0.0018	[177, 187]
HCN/r_{CH}	cc-pVTZ	1.0668	1.0650	0.0018	[178, 187]
HNC/r_{NH}	TZ2Pf	0.9975	0.9970	0.0005	[53, 188]
HNC/r_{NH}	[4321]	0.9960	0.9970	−0.0010	[177, 188]
HCO^+/r_{CH}	cc-pVTZ	1.0936	1.0972	−0.0036	[178, 188]
HOC^+/r_{OH}	cc-pVTZ	0.9916	0.975	0.0166	[178, 189]
H_2O/r_{OH}	cc-pVTZ	0.9594	0.9572	0.0022	[176, 81]
HNO/r_{NH}	cc-pVTZ	1.0759[d]	1.0642[d]	0.0117	[179, 190]
NH_3/r_{NH}	cc-pVTZ	1.0141	1.0116	0.0025	[180, 191]
H_2CO/r_{CH}	cc-pVTZ	1.1033	1.1005	0.0028	[181, 192]
BH_3/r_{BH}	cc-pVTZ	1.1938[d]	1.1900[d]	0.0038	[182, 193]
HOOH/r_{OH}	cc-pVTZ	0.9640	0.965[d]	−0.0010	[176, 194]
HCCH/r_{CH}	cc-pVTZ	1.0637	1.0625	0.0012	[183, 195]
$HCNH^+$/r_{CH}	cc-pVTZ	1.0774	1.0783	−0.0009	[184, 196]
$HCNH^+$/r_{NH}	cc-pVTZ	1.0129	1.0099	0.0030	[184, 196]
CH_4/r_{CH}	cc-pVTZ	1.0890	1.0858±0.001	0.0032	[185, 197]

[a] Bond distances are equilibrium unless indicated otherwise.
Δ = [CCSD(T)−Expt].
[b] ANO basis sets are designated by square brackets,
(e.g., [4321] refers to a $4s$, $3p$, $2d$, $1f$ ANO basis set).
Basis set designation is for the heavy atom. In most cases the H atom basis set contained up through d functions; see the original reference.
[c] First reference is to the CCSD(T) value and the second is to experiment.
[d] Vibrationally averaged. The theoretical value is an r_z value — see Ref. [198].

sets inevitably leads to less accurate predictions, even with the CCSD(T) method. We should also mention that most of the calculations considered here used either atomic natural orbital (ANO) basis sets [85] or Dunning's [79] correlation consistent basis sets. The ANO basis sets were usually those due to Almlöf and Taylor [85], although in a few instances the Lund ANO basis sets [175] were used. For details of the specific basis sets, the interested reader is referred to the original literature.

Presented in Tables III through VII are comparisons of molecular geometries obtained with the CCSD(T) method to experimental values (references to the original literature are given in the tables). Tables III and IV contain X-H bond distances, the former for basis sets that include through f functions and the latter for basis sets that include through g functions. In most instances equilibrium bond lengths are compared, but for those situations where only an experimental vibrationally averaged bond length is known, we have included the theoretical r_z value [198] where possible (i.e., where we had an accurate cubic force field to compute the effect of vibrational averaging). Examination of the Δ [CCSD(T)−Expt] values presented in Table III shows that in most cases theory overestimates the experimental bond distance, which is to be expected since in most situations involving first row atoms, basis set augmentation will lead to shorter bond distances. However, this trend is sometimes different, especially for molecules that do not exhibit neutral Lewis dot structures. For example, it was shown [177] that improvement of the one-particle basis from *spdf* to *spdfg* for HNC slightly increases the N-H bond distance. The negative Δ value for HOOH may be due to the fact that the CCSD(T) distance refers to an equilibrium value, whereas the only bond distance available from experiment is vibrationally averaged (vibrational averaging leads to longer covalent bond distances), or it may be due to an error in the experimental structure. Due to the large number of unique geometrical parameters in HOOH and the presence of the large amplitude torsional mode, the O-H bond distance is not precisely known. In fact, the experimental vibrationally averaged structure of HOOH has been determined by using a set of vibration-torsion-rotation equations to reproduce experimental energy levels using as input a combined *ab initio* /empirical quadratic force field and an assumed O-H bond distance [194]. The other geometrical parameters are considered optimized when the computed energy levels agree with experimental observations. This procedure is performed for several O-H distances and the "best" set of HOOH geometrical parameters are then taken as those that best reproduce the experimental rotational constants. There are clearly several potential pitfalls with this procedure, and the CCSD(T) structure suggest that the HOOH experimental structure needs improvement. For HCO^+, the Δ value is too large and negative, suggesting that the experimental equilibrium structure is in error. In fact Botschwina [199] has recommended a correction to the equilibrium structure of HCO^+ that is in much better agreement with the CCSD(T) *ab initio* calculations.

It is also clear that the Δ values for HOC^+ and HNO are considerably larger than for the other molecules, again suggesting that there is some error in the experimental structure. The situation for HNO is complicated by the fact that there is a very large anharmonicity for the H-N mode leading to a large vibrational averaging effect. Nonetheless, it seems likely that the experimental geometry is in error.

Examination of the Δ values contained in Table IV shows that generally they are smaller than the respective values in Table III by between 0.0010–0.0015Å. Moreover, in every case the assertions made above are supported by the X-H bond distance changes that occur due to improvement of the one-particle basis set.

Part of the difficulty in comparing experimental and theoretical structures is that the experimental values are not direct observables, but rather they are obtained from an analysis of a microwave or high-resolution ro-vibrational spectrum. The situation is further complicated by the fact that experimental equilibrium bond distances are even more difficult to obtain, and that there are several different methods by which to link an equilibrium coordinate and a "vibrationally averaged" coordinate (e.g., see Ref. [198]). Hence, the experimental structure is

TABLE IV
Comparison of CCSD(T) and experimental X-H bond distances (Å), for basis sets containing g functions on the heavy atom.[a]

Molecule/ Parameter	Basis set[b]	CCSD(T)	Expt	Δ	Ref[c]
HF/r_{FH}	cc-pVQZ	0.9162	0.9168	−0.0006	[176, 186]
HF/r_{FH}	[77442]	0.9170	0.9168	0.0002	[58, 186]
HCN/r_{CH}	[54321]	1.0663	1.0650	0.0013	[177, 187]
HCN/r_{CH}	cc-pVQZ	1.0668	1.0650	0.0018	[178, 187]
HNC/r_{NH}	[54321]	0.9961	0.9970	−0.0009	[177, 188]
HCO^+/r_{CH}	cc-pVQZ	1.0935	1.0972	−0.0037	[178, 188]
HOC^+/r_{OH}	cc-pVQZ	0.9900	0.975	0.0150	[178, 189]
H_2O/r_{OH}	cc-pVQZ	0.9579	0.9572	0.0007	[176, 81]
HNO/r_{NH}	cc-pVQZ	1.0740[d]	1.0642[d]	0.0098	[179, 190]
NH_3/r_{NH}	cc-pVQZ	1.0141	1.0116	0.0025	[180, 191]
H_2CO/r_{CH}	cc-pVQZ	1.1022	1.1005	0.0017	[181, 192]
BH_3/r_{BH}	cc-pVQZ	1.1926[d]	1.1900[d]	0.0026	[182, 193]
$HCCH/r_{CH}$	cc-pVQZ	1.0634	1.0625	0.0009	[183, 195]
$HCNH^+/r_{CH}$	cc-pVQZ	1.0772	1.0783	−0.0011	[184, 196]
$HCNH^+/r_{NH}$	cc-pVQZ	1.0118	1.0099	0.0019	[184, 196]
CH_4/r_{CH}	cc-pVQZ	1.0879	1.0858±0.001	0.0021	[185, 197]
CH_4/r_{CH}	[65331]	1.0873	1.0858±0.001	0.0015	[200, 197]

[a] Bond distances are equilibrium unless indicated otherwise.
Δ = [CCSD(T)−Expt].
[b] ANO basis sets are designated by square brackets,
(e.g., [54321] refers to a 5s, 4p, 3d, 2f, 1g ANO basis set).
Basis set designation is for the heavy atom. In most cases the H atom basis set contained up through d functions; see the original reference.
[c] First reference is to the CCSD(T) value and the second is to experiment.
[d] Vibrationally averaged. The theoretical value is an r_s value — see Ref. [198].

subject to systematic errors that are often not well understood, as well as statistical errors, which are usually small and well known but not always, e.g., see CH_4, HOOH, and HOC^+. Therefore, the precise comparison of experimental and theoretical polyatomic molecular geometries is challenging, and it appears that the accuracy of the CCSD(T) method has reached the point where in many instances theory is more reliable than experiment in determining polyatomic molecular geometries.

Tables V and VI contain a similar summary of X-Y bonds. Examining Table V, in this case there is only one instance of a negative Δ value, which is for the O-O linkage in HOOH. Again, part of this is because the CCSD(T) bond distance refers to an equilibrium value whereas the experimental quantity is vibrationally averaged. However, this cannot fully explain the discrepancy since vibrational averaging will not lengthen the O-O bond by anything like 0.006Å (a

TABLE V
Comparison of CCSD(T) and experimental X-Y bond distances (Å), for basis sets containing f functions.[a]

Molecule/ Parameter	Basis set[b]	CCSD(T)	Expt	Δ	Ref[c]
N_2/r_{NN}	[4321]	1.1042	1.0977	0.0065	[49, 186]
CO/r_{CO}	[4321]	1.1361	1.1283	0.0078	[201, 186]
C_2/r_{CC}	[4321]	1.245	1.2425	0.0025	[202, 186]
O_3/r_{OO}	[4321]	1.2749	1.2717	0.0032	[48, 80]
NO_2^+/r_{NO}	[4321]	1.1279[d]	1.125[d]	0.0029	[203, 205]
NNO/r_{NN}	cc-pVTZ	1.1329	1.1273	0.0056	[204, 206]
NNO/r_{NO}	cc-pVTZ	1.1896	1.1851	0.0045	[204, 206]
CO_2/r_{CO}	cc-pVTZ	1.1663	1.1600	0.0063	[204, 207]
HCN/r_{CN}	TZ2Pf	1.1591	1.1532	0.0059	[53, 187]
HCN/r_{CN}	[4321]	1.1604	1.1532	0.0072	[177, 187]
HCN/r_{CN}	cc-pVTZ	1.1601	1.1532	0.0069	[178, 187]
HNC/r_{NC}	TZ2Pf	1.1748	1.1684	0.0064	[53, 188]
HNC/r_{NC}	[4321]	1.1752	1.1684	0.0066	[177, 188]
HCO^+/r_{CO}	cc-pVTZ	1.1128	1.1047	0.0081	[178, 188]
HOC^+/r_{OC}	cc-pVTZ	1.1621	1.1570	0.0051	[178, 189]
HNO/r_{NO}	cc-pVTZ	1.2178[d]	1.2116[d]	0.0062	[179, 190]
H_2CO/r_{CO}	cc-pVTZ	1.2096	1.2033	0.0063	[181, 192]
$HOOH/r_{OO}$	cc-pVTZ	1.4579	1.464[d]	−0.0061	[176, 194]
$HCCH/r_{CC}$	cc-pVTZ	1.2097	1.2024	0.0073	[183, 195]
$HCNH^+/r_{CN}$	cc-pVTZ	1.1378	1.1345	0.0033	[184, 196]

[a] Bond distances are equilibrium unless indicated otherwise.
Δ = [CCSD(T)−Expt].
[b] ANO basis sets are designated by square brackets,
(e.g., [4321] refers to a $4s$, $3p$, $2d$, $1f$ ANO basis set).
See text or original reference for description of other basis sets.
[c] First reference is to the CCSD(T) value and the second is to experiment.
[d] Vibrationally averaged. The theoretical value is an r_z value — see Ref. [198].

typical value would be more like 0.002–0.003Å). Thus it seems very likely that the experimental structure for HOOH is in need of some improvement. From the earlier discussion concerning the O-H bond distance, this is not too surprising, but the non-planar torsional mode in HOOH also complicates the geometrical analysis — see Ref. [208]. All of the other Δ values in Table V seem quite plausible. In addition, the largest Δ values occur for triple bonds (HCO^+, CO, HCCH, HCN, HNC, N_2), which appears consistent since neglected factors such as core-valence correlation and larger basis sets, are expected to have a larger effect on triple X-Y bonds relative to single or double X-Y bonds.

Examination of the Δ values in Table VI shows that in all cases except for O_3 improvement of the basis from *spdf* to *spdfg* leads to better agreement with

TABLE VI
Comparison of CCSD(T) and experimental X-Y bond distances (Å), for basis sets containing g functions.[a]

Molecule/ Parameter	Basis set[b]	CCSD(T)	Expt	Δ	Ref[c]
N_2/r_{NN}	[54321]	1.1007	1.0977	0.0030	[49, 186]
CO/r_{CO}	[54321]	1.1320	1.1283	0.0037	[201, 186]
C_2/r_{CC}	[54321]	1.242	1.2425	−0.0005	[202, 186]
O_3/r_{OO}	[54321]	1.2697	1.2717	−0.0020	[209, 80]
NO_2^+/r_{NO}	[54321]	1.1251[d]	1.125[d]	0.0001	[203, 205]
NNO/r_{NN}	cc-pVQZ	1.1291	1.1273	0.0018	[204, 206]
NNO/r_{NO}	cc-pVQZ	1.1870	1.1851	0.0019	[204, 206]
CO_2/r_{CO}	cc-pVQZ	1.1626	1.1600	0.0026	[204, 207]
HCN/r_{CN}	[54321]	1.1566	1.1532	0.0034	[177, 187]
HCN/r_{CN}	cc-pVQZ	1.1564	1.1532	0.0032	[178, 187]
HNC/r_{NC}	[54321]	1.1721	1.1684	0.0037	[177, 188]
HCO^+/r_{CO}	cc-pVQZ	1.1087	1.1047	0.0040	[178, 188]
HOC^+/r_{OC}	cc-pVQZ	1.1579	1.1570	0.0009	[178, 189]
HNO/r_{NO}	cc-pVQZ	1.2141[d]	1.2116[d]	0.0025	[179, 190]
H_2CO/r_{CO}	cc-pVQZ	1.2066	1.2033	0.0033	[181, 192]
$HCCH/r_{CC}$	cc-pVQZ	1.2065	1.2024	0.0041	[183, 195]
$HCNH^+/r_{CN}$	cc-pVQZ	1.1348	1.1345	0.0003	[184, 196]

[a] Bond distances are equilibrium unless indicated otherwise.
Δ = [CCSD(T)−Expt].
[b] ANO basis sets are designated by square brackets,
(e.g., [54321] refers to a $5s$, $4p$, $3d$, $2f$ $1g$ ANO basis set).
See text or original reference for description of other basis sets.
[c] First reference is to the CCSD(T) value and the second is to experiment.
[d] Vibrationally averaged. The theoretical value is an r_s value — see Ref. [198].

experiment. On average, the largest improvement is for the triple bonds, which is consistent with triple bonds possessing more stringent basis set requirements. Even so, as expected the largest Δ values are still for triple bonds. The Δ values for C_2 and O_3 are negative, which is due to the very large electron correlation effects for these molecules (the $\mathcal{T}_1$ diagnostics are reasonably large for both C_2 and O_3 — see Refs. [48] and [50]). In other words, for these two molecules the basis set limit CCSD(T) bond distances are likely to be shorter than experiment since the effects of higher connected excitations will serve to lengthen the bonds. Because of the unknown contribution from higher connected excitations (which will be more important for X-Y bonds relative to X-H bonds), it is more difficult to assert that particular experimental X-Y bond distances are in error. However, examination of the data contained in Tables V and VI together with consideration of the $\mathcal{T}_1$ diagnostic values (see original references) suggest that the experimental X-Y bond distances for NO_2^+, HOC^+, and probably $HCNH^+$ are somewhat too long. Note that

TABLE VII
Comparison of CCSD(T) and experimental bond angles (degrees), for basis sets containing *f* and *g* functions.[a]

Molecule/ Parameter	Basis set[b]	CCSD(T)	Expt	Δ	Ref[c]
O_3/∠OOO	[4321]	116.93	116.78	0.15	[48, 80]
H_2O/∠HOH	cc-pVTZ	103.58	104.52	−0.94	[176, 81]
HNO/∠HNO	cc-pVTZ	108.09[d]	108.43[d]	−0.34	[179, 190]
NH_3/∠HNH	cc-pVTZ	105.64	106.7	−1.06	[180, 191]
H_2CO/∠HCH	cc-pVTZ	116.19	116.18	0.01	[181, 192]
HOOH/∠OOH	cc-pVTZ	99.55	99.4 [d]	0.15	[176, 194]
HOOH/torsion	cc-pVTZ	113.90	111.8 [d]	2.10	[176, 194]
O_3/∠OOO	[54321]	117.10	116.78	0.32	[209, 80]
H_2O/∠HOH	cc-pVQZ	104.12	104.52	−0.40	[176, 81]
HNO/∠HNO	cc-pVQZ	108.24[d]	108.43[d]	−0.19	[179, 190]
NH_3/∠HNH	cc-pVQZ	106.18	106.7	−0.52	[180, 191]
H_2CO/∠HCH	cc-pVQZ	116.44	116.18	0.26	[181, 192]

[a] Angles are equilibrium unless indicated otherwise.
Δ = [CCSD(T)−Expt].
[b] ANO basis sets are designated by square brackets,
(e.g., [54321] refers to a 5*s*, 4*p*, 3*d*, 2*f* 1*g* ANO basis set).
See text or original reference for description of other basis sets.
[c] First reference is to the CCSD(T) value and the second is to experiment.
[d] Vibrationally averaged. The theoretical value is a θ_z value.

based on a combined experimental/theoretical approach, Botschwina *et al.* [184] have recommended a new equilibrium structure for $HCNH^+$ that is consistent with the above observations. Nevertheless, the agreement between CCSD(T)/*spdfg* and experiment is nothing short of excellent.

Table VII contains a summary of CCSD(T) vs. experiment for bond angles. The effects of basis set improvement are not always systematic for bond angles. In other words, basis set improvement from *spdf* to *spdfg* can increase bond angles or sometimes decrease bond angles, depending on what occurs in the rest of the molecule. For simple bond angles, as in a triatomic, it is generally the case that basis set improvements lead to larger bond angles since the bond distances decrease. As it turns out, all of the examples included in Table VII exhibit slight increases in the bond angle upon going from an *spdf* basis to an *spdfg* basis set, but in all cases this change is fairly small. Indeed, the largest change occurs for H_2O and NH_3 where the bond angle increases by only 0.54°. The Δ value for the torsion angle of HOOH is fairly large (2.10°), but there are two reasons for this discrepancy. Firstly, the experimental value is vibrationally averaged and secondly, as indicated previously, there is some difficulty associated with determination of the experimental HOOH structure due to the nature of the torsional coordinate [208]. The only other Δ value larger than 1.0° is for NH_3 with the *spdf* basis, but

TABLE VIII
Comparison of CCSD(T) and MRCI/MRACPF equilibrium bond distances (Å).

Molecule/ Parameter	Basis set[a]	T_1	CCSD(T)	MRCI/ MRACPF	Δ[b]	Ref
Be_3/r_e	[5321]	0.036	2.232	2.223	0.009	[25, 45]
Be_4/r_e	[421]	0.032	2.148	2.145	0.003	[25, 45]
Mg_3/r_e	[7631]	0.013	3.387	3.372	0.015	[25, 46]
Mg_4/r_e	[531]	0.020	3.202	3.212	−0.010	[25, 46]
Ca_3/r_e	[6521]	–	4.172	4.167	0.005	[47]
Ca_4/r_e	[541]	–	4.131	4.140	−0.009	[47]
$X^3\Pi$ BN	[4321]	–	1.3371	1.3411	−0.0040	[97]

[a] ANO basis sets are designated by square brackets, (e.g., [54321] refers to a $5s$, $4p$, $3d$, $2f$, $1g$ ANO basis set).
[b] CCSD(T)–MRCI/MRACPF.

the Δ value is much smaller for the *spdfg* basis set. This behavior is not surprising as the inversion mode in NH_3 is well known to exhibit a strong dependence on higher angular momentum basis functions. The results in Table VII also show that the bond angle in H_2O is reasonably sensitive to higher angular momentum basis functions.

As shown by the comparisons presented in Tables III–VII, the CCSD(T) level of theory in conjunction with *spdf* or *spdfg* basis sets yields very accurate molecular structures. Even though the remaining errors are small, it is of interest to understand the primary causes of the residual error. Bauschlicher and Partridge [210] have recently investigated the effects of core-valence electron correlation on the spectroscopic constants of N_2 and found that core-valence correlation reduces the equilibrium bond distance by 0.002Å, indicating that this is the major cause of the remaining error. Very recently, Martin and Taylor [211] have performed a similar study of H_2O and HF and found that core correlation reduces the O-H and H-F bond distances by ≈0.00087 and ≈0.00057Å, respectively. Martin and Taylor also investigated relativistic and nonadiabatic effects and found these to be an order of magnitude less important than core correlation effects. Hence, it seems certain that the remaining error in CCSD(T)/*spdfg* geometries is due to the neglect of core-valence correlation.

There are two approaches that may be used to judge the quality of results obtained from an electron correlation method. As discussed above, one approach is to use very large one-particle basis sets and compare the theoretical predictions with accurate experimental data. The second approach is to compare with the results of very high level calculations, and as indicated previously, the series of FCI calculations performed by the Ames group [40] in the mid 1980's demonstrated that full valence reference space MRCI calculations typically give agreement with FCI to within a few microhartrees. Contained in Table VIII is a comparison of CCSD(T) and accurate MRCI equilibrium geometries for some very challenging molecules.

TABLE IX
Errors associated with geometrical parameters from data in Tables III–VII.[a]

Parameter	Basis set	Max error	Avg error	σ^b
r_{XH} [c]	*spdf*	0.0166	0.0029	0.0047
r_{XH} [d]	*spdf*	0.0038	0.0018	0.0015
r_{XY} [c]	*spdf*	0.0081	0.0057	0.0017
r_{XY} [e]	*spdf*	0.0063	0.0050	0.0015
r_{XY} [f]	*spdf*	0.0081	0.0061	0.0017
$\angle$ [g]	*spdf*	−1.06	−0.34	0.54
r_{XH}	*spdfg*	0.0150	0.0021	0.0043
r_{XH} [d]	*spdfg*	0.0026	0.0010	0.0012
r_{XY} [h]	*spdfg*	0.0041	0.0024	0.0015
r_{XY} [e]	*spdfg*	0.0033	0.0020	0.0011
r_{XY} [f]	*spdfg*	0.0041	0.0026	0.0017
$\angle$ [h]	*spdfg*	−0.52	−0.21	0.34

[a] Units are Å and degrees.
[b] One standard deviation (relative to experiment) for geometrical parameters corrected with the average (mean) error.
[c] Excludes HOOH; see text.
[d] Excludes HNO, HCO^+, HOC^+, and HOOH; see text.
[e] Double bonds; NO_2^+, NNO, CO_2, HNO, and H_2CO.
[f] Triple bonds; N_2, CO, C_2, HCN, HNC, HCO^+, HOC^+, HCCH, and $HCNH^+$.
[g] Does not include the HOOH torsion angle.
[h] Excludes O_3; see text.

The $\mathcal{T}_1$ diagnostics (see Table VIII), especially for Be_3 and Be_4, indicate that these metal clusters are highly correlated systems. The agreement between CCSD(T) and MRCI is very good, especially considering the very large electron correlation contribution to their equilibrium structures. In fact, Be_3, Mg_3, and Mg_4 are not bound at the SCF level of theory [44]. Thus the agreement between CCSD(T) and MRCI, especially for Be_3, Mg_3, and Mg_4 must be considered excellent. It is also worth mentioning that CCSD(T) bond lengths obtained in modest basis sets have also been found to be in excellent agreement with MRCI results for challenging cases like FOOF [52].

It is instructive to perform an error analysis on the data contained in Tables III–VII. The summary of such an analysis is contained in Table IX. "Max error" refers to the maximum difference between the CCSD(T) and experimental structures, and "Avg error" is the average error. The σ value given in Table IX is based on a statistical error analysis *after* correcting the CCSD(T) value using the average error. In other words, the average error quantity is subtracted from the CCSD(T) value in order to obtain a "best estimate" geometrical parameter. Since the major part of the error in the *ab initio* values is systematic in nature, it is necessary to correct for this systematic error before performing a statistical error analysis. We have not used the data for HOOH since the *ab initio* and experimental quantities refer to different types of structures. Moreover, we provide analyses of the X-H bond distances in which the HOC^+, HCO^+, and HNO molecules have been excluded since it was noted above that the experimental geometries appear questionable. For the *spdfg* basis, we have also excluded the O_3 molecule as it was noted above that the behavior of this species is atypical due to near degeneracy effects, as evidenced by the fact that O_3 has a much larger $\mathcal{T}_1$ diagnostic than any of the other molecules included in Tables III–VII (see original references). An analysis of all of the X-Y bonds has been given, but also separate analyses for double and triple bonds have been provided.

The first point to note is that the absolute error in the various bond distances and bond angles is substantially reduced upon augmentation of the basis set from *spdf* to *spdfg*. This is evidenced by the fact that the average error for X-H and X-Y bonds is reduced from 0.0018 and 0.0057Å to 0.0011 and 0.0024Å, respectively, while the average error for bond angles is reduced from $-0.34°$ to $-0.21°$. However, it is also interesting to note that the σ values are not substantially reduced with the larger basis set, which is further evidence that the majority of the error is systematic in nature. As expected, the average error for X-Y triple bonds is larger than for double bonds, and also for the σ values. Given the systematic nature of the errors in geometrical structures, especially bond distances, it may be that the most economical approach for determining accurate values will be to use the average error values listed in Table IX to correct CCSD(T) structures obtained with *spdf* quality basis sets. For X-H bond distances, it is recommended that the analyses excluding the HCO^+, HOC^+, and HNO molecules be used, while for X-Y bonds it is recommended to use the double- or triple-bond analyses. The situation for X-Y single bonds is less satisfactory. It may very well be that the X-H (single-bond) correction is appropriate, although on the other hand the X-Y double-bond correction may be more accurate. Clearly some additional studies of X-Y single bonds are needed. We note that Martin [212] has very recently proposed similar corrections to those given here (without the statistical analysis) for molecular geometries obtained with CCSD(T) in conjunction with Dunning's [79] cc-pVDZ, cc-pVTZ, and cc-pVQZ basis sets.

Given that X-H bonds are generally easier to describe with *ab initio* electronic structure theory relative to X-Y bonds, it is somewhat surprising that the σ values for the X-H bond distance are not smaller. There are two probable reasons for this observation. First and most importantly, based on earlier discussion, it is apparent that state-of-the art *ab initio* methods are often capable of predicting equilibrium structures with accuracy that is competitive with the best experimental determinations. Due to resonances, lack of data, and other factors, the analysis of experimental spectra is many times challenging, and it would also seem that X-H bond distances are more prone to these inherent experimental difficulties due to the small mass of hydrogen. Second, it was also noted earlier that in some cases (e.g., HNC) the X-H bond can slightly increase upon basis set augmentation suggesting that for these molecules the systematic error has the opposite sign and that therefore the average error value is not a good estimate of the residual error.

This situation is not as common for X-Y bonds and therefore the average error correction should almost always lead to an improved estimate of the true X-Y bond distance.

There are many more molecules for which CCSD(T)/*spdf* geometry optimizations have been performed, but where the experimental structure is either not known with certainty (e.g., FCCF [213] and HCCF [199]), or is not known at all (e.g., BeH_2 [182] and FONO [209, 214]). Given the discussion above, it seems very likely that these molecular geometries will also be very accurate and the *ab initio* predictions await experimental verification. To summarize, it is evident that CCSD(T) coupled with *spdf* or *spdfg* basis sets is capable of obtaining polyatomic molecular geometries with comparable or even better accuracy than experimental techniques. This degree of accuracy is indeed impressive, but it should be borne in mind that the major difficulty with the experimental analyses is that molecular geometries are not direct observables. In the following section we examine harmonic and fundamental frequencies, the latter being a direct experimental observable. It is shown that while the accuracy of CCSD(T) fundamental vibrational frequencies is also unprecedented among *ab initio* methods, this accuracy is still orders of magnitude less than that achievable from experiment.

3.2 Harmonic and Fundamental Vibrational Frequencies

There have been numerous *ab initio* studies over the last few years in which fundamental vibrational frequencies of polyatomic species have been computed using CCSD(T) quartic force fields. For the sake of brevity, the results of all of these studies cannot possibly be examined in detail here. It is instructive, however, to examine those initial studies which demonstrated the type of accuracy that the CCSD(T) method could provide. As with the molecular structures, this initial group of studies adopted two approaches: a) use large basis sets and compare to experiment; and b) compare to high-level MRCI calculations. Considering the first approach, Table X contains harmonic and fundamental vibrational frequencies for O_3, N_2, and NO_2^+. All of these were computed using *spdf* and *spdfg* basis sets at the CCSD(T) level of theory [48, 49, 203], and are compared to experiment [80, 186, 205].

By 1990 the ozone molecule had become a standard and severe test for high-level electron correlation methods due to the difficulty in correctly describing ω_3, the O-O antisymmetric stretch vibration (see Ref. [48] and references therein). The simplest electron correlation method, MP2 using restricted Hartree-Fock MOs, had been shown to yield an ω_3 value that was about a *factor of two* too large, while one quite sophisticated coupled-cluster method (with a small one-particle basis set) predicted that the equilibrium structure of ozone was asymmetric, i.e., C_s rather than C_{2v}. This latter result was shown to be due to basis set limitations, as larger basis sets predicted the correct C_{2v} minimum. The obvious conclusion from all of the studies prior to 1990 was that a correct description of ω_3/ν_3 would require not only a very sophisticated electron correlation method, but also the use of large one-particle basis sets, including at the very least f functions. These dual requirements essentially eliminated (at that time) the possibility of using a reliable MRCI technique, due to the large number of valence electrons (18), and the relatively

TABLE X
Comparison of CCSD(T) and experimental harmonic and fundamental frequencies (cm^{-1}) for O_3, N_2, and NO_2^+.

Molecule[a]	Mode	CCSD(T)	Expt	Δ[b]
O_3/[4321]	$\omega_1(a_1)$	1152	1135	17
[48, 80]	$\omega_2(a_1)$	717	716	1
	$\omega_3(b_2)$	1057	1089	−32
	$\nu_1(a_1)$	1105	1103	2
	$\nu_2(a_1)$	699	701	−2
	$\nu_3(b_2)$	975	1042	−67
O_3/[5432]	$\omega_1(a_1)$	1153	1135	18
[48, 80]	$\omega_2(a_1)$	718	716	2
	$\omega_3(b_2)$	1053	1089	−36
N_2/[4321]	$\omega_1(\sigma_g)$	2338	2359	−21
[49, 186]	$\nu_1(\sigma_g)$	2310	2330	−20
N_2/[54321]	$\omega_1(\sigma_g)$	2352	2359	−7
[49, 186]	$\nu_1(\sigma_g)$	2324	2330	−6
NO_2^+/[4321]	$\nu_1(\sigma_g)$	1375	1397	−22
[203, 205]	$\nu_2(\pi_u)$	618	639	−21
	$\nu_3(\sigma_u)$	2334	2362	−28
NO_2^+/[54321]	$\nu_1(\sigma_g)$	1384[c]	1397	−13
[203, 205]	$\nu_2(\pi_u)$	625[c]	639	−14
	$\nu_3(\sigma_u)$	2344[c]	2362	−18

[a] See footnote *b* to Table III for ANO basis set designations. First reference is to the CCSD(T) value and the second is to experiment.
[b] Δ = [CCSD(T)−Expt].
[c] Obtained by adding [4321] anharmonic corrections to [54321] harmonic frequencies.

large number of reference configurations that would be needed (at least 12-18). Hence we decided to try the newly developed CCSD(T) method as this approach could easily be used with basis sets containing *f* functions. A summary of this study is presented in Table X. As can been seen, the CCSD(T)/*spdf* method gives ω_1 and ω_2 harmonic frequencies that are accurate to 17 cm^{-1} and ω_3 is accurate to -32 cm^{-1}. At the time, this type of accuracy was virtually unprecedented for any polyatomic molecule, let alone one as difficult to describe as O_3. Anharmonic corrections were computed using a cubic and quartic force field in conjunction with second-order perturbation theory of the nuclear Schrödinger equation [215], and while the ν_1 and ν_2 fundamentals were predicted very well, the agreement for the ν_3 fundamental deteriorated. Nonetheless, this study showed that the CCSD(T) method was capable of achieving "chemical accuracy" in predicting the vibrational spectra of polyatomic species provided that basis sets of *spdf* or *spdfg* quality were used.

In a study of T_d N_4, Lee and Rice [49] investigated the harmonic and fundamental vibrational frequency of the triply-bonded N_2 molecule. Although the ground electronic state of N_2 does not possess a large degree of non-dynamical correlation, it is well known that triple-bonds possess a significant degree of dynamical electron correlation, as evidenced by studies in the early 1980's in which the importance of linked triple excitations (in MP4) was demonstrated [216-219]. Examination of the N_2 results in Table X shows that CCSD(T)/*spdf* and especially CCSD(T)/*spdfg* perform very well for N_2, predicting the fundamental frequency to within 20 and 6 cm^{-1}, respectively. These results showed that the CCSD(T) method was very effective at accounting for dynamical as well as non-dynamical electron correlation effects.

The final set of results in Table X is for the NO_2^+ cation. The CCSD(T) *ab initio* study [203] was motivated by the observation [220] that the available experimental data for NO_2^+, which was from the solid-phase, was not representative of the gas-phase species. Experimental verification of the CCSD(T) predictions was swift [205], however, and examination of the fundamental frequencies in Table X shows that the CCSD(T)/*spdf* and CCSD(T)/*spdfg* levels of theory are very reliable for the prediction of fundamental vibrational frequencies. As is expected from the $\mathcal{T}_1$ diagnostic value for NO_2^+ (≈0.021), none of the fundamentals exhibit the problems found for O_3 ω_3. Thus these three early examples amply demonstrated the degree of accuracy that was possible from the CCSD(T)/*spdf* and CCSD(T)/*spdfg* levels of theory. One final note concerning the results in Table X — examination of the Δ values shows that the error in CCSD(T) vibrational frequencies is much less systematic than is the error in CCSD(T) geometrical parameters. Unfortunately, this means that "average error" corrections will be less effective, but fortunately, the absolute errors are already quite small. This is discussed in more detail later.

Another set of studies that was proceeding simultaneously involved the comparison of CCSD(T) and high-level MRCI harmonic frequencies. Data from these studies [45-47] is collected in Table XI. As indicated previously, small Be, Mg, and Ca clusters are highly correlated systems requiring sophisticated correlation methods just to obtain qualitatively correct results for the trimers. Due to the *sp* near degeneracy in these elements and their metallic nature, both dynamical and non-dynamical electron correlation are very important. The largest difference between CCSD(T) and MRCI for harmonic frequencies is only 15 cm^{-1}, and for the Mg and Ca clusters the largest difference is 5 cm^{-1}. The excellent agreement between CCSD(T) and MRCI shown in Table XI thus provided additional evidence that the CCSD(T) method performs well at describing both dynamical and non-dynamical correlation.

Tables XII and XIII present a comparison of CCSD(T) and experiment for harmonic and fundamental vibrational frequencies for molecules that are reasonably well represented by single-reference based electron correlation methods, since it is instructive to examine the type of accuracy that is possible in this circumstance. Experimental data for the force fields of the NNO and CO_2 molecules are quite extensive including up through sextic force constants, whereas even the quadratic experimental force field for H_2CO is subject to errors due to complications involving multiple Fermi resonances, and no experimental force field for HNO

TABLE XI
Comparison of CCSD(T) and MRCI harmonic frequencies (cm^{-1}).

Molecule[a]	Mode	CCSD(T)	MRCI	Δ[b]
Be_3/[5321]	$\omega_1(a_1')$	480	490	−10
[45]	$\omega_2(e')$	417	427	−10
Be_4/[421]	$\omega_1(a_1)$	602	602	0
[45]	$\omega_2(e)$	436	451	−15
	$\omega_3(t_2)$	527	529	−2
Mg_3/[7631]	$\omega_1(a_1')$	105	110	−5
[46]	$\omega_2(e')$	110	115	−5
Mg_4/[531]	$\omega_1(a_1)$	164	162	2
[46]	$\omega_2(e)$	133	133	0
	$\omega_3(t_2)$	153	151	2
Ca_3/[6521]	$\omega_1(a_1')$	94	95	−1
[47]	$\omega_2(e')$	83	85	−2
Ca_4/[541]	$\omega_1(a_1)$	110	109	1
[47]	$\omega_2(e)$	78	77	1
	$\omega_3(t_2)$	94	94	0

[a] See footnote *b* to Table III for ANO basis set designations.
[b] Δ = [CCSD(T)−MRCI].

exists. Hence the molecules in Table XII represent systems where experimental analyses are well established, whereas only the fundamental vibrational frequencies are well established experimentally for the molecules in Table XIII. Considering Table XII first, note that there is excellent agreement between CCSD(T) and experiment for both the harmonic and fundamental frequencies, indicating that the agreement for the fundamentals is not due to a fortuitous cancellation between harmonic and anharmonic contributions. Interestingly, the errors for NNO are smaller using *spdf* basis relative to using the *spdfg* basis, but the opposite is true for CO_2. This again demonstrates the less systematic nature of the errors associated with vibrational frequencies compared to that observed for geometrical parameters. Also note that the errors in the fundamentals ν are basically the same as those found in the harmonic frequencies, indicating that the anharmonic corrections are very accurate and that the majority of the remaining error is in the *ab initio* harmonic frequencies.

For HNO and H_2CO it is necessary to use variational treatments of the nuclear Schrödinger equation due to Fermi resonances (H_2CO) and the very large anharmonic nature of the H-N stretch in HNO (see Ref. [179]). Nonetheless, the theoretical fundamental vibrational frequencies are again in excellent agreement with experiment. The largest error -24 cm^{-1} is found for the H-N stretch in HNO using the *spdf* basis, but with the *spdfg* basis this error is reduced to less than -1 cm^{-1}. For most of the H_2CO modes contained in Table XIII, the [CCSD(T)-Expt] error tends to be slightly smaller with the *spdf* basis set relative to that

TABLE XII
Comparison of CCSD(T) and experimental harmonic (ω) and fundamental (ν) frequencies (cm^{-1}) for NNO and CO_2.

Molecule[a]	Mode	CCSD(T)	Expt	Δ[b]
NNO/cc-pVTZ	$\omega_1(\sigma)$	2282.8	2282.1	0.7
[204, 206]	$\omega_2(\pi)$	601.1	596.3	4.8
	$\omega_3(\sigma)$	1297.2	1298.3	−1.1
	$\nu_1(\sigma)$	2224.9	2223.8	1.1
	$\nu_2(\pi)$	594.4	588.8	5.6
	$\nu_3(\sigma)$	1271.0	1271.4	−0.4
NNO/cc-pVQZ	$\omega_1(\sigma)$	2287.9	2282.1	5.8
[204, 206]	$\omega_2(\pi)$	602.1	596.3	5.8
	$\omega_3(\sigma)$	1303.5	1298.3	5.2
	$\nu_1(\sigma)$	2230.0[c]	2223.8	6.2
	$\nu_2(\pi)$	595.4[c]	588.8	6.6
	$\nu_3(\sigma)$	1277.3[c]	1271.4	5.9
CO_2/cc-pVTZ	$\omega_1(\sigma_g)$	1346.3	1353.8	−7.5
[204, 207]	$\omega_2(\pi_u)$	660.3	672.9	−12.6
	$\omega_3(\sigma_u)$	2396.5	2396.5	0.0
	$\nu_1(\sigma_g)$	1325.8	1332.9	−7.1
	$\nu_2(\pi_u)$	657.8	672.9	−15.1
	$\nu_3(\sigma_u)$	2350.8	2349.1	1.7
CO_2/cc-pVQZ	$\omega_1(\sigma_g)$	1352.0	1353.8	−1.8
[204, 207]	$\omega_2(\pi_u)$	670.5	672.9	−2.4
	$\omega_3(\sigma_u)$	2396.6	2396.5	0.1
	$\nu_1(\sigma_g)$	1331.5[c]	1332.9	−1.4
	$\nu_2(\pi_u)$	668.0[c]	672.9	−4.9
	$\nu_3(\sigma_u)$	2350.9[c]	2349.1	1.8

[a] First reference is to the CCSD(T) value and the second is to experiment.
[b] Δ = [CCSD(T)−Expt].
[c] Obtained by adding cc-pVTZ anharmonic corrections to cc-pVQZ harmonic frequencies.

obtained with the *spdfg* basis, similar to NNO.

The core correlation studies on N_2, H_2O, and HF mentioned previously [210, 211] also considered the effect of core correlation on the harmonic frequencies. For N_2 and HF, core correlation is found to increase ω by 9 and 4.4 cm^{-1}, respectively. For H_2O core correlation increases ω_1 and ω_3 by 6.6 and 7.1 cm^{-1}, respectively, but decreases the bend ω_2 by 0.8 cm^{-1}. As with the equilibrium geometries, it thus seems likely that for many molecules the primary cause of residual error in CCSD(T)/*spdfg* vibrational frequencies is due to the neglect of core-valence electron correlation.

As indicated above, the large number of *ab initio* studies in which CCSD(T) quartic force fields have been used to compute accurate fundamental vibrational

TABLE XIII
Comparison of CCSD(T) and experimental fundamental frequencies (cm^{-1}) for HNO and H_2CO.

Molecule[a]	Mode	CCSD(T)	Expt	Δ[b]
HNO/cc-pVTZ	$\nu_1(a')$	2660.0	2684.0	−24.0
[179, 221]	$\nu_2(a')$	1496.4	1500.8	−4.4
	$\nu_3(a')$	1560.1	1565.3	−5.2
HNO/cc-pVQZ	$\nu_1(a')$	2683.4	2684.0	−0.6
[179, 221]	$\nu_2(a')$	1502.3	1500.8	1.5
	$\nu_3(a')$	1568.8	1565.3	3.5
H_2CO/cc-pVTZ	$\nu_1(a_1)$	2789.2	2782.5	6.7
[181, 222]	$\nu_2(a_1)$	1749.0	1746.0	3.0
	$\nu_3(a_1)$	1504.3	1500.2	4.1
	$\nu_4(a_1)$	1165.7	1167.3	−1.6
	$\nu_5(a_1)$	2842.0	2843.3	−1.3
	$\nu_6(a_1)$	1247.9	1249.1	−1.2
H_2CO/cc-pVQZ	$\nu_1(a_1)$	2793.5[c]	2782.5	11.0
[181, 222]	$\nu_2(a_1)$	1749.6[c]	1746.0	3.6
	$\nu_3(a_1)$	1498.7[c]	1500.2	−1.5
	$\nu_4(a_1)$	1163.8[c]	1167.3	−3.5
	$\nu_5(a_1)$	2849.8[c]	2843.3	6.5
	$\nu_6(a_1)$	1245.9[c]	1249.1	−3.2

[a] First reference is to the CCSD(T) value and the second is to experiment.
[b] Δ = [CCSD(T) − Expt].
[c] Obtained by adding cc-pVTZ anharmonic corrections to cc-pVQZ harmonic frequencies.

frequencies preclude us from reviewing them here. However, a statistical analysis similar to that given for geometrical parameters is of interest. The results of such an analysis are presented in Table XIV. The molecules included in this analysis are O_3 [48], N_2 [49], NO_2^+ [203], NNO [204], CO_2 [204], HNO [179], H_2CO [181], NH_3 [180], CH_4 [185], BH_3 [182], HCN [177], HNC [177], HCO^+ [178], HOC^+ [178], CO [178], H_2O [176], and HF [176]. Harmonic frequencies are used for H_2O and HF, which should not make any difference since the majority of the residual error is in the harmonic frequencies, as noted above. Also note that other CCSD(T) quartic force fields have been published for HCN [223], NNO [224, 225], and CO_2 [226]. In all, 47 vibrational modes are included for the *spdf* basis while 44 modes are included for the *spdfg* basis set. In the analyses presented in Table XIV, ν_{all} includes all vibrational modes; ν_{XH} denotes that only X-H stretching modes are included; ν_{XY} denotes that only X-Y stretching modes are included; and ν_{bends} includes all other modes (i.e., out-of-plane bends as well as bond angle bends). The σ values have been computed in the same way as for the geometrical parameters — that is, they are one standard deviation determined *after* correcting the frequency

TABLE XIV
Errors associated with CCSD(T) vibrational frequencies (cm^{-1}).[a]

Parameter	Basis set	Max error	Avg abs error	Avg error	σ^b
ν_{all}	*spdf*	87.1	12.4	−2.2	20.6
ν_{XH}	*spdf*	39.1	11.0	1.1	15.4
ν_{XY}	*spdf*	−67.0	14.8	−13.8	17.9
ν_{bends}	*spdf*	87.1	11.5	5.1	23.2
ν_{all} [c]	*spdf*	−28.0	8.8	−3.7	11.3
ν_{XH} [c]	*spdf*	−24.0	8.9	−1.6	11.7
ν_{XY} [c]	*spdf*	−28.0	11.1	−10.0	10.5
ν_{bends} [c]	*spdf*	−21.0	6.7	−0.1	9.7
ν_{all}	*spdfg*	61.6	7.8	1.5	12.5
ν_{XH}	*spdfg*	24.0	8.3	2.5	10.3
ν_{XY}	*spdfg*	−18.0	6.4	−3.2	7.4
ν_{bends}	*spdfg*	61.6	8.6	4.3	16.5
ν_{all} [c]	*spdfg*	−18.0	6.2	−0.5	7.7
ν_{XH} [c]	*spdfg*	17.1	7.1	1.0	8.8
ν_{XY} [c]	*spdfg*	−18.0	6.4	−3.2	7.4
ν_{bends} [c]	*spdfg*	−14.0	5.1	0.5	6.6

[a] Includes either harmonic or fundamental vibrational frequencies from a given molecule. See text for the complete list of molecules included.
[b] One standard deviation (relative to experiment) for frequencies corrected with the average (mean) error.
[c] Excludes $NH_3(\nu_2)$, $O_3(\nu_3)$, $HF(\omega_1)$; see text.

with the average error.

A good indication that the [CCSD(T)-Expt] errors are not very systematic is given by the observation that the average absolute error and the average error are not that consistent, and also by the fact that the average error is usually close to zero. The only exception to this observation is for ν_{XY} because the X-Y vibrations are usually underestimated with both the *spdf* and *spdfg* basis sets.

Since errors for X-H stretching and bending CCSD(T) vibrational frequencies are not very systematic, it is not clear that using the average error to correct these leads to much improvement. However, with a few exceptions, the absolute error of CCSD(T) vibrational frequencies is quite small anyway. Three exceptions have been noted, and these are the O-O antisymmetric stretch in O_3 discussed previously, the inversion mode in NH_3 which is known to need potential energy constants up through octic in order to approach quantitative accuracy (and thus must go beyond second-order perturbation theory in the vibrational analysis), and the H-F stretch in HF. The discrepancy for HF is due to one-particle basis set deficiencies that have been discussed very recently by Martin and Taylor [211]. That is, since fluorine is very electronegative, it is necessary to include diffuse functions in the basis set (i.e., Dunning's [79] augmented correlation consistent basis sets) in order to describe ionic behavior properly. Eliminating these three modes from the analyses dramatically reduces the average errors and the σ values. Indeed, the σ values for the *spdf* basis are around 10 cm^{-1} and for the *spdfg* basis they are all *less* than 10 cm^{-1}.

The analyses in Tables IX and XIV clearly demonstrate that computing ro-vibrational spectra *ab initio* has entered a new age, whereby it is now possible to assign, *for the first time,* the vibrational spectrum of a molecule based on a non-laboratory spectrum. In other words, the ro-vibrational line positions can be predicted with high enough precision using purely *ab initio* data that it is not necessary first to obtain a spectrum from a tightly controlled environment, such as in a laboratory experiment. For example, it should be possible to use a CCSD(T)/*spdfg* simulated ro-vibrational spectrum to determine whether the molecule of interest can be observed in interstellar medium. Another area where this newly developed accuracy has already aided experiment involves vibrational spectra at very high energies, as was recently demonstrated by Bowman *et al.* [227] for the HCN/HNC system.

3.3 Heats of Formation, Binding Energies, and Energy Differences

Perhaps the property of most interest in chemistry is an energy difference. Energy differences are found in many forms such as heats of formation, binding or dissociation energies, or isomeric energy differences to name but a few. However, directly computing energy differences is also one of the most demanding tasks for *ab initio* electronic structure methods. The potential usefulness of the CCSD(T) method in directly computing accurate energy differences was recognized at an early stage. That is, the CCSD(T) method yields an accurate description of electron correlation, but is inexpensive enough to allow the use of very large one-particle basis sets, obviating the need for additivity corrections, at least for small to medium sized molecules, as are used in the G1 and G2 methods [87-89]. This is not to suggest that the G1 and G2 methods are not useful, but if an individual energy difference is desired to very high accuracy, the CCSD(T)/large basis approach provides a viable route. In this section we shall summarize a few specific examples in which the CCSD(T) approach has been shown to yield very accurate directly computed energy differences.

Small Be and Mg clusters are a severe test for electron correlation procedures due to the *sp* near degeneracy in the atoms. Table XV contains a comparison of computed binding energies for the Be_3, Be_4, Mg_3, and Mg_4 clusters [25]. These binding energies refer to atomization energies. In most cases, directly computed atomization energies (using approximate *ab initio* methods) are underestimated since electron correlation is typically more important in the molecule than in the separated atoms, or in other words the differential electron correlation contribution is usually large and positive. Using size-extensive correlation procedures such as CCSD(T) (MRCI in practice can be approximately size-extensive provided that the reference space is large enough, see Ref. [40]) generally means that about the same percentage of the total correlation energy will be recovered in the molecule as in the separated atoms. Thus this same percentage of the differential correlation contribution will be recovered, and when this differential contribution is large in absolute terms, then it is necessary to recover a very large percentage in

TABLE XV
Comparison of CCSD, CCSD(T), and MRCI dissociation energies (kcal/mol).[a]

Molecule[b]	CCSD	CCSD(T)	MRCI	Δ[c]
Be_3/[5321]	11.3	20.4	22.4	−2.0
Be_4/[5321]	63.5	79.4	77.3	2.1
Mg_3/[7631]	1.7	5.7	6.3	−0.6
Mg_3/[531]	8.1	15.9	16.2	−0.3

[a] Results taken from Ref. [25]. All MRCI wave functions used a reference threshold of 0.025, except for Be_4 where a threshold of 0.05 was used; see Ref. [44] for details.
[b] See footnote *b* to Table III for ANO basis set designations.
[c] Δ = [CCSD(T)−MRCI].

order to obtain energy differences to chemical accuracy, which we define here as ±2 kcal/mol. The results in Table XV demonstrate that the CCSD(T) binding energies are indeed a dramatic improvement over the CCSD level of theory and are in excellent agreement with the MRCI values. In fact the CCSD(T) value for Be_4 is even more accurate than the MRCI quantity. This is due to the fact that the reference space for the MRCI wave function was constructed using a rather coarse 0.05 threshold in order to make the MRCI calculation tractable. For the other clusters contained in Table XV a threshold of 0.025 was used, and so the MRCI results are more reliable. Nonetheless, the CCSD(T) binding energies are very accurate, the largest underestimate being 2.0 kcal/mol, and given that the CCSD(T) procedure is exactly size-extensive and considerably less costly than the MRCI technique, CCSD(T) may be applied to much larger clusters than is feasible for MRCI. In a later study [51], direct comparisons with FCI calculations showed that the good agreement between CCSD(T) and MRCI for Be_3 was not due to a fortuitous cancellation of higher-order contributions.

Other studies have also shown that dissociation energies of diatomic molecules can be accurately estimated using the CCSD(T) method in conjunction with large basis sets. For example, employing an *spdfg* basis, Scuseria obtained a CCSD(T) dissociation energy for HF (140.9 kcal/mole) that is only 0.2 kcal/mole smaller than the experimental value [58].

Proton affinities (PAs) represent another energy difference that is of great interest. In general, *ab initio* methods less sophisticated than CCSD(T) are capable of determining PAs to ±2.0 kcal/mol. However, the PA of nitric acid, $HONO_2$, is something of a special case. It is well established [220, 228, 229] that protonation of $HONO_2$ leads to more than one isomer and that the lowest energy isomer involves a complex between H_2O and NO_2^+. In other words, protonation of $HONO_2$ can lead to dissociation of the HO-NO_2 chemical bond. Herein lies the difficulty in accurately computing the PA of $HONO_2$. For example, consider the protonation reaction

$$H^+ + A \rightarrow HA^+ \tag{28}$$

The reason that relatively low-level *ab initio* methods perform so well in determining most PAs is that the bonding in A and HA^+ is very similar (and they both contain the same number of electrons) and thus the cancellation of errors in the directly computed PA is very good. The bonding in $HONO_2$, however, is quite different from that in $H_2O\cdots NO_2^+$, indicating that the cancellation of errors for this system will not be that good. The formation of this complex also complicates the experimental determination of the PA since the complex itself may dissociate under the experimental conditions. In addition, the existence and presence of the less stable protonated forms of $HONO_2$ further complicates the experimental situation. Nevertheless, in 1989 Cacace *et al.* [228, 229] believed that they had been able to unravel these experimental difficulties and proposed that the PA of $HONO_2$ was 168±3 kcal/mol — only slightly larger than H_2O (166 kcal/mol). In 1991 Lee and Rice [220] undertook a CCSD(T) study of the protonation of $HONO_2$ and concluded that the PA of $HONO_2$ was 182.5±4 kcal/mol (0 K), which is considerably different from the experimental value. The final PA was determined from CCSD(T) energies obtained with a [5432] ANO basis set, and it was shown that the MP2 level of theory did not perform very well in comparison to CCSD(T). Lee and Rice also established for the first time that the $H_2O\cdots NO_2^+$ complex is bound by 17.3±2.0 kcal/mol and that the next most stable isomer of protonated $HONO_2$ is about 20 kcal/mol higher in energy than the complex.

When advised of these *ab initio* results, Cacace [230] questioned their validity based on the trend that the PA of HOA usually has a PA that is 8–10 kcal/mol less than the analogous CH_3OA molecule, and the proton affinity of CH_3ONO_2 had been experimentally determined to be 176 kcal/mol [231]. Thus a second CCSD(T) study [232] was undertaken to investigate the PA of CH_3ONO_2. The lowest energy form of protonated CH_3ONO_2 is also a complex, $CH_3OH\cdots NO_2^+$, although in this case the second most stable form is only 4.9 kcal/mol higher in energy [232]. The PA of CH_3ONO_2 was found to be 176.9±5 kcal/mol and the complex was shown to be bound by 19.6±2 kcal/mol (all at 0 K). Hence the CCSD(T) results indicate that the PA of $HONO_2$ is 5.6 kcal/mol *larger* than the PA of CH_3ONO_2. It was shown further that the relative PAs of $HONO_2$ and CH_3ONO_2 could be determined directly by subtracting the difference in the complex binding energies (i.e., 17.3−19.6 kcal/mol) from the enthalpy of the following isodesmic reaction

$$HONO_2 + CH_3OH \rightarrow H_2O + CH_3ONO_2 + \Delta H_1 \quad . \tag{29}$$

Because the complex binding energies are much more easily computed accurately, and because *ab initio* evaluation of the enthalpy of an isodesmic reaction is also an easy task (due to an excellent cancellation of errors), this approach is more reliable than comparing the directly computed PA values. Using this method, it was predicted that the PA of nitric acid is *larger* than that for CH_3ONO_2 by 4.6 kcal/mol, which is in excellent agreement with the value obtained by subtracting the CCSD(T) PAs.

The PA of HOA is generally 8–10 kcal/mol smaller than that for CH_3OA because CH_3 is a much better σ-electron donating group (and π-electron donator as well through hyperconjugation). However, Lee and Rice showed [232] that this reasoning does not take into account situations where the bonding or connectivity in the molecule is changed by protonation — as occurs for $HONO_2$ and CH_3ONO_2. In fact, it was shown that the $HONO_2$ and CH_3ONO_2 relative PAs may be explained by the following bond energies HO–H, CH_3O–H, HO–NO_2, and CH_3O–NO_2, which shall be denoted $D_0(1)$, $D_0(2)$, $D_0(3)$, and $D_0(4)$, respectively. All of these bond energies are either well known experimentally or may be accurately estimated based on experimental data (see Ref. [232] and references therein). The better σ-electron donating ability of CH_3 relative to H suggest that $D_0(1) > D_0(2)$ and $D_0(3) > D_0(4)$, and indeed this is the case. $D_0(1) > D_0(2)$ favors a larger PA for $HONO_2$ whereas $D_0(3) > D_0(4)$ favors a larger PA for CH_3ONO_2, and it turns out that the first difference is considerably larger than the second difference, favoring a larger PA for $HONO_2$ relative to CH_3ONO_2. This latter study was very convincing indeed, and motivated several new experimental investigations [233-235]. The discrepancy between experiment and theory is now resolved and the most recent experimental estimate [235] of the PA of $HONO_2$ is 182.0±2.3 kcal/mol, in excellent agreement with the CCSD(T) value.

The next example concerns the heat of formation of the cis-FONO molecule [214]. Cis-FONO is of interest in the study of the chemistry that takes place in the Earth's atmosphere. As such its heat of formation had been estimated based on some analogous compounds [236]. Many of the molecules in common existence in the Earth's stratosphere are not very stable under the conditions present at the surface, and therefore it is often difficult to perform reliable experiments. Cis-FONO is such a molecule. It is, however, relatively straightforward to compute the heat of formation of cis-FONO by evaluating the enthalpy of the following isodesmic reaction

$$2\mathrm{FONO} + \mathrm{H_2O} \rightarrow 2\mathrm{HONO} + \mathrm{F_2O} \quad , \tag{30}$$

and then combine this with the experimental heats of formation of H_2O, cis-HONO, and F_2O to determine the heat of formation of cis-FONO. Even though isodesmic reactions typically exhibit a large degree of cancellation of errors, it has been noted [237] that when investigating isodesmic reactions involving chlorine or fluorine oxides, it is often better to use a high-level method if high accuracy is desired, such as CCSD(T), since the cancellation of errors is not always very good. Thus combining a CCSD(T) enthalpy of reaction (30) with the experimental data yields +9.5 kcal/mol for $\Delta H^o_{f,298}$(cis-FONO) [214]. This value is expected to be accurate to ±1.0 kcal/mol (not considering possible errors in the experimental heats of formation of H_2O, F_2O, and cis-HONO). The estimate based on analogous compounds was −15±7 kcal/mol, which is clearly in error by almost 25 kcal/mol! Using this approach, the CCSD(T) method has been shown to be very useful in determining accurate heats of formation of many unstable molecules that are thought to be important in atmospheric and/or combustion chemistry (e.g., see Refs. [99,238-242]); molecules for which experimental determination is not a simple task.

TABLE XVI
Comparison of CCSD, CCSD(T), and MRCI dissociation energies (kcal/mol) for N_2.[a]

Basis[b]	no. e^- [c]	CCSD	CCSD(T)	MRCI	Δ[d]
[4321]	10e	206.7	216.0	216.0	0.0
[4321]	14e	209.8	219.2		
[4321]+BSSE	10e	205.6	214.7		
[4321]+BSSE	14e	206.1	215.3		
[54321]	10e	212.8	222.3	222.1	0.2
[54321]	14e	218.5	228.3		
[54321]+BSSE	10e	212.4	221.8		
[54321]+BSSE	14e	213.0	222.7		
[65432]	10e	214.3	223.8	223.6	0.2
[65432]	14e	217.9	227.8		

[a] CCSD and CCSD(T) Results taken from Ref. [243]; MRCI results from Ref. [244]; for comparison, the experimental D_e is 228.4 kcal/mol from Ref. [186].
[b] See footnote *b* to Table III for ANO basis set designations.
[c] Number of electrons included in the correlation procedure.
[d] Δ = [CCSD(T)−MRCI].

The next example concerns the bond energy of N_2. N_2 exhibits one of the largest bond energies in chemistry at 228.4 kcal/mol [186], and very high-level *ab initio* studies have shown that it is extremely difficult to compute this quantity to within 2 kcal/mol [210,243-245]. Soon after development of his open-shell CCSD(T) method [57], Scuseria undertook a study of the N_2 bond energy in order to compare with an earlier high-level MRCI study [243, 244], and these results are summarized in Table XVI. The basis sets and geometry were identical to the earlier MRCI study. Both ten electron (10e) and fourteen electron (14e) calculations were performed and corrected for the basis set superposition energy (BSSE). At the time, the BSSE calculations for the [65432] ANO basis set were too large and were therefore not performed. The first point to note is the excellent agreement between the CCSD(T) and MRCI D_e values. In fact, it could be argued that the CCSD(T) quantities are slightly better, since it is clear that improvements in the n-particle or electron correlation treatment lead to larger D_e values, and the CCSD(T) D_e values are equal to or larger than the respective MRCI quantities. Secondly, as with the Be and Mg clusters, the CCSD(T) level of theory is a substantial improvement over the CCSD level of theory. The difficulty in computing an accurate D_e for N_2 is shown by the fact that even with the [65432] ANO basis set, the 10e CCSD(T) D_e is too small by 4.6 kcal/mol. Bauschlicher and Partridge [210] have recently reinvestigated the D_e of N_2 using basis sets with up through i-functions (and also core-correlating basis functions) in conjunction with the CCSD(T), the internally contracted MRCI (ICMRCI), and the internally contracted ACPF (ICACPF) electron correlation methods. They also found excellent agreement between CCSD(T) and the multireference methods. Their best computed D_e value, including core-valence correlation, is 227.1 kcal/mol, which is in very good agreement with experiment. Interestingly, they estimate that one-particle basis set incompleteness is still the major source of error (0.7±0.2 kcal/mol).

Electron affinities (EAs) of small molecules and atoms are one of the most difficult quantities to compute directly, which is partly a result of the large differential electron correlation contribution. This large differential correlation contribution is unavoidable since the anion possesses one more electron than the neutral species. The EAs of the oxygen and fluorine atoms are two examples of severe cases, especially oxygen. An accurate purely *ab initio* prediction (i.e., no use of extrapolation) of the oxygen EA has been elusive in spite of the use of very large one-particle basis sets and sophisticated electron correlation methods (see Ref. [246] and references therein). Strout and Scuseria [246] investigated the electron affinity of the oxygen atom using the CCSD(T) method and a very large $23s26p10d5f3g$ uncontracted Gaussian basis set. They showed via explicit comparison that for a somewhat smaller basis set, the CCSD(T) method gave an EA in very good agreement with Feller and Davidson's [247] best MRCI value. They also demonstrated that this one-particle basis set was deficient in the p shell and thus arrived at the very large basis noted above. In this basis, the CCSD(T) oxygen EA was computed to be 1.415 eV, which is in good agreement with the experimental value of 1.462 eV. Strout and Scuseria attribute most of the remaining error (0.047 eV = 1.08 kcal/mol) to the lack of higher-angular momentum basis functions (i.e., h, i, etc...), and a much smaller part due to the neglect of higher-order correlation effects [i.e., beyond CCSD(T)].

Scuseria [58] has also investigated the EA of fluorine using the CCSD(T) method. While still difficult, the EA of fluorine is not as difficult to compute accurately as that of oxygen. Hence there have been several studies that have obtained an accurate fluorine EA (see Ref. [58] and references therein). In a small $5s4p2d$ one-particle basis set Scuseria showed that the CCSD(T) method gave an EA that agreed with FCI [248] to better than 0.01 eV. In a large uncontracted $13s9p6d4f2g$ one-particle basis set, CCSD(T) gives an EA of 3.36 eV that is in excellent agreement with the experimental [249] value (3.40 eV). Connected triple excitations were found to contribute +0.17 eV to this value. Given the well known difficulty of these two examples, it is safe to conclude that CCSD(T) is capable of yielding chemical accuracy in computing electron affinities.

As a last example for this section, we consider the energy difference between the linear and bent isomers of C_3^+. In 1991 Scuseria [57], as a first application of his newly developed RHF open-shell CCSD(T) method, investigated this energy difference using a series of basis sets including a large [54321] ANO basis set. The best CCSD(T) result indicated that the $^2B_2$ bent form was 6.8 kcal/mol lower in energy than the $^2\Sigma_u^+$ linear form. There is no experimental value with which to compare, however, later in 1991 Taylor *et al.* [250] investigated this energy difference using sophisticated MRCI techniques. Their best estimate, $5.2^{+1.5}_{-1.0}$ kcal/mol, is very close to Scuseria's value, although they used basis sets that included up through only f-type functions. A more appropriate comparison with the result of Taylor *et al.* is probably Scuseria's CCSD(T)/TZ2Pf energy separation of 4.5 kcal/mol. Using either comparison, it is clear that in this case the MRCI and CCSD(T) methods yield energy separations that agree to about 1 kcal/mol, which is particularly interesting since connected triple excitations are shown to contribute about −11 kcal/mol. There are many other examples in the literature where CCSD(T) has given isomer energy differences to very high accuracy (e.g., see Refs. [53,152,251-253]), and these range from simple cis-trans energy differences to more difficult situations where the connectivity and hence bonding of the atoms is different.

Finally, it is worthy of note that Martin [212, 254] has proposed two semiempirical schemes whereby atomization energies can be obtained to very high accuracy. These are based on CCSD(T) computed atomization energies together with some small empirical correction terms that depend on the basis set and reference geometry used. Nevertheless, the examples reviewed in this section demonstrate that the CCSD(T) method is capable of obtaining directly computed energy differences to better than ±2 kcal/mol, and often to better than ±1 kcal/mol.

3.4 Dipole Moments and Polarizabilities

In order to compute accurate dipole moments μ, polarizabilities α and hyperpolarizabilities, it is important to ensure that the one-particle basis set is diffuse enough as it is the description of the outerlying regions of the wave function that is important — especially for hyperpolarizabilities of small molecules. In fact, for dipole moments it is generally much more important to use basis sets that include enough diffuse functions rather than basis sets that include higher angular momentum functions. The accurate determination of hyperpolarizabilities often requires that diffuse higher angular momentum functions also be included (e.g., see Ref. [54]). Shelton and Rice [255] have recently presented a comprehensive review concerning the comparison of experimental and *ab initio* hyperpolarizabilities, and the interested reader is referred to that article for details. In this section, we concentrate on dipole moments and polarizabilities of molecules at or near their equilibrium structure.

For molecules described well by a single determinant reference function, connected triple excitations are not usually important for computing dipole moments accurate to 0.1 D which is often sufficient accuracy for polyatomics with relatively large dipole moments (e.g., see Ref. [271] for CCSD dipole moments of HF, HCN, HNC, H_2O that are in excellent agreement with experiment). Examination of the theoretical and experimental results for HF, BH, CH, HCN, HNC, FNO, H_2O, H_2S, $ClNO_2$, NH_3, cyclopropenylidene (c-C_3H_2), and dioxirane (c-CH_2O_2) in Table XVII supports this assertion: the CCSD and CCSD(T) dipole moments are very close for all of the molecules except for FNO. Description of the electronic structure of FNO is known to be difficult, however. Indeed, connected triple excitations increase the F-N bond distance by 0.038Å and decrease the F-N stretch by 40 cm^{-1} [237]. The CCSD(T) dipole moments in Table XVII all agree with experiment to about 0.1 D or better with the only exception being NH_3. The discrepancy for NH_3 is no doubt due to vibrational averaging. That is, the inversion potential is known to be very flat and anharmonic, and in the planar structure the dipole moment is zero. Thus vibrational averaging should decrease the dipole and indeed the CCSD(T) value is too large. Further support for this view is given

TABLE XVII
Comparison of CCSD, CCSD(T), and experimental dipole moments (D).[a]

Molecule/Basis	CCSD	CCSD(T)	Expt	Δ[b]	Ref[c]
HF/TZ2P	1.85	1.84	1.80	0.04	[209, 260]
BH/TZ2P	1.34	1.32	1.27	0.05	[33, 186]
CH/TZ2P	–	1.48	1.46	0.02	[256, 186]
HCN/TZ2P	3.02	2.98	2.99	−0.01	[209, 261]
HNC/TZ2P	3.10	3.11	3.05	0.06	[209, 262]
FNO/TZ2P	1.65	1.85	1.81	0.04	[237, 263]
H_2O/TZ2P	1.96	1.95	1.85	0.10	[239, 264]
H_2S/ [d]	1.00	0.98	1.02	0.04	[257, 265]
$ClNO_2$/TZ2P	–	0.45	0.42	0.03	[258, 266]
NH_3/TZ2P	1.67	1.67	1.47	0.20	[32, 267]
c-C_3H_2/TZ2P	–	3.43	3.34[e]	0.09	[209, [e]]
c-CH_2O_2/TZ2P	2.65	2.60	2.48±0.07	0.12	[259, 270]

[a] The *ab initio* dipole moments refer to equilibrium values whereas the experimental quantities are vibrationally averaged except for HF and H_2O.
[b] Δ = [CCSD(T)−Expt].
[c] First reference is to the *ab initio* data and second is to experiment.
[d] See Ref. [55] for description of the basis set.
[e] This is the average of three experimental values: 3.43(2) D (Ref. [268]), 3.32(5) D (quoted in Ref. [269]), and 3.27(1) D (Ref. [269]).

by the fact that connected triple excitations display no effect on the equilibrium dipole moment, and also by the fact that the geometry at which the CCSD(T) dipole moment was calculated is very close to the experimental geometry [191].

The major source of error in the calculations summarized in Table XVII is due to inadequate basis sets. If higher accuracy is desired, as is sometimes the case when making detailed comparisons to experiment or when the total magnitude of the dipole moment is very small, it is necessary to use much larger basis sets, including diffuse higher angular momentum functions. For example, Scuseria *et al.* [272] have examined the dipole moment of CO using CCSD(T) and very large optimized one-particle basis sets. Their desired accuracy was 0.001 D, and using a $10s9p4d2f1g$ basis set they obtain 0.125 D, which is 0.003 D larger than the experimental equilibrium value [273]. Scuseria *et al.* showed that the main deficiency in their calculations was lack of saturation in the g space and lack of h functions. Compared to the desired accuracy, connected triple excitations are found to make a substantial contribution, increasing the dipole moment by 0.060 D in the $10s9p4d2f1g$ basis set. Barnes *et al.* [201] have more recently investigated the dipole moment of CO using CCSD(T) with large ANO basis sets including up to h functions. However, as has been shown by Almlöf and Taylor [274], it is necessary to uncontract the outer primitive functions as well as add diffuse functions to the ANO basis sets in order to obtain high accuracy for electrical properties. In the Barnes *et al.* study, the dipole moment was not the primary focus and it

TABLE XVIII
Comparison of CCSD, CCSD(T), and experimental polarizabilities (a.u.).[a]

Molecule/Basis	CCSD	CCSD(T)	Expt	Δ[b]	Ref[c]
He/9s7p5d4f	1.384	–	1.383[d]	0.001	[275, 277]
Ne/16s11p10d7f	2.628	2.677	2.669	0.008	[55, 278]
Ar/14s12p10d7f	11.08	11.17	11.08	0.09	[275, 278]
Kr/11s10p8d8f	16.86	16.99	16.79	0.20	[275, 278]
Xe/12s11p9d5f	27.25	27.40	27.16	0.24	[275, 278]
HCl/9s8p6d5f	17.08	17.24	17.02	0.22	[276, 279]
CO/8s6p3d	13.00	13.14	13.08[f]	0.06	[33, 280]
NNO/POL[e]	19.52	19.78	19.77	0.01	[33, 281]
CO_2/POL[e]	17.44	17.63	17.50	0.13	[257, 282]
N_2/POL[e]	11.61	11.75	11.8	−0.05	[257, 283]
H_2/3s2p1d	5.22	–	5.18[f]	0.04	[257, 284]
C_2H_4/POL	26.93	27.08	28.70	−1.62	[257, 285]
HF/POL	5.61	5.71	5.52[f]	0.19	[257, 286]
H_2O/POL	9.61	9.79	9.81	−0.02	[257, 282]
NH_3/POL	14.13	14.38	14.56	−0.18	[257, 287]

[a] The *ab initio* polarizabilities do not include a vibrational correction, see text.
[b] Δ = [CCSD(T)−Expt].
[c] First reference is to the *ab initio* data and second is to experiment.
[d] A very accurate calculation by Bishop and Pipin, Ref. [277].
[e] See theoretical reference for description of the basis set.
[f] Vibrational contribution not included.

is clear from examination of their results that the basis sets used are not flexible enough with respect to diffuse functions to attain accuracy approaching 0.001 D, although there is also a geometric effect since they used the theoretical minimum and not the experimental equilibrium bond distance. Nonetheless, the study by Scuseria *et al.* shows that CCSD(T) may be used to obtain highly accurate dipole moments (i.e., approaching ±0.001 D) provided that very large and flexible one-particle basis sets are used.

The accurate determination of polarizabilities α requires more diffuse basis functions than is required for dipole moments. Theoretical studies of polarizabilities are further complicated by two aspects: a) for molecules vibrational contributions can be significant, and these are complicated to include properly (see Refs. [288-290], for example); and b) the direct observables in experiments are frequency dependent polarizabilities, which are significantly more complicated to compute (for example, see Refs. [291-293]). In some cases, experimental frequency independent or static polarizabilities have been determined by extrapolating to zero frequency. We shall be concerned only with static values in this review. The best systems with which to compare experiment and theory are atoms since there is no vibrational contribution. Contained in Table XVIII are CCSD, CCSD(T) and experimental static polarizabilities for the noble gas atoms He, Ne, Ar, Kr, and Xe. The *ab initio* values were obtained with very large basis sets and core-correlation and relativistic effects (for Kr and Xe) were also included. The agreement between CCSD and CCSD(T) is generally good and improves going down the periodic table, indicating that for these atoms, which are described very well by a single determinant reference function, connected triple excitations do not contribute much to the polarizability α. Interestingly, connected triples are much more important for hyperpolarizabilities, even for the noble gas atoms. The agreement between CCSD(T) and experiment is also good, with the largest error being 0.24 au or 0.9%. It is also worthy of note that explicit calculations for the Ne α and γ have shown that CCSD(T) approximates CCSDT very well [55]. Thus very accurate static polarizabilties can be obtained with CCSD(T) provided that very large, flexible and diffuse one-particle basis sets are used.

Also included in Table XVIII are CCSD, CCSD(T) and experimental static polarizabilites for several small molecules. The values quoted are $\bar{\alpha}$ where

$$\bar{\alpha} = \frac{1}{3}(\alpha_{xx} + \alpha_{yy} + \alpha_{zz}) \quad . \tag{31}$$

For convenience, the overbar will be omitted in the following text. Unless otherwise noted, the experimental values include vibrational contributions, whereas the *ab initio* quantities do not. While the basis sets used for the molecules are not as elaborate as those used for the noble gas atoms, they are still reasonably large and in many cases the polarizability consistent basis sets (POL) recommended by Sadlej [294] have been used. Examination of the results shows that both the CCSD and CCSD(T) methods perform well in determining accurate polarizabilities. The contribution from connected triple excitations, while non-negligible for very high accuracy, is not that large either. In comparing to experiment, the largest error is for C_2H_4 and this could be due to several factors, including basis set extension, core correlation, and vibrational effects. Based on the core correlation effects determined for the noble gas atoms, and based on some core correlation calculations by Hammond and Rice [276], it seems clear that core correlation will reduce the α values, which in many cases will bring the CCSD(T) and experimental polarizabilities into better agreement. Even so, the agreement is already good, demonstrating that CC methods are capable of yielding very high accuracy for molecular polarizabilities also. As the calculation of frequency dependent polarizabilities at the CC level, and the inclusion of vibrational effects become more routine, it will be interesting to make detailed comparisons with the experimentally observable quantities.

3.5 Vertical Electronic Excitation Energies

The theoretical study of excited electronic states is an area of intense investigation currently (e.g., see Refs. [295-300]). Single-reference CC theory as applied to ground electronic states is reasonably well defined at this point, with the only exception being the computationally most efficient approach for high-spin open-shell electronic states, as discussed previously.

TABLE XIX
Comparison of LRCCSD and FCI electronic excitation energies (eV).

Molecule/Basis	State	Excitation Type[a]	CCSD	FCI	Δ[b]	Ref[c]
Be/9s9p5d	^1P	1e	5.323	5.318	0.005	[171, 302]
Be/9s9p5d	^1P	1e	7.468	7.462	0.006	[171, 302]
Be/9s9p5d	^1S	1e	6.773	6.765	0.008	[171, 302]
Be/9s9p5d	^1D	2e	7.139	7.089	0.050	[171, 302]
CH^+/4s2p;2s	$^1\Pi$	1e	3.261	3.230	0.031	[301, 302]
CH^+/4s2p;2s	$^1\Delta$	2e	7.889	6.964	0.925	[301, 302]
CH^+/4s2p;2s	$^1\Sigma^+$	2e	9.109	8.549	0.560	[301, 302]
CH^+/4s2p;2s	$^1\Sigma^+$	1e	13.581	13.525	0.056	[301, 302]
H_2O/4s2p;2s	$^1B_1$	1e	8.581	8.696	−0.115	[170, 302]
H_2O/4s2p;2s	$^1A_2$	1e	10.716	10.798	−0.082	[170, 302]
H_2O/4s2p;2s	$^1A_1$	1e	10.737	10.840	−0.103	[170, 302]
H_2O/4s2p;2s	$^1B_2$	1e	13.208	13.272	−0.064	[170, 302]
Be/9s9p5d	^3P	1e	2.729	2.733	−0.004	[169, 302]
Be/9s9p5d	^3S	1e	6.447	6.444	0.003	[169, 302]
Be/9s9p5d	^3DP	1e	7.301	7.295	0.006	[169, 302]
H_2O/4s2p;2s	$^3B_1$	1e	7.790	7.901	−0.111	[170, 303]
H_2O/4s2p;2s	$^3A_1$	1e	9.716	9.898	−0.182	[170, 303]
H_2O/4s2p;2s	$^3A_2$	1e	10.231	10.284	−0.053	[170, 303]
H_2O/4s2p;2s	$^3B_2$	1e	11.813	11.871	−0.058	[170, 303]

[a] 1e denotes a single electron excited state (relative to the ground state) and 2e denotes a two electron excited state.
[b] Δ = [CCSD(T)−FCI].
[c] First reference is to the LRCCSD data and second is to FCI.

There has been considerable work on extending CC theory to excited electronic states, but a "best" general treatment is not yet established. However, for electronic states that correspond to a single excitation from the ground electronic state, the LRCCSD and EOM-CCSD methods appear to be very promising (note: the LRCCSD and EOM-CCSD methods give identical electronic excitation energies). The reason for this optimism is apparent from examination of the results in Table XIX. LRCCSD and FCI vertical excitation energies are compared for singlet excited states of Be, CH^+, and H_2O, and triplet excited states of Be and H_2O. The agreement between LRCCSD and FCI is excellent for all excited electronic states that are a single excitation (i.e., 1e) away from the ground (reference) electronic state. The largest errors occur for the H_2O molecule, with the largest error being −0.182 eV for the $^3A_1$ state. For singly excited states of Be and CH^+, the errors are less than 0.1 eV. Based on these comparisons and additional comparisons between LRCCSD and experiment for CO, N_2, H_2O, H_2CO, C_2H_4, O_3, C_4H_6, C_2H_4O, and C_6H_6 performed by Koch *et al.* [301], Comeau and Bartlett [169], and Head-Gordon and co-workers [170, 171, 304] it appears evident that LRCCSD is capable of achieving $\approx \pm(0.1\text{–}0.2)$ eV accuracy for singly excited states provided that large flexible one-particle basis sets are used. Recall that diffuse basis sets will be needed to describe properly Rydberg states.

It should be noted that excitation energies for electronic states that are more than a single excitation away from the reference state are not nearly as accurate as those from singly excited states. The error for the ^1D Be state (a 2e state; see Table XIX) is almost an order of magnitude greater than found for the singly excited states. The same behavior is found for the two 2e states of CH^+. Hence, it is clear that more sophisticated methods are required in order to obtain an accurate description of these excited electronic states. Nevertheless, the LRCCSD method appears to be very promising for the study of singly excited electronic states, and with the recent development of analytical energy gradients [173, 174] it will be interesting to see how this method performs in characterizing excited electronic state potential energy surfaces.

It is also worthy of note that Head-Gordon and co-workers [170, 171] have shown that, unlike ground state calculations where QCISD and CCSD generally yield similar results, LRQCISD and LRCCSD can exhibit large differences in the prediction of excitation energies. They have determined that the majority of this difference is due to the neglect of the T_1T_2 term in the T_2 equations. Thus although the mathematical manipulations are much simpler, the LRQCISD method is not recommended for a highly accurate description of singly excited electronic states. Head-Gordon *et al.* [304] have also used the LRCCSD method as a model on which to build or define a less costly single-reference perturbation theory approach, which should be applicable to much larger molecules. The initial results appear promising.

4 Future Prospects

We hope that we have convinced the reader of the impressive accuracy that is obtainable for a host of atomic and molecular properties with the CCSD(T) and LRCCSD methods when used with large one-particle basis sets, and the significant formal and implementational developments that have taken place over the last few years, which have made this impressive accuracy possible. However, any field that stagnates will soon be passed by and thus there are many more advances to be made in CC theory. One area of interest is to widen the application of CCSD(T) to transition metal chemistry. Studies by Blomberg *et al.* [305] and by Bauschlicher *et al.* [306], as well as others, suggest that CCSD(T) will perform well for computing ligand binding energies to transition metal atoms provided that the SCF reference function is not too poor a starting point (i.e., the T_1 diagnostic is not too large).

Another area of interest is to apply CC theory to much larger molecules and medium-sized molecules with very large basis sets. As recently discussed by Rendell and Lee [176], there are two major bottlenecks to realizing this goal, a) the disk space and input/output (I/O) requirements, and b) the sheer computational effort (or CPU requirements). For state of the art CCSD(T) programs, it is usually the case that the limiting resource (i.e., that resource which prohibits a given calculation being performed) is the volume of disk space necessary and/or the real time necessary to perform the I/O due to slow access times. The sheer volume of disk space is the major limitation on supercomputers such as the Cray C90, whereas the disk space *and* the I/O time are both limitations on super workstations such as the IBM RS6000 series. Following earlier work [307, 308], Rendell and Lee have proposed approximating specific sets of MO two-electron integrals in order to reduce dramatically the disk space and I/O requirements without substantially affecting the CPU requirements [176]. The results from their initial study appear very promising indeed, and it will be interesting to see if 500 basis function CCSD(T) calculations (their stated goal) are possible with this approach.

Rendell and co-workers [309-311] have also addressed the computational cost limitation by developing CCSD(T) algorithms that are suitable for use on massively parallel, distributed memory, MIMD supercomputers. They showed that very effective and scalable CCSD(T) algorithms could be developed provided that I/O was kept to a minimum, since the I/O subsystem is not really parallel. Hence, incorporation of the approximate MO two-electron integral scheme should significantly improve the performance of these algorithms, and this work continues.

Koch *et al.* [312] have very recently reported an efficient implementation of a direct atomic orbital integral driven CCSD method, which is also designed to eliminate the I/O and disk space limitations. However, the cost of recomputing the integrals each iteration of the CCSD procedure is quite expensive, and it seems likely to us that the scheme proposed by Rendell and Lee [176] will be more useful for large basis set CCSD calculations. Another difficulty concerning the direct atomic orbital integral driven approach is that an efficient (T) step must still be formulated, and due to the nature of this contribution, an efficient formulation will not be a simple task.

Other developments that may prove useful in the future include the use of optimized virtual orbitals in CC theory, as discussed by Adamowicz and Bartlett [313], and also the use of localized orbitals (e.g., see Ref. [314]). The purpose of optimized virtual orbital sets is to obtain a sizeable fraction of the CCSD or CCSD(T) correlation energy with only a small subset of the total number of virtual orbitals, thereby reducing the disk space, I/O, and CPU requirements. The initial comparisons [313] appear interesting, but these are only for relatively small molecules. It will be interesting to see rigorous tests for larger molecules. The purpose for using localized orbitals is similar, in that disk space, I/O, and CPU requirements will all be reduced. The idea is to localize orbitals on specific molecular fragments, and then to eliminate amplitudes that connect fragments that are spatially far removed from one another. Of course, in order to test such an approach a fairly large molecule (by *ab initio* chemists' standards) will need to be examined, and it is probably the current situation that a CCSD calculation cannot easily be performed on a large enough molecule (note that the C_{20} calculations [104, 105] mentioned earlier may constitute a possible test case, but these are quite expensive and at the current limits). Nonetheless, the idea of utilizing localized orbitals is very appealing, and it may be that in the long run this will be a very useful approach, possibly in combination with the approximate two-electron integral scheme mentioned above.

Finally, the recent work of Noga *et al.* [315] is worthy of mention. They have developed a CC-R12 scheme whereby the CC wave function contains linear r_{12} terms. The idea here is to reduce the required size of the one-particle basis set, especially with regard to high angular momentum functions, which are primarily needed to describe correctly the electron-electron cusp. The difficulty with the method is that three-electron integrals appear in the formalism, and in order to avoid computing these a resolution of the identity is used. The basis set requirements for this resolution appear quite stringent, however, and from the results presented thus far it is not evident that the CC-R12 approach will be cost effective. In any case, when accurate predictions of atomic and molecular properties are desired, it is clear that CC theory has several advantages (e.g., the most accurate single-reference based electron correlation procedure; easily extended to include higher-order effects due to its close connection with perturbation theory; wide availability of efficient programs), which suggest that, where applicable, it will be the method of choice for the foreseeable future. The examples presented in the second half of this review, mostly from our own research, certainly demonstrate the very high accuracy that is achievable from the CCSD(T) approach when combined with the use of *spdf* or *spdfg* quality basis sets.

5 Acknowledgements

The authors would like to gratefully acknowledge the many collaborators that have made significant contributions to much of the work reviewed in this article. We would also like to thank Drs. Julia Rice, Peter Taylor, and Alistair Rendell, and colleagues at NASA Ames for reading early versions of this review and making helpful suggestions. The coupled-cluster work at Rice University has been funded by the National Science Foundation, the Welch Foundation, and the Camille and Henry Dreyfus Foundation.

References

1. T. Helgaker and P. R. Taylor, in *Modern Electronic Structure Theory*, Ed. D. R. Yarkony (World Scientific Publishing Company, London, 199x), pp. xxx-xxx.
2. R. J. Bartlett and G. D. Purvis, Int. J. Quantum Chem. **14**, 561 (1978).
3. J. Čižek, Adv. Chem. Phys. **14**, 35 (1969).
4. R. J. Bartlett, Annu. Rev. Phys. Chem. **32**, 359 (1981).
5. J. Paldus, in *New Horizons of Quantum Chemistry*, Eds. P.-O. Lowdin and B. Pullman (Reidel, Dordrecht, 1983), pp. 31-60.
6. V. Knasnička, V. Laurinc, S. Biskupič, and M. Haring, Adv. Chem. Phys. **52**, 181 (1983).
7. R. J. Bartlett, C. E. Dykstra, and J. Paldus, in *Advanced Theories and Computational Approaches for the Electronic Structure of Molecules*, Ed. C. E. Dykstra (Reidel, Dordrecht, 1984), pp. 127-159.
8. S. A. Kucharski and R. J. Bartlett, Adv. Quantum Chem. **18**, 281 (1986).
9. R. J. Bartlett, J. Phys. Chem. **93**, 1697 (1989).
10. K. Raghavachari, Annu. Rev. Phys. Chem. **42**, 615 (1991).
11. J. Paldus, in *Methods in Computational Molecular Physics*, Eds. S. Wilson and G. H. F. Diercksen (Plenum Press, New York, 1992), pp. 99-194.
12. ACES II is a computational chemistry package especially designed for coupled-cluster and many-body perturbation calculations. The SCF, transformation, correlation energy, and gradient codes were written by J. F. Stanton, J. Gauss, J. D. Watts, W. J. Lauderdale, and R. J. Bartlett. The two-electron integrals are taken from the vectorized MOLECULE code of J. Almlöf and P. R. Taylor, and the derivative integrals are taken from the ABACUS program written by T. Helgaker, H. J. Jensen, P. Jørgensen, J. Olsen, and P. R. Taylor. The geometry optimization and vibrational analysis package were written by J. F. Stanton and D. Bernholdt.
13. CADPAC, Issue 5: The Cambridge Analytical Derivatives Package, Cambridge, 1992. A suite of quantum chemistry programs developed by R. D. Amos with contributions from I. L. Alberts, J. S. Andrews, S. M. Colwell, N. C. Handy, D. Jayatilaka, P. J. Knowles, R. Kobayashi, N. Koga, K. E. Laidig, P. E. Maslen, C. W. Murray, J. E. Rice, J. Sanz, E. D. Simandiras, A. J. Stone, and M.-D. Su.
14. GAUSSIAN (Gaussian, Pittsberg, 1992): A quantum chemistry package written by M. J. Frisch, M. Head-Gordon, G. W. Trucks, J. B. Foresman, H. B. Schlegel, K. Raghavachari, M. Robb, J. S. Binkley, C. Gonzalez, D. J. DeFrees, D. J. Fox, R. A. Whiteside, R. Seeger, C. F. Melius, J. Baker, R. L. Martin, L. R. Kahn, J. J. P. Stewart, S. Topiol, and J. A. Pople.
15. MOLCAS-2, a suite of quantum chemistry programs written by K. Andersson, M. R. A. Blomberg, M. P. Fülscher, V. Kellö, R. Lindh, P.-Å. Malmqvist, J. Noga, J. Olsen, B. O. Roos, A. J. Sadlej, P. E. M. Seigbahn, M. Urban, and P.-O. Widmark (Dept. of Theor. Chem., Chem. Centre, Univ. of Lund, Lund, (1992)).
16. MOLPRO (1994) is a suite of programs written by H.-J. Werner and P. J. Knowles with contributions by J. Almlöf, R. D. Amos, M. Deegan, S. T. Elbert, C. Hampel, W. Meyer, K. A. Peterson, R. M. Pitzer, E.-A. Reinsch, A. J. Stone, and P. R. Taylor.
17. PSI, version 2.0, PSITECH Inc., Athens, Georgia (1991). The coupled-cluster programs included in this package were written by G. E. Scuseria.
18. TITAN is a set of electronic structure programs written by T. J. Lee, A. P. Rendell and J. E. Rice.
19. J. F. Stanton, R. J. Bartlett, and C. M. L. Rittby, J. Chem. Phys. **97**, 5560 (1992).
20. A. Balková and R. J. Bartlett, J. Chem. Phys. **99**, 7907 (1993).
21. J. A. Pople, M. Head-Gordon, K. Raghavachari, J. Chem. Phys. **87**, 5968 (1987).
22. G. E. Scuseria and H. F. Schaefer, J. Chem. Phys. **90**, 3700 (1989).
23. J. Paldus, J. Čižek, and B. Jeziorski, J. Chem. Phys. **90**, 4356 (1989).
24. J. A. Pople, M. Head-Gordon, K. Raghavachari, J. Chem. Phys. **90**, 4635 (1989).
25. T. J. Lee, A. P. Rendell, and P. R. Taylor, J. Phys. Chem. **94**, 5463 (1990).
26. R. A. Chiles and C. E. Dykstra, J. Chem. Phys. **74**, 4544 (1981).
27. L. Z. Stolarczyk and H. J. Monkhorst, Int. J. Quantum Chem. Symp. **18**, 267 (1984).
28. G. E. Scuseria and H. F. Schaefer, Chem. Phys. Lett. **142**, 354 (1987).
29. N. C. Handy, J. A. Pople, M. Head-Gordon, K. Raghavachari, and G. W. Trucks, Chem. Phys. Lett. **164**, 185 (1989).

30. K. Raghavachari, J. A. Pople, E. S. Replogle, M. Head-Gordon, and N. C. Handy, Chem. Phys. Lett. **167**, 115 (1990).
31. K. Hirao in *Self-Consistent Field Theory,* Studies in Physical and Theoretical Chemistry, Vol. 70, (1990).
32. T. J. Lee, R. Kobayashi, N. C. Handy, and R. D. Amos, J. Chem. Phys. **96**, 8931 (1992).
33. R. Kobayashi, H. Koch, P. Jørgensen, and T. J. Lee, Chem. Phys. Lett. **211**, 94 (1993).
34. J. F. Stanton, J. Gauss, and R. J. Bartlett, J. Chem. Phys. **97**, 5554 (1992).
35. L. A. Barnes and R. Lindh, Chem. Phys. Lett. **223**, 207 (1994).
36. C. Hampel, K. A. Peterson, and H.-J. Werner, Chem. Phys. Lett. **190**, 1 (1992); Erratum, Chem. Phys. Lett. **192**, 332 (1992).
37. L. A. Curtiss and K. Raghavachari, in *Quantum Mechanical Electronic Structure Calculations with Chemical Accuracy*, Ed. S. R. Langhoff (Kluwer Academic Publishers, Dordrecht, 1995), this volume.
38. G. D. Purvis and R. J. Bartlett, J. Chem. Phys. **76**, 1910 (1982).
39. K. Raghavachari, G. W. Trucks, J. A. Pople, and M. Head-Gordon, Chem. Phys. Lett. **157**, 479 (1989).
40. C. W. Bauschlicher, S. R. Langhoff, and P. R. Taylor, Adv. Chem. Phys. **77**, 103 (1990).
41. S. J. Cole and R. J. Bartlett, J. Chem. Phys. **86**, 873 (1987).
42. J. Noga and R. J. Bartlett, J. Chem. Phys. **86**, 7041 (1987); Erratum, J. Chem. Phys. **89**, 3401 (1988).
43. G. E. Scuseria and H. F. Schaefer, Chem. Phys. Lett. **152**, 382 (1988).
44. T. J. Lee, A. P. Rendell, and P. R. Taylor, J. Chem. Phys. **92**, 489 (1990).
45. A. P. Rendell, T. J. Lee, and P. R. Taylor, J. Chem. Phys. **92**, 7050 (1990).
46. T. J. Lee, A. P. Rendell, and P. R. Taylor, J. Chem. Phys. **93**, 6636 (1990).
47. T. J. Lee, A. P. Rendell, and P. R. Taylor, Theor. Chim. Acta. **83**, 165 (1992).
48. T. J. Lee and G. E. Scuseria, J. Chem. Phys. **93**, 489 (1990).
49. T. J. Lee and J. E. Rice, J. Chem. Phys. **94**, 1215 (1991).
50. G. E. Scuseria and T. J. Lee, J. Chem. Phys. **93**, 5851 (1990).
51. J. D. Watts, I. Cernusak, J. Noga, R. J. Bartlett, C. W. Bauschlicher, T. J. Lee, A. P. Rendell, and P. R. Taylor, J. Chem. Phys. **93**, 8875 (1990).
52. G. E. Scuseria, J. Chem. Phys. **94**, 442 (1991).
53. T. J. Lee and A. P. Rendell, Chem. Phys. Lett. **177**, 491 (1991).
54. P. R. Taylor, T. J. Lee, J. E. Rice, and J. Almlöf, Chem. Phys. Lett. **163**, 359 (1989); Erratum, Chem. Phys. Lett. **189**, 197 (1992).
55. J. E. Rice, G. E. Scuseria, T. J. Lee, P. R. Taylor, and J. Almlöf, Chem. Phys. Lett. **191**, 23 (1992).
56. R. Lindh, J. E. Rice, and T. J. Lee, J. Chem. Phys. **94**, 8008 (1991).
57. G. E. Scuseria, Chem. Phys. Lett. **176**, 27 (1991).
58. G. E. Scuseria, J. Chem. Phys. **95**, 7426 (1991).
59. G. E. Scuseria and H. F. Schaefer, Chem. Phys. Lett. **174**, 501 (1990).
60. G. E. Scuseria, A. C. Scheiner, T. J. Lee, J. E. Rice, and H. F. Schaefer, J. Chem. Phys. **86**, 2881 (1987).
61. T. J. Lee and J. E. Rice, Chem. Phys. Lett. **150**, 406 (1988).
62. G. E. Scuseria, C. L. Janssen, and H. F. Schaefer, J. Chem. Phys. **89**, 7382 (1988).
63. P. Piecuch and J. Paldus, Int. J. Quantum Chem. **36**, 429 (1989).
64. J. F. Stanton, J. Gauss, J. D. Watts, and R. J. Bartlett, J. Chem. Phys. **94**, 4334 (1991).
65. J. Geertsen, S. Eriksen, and J. Oddershede, Adv. Quantum Chem. **22**, 168 (1991).
66. D. Jayatilaka and T. J. Lee, Chem. Phys. Lett. **199**, 211 (1992).
67. S. A. Kucharski and R. J. Bartlett, J. Chem. Phys. **97**, 4282 (1992).
68. P. R. Taylor in *Lecture Notes in Quantum Chemistry,* Vol. 2, Ed. B. Roos (Springer-Verlag, Berlin, 1994), pp. xx-xx.
69. T. P. Zivkovic, Int. J. Quantum Chem. Symp. **11**, 413 (1977).
70. T. P. Zivkovic and H. J. Monkhorst, J. Math. Phys. **19**, 1007 (1978).
71. L. Meissner, A. Balková, and R. J. Bartlett, Chem. Phys. Lett. **212**, 177 (1993).
72. C. Møller and M. S. Plesset, Phys. Rev. **46**, 618 (1934).
73. G. D. Purvis and R. J. Bartlett, J. Chem. Phys. **75**, 1284 (1981).
74. N. Oliphant and L. Adamowicz, Chem. Phys. Lett. **190**, 13 (1992).
75. G. E. Scuseria, T. J. Lee, and H. F. Schaefer, Chem. Phys. Lett. **130**, 236 (1986).
76. P. Pulay, Chem. Phys. Lett. **73**, 393 (1980).
77. P. Pulay, J. Comp. Chem. **3**, 556 (1982).
78. T. P. Hamilton and P. Pulay, J. Chem. Phys. **84**, 5728 (1986).
79. T. H. Dunning, J. Chem. Phys. **90**, 1007 (1989).

80. T. Tanaka and Y. Morino, J. Mol. Spectrosc. **33**, 538 (1970).
81. A. R. Hoy, I. M. Mills, and G. Strey, Mol. Phys. **24**, 1265 (1972).
82. Y. S. Lee, S. A. Kucharski, and R. J. Bartlett, J. Chem. Phys. **81**, 5906 (1984).
83. M. Urban, J. Noga, S. J. Cole, and R. J. Bartlett, J. Chem. Phys. **83**, 4041 (1985).
84. K. Raghavachari, J. A. Pople, E. S. Replogle, and M. Head-Gordon, J. Phys. Chem. **94**, 5579 (1990).
85. J. Almlöf and P. R. Taylor, J. Chem. Phys. **86**, 4070 (1987).
86. D. E. Woon, J. Chem. Phys. **100**, 2838 (1994).
87. J. A. Pople, M. Head-Gordon, D. J. Fox, K. Raghavachari, and L. A. Curtiss, J. Chem. Phys. **90**, 5622 (1989).
88. L. A. Curtiss, C. Jones, G. W. Trucks, K. Raghavachari, and J. A. Pople, J. Chem. Phys. **93**, 2537 (1990).
89. L. A. Curtiss, K. Raghavachari, G. W. Trucks, and J. A. Pople, J. Chem. Phys. **94**, 7221 (1991).
90. T. J. Lee, J. E. Rice, G. E. Scuseria, and H. F. Schaefer, Theor. Chim. Acta. **75**, 81 (1989).
91. A. C. Hurley, *Electron Correlation in Small Molecules* (Academic, London, 1976).
92. T. J. Lee and P. R. Taylor, Int. J. Quantum Chem. Symp. **23**, 199 (1989).
93. D. Jayatilaka and T. J. Lee, J. Chem. Phys. **98**, 9734 (1993).
94. W. D. Laidig, G. D. Purvis, and R. J. Bartlett, Int. J. Quantum Chem. Symp. **6**, 561 (1982).
95. W. D. Laidig, G. D. Purvis, and R. J. Bartlett, Chem. Phys. Lett. **97**, 209 (1983).
96. D. J. Thouless, *The Quantum Mechanics of Many-Body Systems* (Academic, New York, 1972).
97. J. M. L. Martin, T. J. Lee, G. E. Scuseria, and P. R. Taylor, J. Chem. Phys. **97**, 6549 (1992).
98. C. M. Rohlfing and P. J. Hay, J. Chem. Phys. **86**, 4518 (1987).
99. A. L. L. East and W. D. Allen, J. Chem. Phys. **99**, 4638 (1993).
100. P. R. Taylor in *Lecture Notes in Quantum Chemistry,* Vol. 1, Ed. B. Roos (Springer-Verlag, Berlin, 1992), pp. 325-412.
101. P. Pulay, S. Saebø, and W. Meyer, J. Chem. Phys. **81**, 1901 (1984).
102. A. P. Rendell, T. J. Lee, and A. Komornicki, Chem. Phys. Lett. **178**, 462 (1991).
103. P. Čársky, L. J. Schaad, B. A. Hess, M. Urban, and J. Noga, J. Chem. Phys. **87**, 411 (1987).
104. P. R. Taylor, E. Bylaska, J. H. Weare, and R. Kawai, Phys. Rev. Lett., submitted.
105. P. R. Taylor, personal communication.
106. H. Koch, personal communication.
107. N. C. Handy, P. J. Knowles, and K. Somasundrum, Theor. Chim. Acta. **68**, 87 (1985).
108. R. H. Nobes, J. A. Pople, L. Radom, N. C. Handy, and P. J. Knowles, Chem. Phys. Lett. **138**, 481 (1987).
109. R. Murphy, H. F. Schaefer, R. H. Nobes, L. Radom, and R. S. Pitzer, Int. Rev. Phys. Chem. **5**, 229 (1986).
110. J. D. Watts and R. J. Bartlett, J. Chem. Phys. **95**, 6652 (1991).
111. J. Gauss, J. Stanton, and R. J. Bartlett, J. Chem. Phys. **95**, 2639 (1991).
112. W. J. Lauderdale, J. F. Stanton, J. Gauss, J. D. Watts, and R. J. Bartlett, J. Chem. Phys. **97**, 6606 (1992).
113. J. D. Watts, J. Gauss, and R. J. Bartlett, J. Chem. Phys. **98**, 8718 (1993).
114. H. B. Schlegel, J. Phys. Chem. **92**, 3075 (1988).
115. J. F. Stanton, J. Chem. Phys. **101**, 371 (1994).
116. M. Rittby and R. J. Bartlett, J. Phys. Chem. **92**, 3033 (1988).
117. A. D. McLean and B. Liu, J. Chem. Phys. **58**, 1066 (1973).
118. P. J. Knowles, J. S. Andrews, R. D. Amos, N. C. Handy, and J. A. Pople, Chem. Phys. Lett. **186**, 130 (1991).
119. W. J. Lauderdale, J. F. Stanton, J. Gauss, J. D. Watts, and R. J. Bartlett, Chem. Phys. Lett. **187**, 21 (1991).
120. T. J. Lee and D. Jayatilaka, Chem. Phys. Lett. **201**, 1 (1993).
121. T. J. Lee, A. P. Rendell, K. G. Dyall, and D. Jayatilaka, J. Chem. Phys. **100**, 7400 (1994).
122. M. J. O. Deegan and P. J. Knowles, Chem. Phys. Lett., submitted.
123. C. L. Janssen and H. F. Schaefer, Theor. Chim. Acta. **79**, 1 (1991).
124. P. J. Knowles, C. Hampel, and H.-J. Werner, J. Chem. Phys. **99**, 5219 (1993).
125. P. Neogrády, M. Urban, and I. Hubač, J. Chem. Phys. **100**, 3706 (1993).
126. P. Neogrády, M. Urban, and I. Hubač, J. Chem. Phys. **97**, 5074 (1992).
127. H. Nakatsuji and M. Ehara, J. Chem. Phys. **98**, 7179 (1993).

128. I. Hubač and P. Čársky, Phys. Rev. A **22**, 2392 (1980).
129. R. D. Amos, J. S. Andrews, N. C. Handy, and P. J. Knowles, Chem. Phys. Lett. **185**, 256 (1991).
130. C. Murray and E. R. Davidson, Chem. Phys. Lett. **187**, 451 (1991).
131. R. J. Bartlett in *Geometrical Derivatives of Energy Surfaces and Molecular Properties*, Eds. P. Jørgensen and J. Simons (Reidel, Dordrecht, 1986), pp. 35-61.
132. P. Pulay, Mol. Phys. **17**, 197 (1969).
133. N. C. Handy and H. F. Schaefer, J. Chem. Phys. **81**, 5031 (1984).
134. J. Gerratt and I. M. Mills, J. Chem. Phys. **49**, 1719 (1968).
135. L. Adamowicz, W. D. Laidig, and R. J. Bartlett, Int. J. Quantum Chem. Symp. **18**, 245 (1984).
136. J. E. Rice and R. D. Amos, Chem. Phys. Lett. **122**, 585 (1985).
137. J. E. Rice, R. D. Amos, N. C. Handy, T. J. Lee, and H. F. Schaefer, J. Chem. Phys. **85**, 963 (1986).
138. T. J. Lee, N. C. Handy, J. E. Rice, A. C. Scheiner and H. F. Schaefer, J. Chem. Phys. **85**, 3930 (1986).
139. N. C. Handy, R. D. Amos, J. F. Gaw, J. E. Rice, E. D. Simandiras, T. J. Lee, R. J. Harrison, W. D. Laidig, G. B. Fitzgerald and R. J. Bartlett in *Geometrical Derivatives of Energy Surfaces and Molecular Properties*, Eds. P. Jørgensen and J. Simons (Reidel, Dordrecht, 1986), pp. 179-191.
140. N. C. Handy, R. D. Amos, J. F. Gaw, J. E. Rice, and E. D. Simandiras, Chem. Phys. Lett. **120**, 151 (1985).
141. J. E. Rice, T. J. Lee, and N. C. Handy, J. Chem. Phys. **88**, 7011 (1988).
142. T. J. Lee, J. Chem. Phys. **87**, 2825 (1987).
143. T. J. Lee, W. D. Allen, and H. F. Schaefer, J. Chem. Phys. **87**, 7062 (1987).
144. R. Shepard, H. Lischka, P. G. Szalay, T. Kovar, and M. Ernzerhof, J. Chem. Phys. **96**, 2085 (1992).
145. P. Jørgensen and T. Helgaker, J. Chem. Phys. **89**, 1560 (1988).
146. T. Helgaker and P. Jørgensen, Theor. Chim. Acta. **75**, 111 (1989).
147. T. Helgaker, P. Jørgensen, and N. C. Handy, Theor. Chim. Acta. **76**, 227 (1989).
148. P. Jørgensen and J. Simons, J. Chem. Phys. **79**, 334 (1983).
149. G. Fitzgerald, R. J. Harrison, and R. J. Bartlett, J. Chem. Phys. **85**, 5143 (1986).
150. A. C. Scheiner, G. E. Scuseria, J. E. Rice, T. J. Lee, and H. F. Schaefer, J. Chem. Phys. **87**, 5361 (1987).
151. H. Koch, H. J. Aa. Jensen, P. Jørgensen, T. Helgaker, G. E. Scuseria, and H. F. Schaefer, J. Chem. Phys. **92**, 4924 (1990).
152. A. P. Rendell and T. J. Lee, J. Chem. Phys. **94**, 6219 (1991).
153. J. Gauss, J. F. Stanton, and R. J. Bartlett, J. Chem. Phys. **95**, 2623 (1991).
154. J. Gauss, W. J. Lauderdale, J. F. Stanton, J. D. Watts, and R. J. Bartlett, Chem. Phys. Lett. **182**, 207 (1991).
155. R. Kobayashi, N. C. Handy, R. D. Amos, G. W. Trucks, M. J. Frisch, and J. A. Pople, J. Chem. Phys. **95**, 6723 (1991).
156. G. E. Scuseria and H. F. Schaefer, Chem. Phys. Lett. **146**, 23 (1988).
157. J. Gauss and D. Cremer, Chem. Phys. Lett. **163**, 549 (1989).
158. G. Fitzgerald, R. J. Harrison, and R. J. Bartlett, J. Chem. Phys. **85**, 5143 (1986).
159. J. Gauss and D. Cremer, Chem. Phys. Lett. **153**, 303 (1988).
160. E. A. Salter, G. W. Trucks, and R. J. Bartlett, J. Chem. Phys. **90**, 1752 (1989).
161. G. E. Scuseria, J. Chem. Phys. **94**, 442 (1991).
162. T. J. Lee and A. P. Rendell, J. Chem. Phys. **94**, 6229 (1991).
163. T. J. Lee, S. C. Racine, J. E. Rice, and A. P. Rendell, J. Chem. Phys., to be published.
164. R. Kobayashi, R. D. Amos, and N. C. Handy, Chem. Phys. Lett. **184**, 195 (1991).
165. J. D. Watts, J. Gauss, and R. J. Bartlett, Chem. Phys. Lett. **200**, 1 (1992).
166. See the entire issue of *Theor. Chim. Acta,* **80**, No. 6, (1991).
167. H. Koch and P. Jørgensen, J. Chem. Phys. **93**, 3333 (1990).
168. J. F. Stanton and R. J. Bartlett, J. Chem. Phys. **98**, 7029 (1993).
169. D. C. Comeau and R. J. Bartlett, Chem. Phys. Lett. **207**, 414 (1993).
170. R. J. Rico and M. Head-Gordon, Chem. Phys. Lett. **213**, 224 (1993).
171. R. J. Rico, T. J. Lee, and M. Head-Gordon, Chem. Phys. Lett. **218**, 139 (1994).
172. H. Koch, R. Kobayashi, A. Sanchez de Merás, and P. Jørgensen, J. Chem. Phys. **100**, 4393 (1994).
173. J. F. Stanton, J. Chem. Phys. **99**, 8840 (1993).
174. J. F. Stanton and J. Gauss, J. Chem. Phys. **100**, 4695 (1994).

175. P.-O. Widmark, P.-Å. Malmqvist, and B. O. Roos, Theor. Chim. Acta. **77**, 291 (1990).
176. A. P. Rendell and T. J. Lee, J. Chem. Phys. **101**, 400 (1994).
177. T. J. Lee, C. E. Dateo, B. Gazdy, and J. M. Bowman, J. Phys. Chem. **97**, 8937 (1993).
178. J. M. L. Martin, P. R. Taylor, and T. J. Lee, J. Chem. Phys. **99**, 286 (1993); Erratum, J. Chem. Phys. **99**, 9326 (1993).
179. C. E. Dateo, T. J. Lee, and D. W. Schwenke, J. Chem. Phys., in press.
180. J. M. L. Martin, T. J. Lee, and P. R. Taylor, J. Chem. Phys. **97**, 8361 (1992).
181. J. M. L. Martin, T. J. Lee, and P. R. Taylor, J. Mol. Spectrosc. **160**, 105 (1993).
182. J. M. L. Martin and T. J. Lee, Chem. Phys. Lett. **200**, 502 (1992).
183. J. M. L. Martin, T. J. Lee, and P. R. Taylor, to be published.
184. P. Botschwina, Å. Heyl, M. Horn, and J. Flügge, J. Mol. Spectrosc. **163**, 127 (1994).
185. T. J. Lee, J. M. L. Martin, and P. R. Taylor, J. Chem. Phys., submitted.
186. K. P. Huber and G. Herzberg, *Constants of Diatomic Molecules* (Van Nostrand Reinhold, New York, 1979).
187. S. Carter, I. M. Mills, and N. C. Handy, J. Chem. Phys. **97**, 1606 (1992).
188. R. C. Woods, Phil. Trans. R. Soc. Lond. **A 324**, 141 (1988).
189. R. J. Berry and M. D. Harmony, J. Mol. Spectrosc. **128**, 176 (1988).
190. F. W. Dalby, Can. J. Phys. **36**, 1336 (1958).
191. J. L. Duncan and I. M. Mills, Spectrochim. Acta **20**, 523 (1964).
192. J. L. Duncan, Mol. Phys. **28**, 1177 (1974).
193. K. Kawaguchi, J. Chem. Phys. **96**, 3411 (1992).
194. J. Koput, J. Mol. Spectrosc. **115**, 438 (1986).
195. E. Kostyk and H. L. Welsh, Can. J. Phys. **58**, 912 (1980).
196. D.-J. Liu, S.-T. Lee, and T. Oka, J. Mol. Spectrosc. **128**, 236 (1988).
197. D. L. Gray and A. G. Robiette, Mol. Phys. **37**, 1901 (1979).
198. K. Kuchitsu in *Accurate Molecular Structures. Their Determination and Importance,* Eds. A. Domenicano and I. Hargittai (Oxford University Press, Oxford, 1992), pp. 14-46.
199. P. Botschwina, M. Oswald, J. Flügge, Å. Heyl, and R. Oswald, Chem. Phys. Lett. **209**, 117 (1993).
200. J. R. Colt and G. E. Scuseria, Mol. Phys. **75**, 1099 (1992).
201. L. A. Barnes, B. Liu, and R. Lindh, J. Chem. Phys. **98**, 3972 (1993).
202. J. D. Watts and R. J. Bartlett, J. Chem. Phys. **96**, 6073 (1992).
203. T. J. Lee, Chem. Phys. Lett. **188**, 154 (1992).
204. J. M. L. Martin, T. J. Lee, and P. R. Taylor, Chem. Phys. Lett. **205**, 535 (1993).
205. G. Bryant, Y. Jiang, and E. Grant, Chem. Phys. Lett. **200**, 495 (1992).
206. J. L. Teffo and A. Chédin, J. Mol. Spectrosc. **135**, 389 (1989).
207. J. L. Teffo, O. N. Sulakshina, and V. I. Perevalov, J. Mol. Spectrosc. **156**, 48 (1992).
208. A. Willetts, J. F. Gaw, N. C. Handy, and S. Carter, J. Mol. Spectrosc. **135**, 370 (1989).
209. T. J. Lee, unpublished work.
210. C. W. Bauschlicher and H. Partridge, J. Chem. Phys. **100**, 4329 (1994).
211. J. M. L. Martin and P. R. Taylor, Chem. Phys. Lett., in press.
212. J. M. L. Martin, J. Chem. Phys. **100**, 8186 (1994).
213. J. Breidung, W. Schneider, W. Thiel, and T. J. Lee, J. Chem. Phys. **97**, 3498 (1992).
214. T. J. Lee and J. E. Rice, J. Chem. Phys. **97**, 4223 (1992).
215. I. M. Mills, in *Molecular Spectroscopy-Modern Research*, edited by K. N. Rao and C. W. Mathews (Academic, London, 1972), p. 115.
216. R. Krishnan, M. J. Frisch, and J. A. Pople, J. Chem. Phys. **72**, 4244 (1980).
217. M. J. Frisch, R. Krishnan, and J. A. Pople, Chem. Phys. Lett. **75**, 66 (1980).
218. M. F. Guest and S. Wilson, Chem. Phys. Lett. **72**, 49 (1980).
219. S. Wilson and M. F. Guest, Chem. Phys. Lett. **73**, 607 (1980).
220. T. J. Lee and J. E. Rice, J. Phys. Chem. **96**, 650 (1992).
221. J. W. C. Johns, A. R. W. McKellar, and E. Weinberger, Can. J. Phys. **61**, 1106 (1983).
222. D. J. Clouthier and D. A. Ramsay, Annu. Rev. Phys. Chem. **34**, 31 (1983).
223. A. T. Wong and G. B. Bacskay, Mol. Phys. **79**, 819 (1993).
224. W. D. Allen and A. G. Császár, J. Chem. Phys. **98**, 2983 (1993).
225. A. T. Wong and G. B. Bacskay, Chem. Phys. Lett. **207**, 360 (1993).
226. A. G. Császár, J. Phys. Chem. **96**, 7898 (1992).
227. J. M. Bowman, B. Gazdy, J. A. Bentley, T. J. Lee, and C. E. Dateo, J. Chem. Phys. **99**, 308 (1993).
228. F. Cacace, M. Attina, G. de Petris, and M. Speranza, J. Am. Chem. Soc. **111**, 5481 (1989).
229. F. Cacace, M. Attina, G. de Petris, and M. Speranza, J. Am. Chem. Soc. **112**, 1014 (1990).

230. F. Cacace, personal communication.
231. M. Attina, F. Cacace, M. Yanez, J. Am. Chem. Soc. **109**, 5092 (1987).
232. T. J. Lee and J. E. Rice, J. Am. Chem. Soc. **114**, 8247 (1992).
233. Y. Cao, J.-H. Choi, B.-M. Haas, M. S. Johnson, and M. Okumura, J. Chem. Phys. **99**, 9307 (1993).
234. L. S. Sunderlin and R. R. Squires, Chem. Phys. Lett. **212**, 307 (1993).
235. F. Cacace, M. Attina, G. de Petris, and M. Speranza, J. Am. Chem. Soc. **116**, 6413 (1994).
236. W. B. DeMore, S. P. Sander, D. M. Golden, R. F. Hampson, M. J. Kurylo, C. J. Howard, A. R. Ravishankara, C. E. Kolb, M. J. Molina, *Chemical Kinetics and Photochemical Data for Use in Stratospheric Modeling*, Evaluation Number 10, August 15, 1992; NASA JPL Publication 92-20.
237. T. J. Lee, J. Chem. Phys. **99**, 9783 (1993).
238. A. L. L. East, C. S. Johnson, and W. D. Allen, J. Chem. Phys. **98**, 1299 (1993).
239. T. J. Lee, C. M. Rohlfing, and J. E. Rice, J. Chem. Phys. **97**, 6593 (1992).
240. T. J. Lee and A. P. Rendell, J. Phys. Chem. **97**, 6999 (1993).
241. C. W. Bauschlicher and H. Partridge, Chem. Phys. Lett. **208**, 241 (1993).
242. C. W. Bauschlicher and H. Partridge, J. Phys. Chem. **98**, 1826 (1994).
243. G. E. Scuseria, unpublished work.
244. J. Almlöf, B. J. DeLeeuw, P. R. Taylor, C. W. Bauschlicher, and H. Partridge, Int. J. Quantum Chem. Symp. **23**, 345 (1989).
245. H.-J. Werner and P. J. Knowles, J. Chem. Phys. **94**, 1264 (1991).
246. D. L. Strout and G. E. Scuseria, J. Chem. Phys. **96**, 9025 (1992).
247. D. Feller and E. R. Davidson, J. Chem. Phys. **90**, 1024 (1989).
248. C. W. Bauschlicher and P. R. Taylor, J. Chem. Phys. **85**, 2779 (1985).
249. H. Hotop and W. C. Lineberger, J. Phys. Chem. Ref. Data **4**, 539 (1975).
250. P. R. Taylor, J. M. L. Martin, J. P. Francois, and R. Gijbels, J. Phys. Chem. **95**, 6530 (1991).
251. T. J. Lee, Chem. Phys. Lett. **169**, 529 (1990).
252. J. M. L. Martin and T. J. Lee, J. Chem. Phys. **98**, 7951 (1993).
253. T. J. Lee, J. Phys. Chem. **98**, 3697 (1994).
254. J. M. L. Martin, J. Chem. Phys. **97**, 5012 (1992).
255. D. P. Shelton and J. E. Rice, Chem. Rev. **94**, 3 (1994).
256. N. Oliphant and R. J. Bartlett, J. Chem. Phys. **100**, 6550 (1994).
257. H. Sekino and R. J. Bartlett, J. Chem. Phys. **98**, 3022 (1993).
258. T. J. Lee, J. Phys. Chem. **98**, 111 (1994).
259. D. Cremer, J. Gauss, E. Kraka, J. F. Stanton, and R. J. Bartlett, Chem. Phys. Lett. **209**, 547 (1993).
260. J. S. Muenter and J. Klemperer, J. Chem. Phys. **52**, 6033 (1970).
261. W. L. Ebenstein and J. S. Muenter, J. Chem. Phys. **80**, 3989 (1984).
262. G. L. Blackman, R. D. Brown, P. D. Godfrey, and H. I. Gunn, Nature **261**, 395 (1976).
263. D. W. Magnuson, J. Chem. Phys. **19**, 1071 (1951).
264. S. A. Clough, Y. Beers, G. D. Klein, and L. S. Rothman, J. Chem. Phys. **59**, 2254 (1973).
265. R. E. Hillger and M. W. P. Strandberg, Phys. Rev. **83**, 575 (1951).
266. D. Millen and K. M. Sinnot, J. Chem. Soc. **1958**, 350.
267. Y. Ueda and J. Iwahori, J. Mol. Spectrosc. **116**, 191 (1986).
268. H. Kanata, S. Yamamoto, and S. Saito, Chem. Phys. Lett. **140**, 221 (1987).
269. F. J. Lovas, R. D. Suenram, T. Ogata, and S. Yamamoto, Astrophysical J. **399**, 325 (1992).
270. R. D. Suenram and F. J. Lovas, J. Am. Chem. Soc. **100**, 5117 (1978).
271. T. J. Lee, J. Am. Chem. Soc. **111**, 7362 (1989).
272. G. E. Scuseria, M. D. Miller, F. Jensen, and J. Geertsen, J. Chem. Phys. **94**, 6660 (1991).
273. J. S. Muenter, J. Mol. Spectrosc. **55**, 490 (1975).
274. J. Almlöf and P. R. Taylor, J. Chem. Phys. **92**, 551 (1990).
275. J. E. Rice, P. R. Taylor, T. J. Lee, and J. Almlöf, J. Chem. Phys. **94**, 4972 (1991).
276. B. L. Hammond and J. E. Rice, J. Chem. Phys. **97**, 1138 (1992).
277. D. M. Bishop and J. Pipin, J. Chem. Phys. **91**, 3549 (1989).
278. A. Kumar and W. J. Meath, Can. J. Chem. **63**, 1616 (1985).
279. N. J. Bridge and A. D. Buckingham, Proc. R. Soc. Lond. Ser. A **295**, 334 (1966).
280. H.-J. Werner and W. Meyer, Mol. Phys. **31**, 855 (1976).
281. G. R. Alms, A. K. Bunham, and W. H. Flygare, J. Chem. Phys. **63**, 3321 (1975).
282. M. A. Spackman, J. Phys. Chem. **93**, 7594 (1989); this is obtained by analysis of several different experiments.
283. A. D. Buckingham, M. P. Bogaard, D. A. Dunmar, C. P. Hobbs, and B. J. Orr, Trans.

Faraday Soc. **66**, 1548 (1970).
284. A. C. Newell and R. C. Baird, J. Appl. Phys. **36**, 3751 (1965).
285. T. K. Bose and R. H. Cole, J. Chem. Phys. **54**, 3829 (1971).
286. G. H. F. Diercksen and A. J. Sadlej, J. Chem. Phys. **75**, 1253 (1981).
287. G. D. Zeiss and W. J. Meath, Mol. Phys. **33**, 1155 (1977).
288. D. M. Bishop, Rev. Mod. Phys. **62**, 343 (1990).
289. B. Kirtman and M. Hasan, J. Chem. Phys. **96**, 470 (1992).
290. D. M. Bishop, B. Kirtman, H. A. Kurtz, and J. E. Rice, J. Chem. Phys. **98**, 8024 (1993).
291. J. E. Rice and N. C. Handy, J. Chem. Phys. **94**, 4959 (1991).
292. J. F. Stanton and R. J. Bartlett, J. Chem. Phys. **99**, 5178 (1993).
293. R. Kobayashi, H. Koch, and P. Jørgensen, Chem. Phys. Lett. **219**, 30 (1994).
294. A. J. Sadlej, Coll. Czech. Chem. Commun. **53**, 1995 (1988); Theor. Chim. Acta. **79**, 123 (1991).
295. P. J. Bruna and S. D. Peyerimhoff, Adv. Chem. Phys. **67**, 1 (1987).
296. B. O. Roos, Adv. Chem. Phys. **67**, 399 (1987).
297. C. W. Bauschlicher and S. R. Langhoff, Chem. Rev. **91**, 701 (1991).
298. F. B. C. Machado and E. R. Davidson, J. Chem. Phys. **97**, 1881 (1992).
299. M. Peric, B. Engels, and S. D. Peyerimhoff, in *Quantum Mechanical Electronic Structure Calculations with Chemical Accuracy*, Ed. S. R. Langhoff (Kluwer Academic Publishers, Dordrecht, 1995), this volume.
300. B. Roos, M. Fulscher, and P.-Å. Malmqvist, in *Quantum Mechanical Electronic Structure Calculations with Chemical Accuracy*, Ed. S. R. Langhoff (Kluwer Academic Publishers, Dordrecht, 1995), this volume.
301. H. Koch, H. J. Aa. Jensen, P. Jørgensen, and T. Helgaker, J. Chem. Phys. **93**, 3345 (1990).
302. R. L. Graham, D. L. Yeager, J. Olsen, P. Jørgensen, R. J. Harrison, S. Zarrabian, and R. J. Bartlett, J. Chem. Phys. **85**, 6544 (1986).
303. H. Nakatsuji, K. Hirao, and Y. Mizukami, Chem. Phys. Lett. **179**, 555 (1991).
304. M. Head-Gordon, R. J. Rico, M. Oumi, and T. J. Lee, Chem. Phys. Lett. **219**, 21 (1994).
305. M. R. A. Blomberg, P. E. M. Siegbahn, T. J. Lee, A. P. Rendell, and J. E. Rice, J. Chem. Phys. **95**, 5898 (1991).
306. C. W. Bauschlicher, H. Partridge, and G. E. Scuseria, J. Chem. Phys. **97**, 7471 (1992).
307. M. Feyereisen, G. Fitzgerald, and A. Komornicki, Chem. Phys. Lett. **208**, 359 (1993).
308. O. Vahtras, J. Almlöf, and M. Feyereisen, Chem. Phys. Lett. **213**, 514 (1993).
309. A. P. Rendell, T. J. Lee, A. Komornicki, and S. Wilson, Theor. Chim. Acta. **84**, 271 (1993).
310. A. P. Rendell, T. J. Lee, and R. Lindh, Chem. Phys. Lett. **194**, 84 (1992).
311. A. P. Rendell, M. F. Guest, and R. A. Kendall, J. Comp. Chem. **14**, 1429 (1993).
312. H. Koch, O. Christiansen, R. Kobayashi, P. Jørgensen, and T. Helgaker, Chem. Phys. Lett., submitted.
313. L. Adamowicz and R. J. Bartlett, J. Chem. Phys. **86**, 6314 (1987).
314. S. Saebø and P. Pulay, Chem. Phys. Lett. **113**, 13 (1985).
315. J. Noga, W. Kutzelnigg, and W. Klopper, Chem. Phys. Lett. **199**, 497 (1992).

Magnetic Hyperfine Coupling Constants in Free Radicals

Daniel M. Chipman
Radiation Laboratory
University of Notre Dame
Notre Dame, Indiana 46556

August 30, 1994

Abstract. A survey is presented of results from theoretical calculations on electronic spin densities that determine magnetic hyperfine coupling constants experimentally observed in free radicals. Requirements for design of Gaussian basis sets suitable for computation of contact spin densities are described. Focus is primarily on the leading terms in the wave function that provide direct and spin polarization contributions. Effects of nondynamical electron correlation and its spin polarization are also included in those situations where they are significant. Results are presented for a variety of representative hydrocarbons, including nonconjugated π-radicals, σ-radicals, and larger conjugated π-radicals. Appropriate corrections are made in cases where large vibrational or vibronic effects occur. The results are often found to compare favorably with experiment, although some significant systematic deviations are identified that implicate additional important contributions arising from dynamical electron correlation effects. The survey enhances our understanding of the quantitative importance of various contributions to spin densities in free radicals.

S. R. Langhoff (Ed.), Quantum Mechanical Electronic Structure Calculations with Chemical Accuracy, 109–138.

Contents

1 INTRODUCTION

Free radicals act as important, although often transient, intermediate species in many chemical reactions. Study of their properties is therefore essential to fundamental understanding of reaction mechanisms. Since radicals are open-shell systems, one of the most useful experimental methods for their characterization has been that of electron spin resonance (ESR) spectroscopy [1]. This technique provides valuable information about the distribution of unpaired electron spin in a radical by means of highly accurate measurements of the magnetic hyperfine coupling (hfc) constants [2]. These depend [3] on the magnitudes of Fermi contact interactions of unpaired electron spins with any magnetic nuclei that may be present. In hydrocarbons these include the ubiquitous hydrogen and low (approximately 1%) natural abundance carbon-13 nuclei. Isotopic enrichment methods are sometimes employed to enhance observation of ^{13}C hfc.

The ESR technique provides a very useful map of the spin density distribution throughout a radical. This can potentially yield valuable information about the chemical environment of each magnetic nucleus. Proper interpretation of such information can be greatly assisted by theoretical investigations. The challenge for electronic structure theory in this context is then to provide a fundamental understanding of the nature of the effect, particularly to elucidate the origin and magnitude of various physical factors that contribute to determine the net spin density. Together with this, another very important role for theory is to furnish a quantitative tool to aid assignment of experimental spectra in difficult and poorly understood situations. Thus, it is of interest to examine the performance of various theoretical methods that may be applied to calculation of this property.

For radicals tumbling rapidly in solution, only the isotropic hfc due to the Fermi contact interaction is generally experimentally observed. In some instances, it is also possible to observe anisotropic hfc constants, which arise from dipolar interactions. These are generally somewhat easier to calculate than the isotropic couplings, often being described well with simple Hartree-Fock level wave functions, and will not be considered in this work.

Accurate calculation of Fermi contact hfc has proven to be an exceptionally difficult problem for conventional *ab initio* electronic structure methods. Special considerations apply both in the selection of configurations to be included in the wave function and in the choice of one-particle basis set. One reason for the difficulty is that these hfc constants are particularly sensitive to regions of space very near the nuclei, whereas the historical development of electronic structure theory has mostly concentrated instead on obtaining good descriptions of the outer valence regions that are of greatest importance for determination of most other chemical properties. Proper treatment of hfc also requires adequate treatment of the outer valence regions, of course, and additionally requires special consideration of inner valence and, for first-row atoms, $1s$ core regions, all of these in a "balanced" manner. An additional practical complication comes from the fact that most molecular electronic structure calculations utilize Gaussian rather than exponential functions for the one-particle basis, requiring additional measures to ameliorate the potentially poor behavior of Gaussians near the nuclei. This paper will survey recent progress made in our laboratory toward the solution of these problems, with a view to critical evaluation of the current status of our goal of calculating accurate hfc constants for chemically interesting systems.

Having emphasized above the difficulty of hfc calculations, we should point out that some radicals are easier to treat properly than others. In fact, the hfc in certain classes of radicals is reasonably described even with wave functions at the Hartree-Fock level. These are systems in which the total spin density is dominated by the direct contribution of the singly occupied molecular orbital (SOMO), which is usually described well by both spin-restricted open-shell Hartree-Fock (ROHF) [4] and spin-unrestricted (UHF) [5] methods. This category includes many σ-radicals, where the SOMO has substantial s-character at all the nuclei of interest. However, in some σ-radicals the remaining nominally paired electrons may also provide significant contributions to the hfc by indirect mechanisms. These are not described at all by ROHF and are generally not given with the correct magnitude in UHF. Proper spin density determination in such situations then requires use of multiconfiguration wave functions.

Planar π-radicals also constitute a large class of interesting systems. By definition, the SOMO contains no s-character in a π-radical. It therefore has a node in the molecular plane and the ROHF wave function by itself yields zero hfc for all nuclei in the plane. In such cases, the UHF method may still give non-zero values of the hfc constants, but these are generally found to be much too large in magnitude compared to experiment [6]. Annihilation or projection procedures introduced *ad hoc* to remove spin contamination from the UHF wave functions of π-radicals generally lead to hfc constants that are somewhat too small compared with experiment [7]. Such procedures have been useful in qualitative investigations [8]. However, in such instances accurate calculations again require multiconfiguration treatments.

A few words should be said here about the kind of accuracy that one might wish to strive for in hfc calculations. Our goal is to be able to treat moderately large organic radicals of real chemical interest. Even with current computer technology, for any but the smallest systems it is generally not feasible to carry out large-scale calculations that recover most of the electron correlation by treating many configurations, either explicitly or implicitly, with extended basis sets.

Our focus will therefore be instead on methods that utilize a small number of well-chosen configurations together with moderately sized basis sets. These considerations necessarily limit the attainable accuracy. The most generally successful methods within this context correspond to various implementations of the spin polarization model [9]. It will be seen herein that this model is often capable of providing hfc constants within 10-20% of experiment, which is sufficient for many interpretive purposes. In addition, this model usually suffices to identify the sign of the coupling, which is often not determined by the experimental measurements, and it is used for that purpose in this work. To be sure, larger errors are sometimes encountered. Fortunately, it will become apparent that these exceptions are somewhat systematic and therefore predictable. This is of value in pointing the direction for further research in this field.

Other considerations may also limit the accuracy of hfc calculations. For example, vibrational averaging corrections amount to a few percent in typical situations. However, comprehensive evaluation of vibrational averaging effects would be a daunting task in most large polyatomic systems, and is usually not even contemplated. A notable exception is the case of radicals having freedom to allow large amplitude nuclear motions. This may lead to large corrections to the hfc that must be explicitly evaluated for qualitatively meaningful results to be obtained. In such situations, only the large amplitude motions are generally explicitly treated, so that corrections of a few percent from the other remaining untreated modes may still be expected.

Another even more serious consideration limits most investigations. The vast majority of electronic structure calculations are currently carried out on isolated molecules. However, most experimental ESR measurements are carried out either in solution or in low temperature matrices. In those few cases where ESR results are available on the same radical from both gas and condensed phase measurements the hfc constants typically differ by a few percent, ranging up to about 10% in extreme cases. So better agreement than a few percent should not be expected unless solvation effects are explicitly evaluated.

In light of the several factors considered above, deviations of ordinary theoretical calculations on isolated molecules from experiment by less than about 10% should be considered excellent agreement. A more modest goal of agreement to within about 20% would be quite sufficient for most chemical purposes of interpretation and assignment. Higher accuracy is attainable for small systems, of course, and work toward that end should certainly be encouraged. The focus of this paper, however, is on methods that may currently be routinely applied to larger chemically interesting systems. In this context, it will be seen that the goal of consistent agreement within 20% is already a considerable challenge that is being approached but has not yet been fully met.

In the next section some background theoretical material is given, defining the spin density operators and discussing various wave functions that will be used in subsequent applications. After that, some discussion is provided on the design of suitable Gaussian basis sets for hfc calculations in connection with the example of the simple CH radical. Then a variety of hydrocarbon free radicals are considered, each chosen to bring out particular interesting features. In six of the examples the unpaired electron is highly localized at a single center. These include the methynyl, methyl, and ethyl nonconjugated π-radicals and the vinyl, ethynyl, and cyclopropenyl σ-radicals. In the other three examples the unpaired electron is in a conjugated π system delocalized over several centers. These include the allyl, pentadienyl, and cyclohexadienyl π-radicals. The examples therefore cover a broad range of representative organic systems. A concluding section provides a brief summary and evaluation of the results and trends found, and gives a prognosis for future research in this area.

2 THEORETICAL BACKGROUND

The isotropic hfc constant a(N) at a magnetic nucleus N is determined by the Fermi contact interaction and is given in units of the magnetic flux density required to obtain resonance by

$$\mathrm{a}(N) = \frac{8\pi}{3}\frac{g_e}{g_0} g_N \beta_N \mathrm{q}(\mathbf{r}_N).$$

For most organic free radicals, the ratio g_e/g_0 of the actual electron g factor to that of a free electron may be taken as essentially unity. The nuclear g factor g_N and nuclear magneton β_N are known constants. The grouping $(8\pi/3)(g_e/g_0)g_N\beta_N$ can therefore be easily evaluated at the outset and amounts to 1595 Gauss (G) for H and 401 G for ^{13}C. The essential quantity to be calculated is therefore q(**r**), which is the electronic spin density at a point in space **r**. In atomic units this is the effective number of unpaired electron spins per cubic bohr, which is simply given by

$$\mathrm{q}(\mathbf{r}) = \frac{1}{\mathrm{n}}[\rho_\alpha(\mathbf{r}) - \rho_\beta(\mathbf{r})]$$

where $\rho_\alpha(\mathbf{r})$ and $\rho_\beta(\mathbf{r})$ are the separate densities of α and β spin electrons, respectively, and n is the net number of unpaired electrons $(\mathrm{n}_\alpha - \mathrm{n}_\beta)$. The 1/n factor simply normalizes the result to the number of unpaired electrons. This expression may be compared to the total electronic charge density, which is $[\rho_\alpha(\mathbf{r}) + \rho_\beta(\mathbf{r})]$.

If $\Psi(\mathbf{x})$ is the clamped-nucleus electronic wave function for the combined space and spin coordinates **x** of all electrons, the spin density may be evaluated from the expectation value of the spin density operator according to

$$\mathrm{q}(\mathbf{r}) = \frac{1}{\mathrm{n}}\int \Psi^*(\mathbf{x})\sum_j \delta(\mathbf{r}_j - \mathbf{r})2\hat{s}_{zj}\Psi(\mathbf{x})d\mathbf{x}.$$

The delta function operator here picks out the electronic density at a particular point in space **r**. In the absence of the $2\hat{s}_{zj}$ and 1/n factors, this expression would therefore simply give the total electronic charge density. The $2\hat{s}_{zj}$ factor alters it to the spin density, since operating on a wave function $\Psi(\mathbf{x})$ it selects the value +1 if electron j has α spin and −1 if it has β spin. Thus, q(**r**) is called the total electronic spin density, although it is important to bear in mind that despite this name it is actually a function only of spatial coordinates. This function should not be confused with a quantity that is often loosely referred to in the experimental literature as a "spin density" but instead is actually a spin population.

The systems considered here are all doublet free radicals, i.e., each has just one unpaired electron, so for simplicity of presentation we henceforth specialize to that case. The dominant configuration for a doublet radical with $2m$+1 electrons is described in terms of m+1 orbitals in the single Slater determinant

$$\Psi_0 = |\, \psi_1\bar{\psi}_1 \ldots \psi_i\bar{\psi}_i \ldots \psi_m\bar{\psi}_m\psi_{m+1}\,| \,.$$

All orbitals are taken as being real and mutually orthonormal throughout this work. If the orbitals are optimized by a self-consistent-field (SCF) procedure, then Ψ_0 becomes Ψ^{ROHF}. In such a model the doubly occupied orbitals $\{\psi_1 \ldots \psi_m\}$ make no net contribution to the spin density, since each α orbital density is exactly cancelled by an identical β orbital density having the opposite sign. The total spin density is therefore determined only by the direct contribution of the SOMO, which is

$$q^{\mathrm{ROHF}}(\mathbf{r}) = |\psi_{m+1}(\mathbf{r})|^2.$$

Note that this contribution is always nonnegative.

To evaluate contributions from the remaining electrons, it is necessary to include additional terms in the wave function. The most important effects on spin density usually arise from spin polarization of the nominally doubly occupied orbitals. In a spin-restricted approach, this can be described [10] by a configuration-interaction (CI) expansion of the general form

$$\Psi^{\mathrm{SP-SCI}} = c_0\Psi_0 + \sum_i^m c_i\Psi_i$$

where Ψ_i represents the spin polarization arising from a single orbital excitation of the nearly doubly occupied orbital ψ_i to its slightly occupied spin polarization partner orbital ψ_i^*. It is given by

$$\Psi_i = P \mid \psi_1\bar{\psi}_1 \ldots \psi_i^*\psi_i \ldots \psi_m\bar{\psi}_m\bar{\psi}_{m+1} \mid .$$

Suppressing for clarity the doubly occupied orbitals, which are present and unmodified in each of the Slater determinants, the spin polarization projection operator P has the effect

$$P \mid \psi_i^*\psi_i\bar{\psi}_{m+1} \mid = \frac{1}{\sqrt{6}}[2 \mid \psi_i^*\psi_i\bar{\psi}_{m+1} \mid - \mid \psi_i^*\bar{\psi}_i\psi_{m+1} \mid - \mid \bar{\psi}_i^*\psi_i\psi_{m+1} \mid].$$

That is, ψ_i^* and ψ_i are first triplet spin-coupled to one another, then the resultant is coupled with ψ_{m+1} to produce an overall doublet configuration. The other possible doublet spin coupling generally has little influence on the spin density and may be omitted. With respect to the dominant configuration, Ψ_i describes a single excitation of an electron from ψ_i to ψ_i^*, together with a simultaneous "spin flip" of the other unpaired electrons in $\bar{\psi}_i$ and ψ_{m+1}.

In a practical CI calculation, each spin polarization orbital is actually produced as a linear combination of all the "virtual" orbitals obtainable within the given basis set. The linear combination found for any pair i may be collapsed *a posteriori* to the condensed form ψ_i^* [9]. This leads to a compact description having a single spin polarization orbital for each of the nominally doubly occupied pairs. This contraction of the spin polarization orbitals also serves to uniquely define the nearly doubly occupied orbitals, which otherwise would be arbitrary to the extent of a unitary transformation among themselves [11].

The expansion coefficients generally turn out to satisfy the relations $c_0 \approx 1$ and $|c_i| \ll 1$. Assuming this to be true, the spin density obtained from this wave function is given by

$$q^{SP-SCI}(\mathbf{r}) = |\psi_{m+1}(\mathbf{r})|^2 - 2\sqrt{\frac{2}{3}}\sum_{i}^{m} c_i\psi_i(\mathbf{r})\psi_i^*(\mathbf{r}) + \ldots.$$

The first term, representing the direct contribution, is the same as the ROHF result. The next term describes the leading spin polarization corrections. Additional terms also occur, but are not shown explicitly since they are of second order in the small coefficients $\{c_i\}$ and typically make only very small contributions. The integral over all space of the spin polarization contribution approximately vanishes, indicating that globally it consists of nearly balancing positive and negative regions. Accordingly, at any given point in space, such as at a magnetic nucleus where the hfc is experimentally observed, the spin polarization may be large and of either positive or negative sign.

Another common approach to introduce spin polarization is the UHF method, wherein the wave function is kept as a single determinant but the constraint of exactly matching the spatial orbitals describing the α and β partners of each pair is relaxed. This in effect brings in single excitations in a form similar to that shown above, although it does not allow them full variational flexibility. We have recently evaluated arguments for and against this procedure [9] and have concluded that the UHF method, either with or without spin projection to produce a pure spin state, is quantitatively unreliable for most spin density calculations.

Many other implementations are possible that include spin polarization effects within a spin-restricted framework. Richman, Shi, and McCullough [12] and Chipman [13] have suggested use of a wave function of the same generic form as Ψ^{SP-SCI} above, but simultaneously optimizing all orbitals (including those in the dominant configuration Φ_0) and expansion coefficients in a multiconfiguration self-consistent-field procedure, producing what we call a SP–MCSCF wave function. Nakatsuji, Ohta, and Yonezawa [14] have proposed the spin-adapted cluster (SAC) method, which is closely related to SP–MCSCF while also bringing in certain higher-order self-consistency effects. Another possibility emphasized by Chipman [9] and by Nakano, Morihahsi, and Kikuchi [15] is to carry out a conventional all-singles CI calculation (ROHF–SCI), which includes the important spin polarization single excitations described above as well as all other single excitations that generally are of only minor importance. Ellinger *et al.* [16] have utilized low-order perturbation theory approximations to ROHF–SCI. It may be expected that each of these various implementations of the basic spin polarization model will lead to very similar results in most practical calculations.

The major physical effect missing from the above models comes from electron correlation, which requires double and higher excitations to be included in the wave function. These may be broadly classified as dynamical correlations, which have small CI coefficients, or as nondynamical correlations which have large enough coefficients that they really should be included to produce a qualitatively correct wave function. While the presence of nondynamical electron correlations is an unusual situation for closed-shell molecules at their equilibrium geometries, it will be seen in this work that it is a common occurrence in free radicals. This is particularly true for radicals having π electrons delocalized over two or more centers, regardless of whether or not the unpaired electron itself is in the π system.

Nondynamical electron correlations are in fact found in the majority of the systems considered in this work, so it is important to consider methods for their identification. It would be sufficient to carry out a complete-active-space (CAS) treatment correlating all valence electrons, i.e., a full CI calculation among all configurations that can be constructed from a conceptual minimal basis that includes all valence orbitals. However, this would describe many minor dynamical correlations in addition to the desired few nondynamical ones. To make calculations feasible for large systems, it is desirable to keep the configuration list as short as possible. Pulay and Hamilton [17] have suggested a much simpler procedure in which the CAS calculation is limited to only a few active valence orbitals. These are identified as any UHF natural orbitals that have significantly nonintegral occupation numbers, e.g., that differ by more than about 0.02 from the allowed ROHF values of 0, 1, or 2.

However, we have found some instances of nondynamical correlations which are not identified by the Pulay approach. These correspond to double excitations from an orbital of one spatial symmetry to an orbital of a different symmetry. Since the corresponding single excitations are forbidden by symmetry, such configurations are not "sensed" by UHF. We have therefore modified Pulay's idea to allow for such cases. It seems that nondynamical correlations generally involve orbitals that can qualitatively be described within a minimal basis set and that occur in configurations differing by either a single or double excitation from the ROHF determinant. We therefore simply carry out a standard singles+doubles (SDCI) calculation with a minimal basis set, e.g., STO–3G [18]. As with the original Pulay proposal, any natural orbitals having significantly nonintegral occupation numbers are then identified and correlated. A slightly relaxed criterion of differing by more than about 0.03 from an integer is more appropriate for this modified procedure. In most cases, it is obvious which are the important configurations involving these active orbitals. In those rare cases where it is not, a subsequent SDCI calculation over configurations constructed from the natural orbitals clearly identifies the important configurations. It should be emphasized that neither single-reference SDCI wave functions nor minimal basis sets are generally appropriate for spin density calculations. We use the ROHF–SDCI/STO–3G calculations only for qualitative identification of the active orbitals and configurations necessary to describe nondynamical correlations in subsequent calculations.

It might be thought that if configurations describing nondynamical electron correlations should be included in the wave function, then perhaps spin polarization of these configurations may lead to significant contributions to the spin density. That possibility is investigated in this work by a multireference single excitation (MR–SCI) procedure. Here the MR space includes Ψ^{ROHF} and the small number of configurations necessary

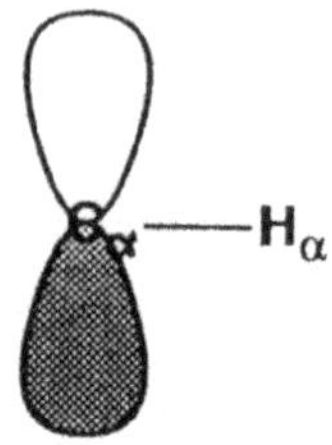

Fig. 1. Sketch of the structure and SOMO of methynyl radical.

to describe the nondynamical correlations. Then all single excitations with respect to *every* configuration in the reference space are included in the final wave function.

It will be seen in the next section that such simple models work well only when a fortuitous cancellation of errors occurs among various higher-order dynamical electron correlation effects. Fortunately, this behavior is apparently systematic enough that the simple calculations often lead to good agreement with experiment. However, there are some instances when it breaks down. One purpose of this survey of representative radicals is then to investigate the question of what circumstances lead to significant dynamical electron correlation corrections to spin densities. This will be evidenced by any findings of large deviations of the present results from experiment.

3 NONCONJUGATED π-RADICALS

3.1 METHYNYL CH

The simple methynyl radical (see Figure 1) serves as a good illustration of the above discussion of wave functions, and will also be used to discuss basis set considerations for spin density evaluation. Calculations were performed at the experimental equilibrium geometry [19]. In the basic electron configuration

$$\Psi_0 = 1\sigma^2\ 2\sigma^2\ 3\sigma^2\ \pi_x$$

the 1σ molecular orbital is essentially the $1s$ core pair localized on carbon, 2σ is similar to the $2s$ pair of a carbon atom but distorted somewhat along the direction of the internuclear axis and could alternatively be called a lone pair on carbon, 3σ is the CH bonding pair, and the unpaired electron resides in a π orbital, say π_x, localized on carbon. Since this is a π-radical, there is no direct contribution from Ψ_0 alone to the isotropic Fermi contact hfc.

As described in the previous section, the leading contact contributions arise from augmenting Ψ_0 with three additional configurations that correspond to $1\sigma \rightarrow 1\sigma^*$,

TABLE I
Comparison of MCSCF calculations to experiment for isotropic hyperfine coupling constants, in Gauss, of CH radical.

Basis set	$a(^{13}C)$				$a(H)$			
	1σ	2σ	3σ	total	1σ	2σ	3σ	total
	Spin polarization wave functions							
6-31G	−2.5	36.1	2.2	35.9	0.0	−24.0	−1.4	−25.4
[42|2]	−36.7	48.6	1.8	13.8	0.0	−17.2	−7.0	−24.3
[631|41]	−37.7	51.7	0.0	14.1	0.0	−10.6	−9.4	−19.9
numerical	−35.8	51.6	0.0	15.8	0.0	−10.0	−11.8	−21.8
	Spin + orbital polarization wave functions							
numerical				17.8				−20.6
	Experiment							
				16.7[a]				−20.6[b]

[a]Reference [21].
[b]Reference [22].

$2\sigma \rightarrow 2\sigma^*$, and $3\sigma \rightarrow 3\sigma^*$ spin polarization single excitations. The best possible wave function of this form, in a variational sense, corresponds to $\Psi^{SP-MCSCF}$ in which both orbitals and CI expansion coefficients are simultaneously optimized. For a diatomic such as CH, it is possible to obtain an essentially exact evaluation of the hfc corresponding to this wave function model by using a seminumerical method [20] for solution of the MCSCF equations. It is seen in Table I that the numerical SP–MCSCF results are in quite good agreement with experiment.

Due to the simplicity of the SP–MCSCF wave function, there are no cross terms and it is further possible to unambiguously examine the separate contribution from each spin-polarized shell. Note that the 1σ shell, i.e., the contributions arising from 1σ and $1\sigma^*$ orbitals which are highly localized in the inner regions near the carbon nucleus, makes a large negative contribution to the carbon coupling, but essentially no contribution to the hydrogen coupling. The 2σ shell makes significant contributions at both nuclei. Finally, the 3σ shell contributes roughly the same as 2σ to the hydrogen coupling, but negligibly to the carbon coupling. This latter result seems to be accidental for the particular case of CH, since the 3σ shell makes significant contributions to the heavy-atom hfc in the analogous NH and OH radicals [11].

One consequence of the high spatial symmetry of this system is that it is also possible to study orbital polarization effects, which arise from $n\sigma \rightarrow n\delta^*$ single excitations, separately from the spin polarization effects that arise from $n\sigma \rightarrow n\sigma^*$ single excitations. Including these in a self-consistent manner similar to that described above leads to a wave function that has also been evaluated [11] with essentially exact seminumerical methods. It can be seen in Table I that the orbital polarization effects on the hfc are not large. Most polyatomic radicals have lower spatial symmetry, thus precluding facile separation of spin and orbital polarization effects. In those cases, we simply understand the term spin polarization to implicitly include both effects.

For polyatomic systems, it is usually necessary to resort to approximate solutions of the equations defining any particular wave function model by the method of linear combination of atomic-like orbitals (LCAO). In this approach each molecular orbital is expanded in terms of a finite number of preordained basis functions. The accuracy of the solution is then limited by the size and suitability of the given basis set of atomic-like orbitals. For the sake of efficiency, most *ab initio* calculations on polyatomic systems utilize groups, i.e., contractions, of Gaussian functions for this purpose. Both numerical and finite basis set calculations can be carried out for simple diatomics such as CH. For a given wave function category such as SP–MCSCF, this provides a convenient means to judge the utility of various Gaussian basis sets for spin density determination by comparing the results to the exact ones found numerically.

Results on CH from a commonly used split valence basis set [23] known as 6–31G are given in Table I. Note particularly that this basis gives a very poor result for the coupling at carbon. The primary reason for this can be traced to the near zero contribution from the 1σ shell, which should be large and negative. The 2σ shell contribution is also underestimated somewhat. Only a fortuitous partial cancellation of these separate large errors prevents the total carbon coupling from being still worse than it actually is. At hydrogen the total coupling is in fair agreement with experiment. Again there is some cancellation of separate errors, here an overestimate of the magnitude of the 2σ shell contribution balancing an underestimate of that from the 3σ shell. This basis is clearly unsuitable for determination of carbon hfc, and just marginally adequate for hydrogen hfc. Addition of polarization functions to form the commonly used 6–31G* and 6–31G** [24] basis sets makes little change in the hfc results, so the lack of polarization functions is not a major reason for the inadequate results from this basis.

One need not look far to understand the poor performance of 6–31G, or of any other split valence or smaller basis set, for hfc calculations. Just as with 1σ, the $1\sigma^*$ spin polarization orbital is localized in the core region of the carbon atom, being just slightly more diffuse and with a node to provide orthogonality. But split valence and smaller basis sets supply only one basis function in the core region. This allows for a reasonable representation of the highly occupied 1σ orbital, but does not provide sufficient variational flexibility to describe the only slightly occupied but nonetheless very important $1\sigma^*$ orbital. That requires a second independent basis function localized in the carbon core region, i.e., the basis must provide at least a double-zeta representation of the core. Similar considerations apply in all the occupied shells, making it clear that an adequate basis set for hfc determination must be of at least full double-zeta quality to describe the relevant spin polarization effects for carbon and other first-row atoms.

The [42|2] basis [25] is a popular double-zeta set. We see in Table I that for the carbon hfc it shows errors in the individual shell contributions that range from about 1-3 G. The total is in good agreement with experiment, being about 3 G too low. At hydrogen there are still significant errors in the separate 2σ and 3σ shell contributions, but these fortuitously cancel to some extent to give a reasonably good total result. It has been found [26, 9] that hfc errors associated with this basis are quite systematic in a number of similar hydrocarbon radicals, allowing them to be largely rectified by empirical scaling procedures.

To allow more confidence in hfc calculations, particularly for radicals that may show unusual bonding effects, it would be preferable to find a basis set that does not depend so heavily on error cancellation to give good results in calculation of spin polarization effects. We have developed the [631|41] basis [27, 11] with this goal in mind. Compared to a double-zeta set, it includes diffuse s(C), p(C) and s(H) functions and a very tight s(H) function, it is uncontracted somewhat to allow additional variational flexibility in the outer core and inner valence regions, and finally d(C) and p(H) orbital polarization functions are added. It is seen in Table I that this basis gives good representations of the separate shell contributions as well as of the total hfc for CH radical, with various errors being about 2 G or less as compared to the exact numerical determination. All spin density results reported in the remainder of this work utilize this specially designed [631|41] basis set.

In place of MCSCF procedures, the calculations reported in the remainder of this work correspond to analogous, but less computationally intensive, single excitation configuration-interaction (SCI) methods. It is seen in Table II that the ROHF–SCI results for CH are very similar to those from SP–MCSCF. Compared to experiment, the relative errors for ROHF–SCI are 22% for carbon and only 4% for hydrogen.

An ROHF–SDCI/STO–3G calculation on CH reveals a nondynamical electron correlation effect, corresponding to the double excitation

$$\Psi_{\rm ND} = \Psi_0(2\sigma^2 \rightarrow \pi_y^2).$$

This nondynamical electron correlation, together with its associated spin polarization, may be treated by MR–SCI calculations. The two configurations Ψ_0 and $\Psi_{\rm ND}$ constituting the reference space are constructed here from the natural orbitals of a full ROHF–SDCI calculation with the [631|41] basis. It should be noted that no attempt is made to impose a restriction forcing π_x (the SOMO) to be spatially equivalent to π_y (the slightly occupied orbital involved in nondynamical correlation). Then spin polarization is described by including all single excitations from each of the two reference configurations.

It is seen in Table II that the MR–SCI result for the carbon coupling is worse than ROHF–SCI, having a large 41% deviation from experiment, while the hydrogen coupling is slightly better, having only a 2% error. This makes it clear that the fairly good spin densities obtained from ROHF–SCI spin polarization calculations for CH do not necessarily imply that electron correlation effects are small. A more extensive investigation [9] on CH has shown that the effect of pair correlations, which is described by double excitations, happens to be nearly cancelled by spin polarization of the pair correlations, which is described by certain triple excitations. As an important consequence, conventional ROHF–SDCI calculations give much worse agreement with experiment than ROHF–SCI in spin density calculations, since the various correlation errors are no longer "balanced". Similar behavior has been observed in several other free radicals and may represent a general trend.

TABLE II
Comparison of calculation to experiment for isotropic hyperfine coupling constants, in Gauss, of representative hydrocarbon π-radicals.

Wave function	$a(^{13}C_\alpha)$	$a(^{13}C_\beta)$	$a(H_\alpha)$	$\bar{a}(H_\beta)$
		Methynyl		
ROHF–SCI	13.0		-19.7	
MR–SCI	9.9		-20.1	
Experiment	16.7[a]		-20.6[b]	
		Methyl		
ROHF–SCI	26.6		-25.6	
Experiment	(27)[c]		(-25)[c]	
		(planar) Ethyl		
ROHF–SCI	30.9	-14.9	-25.9	19.7
Experiment	(29)[d]	-13.6[e]	-22.4[f]	26.9[f]

[a]Reference [21].
[b]Reference [22].
[c]Observed values [29] with corrections [30] to remove the large effect of ν_2 vibrational averaging. See text for details.
[d]Observed value [29] with a correction [32] to remove the large effect of ν_6 vibrational averaging. See text for details.
[e]Reference [29].
[f]Reference [2].

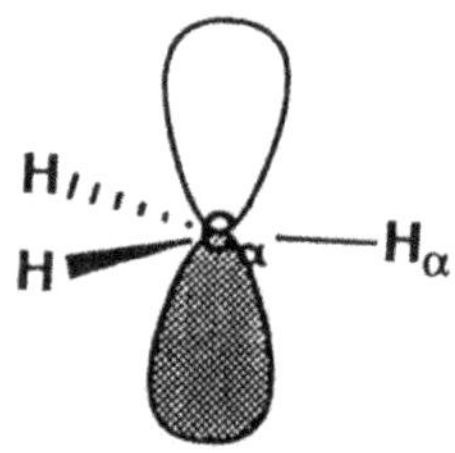

Fig. 2. Sketch of the structure and SOMO of methyl radical. Only symmetry unique positions are explicitly labeled.

3.2 Methyl CH_3

Methyl radical has a planar equilibrium geometry of D_{3h} symmetry (see Figure 2). Calculations were performed at the experimental equilibrium geometry [28]. The basic electron configuration is

$$\Psi_0 = 1a'^2\, 2a'^2\, 1e_x^2\, 1e_y^2\, 1a''$$

where the $1a''$ SOMO is a p_π orbital localized on the carbon atom. Since it has a node in the molecular plane, direct contributions are zero and the hfc arises primarily from spin polarization effects. ROHF–SDCI/STO–3G calculations uncovered no evidence for nondynamical electron correlations in this system.

To be appropriate for a planar radical center, the "experimental" result quoted in Table II for a(^{13}C) actually differs significantly from the observed result [29] of 38.3 G, and the quoted result for a(H) also differs somewhat from the observed value [29] of −23.0 G. The reason for these differences is that there is little resistance to bending of the radical away from its planar equilibrium conformation in the symmetric ν_2 "umbrella" vibrational mode. When the radical bends, the SOMO changes from a pure p_π orbital to become an *sp* hybrid. The *s*-character of the hybrid orbital then allows for a significant direct contribution of the SOMO to the hfc. The carbon coupling is therefore quite sensitive to this geometrical distortion and a significant vibrational correction arises from the ν_2 motion. It turns out that this vibrational motion is highly anharmonic, so some care must be taken [30] to properly evaluate the corrections. The "experimental" estimates of the hfc listed in Table II for a hypothetical nonvibrating planar methyl radical were obtained [30] from analysis of experimental data on the ν_2 potential function and on hfc constants measured in isotopically substituted species.

The spin polarization model is able to account for nearly all of this large vibrational correction to a(^{13}C). ROHF–SCI/[631|41] calculations have predicted [9] a correction of 10 G, as compared to the experimentally derived [30] correction of 11 G. Just as one might expect, calculations have demonstrated [30] that the direct contribution shows a considerably larger variation with bending than does the spin polarization contribution, so that the former is primarily responsible for the large correction. Since the direct contribution is always positive, it is therefore clear that the vibrational correction to the hfc should also be positive. Furthermore, since it becomes more and more positive upon bending further either way from planar, higher vibrational levels will have larger positive corrections. Similar arguments apply to the hydrogen coupling, although there the effects are smaller and the vibrational averaging correction is not as significant.

The ROHF–SCI results in Table II indicate very good agreement of spin polarization calculations with the experimentally derived hfc of methyl radical.

3.3 Ethyl C_2H_5

The ethyl radical may be considered as analogous to methyl, with one hydrogen substituted by a nearly tetrahedral methyl group (see Figure 3). It is depicted in a "staggered" conformation, although there is in fact little resistance to torsion about the CC bond. Calculations were performed at a theoretical MCSCF/6–31G** equilibrium geometry [31] with minor modifications [32] to symmetrize the methyl group and to make the α-carbon planar. In the staggered conformation, C_s symmetry applies and the basic

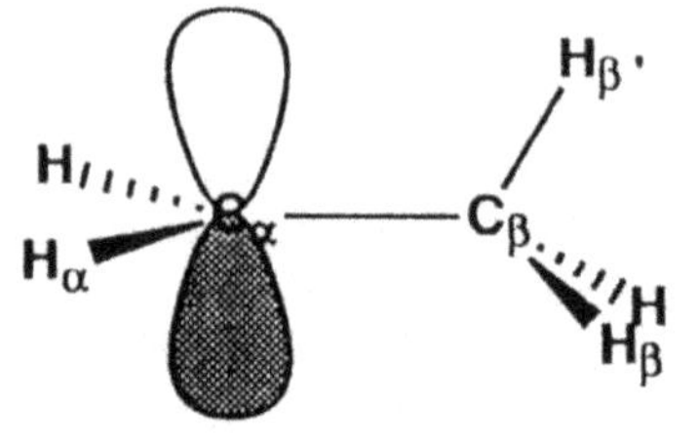

Fig. 3. Sketch of the structure and SOMO of ethyl radical. Only symmetry unique positions are explicitly labeled.

electron configuration is

$$\Psi_0 = 1a'^2 \ldots 6a'^2\ 7a'\ 1a''^2\ 2a''^2.$$

The unpaired electron in $7a'$ is mainly localized on the methylene carbon. If the methylene carbon were perfectly planar and the methyl substituent had local C_{3v} symmetry, then the SOMO would be a strict p_π orbital with a node in the methylene plane. There would then be no direct contributions to the hfc for any of the nuclei in that plane. However, the unpaired electron can delocalize to some extent onto the β-hydrogens of the methyl group via hyperconjugation, leading to direct hfc there. ROHF–SDCI/STO-3G calculations uncovered no evidence for nondynamical electron correlations in ethyl radical.

Just as with methyl, there is little resistance to out-of-plane bending of the methylene group at the radical center. In ethyl this is traditionally labeled the ν_6 mode. The computed equilibrium geometry actually has an out-of-plane bend of nearly 10° at the radical center, together with a staggered methyl conformation. However, the energy surface is very flat near equilibrium for both bending and torsional excursions, so that it is not really very meaningful to speak of a precise equilibrium geometry.

Torsion is practically unhindered when the radical center is exactly planar. The barrier increases somewhat as the radical center bends, but still does not become very large. Torsional reorientations therefore occur rapidly on the ESR time scale, so that experimentally only a single average β-hydrogen hfc is observed. We therefore report only the average calculated $\bar{a}(H_\beta)$ in Table II.

A detailed examination of the interesting hfc dependence on both torsion and out-of-plane bending has been given previously [32]. However, for the present purposes of comparison to experiment, the most important observation is that for any given bend angle at the radical center, $\bar{a}(H_\beta)$ and all other couplings are nearly independent of torsional orientation. Therefore, to a good approximation the coupling between the large amplitude torsional and out-of-plane bending motions can be ignored in hfc determination.

To be appropriate for a planar radical center, the "experimental" result quoted in Table II for a($^{13}C_\alpha$) actually differs significantly from the observed result [29] of 39.1 G. Just as with methyl radical, the reason for this difference is the significant vibrational correction arising from the large amplitude out-of-plane bending motion at the radical center. ROHF–SCI/[631|41] calculations have predicted [32] a correction of 9 G. Recall that in the closely related case of the ν_2 vibration in methyl radical the calculated correction was 10 G as compared to an experimentally derived correction of 11 G. This suggests that the analogous correction for ethyl should be about 10 G, which leads to the estimate of 29 G quoted in Table II for a nonvibrating planar radical center. Note that this correction is not appropriate for the equilibrium geometry, since that is slightly nonplanar. The ν_6 vibrational corrections for hfc at the remaining nuclei are much smaller [32] and have not been applied here.

The ROHF–SCI hfc results in Table II indicate that spin polarization calculations lead to good agreement within 10% of experiment for both carbons, a larger error of 16% for the α-hydrogens, and a significant 28% underestimate for the average β-hydrogen coupling. Comparison with the ROHF results and with experiment indicates that $\bar{a}(H_\beta)$ arises about equally from the direct effect of hyperconjugation, indirect effects from spin polarization, and other effects not treated here, the latter presumably due to dynamical electron correlation corrections.

At room temperature, there is significant population in the first excited vibrational level of the ν_6 out-of-plane bending mode, and even some in the second level. We have calculated [32] that for ν_6=1, a($^{13}C_\alpha$) increases by 20.0 G over its ν_6=0 value, and for ν_6=2 by 28.6 G. One may therefore expect that a($^{13}C_\alpha$) will be sensitive to temperature. Assuming a Boltzmann distribution of vibrational level populations, a($^{13}C_\alpha$) is determined to be 0.9 G higher at 300 K than at 0 K, and its temperature derivative at 300 K is found to be 8 mG/deg. Both of these results are in excellent agreement with direct experimental determination [33], thus validating the assumption that the out-of-plane bend motion is primarily responsible for the temperature dependence.

4 σ-RADICALS

4.1 VINYL C_2H_3

Vinyl is our first example of a σ-radical (see Figure 4). Calculations were performed at a theoretical CASSCF/6–31G** equilibrium geometry [34]. The equilibrium configuration is planar with a calculated CC bond distance of 1.327 Å, which is indicative of a double bond. In the basic electron configuration

$$\Psi_0 = 1\sigma^2 \ldots\ 6\sigma^2\ 7\sigma\ 1\pi^2$$

the 7σ orbital occupied by the unpaired electron is an in-plane *sp* hybrid. Direct contributions to the hfc are therefore possible at all the nuclei.

ROHF–SDCI/STO–3G calculations indicate partial occupations of 1.90 for the bonding 1π orbital and 0.10 for the corresponding antibonding 2π orbital. A multireference

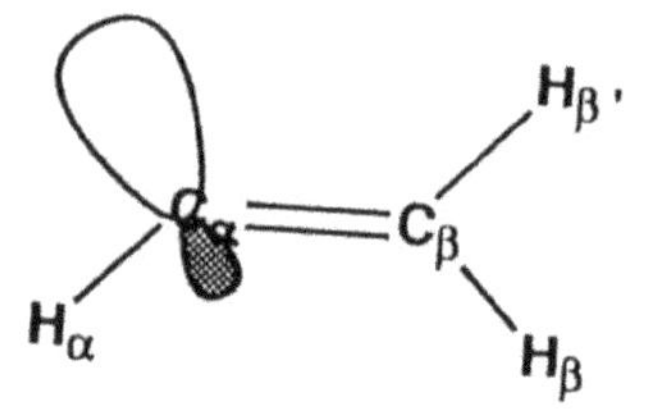

Fig. 4. Sketch of the structure and SOMO of vinyl radical.

space was therefore constructed to describe the nondynamical correlation arising from the dominant configuration and all single and double excitations from 1π into 2π, i.e.,

$$\Psi_{\mathrm{ND}} = \Psi_0(1\pi \rightarrow 2\pi \text{ singles and doubles}).$$

The ROHF result in Table III shows that the α-carbon coupling is dominated by a large direct contribution that is somewhat higher than the experimental result. The ROHF–SCI result indicates that the spin polarization correction to this is positive, taking the total further from experiment to a 28% error. The MR–SCI result shows that there is little further change in hfc upon introduction of the nondynamical electron correlations and their spin polarizations. At the β-carbon, where the experimental coupling is presumably negative, the positive direct contribution is lowered somewhat by the spin polarization correction but again changed little by nondynamical electron correlation, leaving a very large relative error. In summary, neither of the carbons is described satisfactorily by any of the calculations.

The α-hydrogen is described fairly well. The large direct contribution is significantly lowered by spin polarization, and lowered a little further with nondynamical correlation. The errors are 26% with ROHF–SCI results and 9% with MR–SCI. Results are not good, however, for the β-hydrogens. For these nuclei the large direct and spin polarization contributions are both positive and about equal in magnitude, with little further change being brought about with nondynamical correlation. Just as with ethyl radical, these calculations underestimate the β-hydrogen couplings by about 40%.

Experimental evidence [2, 36] indicates that the barrier to inversion of H_α through a linear CCH transition state is ~3-4 kcal/mol. This raises the question of whether or not the H_α wagging motion may be of large amplitude with strong anharmonicity, thereby leading to large vibrational corrections to the hfc. We have carried out (unpublished) exploratory calculations on the magnitude of vibrational corrections to be expected from this source. These indicate that the corrections are not large in any instance, and lead to slight improvement with experiment for all the nuclei except H_α, which drops somewhat below the experimental value. The clear implication is then that dynamical electron

TABLE III

Comparison of calculation to experiment for isotropic hyperfine coupling constants, in Gauss, of representative hydrocarbon σ-radicals.

Wave function	$a(^{13}C_\alpha)$	$a(^{13}C_\beta)$	$a(H_\alpha)$	$a(H_\beta)$	$a(H_{\beta'})$
		Vinyl			
ROHF	128.7	12.7	26.4	21.7	12.3
ROHF–SCI	137.9	−3.3	17.4	42.8	23.5
MR–SCI	138.2	−1.6	15.1	40.5	23.0
Experiment	107.6[a]	−8.6[a]	13.8[b]	65.9[b]	39.6[b]
		Ethynyl			
ROHF	377.1	67.5		5.4	
ROHF–SCI	402.9	78.9		13.0	
MR–SCI	383.6	86.2		15.5	
Experiment	(362)[c]	(76)[c]		(18)[c]	
		Cyclopropenyl			
ROHF	125.2	3.0	35.5	0.2	
ROHF–SCI	136.5	−2.1	36.0	−4.9	
MR–SCI	136.7	−2.4	33.9	−4.8	
Experiment			34.1[d]	−2.5[d]	

[a]Reference [29].
[b]Reference [35].
[c]Observed values [38] with corrections [37] to remove the large effects arising from vibronic coupling. See text for details.
[d]Reference [40].

correlation contributions must be primarily responsible for the remaining significant discrepancies with experiment in the carbon and β-hydrogen hfc constants in vinyl radical.

4.2 Ethynyl C_2H

Ethynyl is another example of a σ-radical (see Figure 5). Calculations were performed at a theoretical MR–CI equilibrium geometry [37]. The equilibrium configuration is linear with a calculated CC bond distance of 1.217 Å that is indicative of a triple bond. In the basic electron configuration

$$\Psi_0 = 1\sigma^2 \ldots\ 4\sigma^2\ 5\sigma\ 1\pi_x^2\ 1\pi_y^2$$

the 5σ orbital occupied by the unpaired electron is an *sp* hybrid oriented along the internuclear axis. Direct contributions to the hfc are therefore possible at all the nuclei.

ROHF–SDCI/STO–3G calculations indicate partial occupations of 1.92 for the equivalent $1\pi_x$ and $1\pi_y$ bonding orbitals and 0.08 for the corresponding equivalent $2\pi_x$ and $2\pi_y$ antibonding orbitals. A multireference space was therefore constructed to describe

Fig. 5. Sketch of the structure and SOMO of ethynyl radical.

nondynamical correlation arising from the dominant configuration and all single and double excitations from $1\pi_x$ into $2\pi_x$ and from $1\pi_y$ into $2\pi_y$, i.e.,

$$\Psi_{\rm ND} = \Psi_0(1\pi_x \rightarrow 2\pi_x \text{ and } 1\pi_y \rightarrow 2\pi_y \text{ singles and doubles}).$$

Vibronic coupling with low-lying excited electronic states leads to complications. Fortunately, hfc corrections due to this effect have been evaluated from very high quality theoretical calculations [37] so that "experimental" couplings appropriate for a hypothetical nonvibrating radical in the ground electronic state can be estimated. The "experimental" values quoted in Table III for $a(^{13}C_\alpha)$, $a(^{13}C_\beta)$, and $a(H_\beta)$ correspond to the actual observed values [38] of 321.9, 55.7, and 15.7 G, respectively, with theoretically determined corrections [37] to remove the large effects arising from vibronic coupling. It is seen that the vibronic correction reduces each of the couplings to some extent, by about 11% for the hydrogen and α-carbon couplings and by a relatively large 28% for the β-carbon coupling.

We see in Table III that the carbon couplings are both largely dominated by the direct contributions, with somewhat smaller corrections from spin polarization and nondynamical electron correlation. On the other hand, spin polarization is the largest contributor at the hydrogen. Comparing to the corrected experimental results, the total calculated couplings are good for carbon and adequate for hydrogen. The ROHF–SCI and MR–SCI relative errors, respectively, amount to 11% and 6% at the α-carbon, to 4% and 13% at the β-carbon, and to 28% and 14% at the hydrogen. The spin polarization model may therefore be said to generally work well for ethynyl radical.

4.3 Cyclopropenyl C_3H_3

Early considerations of cyclopropenyl radical assumed it to have a planar equilateral triangle structure of D_{3h} symmetry. Later studies showed a planar isosceles triangle structure of C_{2v} symmetry to be more stable. However, it turns out that in addition to distortion of the carbon skeleton, pyramidalization also occurs at the α-carbon, so that the most stable form actually corresponds to a nonplanar C_s symmetry structure (see

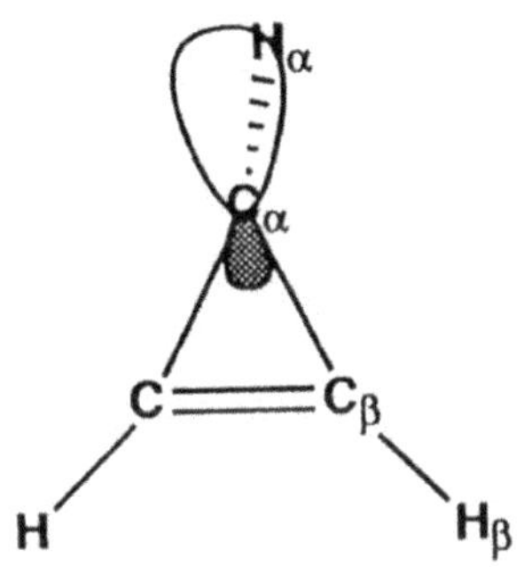

Fig. 6. Sketch of the structure and SOMO of cyclopropenyl radical. Only symmetry unique positions are explicitly labeled.

Figure 6). Calculations were performed at a theoretical MCSCF/[321|2] equilibrium geometry [39]. The calculated distance from the α-carbon to either of the equivalent β-carbons of 1.465 Å is a little shorter than a typical single bond, while the calculated distance between the two β-carbons of 1.315 Å is indicative of a double bond.

In the basic electron configuration

$$\Psi_0 = 1a'^2 \ldots\ 7a'^2\ 8a'\ 1a''^2 \ldots\ 3a''^2$$

the $8a'$ orbital occupied by the unpaired electron is largely an sp hybrid localized at the α-carbon, and oriented to complete the approximately tetrahedral coordination there. While a strict σ-π separation does not apply due to the overall nonplanarity of the radical, the doubly occupied $7a'$ orbital is dominated by p functions perpendicular to the carbon plane and highly localized on the two equivalent β-carbons, clearly approximately representing a π bond between these atoms.

ROHF–SDCI/STO–3G calculations indicate partial occupations of 1.94 for the bonding π-like orbital $7a'$ and 0.07 for the corresponding antibonding π-like orbital $4a''$. A multireference space was therefore constructed to describe the nondynamical correlation arising from the dominant configuration together with a double excitation from $7a'$ into $4a''$, i.e.,

$$\Psi_{\rm ND} = \Psi_0(7a'^2 \rightarrow 4a''^2).$$

The equilibrium form of cyclopropenyl radical has three equivalent conformations which can interconvert through pseudorotation of the CC double bond around the ring. The barrier to pseudorotation, experimentally determined [40] to be ~3-4 kcal/mol, is sufficiently high to allow a specific conformation, and therefore two unique hydrogen couplings, to be observed at low temperature. At higher temperatures, however, pseudorotation becomes facile and only a single effective hydrogen coupling is observed.

The original calculation [39] of the ESR spectrum was reported before this species had been experimentally observed. The results were subsequently used [40] as the major verification that a species found was in fact cyclopropenyl radical. For consistency with other parts of this work, we have recomputed the hfc with the [631|41] basis set. The results, given in Table III, differ only slightly from those obtained earlier [39] with a smaller basis.

The α-carbon and α-hydrogen couplings are both dominated by direct contributions, with small contributions from spin polarization and even smaller from nondynamical correlation. At the β-carbon and β-hydrogen, spin polarization is the largest contributor, and again nondynamical correlation has little further effect. Agreement with experiment is very good for the α-hydrogen, being within 2 G or 6% for all methods considered. Assuming the experimental sign of the β-hydrogen coupling to be negative, the error there of 2.7 G is fairly small in an absolute sense, although it corresponds to a very large relative error due to the rather small magnitude of the coupling. Unfortunately, there is no experimental determination of the ^{13}C hfc with which our results may be compared.

5 CONJUGATED π-RADICALS

5.1 Allyl C_3H_5

Allyl is a planar odd alternant π-radical with C_{2v} symmetry (see Figure 7). Calculations were performed at a theoretical ACPF/MIDI3* equilibrium geometry[41]. The calculated CC bond distances of 1.395 Å are typical of a conjugated system. In the basic electron configuration

$$\Psi_0 = 1a_1^2 \ldots\ 6a_1^2\ 1b_2^2\ \ldots\ 4b_2^2\ 1b_1^2\ 1a_2$$

the unpaired electron is in a nonbonding $1a_2$ π orbital that is delocalized over all three carbons, with a node dictated by symmetry at the central position.

It is well known [42] that allyl radical has nondynamical electron correlation effects in the 3-electron π system that require additional configurations for a proper zeroth-order description. A full ROHF–SDCI/STO–3G calculation showed partial occupations of 1.92 for the bonding orbital $1b_1$ and 0.09 for the corresponding antibonding orbital $2b_1$, which has nodes between each carbon atom. In agreement with previous findings [42], a multireference space was therefore constructed to include nondynamical correlations in the π space arising from the dominant configuration and all single and double excitations from $1b_1$ into $2b_1$, i.e.,

$$\Psi_{ND} = \Psi_0(1b_1 \rightarrow 2b_1 \text{ singles and doubles}).$$

Full SDCI calculations with the final large basis set used for spin density evaluation would be expensive. Fortunately, this is not necessary, since the important nondynamical correlation configurations have already been identified from the small basis calculation. Instead, natural orbitals were determined from a simpler ROHF–SDCI/[631|41]

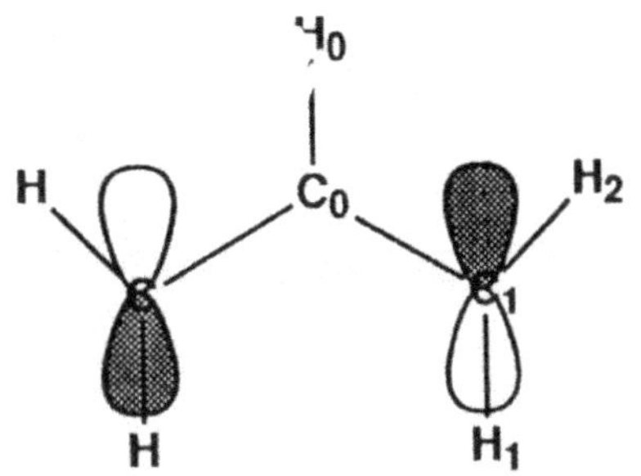

Fig. 7. Sketch of the structure and SOMO of allyl radical. Only symmetry unique positions are explicitly labeled.

calculation restricted to just the π space, i.e., all ten doubly occupied σ orbitals were frozen at their ROHF forms. These natural orbitals were then used as a basis for final MR–SCI spin density calculations, which were otherwise carried out as described previously, i.e., both σ and π single excitations from all reference configurations were included.

As seen in Table IV, the ROHF–SCI spin polarization hfc constants are in fair agreement with experiment, differing by less than 3 G in all cases. These correspond to relative errors of less than 20%, except for a(H_0). There the small 2.1 G absolute error corresponds to a very large relative error due to the small magnitude of the coupling. The MR–SCI procedure leads to small increases in the magnitudes of all the hfc constants, as compared to ROHF–SCI. This produces a notable improvement in a(H_0), which is now in excellent agreement with experiment. It also leads to an improvement in a($^{13}C_1$), but a(H_1), a(H_2), and a($^{13}C_0$) become worse. Thus, even with inclusion of nondynamical electron correlation in the reference space, errors of up to 20% are still seen in the spin polarization hfc constants. These remaining errors are presumably due mainly to dynamical electron correlation effects.

5.2 Pentadienyl C_5H_7

We consider here the W-form of pentadienyl (see Figure 8), another planar odd alternant π-radical with C_{2v} symmetry. Calculations were performed at a theoretical CASSCF/6–31G equilibrium geometry [44]. The calculated interior CC bond distance of 1.421 Å and end CC distance of 1.371 Å alternate around values typical of a conjugated

TABLE IV
Comparison of calculation to experiment for isotropic hyperfine coupling constants, in Gauss, of representative conjugated hydrocarbon π-radicals.

Wave function	$a(^{13}C_0)$	$a(^{13}C_1)$	$a(^{13}C_2)$	$a(^{13}C_3)$	$a(H_0)$	$a(H_1)$	$a(H_2)$	$a(H_3)$
				allyl				
ROHF-SCI	-19.6	18.0			2.1	-14.1	-15.1	
MR-SCI	-20.7	19.5			4.1	-15.8	-16.8	
Experiment	-17.2[a]	21.9[a]			4.2[b]	-13.9[b]	-14.8[b]	
				pentadienyl				
ROHF-SCI	16.7	-15.4	10.9		-13.6	1.9	-7.9	-8.6
MR-SCI	17.8	-17.2	13.3		-13.1	4.0	-11.1	-11.9
Experiment					-11.5[c]	3.4[c]	-9.7[c]	-10.4[c]
				cyclohexadienyl				
ROHF	0.0	0.0	0.0	0.0	0.0	0.0	0.0	16.9
ROHF-SCI	17.9	-14.9	10.9	-9.4	-16.2	1.8	-7.9	34.5
MR-SCI	16.7	-15.4	11.9	-11.7	-14.8	3.5	-10.1	35.8
Experiment	19.3[d]	-12.6[d]	14.1[d]	-12.2[d]	-13.0[b]	2.7[b]	-9.0[b]	47.7[b]

[a]Reference [43].
[b]Reference [2].
[c]Reference [45].
[d]Taken from level crossing μSR measurements on $C_6H_6\mu$, Reference [47].

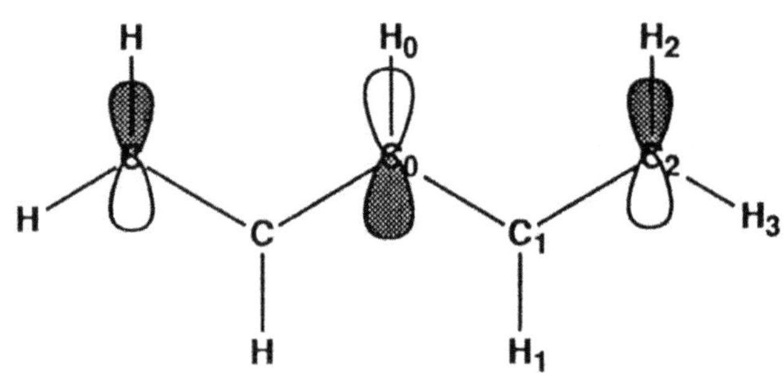

Fig. 8. Sketch of the structure and SOMO of pentadienyl radical. Only symmetry unique positions are explicitly labeled.

system. In the basic electron configuration

$$\Psi_0 = 1a_1^2 \ldots\ 9a_1^2\ 1b_2^2\ \ldots\ 7b_2^2\ 1b_1^2\ 2b_1\ 1a_2^2$$

the unpaired electron is in a nonbonding $2b_1$ π orbital that is delocalized over the entire five-carbon chain, with nodes near two of the interior carbons.

A full ROHF–SDCI/STO–3G calculation showed partial occupations of 1.95 for the bonding orbital $1b_1$ together with 0.05 for the corresponding antibonding orbital $3b_1$, and 1.93 for the largely bonding orbital $1a_2$ together with 0.07 for the corresponding antibonding orbital $2a_2$. A multireference space was therefore constructed to include nondynamical correlations in the π space arising from the dominant configuration, all single and double excitations from $1b_1$ into $3b_1$, all single and double excitations from $1a_2$ into $2a_2$, and double excitations corresponding to simultaneous $1b_1$ to $3b_1$ and $1a_2$ to $2a_2$ single replacements, i.e.,

$$\Psi_{\text{ND}} = \Psi_0(1b_1 \rightarrow 3b_1 \text{ and } 1a_2 \rightarrow 2a_2 \text{ singles and doubles}).$$

Natural orbitals determined from a ROHF–SDCI/[631|41] calculation restricted to just the π space were then used as a basis for the final MR–SCI spin density calculations. This procedure has previously [26] been referred to as MR(π–S′D′D″)–SCI.

The calculated hydrogen hfc constants are compared to experiment in Table IV. Trends are similar to those already discussed for allyl radical. Both ROHF–SCI and MR–SCI calculations are in good overall agreement with experiment, differing by only about 1-2 G. All ROHF–SCI hydrogen couplings have relative errors under 20%, with the exception of the large relative error in the small magnitude a(H_1) coupling. The MR–SCI results again show an increase in the magnitude of all hfc constants. This rectifies, and indeed overcorrects, the problem with the small a(H_1) coupling. Including nondynamical electron correlation in the reference space via MR–SCI gives results that are all within 18% of experiment. Unfortunately, no experimental ^{13}C couplings are available for comparison.

Similar hfc calculations have also been reported [26] for the linear polyene radicals C_9H_{11}, $C_{13}H_{15}$, $C_{17}H_{19}$ and $C_{21}H_{23}$. Important characteristics of the calculated spin density distributions were then extrapolated to the long chain limit. This provided useful insights into the reasons for contradictory interpretations of experimental results on the neutral soliton defect in the important *trans*-polyacetylene conducting polymer.

5.3 Cyclohexadienyl C_6H_7

Cyclohexadienyl is another odd alternant π radical with C_{2v} symmetry (see Figure 9). Calculations were performed at the theoretical CASSCF/6–31G equilibrium geometry [46]. The methylene group is attached to the rest of the planar ring by CC bond distances calculated to be 1.509 Å, indicative of single bonds. The remaining five-carbon segment has CC bond distances calculated to be 1.370 and 1.426 Å, showing a strong similarity to the alternating conjugated framework of pentadienyl radical. In the basic electron configuration

$$\Psi_0 = 1a_1^2 \ldots\ 11a_1^2\ 1b_2^2\ \ldots\ 7b_2^2\ 1b_1^2\ 2b_1^2\ 3b_1\ 1a_2^2$$

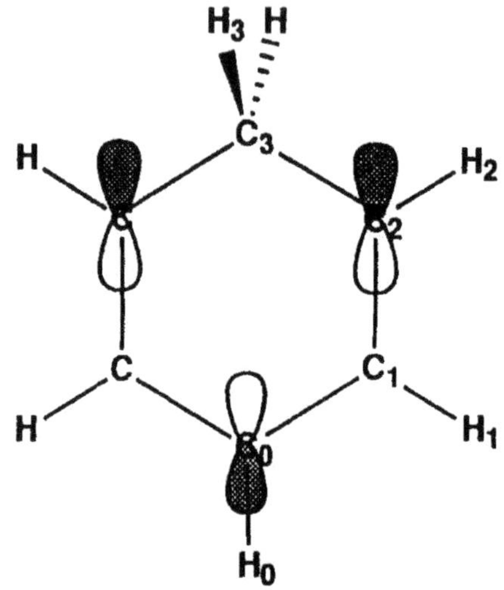

Fig. 9. Sketch of the structure and SOMO of cyclohexadienyl radical. Only symmetry unique positions are explicitly labeled.

the unpaired electron is in a nonbonding $3b_1$ π orbital that is delocalized over the five-carbon conjugated segment, with nodes near two of the interior carbons.

A full ROHF–SDCI/STO–3G calculation showed partial occupations of 1.95 for the bonding orbital $2b_1$ together with 0.05 for the corresponding antibonding orbital $4b_1$, and 1.94 for the largely bonding orbital $1a_2$ together with 0.06 for the corresponding antibonding orbital $2a_2$. Although partial occupations of 1.98 for the $1b_1$ bonding orbital together with 0.01 for the corresponding antibonding orbital $5b_1$ indicate that these are of only marginal importance for treatment of nondynamical correlation, they were included to allow correlation of all the π electrons. A multireference space was therefore constructed in a manner analogous to that already described above in connection with pentadienyl, i. e.,

$$\Psi_{\mathrm{ND}} = \Psi_0(1b_1 \rightarrow 5b_1,\ 2b_1 \rightarrow 4b_1, \text{and } 1a_2 \rightarrow 2a_2 \text{ singles and doubles}).$$

Natural orbitals determined from a ROHF–SDCI/[631|41] calculation restricted to just the π space were then used as a basis for the final MR–SCI spin density calculations.

The calculated hfc constants are compared to experiment in Table IV. Within the conjugated fragment, trends are very similar to those seen in pentadienyl radical. Both

ROHF–SCI and MR–SCI calculations are in fair overall agreement with experiment. Absolute differences from experiment range from about 1-3 G for both methods considered, corresponding to relative errors of up to 33%. The calculated methylene carbon hfc also shows similar behavior. The C_0 and H_0 couplings are exceptions to the previously noted general trend of MR–SCI results being larger in magnitude than those from ROHF–SCI.

The large β-hydrogen methylene coupling $a(H_3)$ in this system has been a subject of considerable interest. The calculations that include spin polarization significantly underestimate this coupling by 12-13 G, corresponding to a relative error of 25-28%. The ROHF calculation indicates that the direct contribution, due to hyperconjugation of the unpaired electron onto the methylene hydrogens, is approximately the same size as the indirect spin polarization contribution. However, this still leaves a considerable 13 G discrepancy with experiment of which only about 1 G is accounted for by nondynamical correlation in the MR–SCI calculation.

Possible reasons for this large error can be examined by considering the smaller H_2CN radical as a model system [48, 49]. This also has a methylene group that can undergo hyperconjugation with an unpaired electron in a nearby π system. Analogous effects are seen there, all magnified by about a factor of two due to the closer proximity of the unpaired electron. ROHF–SCI calculations were found to underestimate the β-hydrogen coupling by about 30%, approximately the same relative error as in cyclohexadienyl. Exploration of the dependence of the hfc on radical structure showed that uncertainties in the equilibrium geometry could not account for more than a small part of the error. Vibrational averaging corrections were found to be on the order of 1 G or less at all nuclei, again a small effect. It was surmised that most of the error must be due to missing dynamical electron correlation effects. However, large scale configuration-interaction calculations with extended basis sets accounted for only part of the discrepancy. It was therefore concluded that high excitation levels and extremely large basis sets would be required to obtain good agreement with experiment. It seems likely that the same should also hold true in cyclohexadienyl radical. Thus hyperconjugation, spin polarization and dynamical electron correlation all apparently play roughly equal roles in contributing to the large methylene hydrogen coupling in H_2CN and, presumably, in cyclohexadienyl radical as well.

Substitution of one methylene hydrogen with an electron withdrawing group such as –OH leads to a significant drop in the hfc of the remaining methylene hydrogen in cyclohexadienyl. It was found [46] that this effect is not due to perturbation of the geometry, which is small and actually counteracts the drop to some extent. It was concluded instead that it is due to disruption of the electronic structure leading to roughly equal drops in each of the contributions from hyperconjugation, spin polarization, and electron correlation.

6 CONCLUSION

We have examined in this work a variety of representative hydrocarbon radicals: the nonconjugated π-radicals methynyl, methyl, and ethyl; the σ-radicals vinyl, ethynyl, and cyclopropenyl; and the conjugated π-radicals allyl, pentadienyl, and cyclohexadienyl. Each has been seen to possess its own unique and interesting features. Here we summarize some conclusions that can be drawn from the general trends found in the hfc calculations.

Properly prepared Gaussian basis sets can be suitable for contact spin density calculations. The [631|41] basis utilized here seems adequate for evaluation of direct, spin polarization, and nondynamical electron correlation hfc contributions. Inclusion of dynamical electron correlation effects would require larger basis sets, of course, particularly with more high angular momentum functions. Truly quantitative work would also require inclusion of additional very tight s-functions in the basis set [11].

The spin polarization method as implemented in the ROHF–SCI computational procedure usually provides hfc constants within about 25% of experiment, and often much better than that. Two systematic exceptions to this have been noted, both corresponding to underestimations of hydrogen hfc constants. In the case of the small hfc constants of hydrogens located at nodal positions within the odd alternant network of a conjugated π-radical, this systematic error seems to be largely rectified by including nondynamical π electron correlations in the reference space, e.g., via the MR–SCI procedure. However, that is still not sufficient for the case of the large hfc constants of β-hydrogens such as in the cases of ethyl, vinyl, and the methylene group of cyclohexadienyl, although not of ethynyl. Proper description in that situation presumably requires inclusion of dynamical electron correlation effects.

Several research groups are currently engaged in pursuing promising approaches to the important matter of the *ab initio* evaluation of dynamical electron correlation contributions to hfc constants. We mention some of them here and give representative references to lead the interested reader into the recent literature, although no attempt is made to be comprehensive.

Many workers utilize more sophisticated spin-restricted wave functions together with extended basis sets in an attempt to recover a large fraction of the dynamical electron correlation. Kristiansen and Veseth [50] have used many-body perturbation theory. Nakatsuji and Izawa [51] have tested a SAC-CI formulation. Sundholm and Olsen [52] have used a MCSCF procedure. Chong, Langhoff, and Bauschlicher [53] have compared CASSCF, MR–CI, and coupled-pair functional methods to full CI. Suter and Ha [54] used MR-CI procedures. Feller and Davidson [55] have recently reviewed their considerable work involving MR–CI calculations. Engels *et al.* [56] have utilized MR–CI methods, while Engels [57] has also proposed a perturbation theory approximation to MR–CI based on the interesting observation that the effect of dynamic correlations arises not so much from their direct participation in the wave function, but rather from their presence indirectly altering the coefficients of the spin polarization terms.

Although UHF itself is usually unsatisfactory for hfc calculations, finite-field spin-unrestricted approaches that include correlation effects often perform well. Botschwina, Flesch, and Meyer [58] used self-consistent electron pair methods. Barone *et al.* [59] used MP2 perturbation theory. Sekino and Bartlett [60] have studied a coupled-cluster formulation. Carmichael [61] has used the closely related QCISD(T) method. Fernandez *et al.* [62] have proposed a novel hybrid procedure utilizing a field-free spin-restricted reference and finite-field spin-unrestricted response.

These studies have all shed light on the effects of electron correlation on hfc in particular situations. Unfortunately, most have not yet been consistently and systematically applied to enough different polyatomic systems to allow general conclusions to be drawn.

In conclusion, it may be said that the spin polarization model provides a useful qualitative description of the hfc in hydrocarbon radicals. It is therefore a desirable first step in any hfc investigation. It approaches but does not fully meet the goal of consistent quantitative agreement to within 20% of experiment. That requires more sophisticated computational procedures, which are a fruitful active area of current research. To make such studies feasible on large systems, it will be necessary to identify the nature of the important dynamical electron correlation contributions and to develop economical procedures for their evaluation. Only then will theory be able to reliably provide quantitative agreement with experimental hfc results on large systems of chemical interest.

ACKNOWLEDGEMENT

This work was supported by the Office of Basic Energy Sciences of the U. S. Department of Energy. This is Contribution No. SR-157 of the Notre Dame Radiation Laboratory.

References

1. R. W. Fessenden and R. H. Schuler, in *Advances in Radiation Chemistry*, Vol. 2, edited by M. Burton and J. L. Magee (Interscience, New York, 1970), p. 1.
2. R. W. Fessenden and R. H. Schuler, J. Chem. Phys. **39**, 2147 (1963).
3. A. Carrington and A. D. McLachlan, *Introduction to Magnetic Resonance with Applications to Chemistry and Chemical Physics* (Harper and Row, New York, 1967).
4. C. C. J. Roothaan, Rev. Mod. Phys. **32**, 179 (1960).
5. J. A. Pople and R. K. Nesbet, J. Chem. Phys. **22**, 571 (1954).
6. D. M. Chipman, J. Chem. Phys. **78**, 4785 (1983).
7. D. M. Chipman, J. Chem. Phys. **71**, 761 (1979).
8. D. M. Chipman, Rad. Res. **85**, 257 (1981).
9. D. M. Chipman, Theor. Chim. Acta **82**, 93 (1992).
10. H. M. McConnell and D. B. Chesnut, J. Chem. Phys. **28**, 107 (1958).
11. D. M. Chipman, J. Chem. Phys. **91**, 5455 (1989).
12. K. W. Richman, Z. Shi, and E. A. McCullough, Chem. Phys. Lett. **141**, 186 (1987).
13. D. M. Chipman, Phys. Rev. A **39**, 475 (1989).
14. H. Nakatsuji, K. Ohta, and T. Yonezawa, J. Phys. Chem. **87**, 3068 (1983).
15. T. Nakano, K. Morihahsi, and O. Kikuchi, Chem. Phys. Lett. **186**, 572 (1991).
16. Y. Ellinger, F. Pauzat, V. Barone, J. Douady, and R. Subra, J. Chem. Phys. **72**, 6390 (1980).
17. P. Pulay and T. P. Hamilton, J. Chem. Phys. **88**, 4926 (1988).
18. W. J. Hehre, R. F. Stewart, and J. A. Pople, J. Chem. Phys. **51**, 2657 (1969).
19. K. P. Huber and G. Herzberg, *Molecular Spectra and Molecular Structure. IV. Constants of Diatomic Molecules* (Van Nostrand Reinhold, New York, 1979), p. 140.

20. E. A. McCullough, Comp. Phys. Rep. **4**, 265 (1986).
21. T. C. Steimle, D. R. Woodward, and J. M. Brown, J. Chem. Phys. **85**, 1276 (1986).
22. C. R. Brazier and J. M. Brown, Can. J. Phys. **62**, 1563 (1984).
23. W. J. Hehre, R. Ditchfield, and J. A. Pople, J. Chem. Phys. **56**, 2257 (1972).
24. P. C. Hariharan and J. A. Pople, Theor. Chim. Acta **28**, 213 (1973).
25. T. H. Dunning, J. Chem. Phys. **53**, 2823 (1970).
26. B. Kirtman, M. Hasan, and D. M. Chipman, J. Chem. Phys. **95**, 7698 (1991).
27. D. M. Chipman, Theor. Chim. Acta **76**, 73 (1989).
28. G. Herzberg, Proc. Roy. Soc. (London) **A262**, 291 (1961).
29. R. W. Fessenden, J. Phys. Chem. **71**, 74 (1967).
30. D. M. Chipman, J. Chem. Phys. **78**, 3112 (1983).
31. D. Feller and E. R. Davidson, Theor. Chim. Acta **68**, 57 (1985).
32. D. M. Chipman, J. Chem. Phys. **94**, 6632 (1991).
33. D. Griller, P. R. Marriott, and K. F. Preston, J. Chem. Phys. **71**, 3703 (1979).
34. A theoretical equilibrium geometry for vinyl radical was obtained from a CASSCF/6–31G** optimization, where the active space included two π electrons correlated in two π orbitals. This gave an all planar structure having R(CC) = 1.327 Å, $R(CH_\alpha)$ = 1.073 Å, $R(CH_\beta)$ = 1.079 Å, $R(CH_{\beta'})$ = 1.076 Å, $\theta(CCH_\alpha) = 133.4°$, $\theta(CCH_\beta) = 121.6°$, and $\theta(CCH_{\beta'}) = 121.4°$.
35. P. H. Kasai, J. Am. Chem. Soc. **94**, 5950 (1972).
36. H. Kanamori, Y. Endo, and E. Hirota, J. Chem. Phys. **92**, 197 (1990).
37. M. Perić, B. Engels, and S. D. Peyerimhoff, J. Mol. Spect. **150**, 56,70 (1991).
38. W. R. M. Graham, K. I. Dismuke, and W. Weltner, J. Chem. Phys. **60**, 3817 (1974).
39. D. M. Chipman and K. E. Miller, J. Am. Chem. Soc. **106**, 6236 (1984).
40. G. L. Closs and O. D. Redwine, J. Am. Chem. Soc. **108**, 506 (1986).
41. P. G. Szalay, A. G. Császár, G. Fogarasi, A. Karpfen, and H. Lischka, J. Chem. Phys. **93**, 1246 (1990).
42. O. Kikuchi, Chem. Phys. Lett. **72**, 487 (1980).
43. H. J. McManus, R. W. Fessenden, and D. M. Chipman, J. Phys. Chem. **92**, 3778 (1988).
44. H. O. Villar and M. Dupuis, Chem. Phys. Lett. **142**, 59 (1987).
45. J. Foxall, B. C. Gilbert, H. Kazarians-Moghaddam, R. O. C. Norman, W. T. Dixon, and G. H. Williams, J. Chem. Soc., Perkin Trans. II, 273 (1980).
46. D. M. Chipman, J. Phys. Chem. **96**, 3294 (1992).
47. R. F. Kiefl, P. W. Percival, J.-C. Brodovitch, S.-K. Leung, D. Yu, K. Venkateswaran, and S. F. J. Fox, Chem. Phys. Lett. **143**, 613 (1988).
48. H. J. McManus, R. W. Fessenden, and D. M. Chipman, J. Phys. Chem. **92**, 3781 (1988).
49. D. M. Chipman, I. Carmichael, and D. Feller, J. Phys. Chem. **95**, 4702 (1991).
50. P. Kristiansen and L. Veseth, J. Chem. Phys. **84**, 2711 (1986).
51. H. Nakatsuji and M. Izawa, J. Chem. Phys. **91**, 6205 (1989).
52. D. Sundholm and J. Olsen, J. Chem. Phys. **94**, 5051 (1991).
53. D. P. Chong, S. R. Langhoff, and C. W. Bauschlicher, J. Chem. Phys. **94**, 3700 (1991).
54. H. U. Suter and T.-K. Ha, Chem. Phys. **154**, 227 (1991).
55. D. Feller and E. R. Davidson, in *Molecular Spectroscopy, Electronic Structure, and Intramolecular Interactions* (Springer-Verlag, New York, 1991), p. 429.
56. B. Engels, M. Perić, W. Reuter, S. D. Peyerimhoff, and F. Grein, J. Chem. Phys. **96**, 4526 (1992).
57. B. Engels, Chem. Phys. Lett. **179**, 398 (1991).
58. P. Botschwina, J. Flesch, and W. Meyer, Chem. Phys. **74**, 321 (1983).
59. V. Barone, C. Minichino, H. Faucher, R. Subra, and A. Grand, Chem. Phys. Lett. **205**, 324 (1993).
60. H. Sekino and R. J. Bartlett, J. Chem. Phys. **82**, 4225 (1985).
61. I. Carmichael, J. Phys. Chem. **95**, 108 (1991).
62. B. Fernandez, P. Jørgensen, J. Byberg, J. Olsen, T. Helgaker, H. Jørgen, and Aa. Jensen, J. Chem. Phys. **97**, 3412 (1992).

Calculation of Accurate Bond Energies, Electron Affinities, and Ionization Energies

Larry A. Curtiss
Chemical Technology Division/Materials Science Division, Argonne National Laboratory, Argonne, Illinois 60439

and

Krishnan Raghavachari
AT&T Bell Laboratories, Murray Hill, New Jersey 07974

August 31, 1994

Abstract Gaussian-2 theory (G2 theory) is a general procedure, based on *ab initio* molecular orbital theory, for the accurate calculation of energies such as dissociation energies, electron affinities, ionization energies, proton affinities, appearance potentials, and enthalpies of formation. In this chapter we review the theoretical basis of G2 theory and its performance on a set of 125 molecular energies used for testing it. The validity of various aspects of G2 theory is reviewed including the additivity assumptions, zero-point energy correction, geometries, and the higher level correction. Comparisons to other methods for calculation of thermochemical data based on *ab initio* molecular orbital theory are made. Finally, some of the applications of G2 (and G1) theory which have been reported in the literature are reviewed.

S. R. Langhoff (Ed.), Quantum Mechanical Electronic Structure Calculations with Chemical Accuracy, 139–171.

Contents

1 INTRODUCTION

Since the early 1970s when *ab initio* molecular orbital calculations became "routine," one of the major goals of quantum chemistry has been the calculation of molecular energies, such as bond dissociation energies to chemical accuracy, i.e. on the order of 1-2 kcal/mol. Over the last 20 years dramatic progress[1] has been made in achieving this goal with recent work proving to be extremely useful in the interpretation and evaluation of experimental data as well as providing data where experiment is not possible.

The initial work in this area in the 1970s was at the Hartree-Fock (HF) level and used small basis sets such as STO-3G and 4-31G. These methods depended on the cancellation of errors, such as neglect of correlation effects, by suitable choice of reactions. Pople, Radom, Hehre, and Ditchfield[2,3] introduced the idea of isodesmic reactions which are bond separation reactions having reactants and products with the same number of each type of bond. Along the same lines, George *et al.*[4,5] introduced homodesmotic reactions, which had the additional constraint that the structural groups be the same.

Since the early 1980s most approaches for accurate calculation of energies have included correlation treatments. This has been made possible by new developments in theoretical methodology as well as dramatic advances in computer technology. These approaches have generally been based on a moderate level of correlation treatment with large basis sets, which are supplemented by corrections from a high level of correlation treatment using a moderate-sized basis set and/or some form of experimental input. An example of this approach is the work of Pople *et al.*[6,7] who used a moderate level of correlation (Møller-Plesset perturbation theory to fourth order, MP4) with large basis sets to study the atomization energies of AH_n hydrides (A=Li-Cl). The remaining correlation effects were estimated by determining the energy of each A-H bond relative to that of the hydrogen molecule and then converting to atomization energies using the exact energy of dissociation of the hydrogen molecule. This was approximately 4 kcal/mol per hydrogen molecule or electron pair, and was referred to as the isogyric energy correction. This method was quite successful for the AH_n hydrides[7] and their ionization potentials.[8-10] It has also been applied to third-row nontransition metal hydrides AH_n.[11,12]

The work on AH_n hydrides[6-10] led to our development, in collaboration with John Pople of Gaussian-2 theory[13] (G2 theory). This is a general procedure, based on *ab initio* molecular orbital theory, for the accurate calculation of energies such as dissociation energies, electron affinities, ionization energies, proton affinities, appearance potentials, and enthalpies of formation. It has been tested on a large number of systems, containing first and second row atoms, that have reliable experimental energies. G2 theory has also been applied to numerous systems for which the experimental data are less well known or nonexistent.

In this chapter, various aspects of G2 theory are covered. In Section 2 the theoretical basis of G2 theory and its predecessor, G1 theory,[14] are described. In Section 3 the performance of G2 theory for calculation of bond energies, ionization energies, and electron affinities is assessed. The results of calculations on a set of 125 molecular energies used for testing the methods are summarized. The validity of the various aspects of G2 theory is discussed, including the additivity assumptions, zero-point energy correction, geometries, and the higher level correction. Comparisons to other methods for calculation of thermochemical data based on *ab initio* molecular orbital theory are made. In Section 4 some of the applications of G2 (and G1) theory which have been reported in the literature, including our own, are reviewed. In Section 5 various modified versions of G2 theory are discussed. Finally, concluding remarks and possible areas for future research are presented in Section 6.

2 ELEMENTS OF G2 THEORY

2.1 G1 THEORY

In G1 theory[14] an equilibrium structure, a total electronic energy E_e, and a set of harmonic frequencies are calculated for each structure associated with a local minimum on a potential energy surface. The harmonic frequencies for each structure are used to obtain a zero-point correction $\Delta(ZPE)$ and, hence, the energy of the lowest vibronic state of the molecular system

$$E_o = E_e + \Delta(ZPE). \tag{1}$$

The E_o is the energy of the system at 0 K. The principal steps in G1 theory are as follows.

1. Initial equilibrium structures are obtained at the Hartree-Fock level with the 6-31G(d) basis.[15,16] Spin-restricted (RHF) theory is used for singlet states and spin-unrestricted (UHF) for others.[17]
2. The HF/6-31G(d) model is used to calculate harmonic frequencies, which are then scaled by a factor of 0.8929 to take account of known deficiencies at this level.[18] These frequencies give the zero-point energy $\Delta(ZPE)$ used to obtain E_o.
3. The equilibrium geometries are refined to the MP2/6-31G(d) level (Møller-Plesset perturbation theory to second order with the 6-31G(d) basis set), using all electrons (MP=full) for the calculation of correlation energies. These are then the final equilibrium geometries in the theory and are used in all single-point calculations at higher levels of theory. All of these subsequent calculations treat only valence electrons in the calculation of electron correlation, i.e. the frozen core approximation (FC).
4. The first higher level calculation is at full fourth-order MP theory,[19] with the 6-311G(d,p) basis set,[20] i.e., MP4(SDTQ)/6-311G(d,p) [also referred to as MP4/6-311G(d,p)]. The MP4/6-311G(d,p) energy is then modified by the

corrections from higher level calculations in steps 5-7.

5. A correction for diffuse functions[21,22] is evaluated as

$$\Delta E(+) = E[\mathrm{MP4/6\text{-}311{+}G}(d,p)] - E[\mathrm{MP4/6\text{-}311G}(d,p)]. \quad (2)$$

6. A correction for higher polarization functions[22] on non-hydrogen atoms is evaluated as

$$\Delta E(2df) = E[\mathrm{MP4/6\text{-}311G}(2df,p)] - E[\mathrm{MP4/6\text{-}311G}(d,p)]. \quad (3)$$

7. A correction for correlation effects beyond fourth order perturbation theory is evaluated as

$$\Delta E(\mathrm{QCI}) = E[\mathrm{QCISD(T)/6\text{-}311G}(d,p)] - E[\mathrm{MP4/6\text{-}311G}(d,p)]. \quad (4)$$

   This correction uses the method of quadratic configuration interaction.[23]

8. These three corrections are combined in an additive manner:

$$E(\mathrm{combined}) = E[\mathrm{MP4/6\text{-}311G}(d,p)] + \Delta E(+) + \Delta E(2df) + \Delta E(\mathrm{QCI}). \quad (5)$$

9. A final "higher level correction" is added to give the energy for stationary nuclei:

$$E_e(\mathrm{G1}) = E(\mathrm{combined}) + \Delta E(\mathrm{HLC}) \quad (6)$$

   where

$$\Delta E(\mathrm{HLC}) = -A n_\beta - B n_\alpha. \quad (7)$$

   The n_α and n_β are the number of α and β electrons, respectively, with $n_\alpha \geq n_\beta$. For G1 theory, A = 5.95 millihartrees and B = 0.19 millihartrees (equivalent to 6.14 millihartrees per electron pair). These numbers are chosen so that E_e is exact for the hydrogen atom and the hydrogen molecule.

10. Finally, the total energy at zero temperature is obtained by adding the zero-point correction, obtained from the frequencies of step 2:

$$E_o(\mathrm{G1}) = E_e + \Delta(\mathrm{ZPE}). \quad (8)$$

The energy E_o is referred to as the G1 energy.

2.2 G2 THEORY

In G2 theory[13] a correction, Δ, is added to the G1 energy. This is composed of two parts, both computed at the MP2 level. The first part, Δ_1, corrects for nonadditivity caused by the assumption of separate basis set extensions for diffuse-*sp* functions (+) and higher polarization functions (2*df*) in G1 theory:

$$\Delta_1 = \Delta(+2df) - \Delta(+) - \Delta(2df) \tag{9}$$

where $\Delta(+2df) = \text{E[MP2/6-311+G}(2df,p)] - \text{E[MP2/6-311G}(d,p)], \Delta(+) = \text{E[MP2/6-311+G}(d,p)] - \text{E[MP2/6-311G}(d,p)]$, and $\Delta(2df) = \text{E[MP2/6-311G}(2df,p)] - \text{E[MP2/6-311G}(d,p)]$. The second part, Δ_2, is a correction for addition of a third d-function[22] to the nonhydrogen atoms and a second p-function[22] to the hydrogens:

$$\Delta_2 = \text{E[MP2/6-311+G}(3df,2p)] - \text{E[MP2/6-311+G}(2df,p)] \tag{10}$$

The total correction Δ is then

$$\begin{aligned}\Delta = \Delta_1 + \Delta_2 = \; & \text{E[MP2/6-311+G}(3df,2p)] - \text{E[MP2/6-311G}(2df,p)] \\ & - \text{E[MP2/6-311+G}(d,p)] + \text{E[MP2/6-311G}(d,p)].\end{aligned} \tag{11}$$

The G2 energy is then obtained as follows:

$$E_o(\text{G2}) = E_o(\text{G1}) + \Delta \; + 1.14\text{npair} \tag{12}$$

where npair is the number of valence pairs. The last term in Eqn. (12) lowers the higher level correction ΔE(HLC) of 6.14 millihartrees per electron pair in G1 theory by 1.14 millihartrees to 5.00 millihartrees per electron pair in G2 theory (A=4.81 in Eqn (7)). The value of 1.14 millihartrees was determined to give zero mean deviation from experiment of the calculated atomization energies of 55 molecules having well-established experimental values (cf. Table I). G2 theory requires only one additional calculation, MP2/6-311+G($3df,2p$), the other MP2 values in Eqn. (11) already being done in G1 theory as part of the MP4 calculations. All of the calculations required for G2 theory are available in the quantum chemical computer program Gaussian90.[24] A more detailed discussion of the correlation and basis set requirements for accurate quantum chemical thermochemical predictions has been presented in another publication.[25]

3 ASSESSMENT OF G2 THEORY

3.1 Comparison with Experiment

G2 theory has been tested against experiment by comparing theoretical and experimental values for zero-temperature (0 K) atomization energies, ionization energies, electron affinities, and proton affinities.[13] The comparisons have been limited to accurate experimental data so that any differences should be largely due to error in G2 theory. To achieve this goal only experimental data believed to be accurate to ±1 kcal/mol were used. The experimental data were taken largely from compilations such as the JANAF tables[26] and Huber and Herzberg's book on diatomics.[27] A total of 125 energies were included in the comparison: 38 ionization potentials,

TABLE I

Ionization potentials, electron affinities, and proton affinities calculated from G2 theory. Values in parentheses are differences between G2 theory and experiment (i.e., G2 - experiment).[a]

	Species		Species		Species	
Ionization Potentials	Li	123.1(-1.2)	Cl	296.3(-2.8)	ClH	293.1(-0.9)
	Be	217.0(2.1)	CH_4	292.4(1.4)	C_2H_2	263.3(0.5)
	B	189.1(-2.3)	NH_3	235.0(0.2)	C_2H_4	244.0(1.6)
	C	257.8(-1.8)	OH	299.3(-0.7)	CO	323.1(0.0)
	N	333.9(-1.4)	OH_2	291.2(0.2)	$N_2(^2\Sigma_g^+)$	358.8(-0.5)
	O	311.8(-2.1)	FH	370.8(0.9)	$N_2(^2\Pi_u)$	384.4(-0.7)
	F	401.0(-0.7)	SiH_4	253.9(0.2)	O2	280.6(2.3)
	Na	114.1(-4.4)	PH	232.7(-1.4)	P2	243.1(0.2)
	Mg	176.4(0.1)	PH_2	224.1(-2.3)	S2	214.0(-1.8)
	Al	136.7(-1.2)	PH_3	227.6(0.0)	Cl2	265.4(0.2)
	Si	186.8(-1.2)	SH	237.7(-1.4)	ClF	291.7(-0.2)
	P	240.7(-1.2)	$SH_2(^2B_1)$	240.5(-0.9)	SC	263.3(2.1)
	S	235.2(-3.7)	$SH_2(^2A_1)$	294.0(-0.7)		
Electron Affinities	C	27.4(-1.6)	CH_3	0.9(-0.9)	HS	53.0(-0.3)
	O	32.3(-1.4)	NH	6.9(-1.8)	O_2	10.8(0.7)
	F	77.5(-0.9)	NH_2	17.8(0.7)	NO	-1.6(-2.1)
	Si	31.1(-0.8)	OH	43.1(0.9)	CN	91.5(3.5)
	P	14.8(-2.4)	SiH	27.2(-2.2)	PO	24.0(-1.2)
	S	46.1(-1.8)	SiH_2	22.8(-3.1)	S_2	38.3(-0.1)
	Cl	83.0(-0.3)	SiH_3	32.7(-0.5)	Cl_2	54.9(-0.2)
	CH	26.1(-2.5)	PH	22.1(-0.9)		
	CH_2	15.2(0.2)	PH_2	28.8(-0.2)		
Proton Affinities	NH_3	202.5(0.0)	SiH_4	153.0(-1.0)	ClH	133.0(-0.6)
	OH_2	163.1(-2.0)	PH_3	186.2(-0.9)		
	C_2H_2	153.6(1.3)	SH_2	167.7(-1.1)		

[a] All quantities in kcal/mol. References for experimental values are listed in Ref. 14.

25 electron affinities, and 7 proton affinities listed in Table I and 55 atomization energies listed in Table II. These tables contain the G2 values for these energies and their deviations from experiment.

Table III contains a summary of the average absolute deviations of the G2 theoretical values from experiment for the 125 energies grouped by the type of energy quantity and the type of molecule (i.e., first-row vs. second-row). Also given in Table III are the average absolute deviations for G1 theory and two

TABLE II
Atomization energies calculated from G2 theory. Values in parentheses are differences between G2 theory and experiment (i.e., G2 - experiment).[a]

Species		Species		Species	
LiH	56.6(0.6)	SH_2	173.0(-0.2)	CO_2	384.6(2.7)
BeH	45.5(-1.4)	ClH	102.6(0.4)	Na_2	19.2(2.6)
CH	80.5(0.6)	Li_2	25.9(1.9)	Si_2	73.6(-0.4)
$CH_2(^3B_1)$	178.6(-1.0)	LiF	137.5(-0.1)	P_2	114.7(-1.4)
$CH_2(^1A_1)$	172.0(1.4)	HCCH	387.2(-1.7)	S_2	97.4(-3.3)
CH_3	289.1(-0.1)	C_2H_4	531.7(-0.2)	Cl_2	55.8(-1.4)
CH_4	393.2(0.7)	C_2H_6	666.6(0.3)	NaCl	98.8(1.3)
NH	77.9(-1.1)	CN	176.0(-0.6)	SiO	188.8(-1.7)
NH_2	170.1(0.1)	HCN	302.8(1.0)	SC	170.5(1.0)
NH_3	276.5(-0.2)	CO	258.0(1.8)	SO	120.8(-2.7)
OH	101.6(0.3)	HCO	271.4(1.1)	ClO	61.2(-2.1)
OH_2	219.6(0.3)	H_2CO	359.3(2.1)	ClF	61.0(0.7)
FH	136.3(1.1)	H_3COH	482.3(1.5)	Si_2H_6	503.0(2.9)
$SiH_2(^1A_1)$	147.1(2.7)	N_2	223.8(-1.3)	CH_3Cl	372.1(1.1)
$SiH_2(^3B_1)$	123.8(0.4)	H_4N_2	404.4(-1.0)	CH_3SH	445.0(-0.1)
SiH_3	213.5(-0.5)	NO	150.6(0.5)	HOCl	156.8(0.5)
SiH_4	304.8(2.0)	O_2	115.6(-2.4)	SO_2	248.9(-5.1)
PH_2	144.9(0.2)	HOOH	252.1(-0.2)		
PH_3	226.4(-1.0)	F_2	36.6(-0.3)		

[a] All quantities in kcal/mol. References for experimental values are listed in Ref. 14

modifications of G2 theory [G2(MP2) and G2(QCI) theory, which are discussed in Section 5].

For the 125 energies, the average absolute deviation of G2 theory from experiment is 1.21 kcal/mol. The 55 G2 atomization energies have an average absolute deviation of 1.16 kcal/mol. G2 theory performs somewhat better for atomization energies of molecules containing first-row atoms (0.92 kcal/mol) than for those containing second-row atoms (1.47 kcal/mol). The 38 ionization potentials have an average absolute deviation of 1.24 kcal/mol (first-row: 1.13 kcal/mol, second-row: 1.34 kcal/mol). The 25 electron affinities have an average absolute deviation of 1.29 kcal/mol. In this case, G2 theory performs better for the second-row systems (1.08 kcal/mol) than for the first-row systems (1.52 kcal/mol). The seven G2 level proton affinities have an absolute average deviation of 1.04 kcal/mol.

Some of the larger deviations from experiment in Tables I and II are systematic. A serious deviation has to do with triplet states where the higher level correction may be failing. The G2 singlet-triplet energy separations of CH_2 and SiH_2 differ from experiment by more than 2 kcal/mol. The experimental separations are 9.0 and 21.0 kcal/mol, respectively, compared to the theoretical values of 6.6 and 23.3 kcal/mol. The calculated dissociation energies of the triplet molecules O_2, S_2, and SO are too low by 2.4, 3.3, and 2.7 kcal/mol, respectively. G2 theory also does poorly on some atomic ionization energies. The calculated ionization energies of five atoms (Be, B, Na, S, Cl) deviate from experiment by more than 2 kcal/mol.

Twenty-four of the energies in Tables I and II differ by more than 2 kcal/mol from experiment. Of these only five deviate by more than 3 kcal/mol: the dissociation energies of SO_2 (−5.1 kcal/mol) and S_2 (−3.3 kcal/mol), the ionization energy of Na (−4.4 kcal/mol), and the electron affinities of CN (+3.5 kcal/mol) and SiH_2 (−3.1 kcal/mol). The error in the ionization energy of Na is due to neglect of inner shell correlation,[10] the error in the electron affinity of SiH_2 is partially accounted for by the additivity approximation,[28] and the error in the dissociation energy of S_2 was discussed above. The two largest deviations (CN, SO_2) are largely due to geometry problems and are discussed in Section 3.6.

Hence, G2 theory is quite successful for a wide range of systems and energy quantities with only a *single molecule-independent* semi-empirical parameter, Δ(HLC). All of the 125 test cases involved systems containing one or two heavy nonhydrogen atoms, with the exception of CO_2 and SO_2. The performance of the theory has not been thoroughly assessed for larger molecules (three or more nonhydrogen atoms). However, there have been numerous applications to larger systems which have generally been successful, and some are discussed in Section 4.

3.2 G1 Theory vs. G2 Theory

The average absolute deviations summarized in Table III indicate that G2 theory is a significant improvement over G1 theory for atomization energies, ionization energies, and electron affinities while proton affinities remain about the same. Overall, the average absolute deviation improves from 1.53 kcal/mol to 1.21 kcal/mol.

The most dramatic improvement in atomization energy is that of LiF where the dissociation energy decreases from 141.3 kcal/mol (G1) to 137.5 kcal/mol (G2). The improvement occurs because the nonadditivity, Δ_1, of the (+) and (2*df*) basis set expansions is large (6.83 millihartrees). Apparently, in ionic systems the polarization extension (2*df*) on the positive center (Li) duplicates some of the contribution of the diffuse function (+) on the negative center (F). This is corrected in G2 theory.

The addition of the third *d*-function and the second *p*-function to the basis set contributes to improvement in the atomization energies at the G2 level. The increase in basis set size, along with the modification in the higher level correction ΔE(HLC), has the largest effect on hydrides of the first-row compounds. In the majority of these cases, the G2 value is in better agreement with experiment and contributes to its overall improvement relative to G1 theory. The atomization

TABLE III
Average Absolute Deviations of G1, G2, G2(MP2), G2(QCI) Results Compared with Experiment.

Quantity[a]	Species	Average Absolute Deviation G1	G2	G2(QCI)	G2(MP2)	G2(NoQCI)[b]
ΣD_o	1st-row	1.49	0.92	0.97	1.25	1.90
	2nd-row	1.71	1.47	1.47	1.40	1.63
	1st- + 2nd-row	1.56	1.16	1.19	1.32	1.78
IP_o	1st-row	1.27	1.13	1.08	1.50	1.97
	2nd-row	1.59	1.34	1.27	2.19	2.36
	1st- + 2nd-row	1.43	1.24	1.18	1.86	2.18
EA_o	1st-row	1.57	1.52	1.41	2.04	2.46
	2nd-row	1.94	1.08	0.87	1.94	1.44
	1st- + 2nd-row	1.75	1.29	1.13	1.99	1.93
PA_o	1st-row	.73	1.10	1.10	1.13	1.10
	2nd-row	1.10	.90	1.12	0.28	1.08
	1st- + 2nd-row	.95	.99	1.11	0.64	1.09
All	1st-row	1.41	1.10	1.09	1.46	1.99
All	2nd-row	1.68	1.31	1.25	1.70	1.79
All	1st- + 2nd-row	1.53	1.21	1.17	1.58	1.91

[a] ΣD_o = atomization energy, IP_o = ionization potential, EA_o = electron affinity, PA_o = proton affinity

[b] G2 theory without the $\Delta E(QCI)$ correction.

energies of C_2H_6 [2.8 kcal/mol too low (G1) compared to 0.3 kcal/mol too high (G2)], NH_3 [2.5 kcal/mol too low (G1) compared to 0.2 kcal/mol too low (G2)], N_2H_4 [4.0 kcal/mol too low (G1) compared to 1.0 kcal/mol too low (G2)], and H_2O_2 [2.0 kcal/mol too low (G1) compared to 0.2 kcal/mol too low (G2)] are notable for their improved agreement with experiment at the G2 level. Also, the addition of the third *d*-function is largely responsible for the increase by 2 kcal/mol of the dissociation energy of SO_2 [from 246.6 (G1) to 248.9 kcal/mol (G2)]. However, it still remains 5.1 kcal/mol below the experimental value. This is discussed further in Section 3.6. Much of the improvement in the G2 electron affinities compared to those of G1 theory can be traced to the correction for the nonadditivity of the diffuse-*sp* and 2*df* basis set extensions in the G1 procedure.

3.3 Importance of the Quadratic Configuration Interaction (QCI) Correction

The importance of including correlation contributions beyond fourth-order Møller-Plesset perturbation theory by using quadratic configuration interaction in G2 theory has been investigated for all 125 test cases by Curtiss, Raghavachari, and Pople.[29] The average absolute deviations for G2 theory without the QCI correction [Eqn. (4)] are summarized as G2(NoQCI) in Table III. The deviation increases from 1.21 to 1.91 kcal/mol when the QCI correction is not included. Agreement with experiment is very poor for a significant number of test energies: 19 energies deviate by more than 3 kcal/mol with a maximum deviation of 19.5 kcal/mol. For example, the atomization energies of CN, CO, H_2CO, and CO_2 deviate by 11.6, 6.1, 3.8, and 10.0 kcal/mol, respectively. A post MP4-correction is necessary for unsaturated molecules. This has previously been shown for molecules such as CN and N_2.[30,31]

3.4 Higher Level Correction

A major difficulty in calculating accurate energies is that the correlation energy between two spin-paired electrons converges slowly as increasing angular quantum numbers are included in the basis set. It is difficult to correct this problem by additional calculations. In G1 and G2 theory the residual basis set effect is approximated by an overall higher level correction, ΔE(HLC). In G1 theory this quantity is chosen to give the exact energies for the hydrogen atom and the hydrogen molecule, whereas in G2 theory it is chosen to give zero mean deviation from experiment of the calculated atomization energies of the 55 molecules.

The higher level correction in G2 theory is a *single molecule-independent* semi-empirical parameter [eqn. (7) with $A = 4.81$] used to account for the residual basis set effect. Uniform application of this type of correction is reasonable if the basis set used is reasonably balanced so that residual errors per electron are approximately constant over a wide range of molecules. This cannot be demonstrated in an *ab initio* sense. It can only be validated if it succeeds in reproducing well-established experimental data within a certain accuracy. One of the areas for improvement in G2 theory is a method for determining ΔE(HLC) that does not depend on fitting it to a set of experimental energies.

3.5 Additivity Assumptions

The total energies obtained in G1 and G2 theory [before addition of the ΔE(HLC) and Δ(ZPE) terms] are *effectively* at the QCISD(T)/6-311+G(2*df*,*p*) and QCISD (T)/ 6-311+G(3*df*,2*p*) level, respectively, making certain assumptions about additivity. Curtiss *et al.*[28] examined the validity of the three "additivity approximations" (also referred to as "corrections") in G2 theory: (1) use of second order Møller-Plesset perturbation theory to account for the nonadditivity of the diffuse functions and additional polarization functions [Δ_1, Eqn. (9)]; (2) use of second order Møller-Plesset perturbation theory to obtain the energy change of the $(2df, p)$ to $(3df, 2p)$ basis set extension [Δ_2, Eqn. (10)], and (3) use of the 6-311G(d,p) basis to evaluate correlation effects beyond fourth order Møller-Plesset perturbation theory in the quadratic configuration interaction calculation [Eqn. (4)]. These approximations are used in G2 theory because they provide savings in computational time and disk storage on large molecular systems. The three additivity approximations were assessed by carrying out full QCISD(T)/6-311+G$(3df, 2p)$ calculations[28] for the set of 125 energies used in testing G2 theory. The E_o values were obtained as follows

$$E_o = E[\text{QCISD(T)/6-311+G}(3df, 2p)] + \Delta E(\text{HLC}) + \Delta(\text{ZPE}) \tag{13}$$

with A = 5.13 millihartrees in the expression for ΔE(HLC) [Eqn. (7)]. The value for A was derived in a similar manner as for G2 theory, i.e., to give zero mean deviation from experiment of the calculated atomization energies of 55 molecules. Similar values for A were derived using slightly different methods of fitting the set of data including adjustment to obtain the smallest average absolute deviation from experiment for the 55 atomization energies (A=5.13) and the entire set of 125 energies (A=5.10). The total energy obtained from Eqn. (13) is referred to as the G2(QCI) energy.

The comparisons of the G2(QCI) energies with experiment for the set of 125 energies are summarized in Table III. The average absolute deviation for all 125 energies is 1.17 kcal/mol, only slightly better than that for G2 energies (1.21 kcal/mol). The average absolute deviation of the G2(QCI) energies from the G2 energies is only 0.30 kcal/mol (0.013 eV). Thus, the three approximations that are used in G2 theory to obtain the *effective* QCISD(T)/6-311+G$(3df, 2p)$ energies work very well for the 125 test cases. Only two G2(QCI) energies differ by more than 1.0 kcal/mol from the G2 energies: the electron affinities of OH (2.3 kcal/mol) and NH_2 (1.2 kcal/mol). The reason for the large discrepancies in the OH and NH_2 electron affinities can be traced to the use of the QCISD(T)/6-311G(d,p) calculation in G2 theory to obtain the QCI correction (Eqn (4)). The diffuse functions make an important contribution at the QCISD(T) level for OH^- and NH_2^- which is included in the G2(QCI) energy, but not in the G2 energy.

The saving in cpu time and disk storage that are gained from the use of the three additivity approximations in G2 theory can be significant. For the 125 systems use of the approximations reduces the cputime by a factor of about 2-3 and disk space by factor of 2-10.

3.6 Geometries

The MP2(FULL)/6-31G(d) geometries used in G2 theory are generally in reasonable agreement with experimental geometries and any discrepancies have only small effects on the calculated energies in most cases for which it has been tested.[13] For the 26 diatomics in Tables I and II, the bond lengths have an average absolute deviation of 0.027 Å compared to experiment. For example, the bond lengths of SiO and SO are too long by 0.03 and 0.04 Å, respectively. Use of the experimental distances for SiO and SO in G1 theory, instead of the MP2 values, increases their bond dissociation energies by only 0.3 and 0.4 kcal/mol, respectively.[14b] However, there are also a few cases where the geometries have a significant effect on the energies. The two cases in poorest agreement with experiment in Ref. 13 (dissociation energy of SO_2 and electron affinity of CN) are ones that are, in part, due to problems with geometry.

One of the major failures of G2 theory is the atomization energy of SO_2.[13] The G2 value of 248.9 kcal/mol is 5.1 kcal/mol less than the experimental atomization energy of 254.0 kcal/mol. No improvement is found when the full QCISD(T)/6-311+G($3df, 2p$) calculation was carried out: the G2(QCI) atomization energy is 248.7 kcal/mol.[28] The MP2/6-31G(d) geometry of SO_2 is poor (SO distance of 1.477 Å compared to the experimental value of 1.432 Å). Calculation of the atomization energy of SO_2 at the experimental geometry leads to G2 and G2(QCI) atomization energies of 249.6 and 250.4 kcal/mol, respectively. The latter value differs from experiment by 3.6 kcal/mol compared to 5.1 kcal/mol for G2 theory. Hence, improvement is obtained by use of a better geometry and G2(QCI) energies. The remaining discrepancy may be due to a problem with the higher level correction.

The G2 electron affinity of CN, 3.97 eV, is 0.15 eV (3.5 kcal/mol) higher in energy than the experimental value of 3.82 $\pm$ 0.02 eV from Berkowitz et al.[32] Neumark *et al.*[33] recently reported a value of 3.86 $\pm$ 0.004 eV for the electron affinity of CN from a photoelectron study. This is within 0.1 eV of the G2 value. In addition, the G2 value is lowered to 3.89 eV when QCISD/6-31G(d) geometries are used instead of MP2/6-31G(d) geometries.[34] The bond distance in neutral CN is especially sensitive to higher level correlation treatment as it increases from 1.135 [MP2/6-31G(d)] to 1.181 Å [QCISD(T)/6-31G(d)]. The experimental value[26] is 1.175 Å.

There are also some systems where the MP2/6-31G(d) level of theory may be totally inadequate for geometries in which case G2 theory will fail unless some other level of theory is used for the geometry. McKee[35] has pointed out such a case in the dimethyl sulfide - OH adduct where a basis set larger than 6-31G(d) is required to get a stable adduct.

3.7 Zero-Point Energies

The level of calculation used for zero-point energies in G2 theory (scaled HF/6-31G(d) frequencies) was set by the computational resources available at the time of development of G1 theory. It would now be feasible to calculate the frequencies for all the species considered in Ref. 13 at a higher level, such as MP2/6-31G(d). A study of the effect of using better estimates of the zero-point energy in G2 theory on the 125 test energies would be useful.

Ma, Smith, and Radom[36] have examined the energies of CH_3OH, CH_3OH^+, and

$CH_2OH_2^+$ using G2 theory with MP2/6-31G(d) frequencies scaled by 0.93 instead of HF/6-31G(d) frequencies scaled by 0.8929. They found that the ionization energy is decreased by 0.6 kcal/mol (0.02 eV). The energy difference between CH_3OH^+ and $CH_2OH_2{}^+$ is decreased by 0.7 kcal/mol (0.03 eV).

Grev, Janssen, and Schaefer[37] have recently recommended using a scale factor of 0.91 instead of 0.8929 for HF/6-31G(d) vibrational frequencies. The currently used scale factor, 0.8929, is derived from scaling HF/6-31G(d) frequencies to reproduce experimentally observed fundamental frequencies.[18] Grev *et al.*[37] argue that the correct zero-point vibrational energy is not one-half the sum of the fundamental frequencies and suggest that 0.91 is a more appropriate scale factor until better methods for determining those energies are developed. We have examined the use of a scale factor of 0.91 in the calculation of G2(QCI) energies discussed in Section 3.5. For the 55 atomization energies in Table I, use of the 0.91 scale factor gives an average absolute deviation of 1.19 kcal/mol, exactly the same as the deviation obtained with the 0.8929 scale factor. When the constant A in the $\Delta E(HLC)$ expression [eqn. (7)] is adjusted to give the smallest average absolute deviation for the 55 atomization energies, the average absolute deviation is reduced to 1.17 kcal/mol. Hence, this suggests that use of a scale factor of 0.91 improves agreement with experiment only slightly, not enough to justify its use in G2 theory.

3.8 Relationship to Other Methods for Calculations of Accurate Energies

Since the early 1980s there have been a variety of methods[6,7,38–51] used for the calculation of dissociation energies and enthalpies of formation of molecules. These methods have been of varying degrees of accuracy and some have been on only one molecule, others on a series of related systems. Generally, the methods are based on a moderate level of correlation treatment that is calibrated in most cases against either some higher level of theory or experimental input (or both).

The isogyric approach by Pople *et al.*[6,7] in which one parameter, the bond energy of H_2, is used as a correction factor to predict bond energies, ionization energies, and proton affinities of AH_n species was discussed in the introduction. Martin *et al.*[38] used a similar isogyric approach to study dissociation energies of AH_n molecules and extended the accuracy with the inclusion of sp bond functions. Sana *et al.*[39] have also used a similar isogyric approach for a series of AH_n and ABH_n compounds. As discussed in the introduction, G2 theory evolved out of the isogyric approach for calculating accurate thermochemical data.

A number of researchers[47–49] have used specially chosen reactions (isodesmic, hydrogenation, exchange, etc.) in high level correlation calculations to obtain accurate thermochemical data for selected species. In this approach an appropriate reaction is chosen for which all species have accurate experimental thermochemical data available except the species of interest. A high level of theory is used to calculate the reaction energy and the enthalpy of formation of the unknown species is then extracted. Often more than one reaction is used to assess the reliability of the calculation. This approach has also been used in some applications of G2 theory. Methods based on specially chosen reactions can give quite accurate enthalpies of formation because they take advantage of cancelation of correlation effects and accurate experimental data. However, suitable reference molecules are often not available, making this method inapplicable in many cases.

Recently, methods using experimental information for *molecule-dependent* empirical corrections have been quite successful. Ho *et al.*[40] have used MP4/6-31G(d) calculations, along with empirical bond additivity corrections based on bond types and bond distances (BAC-MP4). The correction factors are determined by fitting to well-known enthalpies of formation. They have applied the method to a series of silicon hydrides. Along the same lines, Sax and Kalcher[41] used multiconfiguration self-consistent field (MCSCF) calculations with experimental input (enthalpies of formation of Si_2, SiH_4, Si_2H_6) to obtain enthalpies of formation of Si_nH_m (m=1-5) hydrides. While these methods have been successful, they may have problems in certain cases, such as clusters where the bonding may be unconventional or systems where the bonding is delocalized so that a local correction may not be valid. Also such methods cannot be used for pairs of atoms which have not been calibrated.

Bauschlicher, Langhoff, Taylor, and coworkers[1,43,46] have very successfully applied a variety of methods to obtain accurate thermochemical data for individual molecules. Their work has been based on converging to the one-particle limit through the use of atomic natural orbitals (ANO) at a moderate level of correlation treatment [such as complete active space self-consistent-field multi-reference configuration interaction (CASSCF/MRCI)]. The correlation treatment is calibrated against full configuration interaction (FCI) calculation on smaller systems or against accurate experimental data, in some cases using isodesmic reactions, to determine probable error in the one-particle basis set. They have achieved accuracies of 1-2 kcal/mol using these methods and have applied them to investigate C-H bond energies in molecules such as C_2H_2 as well as bond dissociation energies in other molecules.

Petersson *et al.*[42] have developed a scheme referred to as the complete basis set (CBS) method to extrapolate MP2 energies to estimate residual basis set effects. They have used this scheme in combination with correlation treatment at the QCISD(T) level to study the dissociation energies of a series of molecules including NF, O_2, C_2H_2, etc. Grev and Schaefer[44] have used coupled cluster methods [CCSD(T)] with large ANO basis sets to study the thermochemistry of CH_n, SiH_n hydrides and some of their cations. They achieved bond energies accurate to 0.5 kcal/mol without any corrections.

G2 theory uses a moderate level of correlation treatment with large basis sets, augmented by a higher level of correlation treatment with a moderate basis set [QCISD (T)/6-311G(d,p)]. Unlike most of the other approaches, it is not dependent on calibration with experimental data for related species either through isodesmic reactions or in some other manner. It does have the *single molecule-*

independent semi-empirical parameter, $\Delta E(HLC)$, which was chosen by fitting to a large set of well-established experimental data. Although other approaches are more accurate for individual bond energies, usually because of specific calibration with experiment or higher level of correlation treatment, G2 theory is the most general approach, considering the variety of systems and energy quantities that can be reliably handled. It has also been quite thoroughly tested on small molecules and computationally feasible on computers available to most scientists.

4 APPLICATIONS OF G2 THEORY

G2 theory and its predecessor, G1 theory, have been applied to many molecular systems[52–103] and have in most cases been quite successful. They have been used to make predictions of bond dissociation energies, ionization energies, electron affinities, appearance energies, proton affinities, and enthalpies of formation. We now review some of these applications.

4.1 C_2H_n

G1 and G2 theory have been systematically applied to C_2H_n hydrides by Curtiss *et al.*[13,52] The G2 bond dissociation energies of the hydrides are summarized in Table IV. G1 theory has also been used to calculate the ionization energies of most of the C_2H_n hydrides as well as appearance potentials of the cations.[52] Agreement with experiment has generally been very good.

The energy for breaking the first CH bond of acetylene has been the subject of numerous theoretical and experimental studies. For acetylene the experimental D_o(HCC-H) values have fallen in two groups,[104] one at around 126 kcal/mol and the other at 131-132 kcal/mol. The G1 value[52b] published in 1989 is 133.4 kcal/mol and supports the higher experimental values. Other theoretical work reported since then has also supported the higher values. Montgomery and Petersson[42a] reported 131.5 kcal/mol, Wu and Carter[45] reported 129.7 kcal/mol, and Bauschlicher et al[43b] reported 130.1±1.0 kcal/mol. The G2 value of 133.4 kcal/mol is the same as the G1 value; however, use of QCISD(T)/6-31G(*d*) geometries (which is important for C_2H radical) reduces the G2 value to 132.3 kcal/mol.[13] This value is in good agreement with the recent experimental values of 131.3±0.7[105] and 131.1 ± 0.7[106] kcal/mol.

The experimental values of $D_o(H_2CCH\text{-}H)$ for ethylene have fallen in the range 98-117 kcal/mol.[107] The G1 value[52a] published in 1988 is 110.1 kcal/mol. Subsequent theoretical studies have been in agreement. Wu and Carter[45] reported a value of 109.5 kcal/mol and Bauschlicher *et al.*[43c] reported a value of 110.4 ± 2 kcal/mol. The G2 value[13] of 110.4 kcal/mol is nearly the same as the G1 value. The experimental study of Berkowitz *et al.*[108] based on photoionization experiments gave a lower limit of 106.6 ± 1 kcal/mol. This value was derived from the appearance potential of $C_2H_3^+$ from C_2H_4 (13.22 ± 0.02 eV) which is well known and the ionization potential of C_2H_3 (8.59) which is not as well known. On the

TABLE IV
G2 Bond Dissociation Energies[a] (in kcal/mol) of X_2H_n (X = C, Si) Compounds at 0 K

	X = C	X = Si
X-H bonds		
$X_2H_6 \rightarrow X_2H_5 + H$	100.8	88.1
$X_2H_5 \rightarrow X_2H_4 + H$	34.1	61.9
$X_2H_4 \rightarrow X_2H_3 + H$	110.4	83.7
$X_2H_3 \rightarrow X_2H_2 + H$	34.0	45.7
$X_2H_2 \rightarrow X_2H + H$	88.3	77.7
$X_2H \rightarrow X_2 + H$	106.2	72.3
X-X bonds		
$X_2H_6 \rightarrow 2XH_3$	88.3	76.0
$X_2H_4 \rightarrow 2XH_2$	174.4	58.8
$X_2H_2 \rightarrow 2XH$	226.3	82.1
$X_2 \rightarrow 2X$	146.9	73.6
Hydrogenation Energies		
$X_2H_6 \rightarrow X_2H_4 + H_2$	30.5	45.5
$X_2H_4 \rightarrow X_2H_2 + H_2$	40.1	25.1
$X_2H_2 \rightarrow X_2 + H_2$	135.3	45.6

[a] From G2 energies in Ref. 13 and 54. Results are for lowest energy structures. Note that the G2 bond energy for C_2 is incorrect in Ref. 13. The correct value is listed here and in the text (G2 total energy is −75.80270 hartrees.

basis of G1 theory, it was suggested [52a] that the true origin of the photoionization process ($C_2H_3 \rightarrow C_2H_3^+ + e^-$) has not been observed due to a Franck-Condon effect. If so, the value of 106.6 kcal/mol would be increased by one or two quanta (1.8 kcal/mol). The theoretical $D_o(H_2CCH\text{-}H)$ values are also consistent with the value of 109.7 ± 0.8 kcal/mol reported by Ervin *et al.*[105] from negative ion photoelectron spectroscopy. The theoretical values are not consistent with Shiromaru *et al.*[107] who obtained $D_o(H_2CCH\text{-}H)$ = 116.7 kcal/mol from the threshold energy of the process $C_2H_4 + h\upsilon \rightarrow C_2H_3 + H^+ + e$ or Russell et al.[109] who obtained 106.5±0.3 kcal/mol at 298 K (105.1 kcal/mol at 0 K) from a kinetic study.

The experimental values for the dissociation energy of $C_2(^1\Sigma_g{}^+)$ have fallen in the range 141-145 kcal/mol. The JANAF tables[26] list a value of 141.0±1 kcal/mol from a spectroscopic study. However, Huber and Herzberg[27] list a value of 143.2 kcal/mol which is the average of two thermochemical values of 142.3±2.5 and 144.1 kcal/mol. A recent experimental study by Urdahl *et al.*[110] based on ro-vibrational cutoffs in the LIF spectrum of C_2 gave a $D_o(C_2)$ of 145.2 ± 0.5 kcal/mol. The G2 value for the dissociation energy is 146.9 kcal/mol.[13] Bauschlicher *et al.*[46d] reported a theoretical value of 140.9 kcal/mol, but indicated that remaining basis set saturation would increase the value. The G2 value may be too large because the higher level correction overestimates the basis set deficiency in singlet C_2 molecule relative to two triplet carbon atoms. Further study of the dissociation energy of C_2 is required, both by experiment and theory, to determine a more accurate value for its D_o.

The experimental $D_o(CC\text{-}H)$ values range from 116.3 to 112.4.[110] Bauschlicher *et al.*[43a] reported a value of 112.5 ± 2 kcal/mol using large basis sets and high levels of correlation treatment. This is in agreement with the recently measured value of 112.0 ± 0.8 kcal/mol by Urdahl *et al.*[110] from ro-vibrational cutoffs. The G2 value[13] for $D_o(CC\text{-}H)$ is considerably lower at 106.2 kcal/mol due to problems with the energies of C_2 and C_2H. This is reflected in the bond dissociation energy of C_2 and the energy for $C_2H_2 \rightarrow C_2H + H$ (see above).

Smith *et al.*[53] have examined the vinylidene-acetylene rearrangement using G2 theory. They found an energy difference between acetylene and vinylidene of 42.8 kcal/mol and a barrier of 0.6 kcal/mol. They went beyond G2 theory with a QCISD(T)/6-311++G(3*df*, 2*pd*)//QCISD/6-311G(*d*,*p*) calculation and found an energy difference of 42.3 kcal/mol and a barrier of 0.8 kcal/mol, both in good agreement with G2 theory.

4.2 Si_2H_n

G2 theory has been applied systematically to Si_2H_n hydrides and their cations by Curtiss *et al.*[54] The G2 results are in agreement with the appearance potentials of $Si_2H_n{}^+$, n = 2-5 and the adiabatic ionization potentials of Si_2H_n, n = 2, 4-6 reported recently by Ruscic and Berkowitz[111] in their photoionization study. The theoretical work supported the observation of the species Si_2H_2, Si_2H_3, Si_2H_4 and Si_2H_5 for the first time.[111] The G2 enthalpies of formation are in agreement with the values derived by Ruscic and Berkowitz[111] for Si_2H_n, n = 2-5 and $Si_2H_n{}^+$, n = 2-6. They are also in agreement with previous theoretical results[40,41] on the enthalpies of the neutrals in most cases.

G2 bond dissociation energies are listed in Table IV. The results indicate that breaking an Si-H bond is easier in the open-shell species, i.e., n = 1, 3, 5 than in the closed-shell systems, i.e., n = 2, 4, 6. This is similar to the trend for the C-H bonds listed in Table IV, although the magnitudes differ. Breaking the Si-Si bonds in Si_2H_6, Si_2H_4, and Si_2H_2 requires 76.0, 58.8, and 82.1 kcal/mol, respectively. There is no significant increase in bond strength with increase in bond order as occurs for C-C bonds. The Si-Si double bond in Si_2H_4, in fact, appears to be weaker than the Si-Si single bond in Si_2H_6. However, this is principally due to the fact that SiH_2 has a singlet ground state, whereas CH_2 has a triplet ground state. Dissociation of Si_2H_4 to give two triplet SiH_2 groups requires 105.5 kcal/mol, significantly larger than the Si-Si bond energy in Si_2H_6. The hydrogenation energies of the silicon species are all around 20-50 kcal/mol. The most significant difference with the carbon species is the much greater stability of acetylene with respect to elimination of H_2 (135.3 kcal/mol) than disilyne (45.6 kcal/mol).

4.3 B_2H_n

The B_2H_n hydrides and their cations have been studied by Curtiss and Pople[55] using G1 theory. The $B_2H_5{}^+$ cation is found to have a D_{3h} structure with three hydrogens bridging the borons.[55f] The calculated appearance potential for $B_2H_5{}^+$ from B_2H_6 is in agreement with the value measured by Ruscic et al[112a]. The calculations from G1 theory showed $B_2H_6{}^+$ to be a weak complex between $B_2H_5{}^+$ and H that is likely to exist only in a metastable state, consistent with experimental observations.[112a] The G1 dimerization energy of BH_3 is -36.2 kcal/mol at 0 K, in good agreement with recent experimental[112a] and previous theoretical results.[113]

In the same series of studies[55] the G1 appearance potential of $B_2H_4{}^+$ from B_2H_6 was found to be in disagreement with experiment[112a] by about 0.5 eV. It was suggested that the experiment might be detecting a triply-bridged $B_2H_4{}^+$ structure which G1 theory predicted to be 0.49 eV less stable than the doubly-bridged equilibrium structure.[55d] In subsequent work, Ruscic et al[112b] observed B_2H_4 and a doubly-bridged $B_2H_4{}^+$ cation and confirmed the G1 predictions. The experimental ionization potential of B_2H_4 (to the doubly-bridged $B_2H_4{}^+$ cation) is in good agreement with G1 theory.[55e] The $B_2H_3{}^+$, $B_2H_2{}^+$, and B_2H^+ cations were also studied at the G1 level.[55c] They do not contain hydrogen bridges as in the $B_2H_6{}^+$, $B_2H_5{}^+$, and B_2H^+ cations. Agreement was found with photoionization studies of these cations.[112]

In related work, G1 theory was used to study BH_n molecules and their cations.[55a] This work found a D_e value for BH of 82.8 kcal/mol in disagreement with the spectroscopic value[27] of 78.9 kcal/mol. In subsequent calculations[114] at the QCISD(T) level with a very large basis set, the D_o was found to be 81.5±1 kcal/mol. The experimental value was obtained from a predissociation limit in the excitation spectrum to the lowest Π state with a correction for the repulsive hump in the Π potential curve. It was suggested that the observed dissociation limit of 82.6 kcal/mol might be close to the true value of D_o because the tunneling effect had not been observed.[114] This was confirmed by further high level theoretical calculations by Bauschlicher *et al.*[115]

4.4 COH_n

G2 theory was used to study the energies of CH_2OH, CH_3O and related compounds (CH_3OH, CH_3O^+, CH_2OH^+, H_2CO) by Curtiss *et al.*[56] The G2 result for the adiabatic ionization potential of CH_3O (10.78 eV) provided support for the value of 10.726 ± 0.008 eV reported at the same time by Ruscic and Berkowitz[116a] from a photoionization study. It was concluded that a previous photoelectron result[117] of 7.37±0.03 eV was incorrect.[56,116a] The G2 values for the O-H and C-H bond dissociation energies of CH_3OH are 105.0 and 96.2 kcal/mol, respectively. It was suggested that the D_o(H-CH_2OH) of 92.6 kcal/mol from kinetic measurements[118] may be too low. A theoretical study by Bauschlicher *et al.*[43e] gave O-H and C-H bond dissociation energies of 105.0±2 and 96.2±2 kcal/mol, respectively, in agreement with the G2 results. Espinosa-Garcia and Olivares del Valle[51] reported an enthalpy of formation for CH_2OH radical based on MP4 calculations (with post MP4 corrections). Their enthalpy value corresponds to a D_o(H-CH_2OH) of 94.8±0.4 kcal/mol, although these error limits are probably too optimistic. Ruscic and Berkowitz[116b] reported a more accurate measurement of the appearance potential of CH_2OH^+ from CH_3OH, from which they derived a value for D_o(H-CH_2OH) of 94.6±0.1 kcal/mol, which falls within the error limits of all three theoretical values.

Ma *et al.*[57] have reported a detailed study of the rearrangement and fragmentation reactions of methanol cation (CH_3OH^+) using G1 and G2 theory. They found that in most cases agreement between theory and experiment is very good. However, they found a discrepancy of 0.12 eV for the ionization of methanol. In subsequent work[36] they found that this was reduced to 0.04 eV by avoiding the additivity approximations and by employing QCISD(T) geometries and MP2/6-31G(*d*) zero-point energies.

4.5 CSH_n

The organosulfur systems CSH_n (n=0-4) and $CSH_n{}^+$ (n=0-5) have been studied using G2 theory.[58,59] Dissociation energies, ionization energies and enthalpies of formation were reported. The calculated enthalpy of formation of CH_3S at 0 K (31.6 kcal/mol) is in agreement with the value obtained from a recent kinetic study[119] (31.44 kcal/mol), but in disagreement with values from recent photofragment studies[120] (34.2-35.5 kcal/mol). G2 theory predicts the CH_2SH^+ cation to lie 33.1 kcal/mol lower in energy than the CH_3S^+ cation, close to the value of 33.5±2 kcal/mol deduced by Ruscic and Berkowitz,[121] from experiment, but significantly less than the value of 41-43 kcal/mol obtained from experiment by Nourbaksh *et al.*[120] Finally, G2 theory energies were used to assess enthalpies of formation for organosulfur compounds in the thermochemical compilation of Lias *et al.*[122] It was suggested that three of the values (for CH_3S^+, $CH_2SH_2^+$, and CH_2SH^+) may have to be revised.

Chiu *et al.*[60] used G2 theory to study the energies of CH_2SH, CH_2S^-, CH_3S^-, CH_2SH^-, CH_3SH^-, CH_3^+, and CH_3SH^+. They found good agreement with experiment for the ionization energies and electron affinities of CH_3S and CH_2SH. Ma *et al.*[61] have reported G2 predictions for the enthalpies of formation of various isomers of C_2H_5S and $C_2H_5S^+$. The results were in agreement with their measured values from photoionization and photodissociation studies.

4.6 Proton Affinities and Gas Phase Acidities

G2 theory was tested on seven proton affinities by Curtiss *et al.*[13] and found to be in good agreement with experiment (average absolute deviation of 1.04 kcal/mol). Smith and Radom[62] evaluated G2 theory in the calculation of gas phase acidities of 23 molecules. They found it to perform very well with a mean error of 1.24 kcal/mol and consistently within 2 kcal/mol of reliable experimental data. They found in a small number of cases that there is a larger discrepancy between experiment and theory and suggested that in these cases a reexamination of the experimental data may be warranted.

Curtiss *et al.*[63] used G1 and G2 theory to predict accurate proton affinities of H_3SiO^-, H_3SiOH, $H_3SiOSiH_3$ and $H_3SiOAlH_3$, species that are used in modeling of zeolite frameworks. Maclagan[64] used G2 theory to predict the proton affinities of NCO and CNO. Ekern *et al.*[65] used G1 and G2 theory to calculate the proton affinity of N_2O in a study of isomers of protonated N_2O. The site of protonation is predicted to be the oxygen.

Smith and Radom[66] have recently used G2 theory to obtain proton affinities for 31 small molecules. The theoretical values are in good agreement for a range of bases with proton affinities spanning some 120 kcal/mol. They found a small number of discrepant cases which could be reconciled if the currently accepted proton affinity of isobutene was lowered by 2-5 kcal/mol. On the basis of the G2 results Smith and Radom[66] suggested that the current experimental $\Delta H^\circ_{f\ 298}$ for t-butyl cation may be in error, which could explain the discrepancy between theory and experiment on the proton affinity of isobutene.

4.7 Nitrogen Containing Compounds

G2 theory has been used to study the energies of N_2H_n (n=0-4) molecules and their cations by Pople and Curtiss.[67] Overall, good agreement was found between theory and experiment. The G2 results support recent experimental results[123] for the enthalpy of formation of N_2H_2 and the ionization potential of N_2H_3. Smith[68] used G2 theory to calculate the relative energies of five stationary points on the N_2H_2 potential surface including two transition states. The results are in excellent agreement with density functional methods that included nonlocal corrections.[68]

Schlegel and Skancke[69] used G2 theory to calculate thermodynamic data for hydrazine [H_2NNH_2], triazane [$HN(NH_2)_2$] and triaminoammonia [$N(NH_2)_3$] and obtained NN bond energies for this series.

Carbon-nitrogen compounds have been studied by several groups using G2 theory. Armstrong *et al.*[70] calculated G2 enthalpies of formation for $H_2NCH_2^+$, H_2NCH_2, H_2NCH_3, $H_2NCH_3^+$, and $H_3NCH_2^+$. The results are in good agreement with experiment where comparisons were possible. Cacace *et al.*[71] calculated G1 energies for cyanamide (H_2NCN) and two related ions H_2NCNH^+ and $HNCN^-$. The proton affinities of H_2NCN and $HNCN^-$ are in agreement with values they measured using mass spectrometry. Smith et al.[72] used G2 theory to predict a new value for the enthalpy of formation of formaldimine ($CH_2{=}NH$). Smith and Radom[92] used G2 theory and experimental enthalpies of formation for reference molecules to calculate enthalpies of formation of ethynol ($HC{\equiv}COH$), vinyl alcohol ($CH_2{=}CHOH$), ketene ($CH_2{=}C{=}O$), and acetaldehyde ($CH_3CH{=}O$) from bond separation energies. Very good agreement was found with experiment for these four systems. They used G2 theory in the same way to predict enthalpies of formation of the nitrogen analogs $HC{\equiv}CNH_2$, $CH_2{=}CHNH_2$, $CH_2{=}C{=}NH$, and $CH_3CH{=}NH$.

Grandinetti *et al.*[73] used G1 theory to study several nitrogen fluoride ions including NF_2H^+, NH_2F^+, F_3NH^+, and F_2NFH^+. The results were used to help interpret mass spectrometric experiments on these cations. Wiberg[74] has used G1 theory to calculate bond dissociation energies for H_2NX (X = Me, H_2N, F) and compared them to bond dissociation energies of MeX, HOX, and FX derivatives.

4.8 Silicon Containing Compounds

The silicon-oxygen systems $SiOH_n$ (n=0-4), SiO_2H_n (n=0-4), and $SiOH_n^+$ (n=0-5) have been studied using G2 theory by several researchers.[75–78] There is little experimental thermochemical data for these systems. The G2 results provide accurate estimates of the enthalpies of formation, ionization energies, and bond energies of these species. Ding and Marshall[83] used G2 theory to calculate the barrier for the reaction $O + SiH_4 \rightarrow OH + SiH_3$. They found that the G2 barrier lies within 1.5 kcal/mol of the experimental value measured using a flash photolysis resonance-fluorescence technique.

Su and Schlegel[80] used G2 theory in a study of the reactants, transition structures, and products for various channels for the thermal decomposition of $SiH_{4-n}Cl_n$ (n=0-4). Luna and Yanez[78] used G2 theory to study the potential energy surface for the reaction of $Si^+(^2P)$ with methanol. Calculated enthalpies of formation of the possible products of this reaction are in close agreement with experiment, where available. An exception is $SiCH_2^+$ which differs by about 30 kcal/mol. This is much more than expected and should be further investigated. Goldberg *et al.*[81] used G2 theory to study $SiNH_n$ and $SiNH_n^+$ (n=0-2) and several transition states. They found $SiNH_2^+$ to be the most stable isomer on the $[H_2SiN]^+$ potential energy surface, in keeping with their mass spectrometric study. Boatz and Gordon[79] have predicted enthalpies of formation of silaethylene, disilene, and their silylene isomers using G1 theory.

Michels and Hobbs[82] used G2 theory to study the thermochemistry of SiH_n^- and SiF_n^- anions and compared the results to calculations based on isodesmic reactions and experiment. Significant differences (≈10 kcal/mol) were found for SiF and SiF_2, suggesting that new experimental investigations are warranted. The G2 atomization energy of SiF_4 is too low by 7.1 kcal/mol. Michels and Hobbs also investigated CF_4 and found its G2 atomization energy too high by 5.6 kcal/mol. These latter two cases represent failures of G2 theory on the order of the failure for SO_2 (see section 3.6), although in the opposite direction. The reason for the failure of G2 theory for SiF_4 and CF_4 needs further investigation.

4.9 Clusters

G2 theory has been used to calculate accurate electron affinities and binding energies of silicon clusters containing up to five silicon atoms.[84] This work is discussed in more detail in another chapter of this book. In other work on clusters, Martin *et al.*[85] have used a modified G1 theory to calculate total atomization energies of B_2, B_3, and B_4 clusters and several carbon clusters (C_4, C_5) for which they proposed revised enthalpies of formation.

4.10 Ring Compounds

G2 theory has been used in the calculation of a number of ring compounds.[86–88] The largest study of ring compounds has been carried out by Seasholtz *et al.*[87] They used G2 theory to calculate the enthalpies of formation of a series of imines (compounds containing a carbon-nitrogen double bond) including the ring structures 1,4,5,6-tetrahydropyrimidine (THP), pyroline, and oxazoline. The enthalpy of oxazoline is in agreement with experiment while that of THP differed by about 6 kcal/mol. In the latter case the difference between theory and experiment was not reconciled. Seasholtz *et al.*[87] also used G2 theory to calculate enthalpies of formation for a series of chlorinated olefins and obtained good agreement with experiment.

4.11 Transition States

G2 theory has been used for calculation of some transition states.[57,80,81,83,103] However, in the original formulation of G2 theory[13] no transition states were tested and its use is questionable in cases where there is a change in the number of electron pairs because of the way in which the higher level correction is determined. Durant and Rohlfing[95] recently tested out G2 theory on a number of small open-shell systems which have been well characterized in other theoretical calculations. They found that G2 theory performed well in predicting energies of transition states including some nonisogyric reactions with an average absolute deviation of 1.5 kcal/mol in the classical barrier heights. They found improved results when QCISD/6-311G(d,p) geometries and frequencies were used.

4.12 Electron Affinities

Yu *et al.*[100] have used G2 theory to calculate the electron affinities of HCO_2 and the triatomic species N_3, OCN, SCN, NO_2, O_3, SO_2, ClO_2, and N_2O. They found that the G2 electron affinities for N_3, NO_2, O_3 and SO_2 were in agreement with accurate experimental measurements with an average error of 0.06 eV. The G2 results were used to provide evidence for preferred values for the other species where the experimental data were less accurate or less extensive.

4.13 Other

Methyl hypofluorite (CH_3OF) has been the subject of a study using G2 theory.[90] Its calculated enthalpy of formation is in agreement with a recent experimental determination[124] for this recently synthesized compound. The G2 energies for fragmentation processes of CH_3OF are in good agreement with the experimental results from photoionization data. The energy of hypofluorous acid, HOF, has also been the subject of studies using G1 and G2 theory.[90,91] This work has helped lead to a revised experimental value for the O-F bond dissociation energy of HOF.[124] Hrusak and Schwarz[103] used G1 theory to explore the potential energy surface of $CHOF^+$. Five isomers are predicted with the most stable being trans $FCOH^+$. Three of the isomers are separated by large enough barriers to allow them to be observed experimentally. The calculated thermochemistry of $CHOF^+$ is in good agreement with the available experiment.

Esseffar *et al.*[93a] used G2 theory to calculate thermochemical data for PNH_n (n=0-2) and PNH_n^+ (n=0-3). They reported relative stabilities, ionization potentials, enthalpies of formation and proton affinities. There is almost no experimental information on these quantities. Esseffar *et al.*[93b] have also reported enthalpies of formation of POH and HPO (singlet and triplet states), along with their ionization potentials.

Ma *et al.*[98] used G2 theory to study van der Waals dimers R_1R_2 and their cations $R_1R_2^+$ (R_1R_2 = He, Ne, Ar, Kr). They found good agreement between theory and the available experiment on ionization energies of the neutral dimers and the dissociation energies of the cations. Since the dissociation energies of the neutral dimers are less than 2 kcal/mol, G2 theory is not expected to give an accurate account of these quantities. McKee[96] studied three adducts of atomic chlorine with CS_2 using G1 theory. The largest binding energy was found for a symmetrical complex with Cl attached to C. The binding energy is in agreement with experiment.

5 MODIFICATIONS TO G2 THEORY

5.1 G2(MP2)

A variant of G2 theory, referred to as G2(MP2) theory, using reduced Møller-Plesset perturbation theory has been proposed[125] to handle larger systems than are possible with G2 theory. In G2(MP2) theory the basis set extension corrections in Eqns. (2), (3), and (11) are replaced by a single correction obtained at the MP2 level:

$$\Delta_{\mathrm{MP2}} = \mathrm{E}[\mathrm{MP2/6\text{-}311{+}G}(3df,2p)] - \mathrm{E}[\mathrm{MP2/6\text{-}311G}(d,p)]. \tag{14}$$

The total G2(MP2) energy is thus given by

$$\mathrm{E}_o = \mathrm{E}[\mathrm{QCISD(T)/6\text{-}311G}(d,p)] + \Delta_{\mathrm{MP2}} + \Delta\mathrm{E(HLC)} + \Delta\mathrm{(ZPE)}. \tag{15}$$

The value of A (4.81 millihartrees) used for ΔE(HLC), [Eqn. (7)] in G2(MP2) theory was chosen to give the smallest average absolute deviation from experiment for the entire set of 125 energies. This value turns out to be the same as that used for ΔE(HLC) in G2 theory. The calculation of the G2(MP2) energy requires only two calculations, QCISD(T)/6-311G(d,p) and MP2/6-311+G($3df,2p$). G2(MP2) theory eliminates the MP4/6-311G($2df,p$) calculation which is the largest calculation in G2 theory in terms of disk storage. Use of direct calculations makes the MP2/6-311+G($3df,2p$) calculation in Eqn. (14) feasible. In G2(MP2) theory the limiting step is now the QCISD(T)/6-311G(d,p) calculation.

The average absolute deviation of the G2(MP2) energies from experiment of the set of 125 energies in Tables I and II is 1.58 kcal/mol compared to the corresponding G2 value of 1.21 kcal/mol. The average absolute deviation of the G2(MP2) energies from the G2 energies is 0.74 kcal/mol. G2(MP2) theory works best for dissociation energies and proton affinities. Of the 55 dissociation energies in Table II the average absolute deviation from experiment is 1.31 kcal/mol compared to 1.16 kcal/mol for G2 theory. For the most part the molecules which deviate by more than 2 kcal/mol from experiment are similar in G2 and G2(MP2) theories. G2(MP2) theory does not do as well on ionization energies and electron affinities as G2 theory. This is largely due to the results for the atoms where the higher levels of correlation are apparently important.

The absence of the MP4/6-311G($2df,p$) calculation in G2(MP2) theory provides significant savings in computational time and disk storage such that larger systems can be handled than in G2 theory. There is some loss in accuracy caused by using the reduced order of Møller-Plesset theory. Further testing of G2(MP2) theory on larger systems is required to better assess its reliability. A modification of G2 theory based on MP3 energies, G2(MP3), was also presented in Ref. 125. It had an average absolute deviation from experiment of 1.52 kcal/mol (compared to 1.21 kcal/mol for G2 theory). It was concluded that G2(MP3) had little advantage over G2(MP2), so it was not recommended.

G2(MP2) and G2 thermochemical data have been compared in a number of other studies on N_2H_2 and some nitrogen containing ring compounds.[68,87] The agreement between the two methods in these studies is good. In addition, G2(MP2) theory has been used to calculate dissociation energies for $HAlF_4$, $LiAlF_4$, and $NaAlF_4$.[94] Theory is in good agreement with experiment for $LiAlF_4$ and in poor agreement for $NaAlF_4$. The reason for the poor agreement in the latter case is unclear.

5.2 G2(QCI)

G2 theory based on QCISD(T)/6-311+G(3*df*, 2*p*) calculations was described in Section 3.5. In this modified G2 theory, referred to as G2(QCI),[28] the energy is evaluated via Eqn. (13). The G2(QCI) energies for the 125 systems have been given in Ref. 28. The average absolute deviation of 1.17 kcal/mol is about the same as that for G2 theory (1.21 kcal/mol). In view of the relatively modest increase in cost, calculation of G2(QCI) energies instead of G2 energies maybe a viable alternative in some studies, although mixing of the two energies is not recommended.

6 CONCLUDING REMARKS

G2 theory is a general predictive procedure based on *ab initio* molecular orbital theory for calculation of energies of molecular systems containing first- and second-row atoms which has been tested on a large number of small molecules, most containing one or two non-hydrogen atoms. Since its publication G2 theory has been applied to numerous other species, some of which have been discussed in Section 4. Generally G2 theory (and its predecessor G1 theory) have proved quite accurate in calculations of appearance potentials, ionization energies, electron affinities, and proton affinities. G2 theory is one of several theoretical methods being used for interpretation of experimental data and for providing accurate predictions of thermochemical data where no reliable experimental values are available. G2(MP2) theory,[125] which uses reduced Møller-Plesset perturbation orders, should also find uses in calculation of energies in larger systems, i.e. ones that cannot be handled by G2 theory.

There are some species such as SO_2, SiF_4, and CF_4 for which G2 theory does poorly in reproducing known experimental energies (see Sections 3.6 and 4.8). Testing of G2 theory on larger systems (more than two nonhydrogen atoms) having accurate thermochemical data is clearly an important area for future research and will hopefully lead to improved versions of G2 theory.

G2 theory is being extended to compounds containing third-row nontransition metal atoms (K, Ca, Ga-Br). McGrath and Radom[126] have already extended it to compounds containing Br and found good agreement with experiment on such molecules as Br_2 and CH_3Br. It is currently being extended to include the other third-row nontransition atoms.[127] Spin-orbit effects become important in these compounds and must be taken into account.[11,12,126] A better method for calculation of the higher level correction is desirable and work is in progress[128] on this aspect based the complete basis set (CBS) method of Petersson *et al.*[129] The use of better geometries and zero-point energies is another area where systematic studies of G2 theory would be fruitful.

Finally, density functional methods[130,131,132] have recently been tested on the first- and second-row compounds used to evaluate G2 theory. While not as accurate as G2 theory, these methods look promising for larger systems because of their computational efficiency.

7 ACKNOWLEDGEMENTS

The authors are indebted to Professor John A. Pople for many enlightening discussions through the years. This work was supported by the U.S. Department of Energy, Office of Basic Energy Sciences, Division of Materials Sciences, under Contract No. W-31-109-ENG-38.

REFERENCES

1. C. W. Bauschlicher and S. R. Langhoff, Science **254**, 394 (1991).
2. R. Ditchfield, W. J. Hehre, J. A. Pople, and L. Radom, Chem. Phys. Lett. **5**, 13 (1970).
3. W. J. Hehre, R. Ditchfield, L. Radom, and J. A. Pople, J. Am. Chem. Soc. **92**, 4796 (1970).
4. P. George, M. Trachtman, C. W. Bock, and A. M. Brett, Theor. Chim. Acta **38**, 121 (1975).
5. P. George, M. Trachtman, C. W. Bock, and A. M. Brett, Tetrahedron **32**, 317 (1976).
6. J. A. Pople, M. J. Frisch, B. T. Luke, and J. S. Binkley, Int. J. Quant. Chem. **S17**, 307 (1983).
7. J. A. Pople, M. J. Frisch, B. T. Luke, and J. S. Binkley, J. Phys. Chem. **89**, 2198 (1985).
8. J. A. Pople and L. A. Curtiss, J. Phys. Chem. **91**, 155 (1987).
9. J. A. Pople and L. A. Curtiss, J. Phys. Chem. **91**, 3637 (1987).
10. L. A. Curtiss and J. A. Pople, J. Phys. Chem. **92**, 894 (1988).
11. R. C. Binning, Jr. and L. A. Curtiss, J. Chem. Phys. **92**, 1860 (1990).
12. R. C. Binning, Jr. and L. A. Curtiss, J. Chem. Phys. **92**, 3688 (1990).
13. L. A. Curtiss, K. Raghavachari, G. W. Trucks, and J. A. Pople, J. Chem. Phys. **94**, 7221 (1991).
14. (a) J. A. Pople, M. Head-Gordon, D. J. Fox, K. Raghavachari, and L. A. Curtiss, J. Chem. Phys. **90**, 5622 (1989); (b) L. A. Curtiss, C. Jones, G. W. Trucks, K. Raghavachari, and J. A. Pople, J. Chem. Phys. **93**, 2537 (1990).
15. (a) P. C. Hariharan and J. A. Pople, Theor. Chim. Acta **28**, 213 (1973); (b) M. M. Francl, W. J. Pietro, W. J. Hehre, J. S. Gordon, D. J. Defrees, and J. A. Pople, J. Chem. Phys. **77**, 3654 (1982).
16. The notation 6-31G(d) is synonymous with 6-31G*. Similarly 6-311G**, 6-311+G** and 6-311G**($2df$) are synonymous with 6-311G(d,p), 6-311+G(d,p), and 6-311G($2df,p$) respectively.
17. W. J. Hehre, L. Radom, J. A. Pople, and P. von R. Schleyer, *Ab Initio* Molecular Orbital Theory" (John Wiley, New York, 1987).
18. J. A. Pople, H. B. Schlegel, R. Krishnan, D. J. Defrees, J. S. Binkley, M. J. Frisch, R. A. Whiteside, R. F. Hout, and W. J. Hehre, Int. J. Quantum Chem. **S15**, 269 (1981).
19. R. Krishnan, M. J. Frisch, and J. A. Pople, J. Chem. Phys. **72**, 4244 (1980).
20. R. Krishnan, J. S. Binkley, R. Seeger, and J. A. Pople, J. Chem. Phys. **72**, 650 (1980).
21. T. Clark, J. Chandrashekar, G. W. Spitznagel, and P. von R. Schleyer, J. Comput. Chem. **4**, 294 (1983).
22. M. F. Frisch, J. A. Pople, and J. S. Binkley, J. Chem. Phys. **80**, 3265 (1984).
23. J. A. Pople, M. Head-Gordon, and K. Raghavachari, J. Chem. Phys. **87**, 5968 (1987).

24. M. J. Frisch, M. Head-Gordon, P. M. W. Gill, G. W. Trucks, J. B. Foresman, H. B. Schlegel, K. Raghavachari, M. A. Robb, J. S. Binkley, C. Gonzalez, D. J. Defrees, D. J. Fox, R. A. Whiteside R. Seeger, C. F. Melius, J. Baker, R. L. Martin, L. R. Kahn, J. J. P. Stewart, S. Topiol, and J. A. Pople, Gaussian, Inc., Pittsburgh PA 1992.
25. K. Raghavachari and L. A. Curtiss, in "Modern Electronic Structure Theory," edited by D. Yarkony, in press.
26. M. W. Chase, Jr., C. A. Davies, J. R. Downey, Jr., D. J. Frurip, R. A. McDonald, and A. N. Syverud, J. Phys. Chem. Ref. Data **14**, Suppl. No. 1 (1985).
27. H. Huber and G. Herzberg, "Molecular Spectra and Molecular Structure 4. Constants of Diatomic Molecules" (van Nostrand, Princeton 1979).
28. L. A. Curtiss, J. E. Carpenter, K. Raghavachari, and J. A. Pople, J. Chem. Phys. **96**, 9030 (1992).
29. L. A. Curtiss, K. Raghavachari, and J. A. Pople, Chem. Phys. Lett. **214**, 183 (1993).
30. K. Raghavachari, J. Chem. Phys. **82**, 4607 (1985).
31. R. H. Nobes, J. A. Pople, L. Radom, N. C. Handy, and P. J. Knowles, Chem. Phys. Lett. **138**, 481 (1987).
32. J. Berkowitz, W. A. Chupka, and T. A. Walter, J. Chem. Phys. **50**, 1497 (1969).
33. S. E. Bradforth, E. H. Kim, D. W. Arnold, and D. M. Neumark, J. Chem. Phys. **98**, 800 (1993).
34. L. A. Curtiss, unpublished work.
35. M. L. McKee, J. Chem. Phys. **97**, 10971 (1993).
36. N. L. Ma, B. J. Smith, and L. Radom, J. Phys. Chem. **96**, 5804 (1992).
37. R. S. Grev, C. L. Janssen, and H. F. Schaefer III, J. Chem. Phys. **95**, 5128 (1991).
38. J. M. L. Martin, J. P. Francois, and R. Gijbels, Chem. Phys. Lett. **163**, 387 (1989).
39. M. Sana and G. Leroy, J. Mol. Struct. (Theochem) **72**, 309 (1991).
40. (a) P. Ho and C. F. Melius, J. Phys. Chem. **94**, 5120 (1990); (b) P. Ho, M. E. Coltrin, J. S. Binkley, and C. F. Melius, J. Phys. Chem. **90**, 3399 (1986).
41. A. F. Sax and J. Kalcher, J. Phys. Chem. **95**, 1768 (1991).
42. (a) J. A. Montgomery and G. A. Petersson, Chem. Phys. Lett. **168**, 75 (1990); (b) J. A. Montgomery, G. A. Petersson, M. A. Al-Latham, and J. Mantzaris, Chem. Phys. Lett. **169**, 497 (1990).
43. (a) C. W. Bauschlicher and S. R. Langhoff, Chem. Phys. Lett. **173**, 367 (1990); (b) C. W. Bauschlicher, S. R. Langhoff, and P. R. Taylor, Chem. Phys. Lett. **171**, 42 (1990); (c) C. W. Bauschlicher and S. R. Langhoff, Chem. Phys. Lett. **177**, 133 (1991); (d) S. R. Langhoff, C. W. Bauschlicher, and P. R. Taylor, Chem. Phys. Lett. **180**, 88 (1991); (e) C. W. Bauschlicher, S. R. Langhoff, and S. P. Walch, J. Chem. Phys. **96**, 450 (1992).

44. R. S. Grev and H. F. Schaefer III, J. Chem. Phys. **97**, 8389 (1992).
45. C. J. Wu and E. A. Carter, J. Am. Chem. Soc. **112**, 5893 (1990).
46. (a) C. W. Bauschlicher, B. H. Lengsfield, and B. Liu, J. Chem. Phys. **77**, 4084 (1982); (b) S. R. Langhoff, C. W. Bauschlicher, and H. Partridge, J. Chem. Phys. **84**, 1687 (1986); (c) C. W. Bauschlicher and S. R. Langhoff, J. Chem. Phys. **87**, 2919 (1987); (d) C. W. Bauschlicher, S. R. Langhoff, and P. R. Taylor, Astrophys. J. **332**, 531 (1988).
47. E. W. Ignacio and H. B. Schlegel, J. Phys. Chem. **96**, 5830 (1992).
48. J. S. Francisco and S. P. Sander, J. Chem. Phys. **99**, 6219 (1993).
49. (a) A. L. L. East and W. D. Allen, J. Chem. Phys. **99**, 4638 (1993); (b) A. L. L. East, C. S. Johnson, and W. D. Allen, J. Chem. Phys. **98**, 1299 (1993).
50. M. W. Wong and L. Radom, J. Phys. Chem. **94**, 638 (1990).
51. J. Espinosa-Garcia, F. J. Olivares del Valle, J. Phys. Chem. **97**, 3377 (1993).
52. (a) L. A. Curtiss and J. A. Pople, J. Chem. Phys. **88**, 7405 (1988); (b) L. A. Curtiss and J. A. Pople, J. Chem. Phys. **91**, 2420 (1989); (c) B. Ruscic, J. Berkowitz, L. A. Curtiss, and J. A. Pople, J. Chem. Phys. **91**, 114 (1989).
53. B. J. Smith, R. Smernik, and L. Radom, Chem. Phys. Lett. **188**, 589 (1992).
54. L. A. Curtiss, K. Raghavachari, P. W. Deutsch, and J. A. Pople, J. Chem. Phys. **95**, 2433 (1991).
55. (a) L. A. Curtiss and J. A. Pople, J. Chem. Phys. **89**, 614 (1988); (b) L. A. Curtiss and J. A. Pople, J. Chem Phys. **89**, 4875 (1988); (c) L. A. Curtiss and J. A. Pople, J. Chem. Phys. **91**, 4809 (1989); (d) L. A. Curtiss and J. A. Pople, J. Chem. Phys. **90**, 4314 (1989); (e) L. A. Curtiss and J. A. Pople, J. Chem. Phys. **91**, 5118 (1989); (f) L. A. Curtiss and J. A. Pople, J. Chem. Phys. **91**, 4189 (1989).
56. L. A. Curtiss, L. D. Kock, and J. A. Pople, J. Chem. Phys. **95**, 4040 (1991).
57. N. L. Ma, B. J. Smith, J. A. Pople, and L. Radom, J. Am. Chem. Soc. **113**, 7903 (1991).
58. L. A. Curtiss, R. H. Nobes, J. A. Pople, and L. Radom, J. Chem. Phys. **97**, 6766 (1992).
59. R. H. Nobes and L. Radom, Chem. Phys. Lett. **189**, 554 (1992).
60. S. W. Chiu, W. K. Li, W. B. Tzeng, and C. Y. Ng, J. Chem. Phys. **97**, 6557 (1992).
61. Z.-X. Ma, C.-L. Liao, H.-M. Yin, C. Y. Ng, S.-W. Chiu, N. L. Ma, and W.-K. Li, Chem. Phys. Lett. **213**, 250 (1993).
62. B. J. Smith and L. Radom, J. Phys. Chem. **95**, 10549 (1991).
63. L. A. Curtiss, H. Brand, J. B. Nicholas, and L. E. Iton, Chem. Phys. Lett. **184**, 215 (1991).
64. R. Maclagan, J. Phys. Chem. **97**, 4614 (1993).
65. S. P. Ekern, A. J. Illies, and M. L. McKee, Mol. Phys. **78**, 263 (1993).
66. B. J. Smith and L. Radom, J. Am. Chem. Soc. **115**, 4885 (1993).
67. J. A. Pople and L. A. Curtiss, J. Chem. Phys. **95**, 4385 (1991).
68. B. J. Smith, J. Phys. Chem. **97**, 10513 (1993).

69. H. B. Schlegel and A. Skancke, J. Am. Chem. Soc. **115**, 7465 (1993).
70. D. A. Armstrong, A. Rauk, and D. K. Yu, J. Am. Chem. Soc. **115**, 666 (1993).
71. F. Cacace, G. Petris, F. Grandinetti, and G. Occhiucci, J. Phys. Chem. **97**, 4239 (1993).
72. B. J. Smith, J. A. Pople, L. A. Curtiss, and L. Radom, Aust. J. Chem. **45**, 285 (1992).
73. (a) F. Grandinetti, J. Hrusak, and H. Schwarz, Chem. Phys. Lett. **204**, 53 (1993); (b) F. Grandinetti, J. Hrusak, D. Schroder, S. Karrass, and H. Schwarz, J. Am. Chem. Soc. **114**, 2806 (1992).
74. K. Wiberg, J. Phys. Chem. **96**, 5800 (1992).
75. D. J. Lucas, L. A. Curtiss, and J. A. Pople, J. Chem. Phys. **99**, 6697 (1993).
76. J. R. Flores, J. Phys. Chem. **96**, 4414 (1992).
77. C. L. Darling and H. B. Schlegel, J. Phys. Chem. **97**, 8207 (1993).
78. A. Luna and M. Yanez, J. Phys. Chem. **97**, 10659 (1993).
79. J. A. Boatz and M. S. Gordon, J. Chem. Phys. **94**, 7331 (1990).
80. M.-D. Su and H. B. Schlegel, J. Phys. Chem. **97**, 9981 (1993).
81. N. Goldberg, J. Hrusak, M. Iraqi, and H. Schwarz, J. Phys. Chem. **97**, 10687 (1993).
82. H. H. Michels and R. H. Hobbs, Chem. Phys. Lett. **207**, 389 (1993).
83. L. Y. Ding and P. Marshall, J. Chem. Phys. **98**, 8545 (1993).
84. L. A. Curtiss, P. W. Deutsch, and K. Raghavachari, J. Chem. Phys. **96**, 6868 (1992).
85. (a) J. M. L. Martin, J. P. Francois, and R. Gijbels, Chem. Phys. Lett. **193**, 243 (1992); (b) J. M. L. Martin, J. P. Francois, and R. Gijbels, Chem. Phys. Lett. **189**, 529 (1992); (c) J. M. L. Martin, J. P. Francois, and R. Gijbels, J. Chem. Phys. **94**, 3753 (1991); (d) J. M. L. Martin, J. P. Francois, and R. Gijbels, J. Chem. Phys. **95**, 9420 (1991).
86. M. W. Wong, R. Leuntoung, and C. Wentrup, J. Am. Chem. Soc. **115**, 2465 (1993).
87. M. B. Seasholtz, N. G. Rondan, J. M. Ruiz, B. T. Colegrove, and T. B. Thompson, manuscript in preparation.
88. M. W. Wong and L. Radom, J. Am. Chem. Soc. **115**, 1507 (1993).
89. K. Raghavachari, G. W. Trucks, J. A. Pople, and E. Replogle, Chem. Phys. Lett. **158**, 207 (1989).
90. L. A. Curtiss and J. A. Pople, J. Chem. Phys. **95**, 7962 (1991).
91. J. A. Pople and L. A. Curtiss, J. Chem. Phys. **90**, 2833 (1989).
92. B. J. Smith and L. Radom, J. Am. Chem. Soc. **114**, 36 (1992).
93. (a) M. Esseffar, A. Luna, O. Mo, and M. Yanez, J. Phys. Chem. **97**, 6607 (1993); (b) M. Esseffar, A. Luna, O. Mo, and M. Yanez, Chem. Phys. Lett. **209**, 557 (1993).
94. (a) L. A. Curtiss and G. Scholz, Chem. Phys. Lett. **205**, 550 (1993); (b) L. A. Curtiss, Proceedings of the International Symposium on Molten Salt

and Technology, Edited by M. L. Saboungi and H. Kojima, (Electrochemical Society, Pennington, N. J.) p. 31 1993.
95. J. Durant and C. Rohlfing, J. Chem. Phys. **98**, 8031 (1993).
96. M. L. McKee, Chem. Phys. Lett. **209**, 195 (1993).
97. N. Nartz, G. Rasul, and G. Olah, J. Am. Chem. Soc. **115**, 1277 (1993).
98. N. L. Ma, W. K. Li, and C. Y. Ng, J. Chem. Phys. **99**, 3617 (1993).
99. A. Rauk, E. Tschuikow-Roux, Y. Chen, M. P. McGrath and L. Radom, J. Phys. Chem. **97**, 7947 (1993).
100. D. Yu, A. Rauk, and D. A. Armstrong, J. Chem. Phys. **96**, 6031 (1992).
101. J. M. Seminario and P. Politzer, Int. J. Quant. Chem. **S26**, 497 (1992).
102. A. Luna, O. Mo, and M. Yanez, Chem. Phys. Lett. **197**, 581 (1992).
103. J. Hrusak and H. Schwarz, J. Phys. Chem. **97**, 4659 (1993).
104. P. G. Green, J. L. Kinsey, and R. W. Field, J. Chem. Phys. **91**, 5160 (1990).
105. K. M. Ervin, S. Gronert, S. E. Barlow, M. K. Gilles, A. G. Harrison, V. W. Bierbaum, C. H. Depuy, W. C. Lineberger, and G. B. Ellison, J. Am. Chem. Soc. **112**, 5750 (1990).
106. B. Ruscic and J. Berkowitz, J. Chem. Phys. **93**, 5586 (1990).
107. H. Shiromaru, Y. Achiba, K. Kimura, and Y. T. Lee, J. Phys. Chem. **91**, 17 (1987).
108. J. Berkowitz, C. A. Mayhew, and B. Ruscic, J. Chem. Phys. **88**, 7396 (1988).
109. J. J. Russell, J. A. Seetula, D. Gutman, and S. M. Senkan, J. Phys. Chem. **93**, 5184 (1989).
110. R. S. Urdahl, Y. Bao, and W. M. Jackson, Chem. Phys. Lett. **178**, 425 (1991).
111. (a) B. Ruscic and J. Berkowitz, J. Chem. Phys. **95**, 2407 (1991); (b) B. Ruscic and J. Berkowitz, J. Chem. Phys. **95**, 2416 (1991).
112. (a) B. Ruscic, C. A. Mayhew, and J. Berkowitz, J. Chem. Phys. **88**, 5580 (1988); (b) B. Ruscic, M. Schwarz, and J. Berkowitz, J. Chem. Phys. **91**, 4576 (1989).
113. M. Page, G. F. Adams, J. S. Binkley, and C. F. Melius, J. Phys. Chem. **91**, 2675 (1987).
114. L. A. Curtiss and J. A. Pople, J. Chem. Phys. **90**, 2522 (1989).
115. C. W. Bauschlicher, S. R. Langhoff, and P. R. Taylor, J. Chem. Phys. **93**, 502 (1990).
116. (a) B. Ruscic and J. Berkowitz, J. Chem. Phys. **95**, 4033 (1991); (b) B. Ruscic and J. Berkowitz, J. Phys. Chem. **97**, 11451 (1993).
117. J. M. Dyke, J. Chem. Soc. Faraday Transactions 2 **83**, 69 (1987).
118. D. M. Golden and S. W. Benson, Chem. Rev. **69**, 125 (1969).
119. J. M. Nicovich, K. D. Kreutter, C. A. van Dijk, and P. H. Wine, J. Chem. Phys. **96**, 2518 (1992).
120. S. Nourbaksh, K. Norwood, C.-L. Liao, and C. Y. Ng, J. Chem. Phys. **95**, 5014 (1991), and references therein.
121. (a) B. Ruscic and J. Berkowitz, J. Chem. Phys. **97**, 1818 (1992); (b) B.

Ruscic and J. Berkowitz, J. Chem Phys. **98**, 2568 (1993).
122. S. Lias, J. E. Bartmess, J. F. Liebman, J. L. Holmes, R. D. Levin, and W. G. Mallard, J. Phys. Chem. Ref. Data **17**, Suppl. No. 1 (1988).
123. B. Ruscic and J. Berkowitz, J. Chem. Phys. **95**, 4378 (1991).
124. B. Ruscic, E. Appelman, and J. Berkowitz, J. Chem. Phys. **95**, 7957 (1991).
125. L. A. Curtiss, K. Raghavachari, and J. A. Pople, J. Chem. Phys. **98**, 1293 (1993).
126. M. P. McGrath and L. Radom, J. Chem. Phys. **94**, 511 (1991).
127. L. A. Curtiss, N. Davis, M. P. McGrath, and L. Radom, work in progress.
128. M. A. Al-Laham, K. Raghavachari, J. A. Pople, and L. A. Curtiss, to be published.
129. (a) G. A. Petersson, A. Bennett, T. G. Tensfeldt, M. A. Al-Laham, W. A. Shirley, and J. Montzaris, J. Chem. Phys. **89**, 2193 (1988); (b) G. A. Petersson and M. A. Al-Laham, J. Chem. Phys. **94**, 6081 (1991); (c) G. A. Petersson, T. G. Tensfeldt, and J. A. Montgomery, J. Chem. Phys. **94**, 6091 (1991); (d) G. A. Petersson and M. Braunstein, J. Chem. Phys. **83**, 5129 (1985); (e) G. A. Petersson, A. K. Lee, and A. Bennett, J. Chem. Phys. **83**, 5105 (1985).
130. (a) P. M. W. Gill, B. G. Johnson, J. A. Pople, and M. J. Frisch, Int. J. Quant. Chem. **S26**, 319 (1992); (b) B. G. Johnson, P. M. W. Gill, and J. A. Pople, J. Chem. Phys. **97**, 7846 (1992); (c) P. M. W. Gill, B. G. Johnson, J. A. Pople, and M. J. Frisch, Chem. Phys. Lett. **197**, 499 (1992).
131. (a) A. D. Becke, J. Chem. Phys. **96**, 2155 (1992); (b) A. Becke, J. Chem. Phys. **97**, 9173 (1992); (c) A. Becke, J. Chem. Phys. **98**, 1372 (1993).
132. J. M. Seminario, Chem. Phys. Lett. **206**, 547 (1993).

Accurate Theoretical Studies of Small Elemental Clusters

Krishnan Raghavachari
AT&T Bell Laboratories, Murray Hill, New Jersey 07974

and

Larry A. Curtiss
Chemical Technology Division, Argonne National Laboratory, Argonne, Illinois 60439

August 31, 1994

Abstract

This article reviews the applications of accurate quantum chemical techniques to the study of atomic clusters. In particular, we focus on clusters of the main group elements boron, aluminum, carbon, silicon, phosphorus, and sulfur. Only smaller cluster systems are included since calculations with chemical accuracy are available only for these systems. Particular emphasis is placed on the applications of Gaussian-1 and Gaussian-2 theories. Comparisons with recent experimental values for the binding energies of carbon clusters (C_2-C_7) and silicon clusters (Si_2-Si_7) indicate an accuracy of about 0.1-0.2 eV for these methods.

S. R. Langhoff (Ed.), Quantum Mechanical Electronic Structure Calculations with Chemical Accuracy, 173–207.

Contents

1 Introduction

The field of *ab initio* quantum chemistry has witnessed dramatic developments in the last decade which have led to the emergence of novel theoretical techniques with unprecedented accuracy [1-3]. When applied to small molecules, these methods appear to realize the goal of attaining "chemical accuracy" ($\pm 1 - 2$ kcal/mol). Calculations with chemical accuracy are now being performed by several groups using different theoretical techniques on a fairly wide variety of small molecules. One of the most important advantages of theoretical work is that it can be applied to species for which experimental data are difficult to obtain. An area of research that has witnessed dramatic growth in the last decade has been the study of atomic clusters [4-14]. Clusters bridge the gap between atoms and solids and have a variety of bonding environments ranging from van der Waals to covalent to metallic interactions. While many cluster systems can be generated relatively easily experimentally, they are frequently unstable or reactive so that many important properties such as geometrical structures and binding energies are difficult to obtain and remain uncertain. Theory is particularly suited to provide the necessary complementary information to understand and interpret the experimental observations for these systems.

In this review, we consider the applications of accurate quantum chemical techniques to the study of small elemental clusters. We place particular emphasis on the applications of Gaussian-1 (G1) [15] and Gaussian-2 (G2) [16] theories developed by Pople, Curtiss, Raghavachari and coworkers, though applications of other methods are also included. Detailed descriptions of the G1 and G2 methods have been given in another chapter in this book [2]. Briefly, they are general predictive procedures applicable to any molecular system in an unambiguous manner, which can reproduce known experimental data to a prescribed accuracy of ± 2 kcal/mol, and can be applied with similar accuracy to other species where the experimental data are unknown or uncertain. These are composite techniques where a sequence of well-defined calculations is performed to arrive at the total energy of a given molecular species. A series of additivity approximations are made to make these techniques fairly widely applicable. A simpler variant of G2 theory, termed G2(MP2) [17], is particularly suited to the study of clusters containing $5 - 10$ atoms. A more accurate version of G2 theory without the additivity approximations, termed G2(QCI) [18], has been applied to some of the smaller cluster systems.

We restrict our review to main group clusters in the first and second rows of the periodic table. Transition metal clusters have been excluded since accurate theoretical procedures that can reach chemical accuracy are still not available for these systems. We also do not consider weakly bound van der Waals clusters such as those of rare gases (e.g., Ar). We have also excluded the alkali and alkaline earth clusters (Groups I and II) which have been comprehensively reviewed recently by Bonacic-Koutecky *et al.* [8]. Finally, our emphasis is on smaller clusters, since accurate calculations with chemical accuracy are available only for those systems.

2 Boron Clusters

Boron compounds are very well known for exhibiting three-dimensional networks of boron clusters [19]. For example, boron hydrides, carboranes, and metal borides all exhibit cluster structures based on polyhedral geometrical arrangements of boron atoms. However, relatively few studies have been performed on isolated boron clusters [20-35]. Experimentally, Berkowitz and Chupka employed mass spectrometry to study $B_2 - B_5$ ejection from solid boron by focused ruby laser light [21]. Laser ablation of boron nitride appears to produce boron clusters containing up to 50 atoms [25]. The most detailed experimental work has been performed by Anderson and coworkers [24] who carried out collision induced dissociation studies of boron clusters (up to B_{13}) and also investigated the chemical reactivity of boron clusters. Theoretical investigations have been carried out by density functional methods [33] up to B_{14} and by *ab initio* Hartree-Fock methods [32] up to B_{12}. However, calculations with chemical accuracy have been performed only on B_2 – B_4 using a modified version of G1 theory by Martin *et al.* [28].

2.1 B_2

B_2 has two low-lying electronic states, a $^3\Sigma_g^-$ $(\sigma_g^2\sigma_u^2\pi_u^2)$ and a $^5\Sigma_u^-$ $(\sigma_g^2\sigma_u^1\pi_u^2\sigma_g^1)$. Dupuis and Liu [26a] have performed large scale configuration-interaction studies to consider the relative energy between the two states and have identified the triplet as the ground state. The quintet state occurs only 0.16 eV higher in energy. A similar value of 0.21 eV has been obtained by Langhoff and Bauschlicher [26c] using MRCI techniques. A larger energy difference of 0.33 eV has been obtained by the recent G1 calculations of Deutsch *et al.* [27]. The theoretical results are consistent with the experimental assignment of a triplet ground state in the spectrum of B_2 produced in a discharge by Douglas and Herzberg [28].

The dissociation energy of B_2 has been computed by Deutsch *et al.* with G1 theory to be 2.71 eV [27]. The same value was obtained using a modified G1 theory by Martin *et al.* [28]. A slightly larger value of 2.85 eV has been estimated by Langhoff and Bauschlicher [26c]. The experimental value of 3.0 eV listed in the JANAF tables [36] and by Huber and Herzberg [37] is likely to be an overestimate though the reported uncertainty (0.5 eV) is fairly large.

Deutsch *et al.* [27] have also considered the positive ion B_2^+ with G1 theory. They have obtained a $^2\Sigma_g^+$ ground state, in agreement with the earlier work of Bruna and Wright [26b]. A $^2\Pi_u$ state occurs 0.29 eV higher in energy at the G1 level. The first two ionization potentials of B_2 at the G1 level are 8.99 eV and 9.27 eV, in excellent agreement with the values calculated by Bruna and Wright [26b]. An experimental value of 10.3 eV derived by Hanley *et al.* [24] appears to be too large and may be partly due to large uncertainties in the measured appearance potentials.

2.2 B_3

B_3 has a D_{3h} geometry with a $^2A_1'$ electronic state. This is similar to the ground state of Al_3 seen in the next section. However, unlike Al_3, which has several low-lying states, other electronic states of B_3 are significantly higher in energy. Martin *et al.* [28] obtain a total binding energy of 8.36 and 8.20 eV for B_3 using regular G1 theory and a modified version of G1 theory, respectively. The difference between the two appears to result from an approximate treatment of the effects of spin contamination in this molecule.

2.3 B_4

Many different isomers of B_4 have been considered carefully by Martin *et al.* [28]. They obtain a singlet rhombus (D_{2h}, ^1A_g) as the ground state. It is a pseudo Jahn-Teller distorted form of a singlet square ($^1A_{1g}$), although the energy stabilization from the distortion appears to be very small. Using large basis sets and higher levels of correlation, Martin *et al.* [28] obtain an energy difference between the two of only 0.02 eV. This implies that the structure of B_4 would be effectively a square at practical temperatures.

The total binding energy of B_4 has been obtained by Martin *et al.* to be 13.56 eV with the G1 method [28]. The dissociation energy of $B_4 \rightarrow B_3 + B$ is thus about 5.3 eV, indicating that B_4 is a very stable cluster.

As mentioned earlier, calculations with chemical accuracy are not available for the larger boron clusters. An interesting qualitative observation for these clusters is that planar or quasi-planar structures appear to be very stable for boron clusters at least up to B_{12} [33]. Kawai and Weare have reported that an icosahedral cage becomes stable for B_{14} using density functional techniques [33].

3 Aluminum Clusters

In contrast to boron which is a large band gap semiconductor, aluminum is a nearly free electron metal. The presence of three valence electrons per atom makes aluminum clusters more complicated and more interesting than those of simple alkali metals [8]. For metal clusters the simple electronic shell model [38] predicts "magic numbers" at 20 and 40 valence electrons. It is much easier to test the validity of such a model in the case of aluminum clusters (7 – 14 atoms) than for the alkali elements (20 – 40 atoms). A variety of experimental measurements such as negative ion photodetachment, photofragmentation, collision induced dissociation, chemical reactivity studies, etc. are available on Al clusters [39-53]. Since Al is a main group element, detailed theoretical studies of the structures and stabilities of small Al clusters are feasible and have been performed by several groups using techniques ranging from empirical force fields to *ab initio* quantum chemistry [54-66]. The most detailed and comprehensive theoretical study on aluminum clusters has been performed by Jones [65] using local density functional techniques. However, since the local density method substantially overestimates bond energies, in much of the following discussion we use the results from accurate quantum chemical studies which we have performed using G2 and G2(MP2) methods [66]. As in other systems, we focus on the small clusters Al_2–Al_7 where accurate theoretical results are available.

3.1 Al_2 AND Al_2^+

Al_2 has two very low-lying electronic states, which are almost isoenergetic, a $^3\Pi_u$ state ($\sigma_g^2\sigma_u^2\sigma_g^1\pi_u^1$) and a $^3\Sigma_g^-$ state ($\sigma_g^2\sigma_u^2\pi_u^2$). In spite of the very small energy difference of less than 0.1 eV, Bauschlicher *et al.* [54] performed accurate theoretical calculations comparing the two electronic states and made a definitive prediction that the ground state of Al_2 is $^3\Pi_u$. Similar results were also reported by Sunil and Jordan [55]. This prediction has been confirmed by the experimental results of Cai *et al.* [51]. The bond length in the $^3\Pi_u$ state (2.697 Å at MP2/6-31G*) is considerably larger than that in the $^3\Sigma_g^-$ state (2.469 Å at MP2/6-31G*), consistent with the presence of more π-electrons in the latter. The energy difference between the two states obtained by Bauschlicher and coworkers is 0.03 eV [54]. At the G2 level, the energy difference is very similar, 0.07 eV.

The binding energy of Al_2 at the G2 level is 1.38 eV [66]. This is in excellent agreement with the value of 1.40 eV derived by Bauschlicher *et al.* [54]. The binding energy per atom for Al_2 is 0.70 eV, considerably smaller than the cohesive energy for bulk Al which is 3.43 eV. A large part of this difference is directly attributable to the much larger number of nearest neighbors (12) of Al in the bulk fcc lattice.

Al_2^+ also has two possible candidates for its ground state, $^2\Sigma_g^+$ and $^2\Pi_u$. Among the two, the $^2\Sigma_g^+$ state is more stable and the adiabatic ionization potential of Al_2 at the G2 level is 5.91 eV. This is slightly smaller than the ionization potential of Al atom (5.93 eV at G2, 5.98 eV experiment). The lower ionization potential of Al_2 is consistent with the dissociation of Al_3^+, which yields Al_2^+ + Al predominantly [41,42].

3.2 Al_3 AND Al_3^+

Al_3 has several low-lying electronic states which have been considered by several groups [57-60,64-66]. The nature of these states are easily understood by considering a D_{3h} framework for the molecule. Among its nine valence electrons, six of them occupy molecular orbitals that are principally composed of 3*s* atomic orbitals. The remaining three electrons have to be distributed among four possible molecular orbitals, which are different *p*-bonding combinations, an out-of-plane π-bonding orbital (a_2''), an in-plane radial bonding orbital (a_1'), and a pair of degenerate in-plane tangential bonding orbitals (e'). This results in three low-lying electronic states, $^2A_1'$, $^2A_2''$, and $^4A_2$, where the quartet is a Jahn-Teller distorted state (C_{2v}) due to partial occupation of the degenerate e' orbital [66].

The presence of several low-lying states suggests that high levels of theory with large basis sets are necessary to identify the ground state of Al_3. At the G2 level, the $^2A_1'$ is the ground state and occurs 0.13 and 0.28 eV lower in energy than the $^2A_2''$ and $^4A_2$ states, respectively [66]. The identification of the $^2A_1'$ ground state with G2 theory is in agreement with the results from previous high level calculations [58-60,64,65]. Experimentally, magnetic deflection measurements [39] indicate a doublet ground state while matrix electron spin resonance (ESR) measurements [52] suggest a quartet ground state. Our theoretical results clearly support a doublet ground state.

The G2 binding energy of Al_3 is 3.69 eV [66]. The binding energy per atom of 1.23 eV is much larger than that of Al_2, but still only about a third of the cohesive energy for bulk Al. The dissociation energy of $Al_3 \rightarrow Al_2 + Al$ has been measured by Fu *et al.* [53] by resonant multiphoton ionization spectroscopy as 2.4 eV. An earlier collision induced dissociation measurement of Hanley *et al.* [42] yielded a much lower value of 1.4 – 1.8 eV. Our calculated value of 2.32 eV [66] at the G2 level clearly supports the spectroscopic measurements of Fu *et al.* [53].

Unlike neutral Al_3, Al_3^+ has a well-defined $^1A_1'$ ground state (D_{3h} symmetry). The calculated adiabatic ionization potential of 6.63 eV at the G2 level is significantly larger than that of Al atom. This is consistent with the dissociation of Al_4^+, which yields $Al^+ + Al_3$ with the charge being carried by the species with the lower ionization potential [41,42]. The larger ionization potential of Al_3 is a consequence of the significant bonding interactions of the $3p$ orbitals [59].

3.3 Al_4 and Al_4^+

The size at which aluminum clusters become three-dimensional can be studied by considering the isomers of Al_4. Simple considerations of the close-packing nature of metallic systems suggest that Al_4 may have a tetrahedral or distorted tetrahedral structure. However, theoretical calculations clearly show that Al_4 has a planar rhombus geometry [57,60,65,66]. However, as in the case of Al_2 and Al_3, there are several low-lying electronic states for Al_4. Again this results from the availability of three different bonding molecular orbitals (an out-of-plane π-orbital b_{3u}, an in-plane tangential bonding orbital b_{3g}, and an in-plane symmetric bonding orbital a_g) for the four "p"-electrons [66]. We have obtained three low-lying electronic states, ^1A_g, $^3B_{3u}$, and $^3B_{3g}$. In agreement with previous studies, G2 theory yields the ^1A_g as the ground state [66]. However, we find that the $^3B_{3g}$ state, which has not been considered previously, is only 0.06 eV less stable at the G2 level. Since G2 theory tends to underestimate the stability of triplet states slightly, it is possible that the ground state of Al_4 is a triplet. The other triplet, $^3B_{3u}$, is higher in energy than $^3B_{3g}$ by 0.09 eV. The total binding energy of Al_4 at the G2 level is 5.83 eV.

There are several low-lying electronic states of Al_4^+ also. The three obvious candidates are ^2A_g, $^2B_{3g}$, and $^2B_{3u}$ corresponding to an unpaired electron in each of the low-lying orbitals. Among the three, the $^2B_{3u}$ state collapses to a square D_{4h} geometry ($^2A_{2u}$ state). At the G2 level of theory, the ground state of Al_4^+ is $^2B_{3g}$ with the ^2A_g and the $^2A_{2u}$ states being only 0.14 and 0.13 eV higher in energy, respectively.

The adiabatic ionization potential of Al_4 at the G2 level is 6.63 eV, close to that of Al_3 and larger than that of Al atom. This is again consistent with the formation of $Al^+ + Al_4$ from the collision induced dissociation of Al_5^+ [41,42].

3.4 Al_5 and Al_5^+

Al_5 appears to be at the threshold of the transition of aluminum clusters to three-dimensional structures [65,66]. Our results indicate that the ground state of neutral Al_5 is probably planar while that of the positive ion is three-dimensional (Figure

1). Pettersson *et al.* [64] also obtained a planar ground state while Jones [65] has reported that a planar and a non-planar isomer of Al_5 are almost isoenergetic using density functional methods.

The most stable structure of Al_5 that we have calculated in our study is a planar edge-capped rhombus (C_{2v} symmetry) with a $^2B_2$ electronic state. However, there is some uncertainty because of the high degree of spin contamination in the starting UHF wave function. The G2(MP2) binding energy of this isomer of Al_5 is 8.23 eV. The lowest energy fragmentation channel of Al_5 is Al_4 + Al, which requires 2.40 eV.

Removal of an electron from the planar structure to give Al_5^+ yields an ionic structure that is not a local minimum. Following a series of distortions, Al_5^+ eventually ends up as a non-planar C_{2v} singlet isomer, which can be considered as a capped distorted rhombus where the capping atom is out-of-plane (Figure 1). This transformation on ionization indicates that the Al_5 cluster is very close to the transformation from two-dimensional to three-dimensional structures. In fact, all the larger aluminum clusters adopt three-dimensional structures [65,66].

The adiabatic ionization potential for Al_5 is 6.05 eV at the G2(MP2) level of theory. However, since the neutral and ionic forms are very different from each other, it is unlikely that Franck-Condon factors will permit this to be experimentally observed. Indeed, the experimentally observed ionization potential of Al_5 is somewhat larger, around 6.4 eV [48].

3.5 Al_6

Al_6 clearly adopts a three-dimensional structure. Simple expectations from the nature of metallic systems clearly suggest a close-packed octahedral structure for this cluster. Indeed, Upton [59] and Pettersson *et al.* [64] have studied an octahedral isomer for this cluster. As in other clusters considered earlier, the levels derived from *s*-atomic orbitals are considerably below those derived from *p*-atomic orbitals. For the six highest "*p*"-electrons, there are two available orbitals, a symmetric a_{1g} orbital and a triply degenerate t_{2g} orbital in O_h symmetry. We find that the lowest energy is obtained by the $(a_{1g})^2(t_{2g})^4$ configuration, which goes to lower symmetry due to Jahn-Teller distortions. Following a series of lower symmetry pathways, we have identified three different low energy forms, two singlets and a triplet. Among the three, a D_{3d} singlet (Figure 1) is found to be the most stable form, in agreement with the results reported earlier by Jones [65]. A different singlet isomer with only C_s symmetry (Figure 1) has also been found to be a true minimum at both the HF/6-31G* and MP2/6-31G* levels of theory. It occurs only 0.21 eV higher in energy than the D_{3d} form. The low symmetry and relatively open forms of this isomer are somewhat surprising and indicate the unexpected pathways of evolution of cluster structures. Both isomers, however, are related to a stable 20 valence electron structure seen for Al_7^+. Finally, a triplet isomer also occurs very close in energy to the C_s form at the G2(MP2) level of theory. Since G2 theory tends to underestimate the stability of triplets, it is likely that the triplet and the best singlet are almost isoenergetic. Jones has reached a similar conclusion [65].

The binding energy of Al_6 at the G2(MP2) level of theory is 11.65 eV. As in all the previous clusters, loss of an atom is the lowest energy dissociation channel.

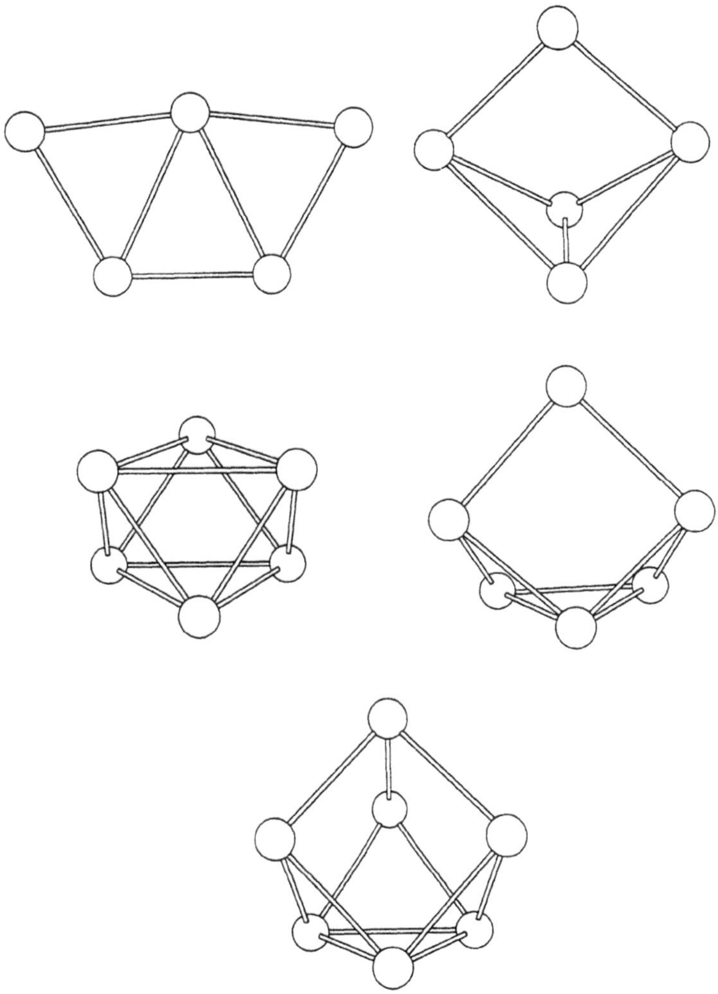

Fig. 1. Structures of aluminum clusters. Shown from the top are Al_5 (C_{2v}), Al_5^+ (C_{2v}), Al_6 (D_{3d}), Al_6 (C_s), and Al_7^+ (C_{3v}) forms.

The energy for the process $Al_6 \rightarrow Al_5 + Al$ is 3.42 eV at the G2(MP2) level of theory.

3.6 AL$_7$ AND AL$_7^+$

A simple theoretical model which has proved to be very successful in explaining the properties of metallic clusters is the electronic shell model [38]. For the spherical shell model, the electronic energy levels show a shell structure with stable arrangements being indicated at shell closings occurring at 8, 20, 40, ... valence electrons. Experimental works on aluminum clusters clearly show enhanced stability at Al_7^+ which has 20 valence electrons [41,42,48].

Detailed investigations of the geometric and electronic structure of Al_7^+ has been carried out to understand the consequences of this shell model [66]. The geometrical arrangement of seven atoms, which can give rise to a spheroidal electronic structure is not obvious. The most compact geometrical arrangement of seven atoms is a pentagonal bipyramid, which is an obvious candidate. However, this D_{5h} structure is found not to be a local minimum for Al_7^+. Consideration of lower symmetry structures finally yields a C_{3v} isomer (Figure 1), which appears to be the ground state of Al_7^+. This structure is less compact than the D_{5h} form in spite of its greater stability. It can be considered as a face-capped octahedron where the face being capped has opened out considerably. If we calculate the distance of all the atoms from its center of mass, all the distances occur in a narrow range of 2.15 ±0.15 Å for the MP2/6-31G* geometry. In other words, all the atoms in this cluster are roughly on the surface of a hollow sphere. Consideration of the one-electron levels in this structure suggests obvious connections to the shell model. Similar observations have been made previously by Jones based on local density functional calculations [65].

The neutral cluster Al_7 also has C_{3v} symmetry and a $^2A_1$ state. A clear indication of the extra stability of Al_7^+ is the ionization potential of Al_7. At the QCISD(T)/6-311G* level, the ionization potential is 5.12 eV, significantly smaller than those of Al_4 and Al_5, and in good agreement with the value obtained by Jones [65]. The low energy isomers of Al_6 including the low symmetry C_s form can be clearly related to the structure of Al_7^+ and indeed can be obtained by removing the more weakly bound atoms in this structure (Figure 1).

4 Carbon Clusters

Carbon clusters are by far the most intensively studied cluster system [67-70]. Experimentally, a rich variety of studies including photodissociation, metastable dissociation, photoelectron spectroscopy, vibrational spectroscopy, ESR, thermodynamic measurements, gas-phase ion chromatography, etc. have been performed on carbon clusters [71-92]. Recent work in this field has accelerated dramatically due to the emergence of the hollow spheroidal carbon clusters referred to as fullerenes (C_{60}, C_{70}, etc.) [68-70,72,77]. The evolution of the structures of carbon clusters from chains to monocyclic rings to polycyclic rings to fullerenes has been the subject of several recent investigations [82,83,88]. Numerous theoretical papers on the smaller clusters $C_2 - C_{10}$ using a wide range of theoretical methods have also been published [93-108]. In this work, we only include the results on $C_2 - C_7$ where theoretical calculations with chemical accuracy have been performed.

Before the specific cluster systems are discussed, several general points can be noted. From a general structural point of view, small carbon clusters are remarkably simple [93]. All of them have either linear or monocyclic ground states involving multiple bonding. This is clearly due to the presence of strong π-bonding involving carbon. Three-dimensional structures similar to those seen in the case of small silicon clusters (*vide infra*) are found to be considerably higher in energy [94]. Fused-ring structures closely related to the graphite fragment (e.g., naphthalene like form of C_{10}) are also very unstable for the smaller clusters [94].

The basic features of the chains and rings can be understood from qualitative considerations [93]. In a simple analysis of all the carbon clusters as linear structures (with 2n-2 π electrons for C_n), the odd numbered clusters can be expected to be stable singlet ground states, while the even-numbered clusters can be expected to be triplet states, which are comparatively less stable. Theoretical studies have revealed that these even-numbered clusters such as C_4, C_6 and C_8 have alternative ring isomers, which are comparably stable [94]. The ring structures benefit from the additional bonding resulting from ring closure. However, the energy gained from such bond formation has to be weighed against the strain energy that may be present in the more compact ring structures. This balance between two opposing factors makes it possible for some of the even-numbered clusters which are comparatively less stable as linear forms (e.g., C_6) to have monocyclic ground states. As the rings get larger, the strain energy gets smaller and the rings become significantly more stable than chains. Similar considerations are expected to hold until the cluster gets large enough to form closed three-dimensional fullerenes.

4.1 C_2, C_2^+, AND C_2^-

C_2 and its ions have been reviewed comprehensively by Weltner and Van Zee [67], and we consider only the results from G2 theory in this subsection. The binding energy of C_2 at the G2 level of theory is 6.37 eV. This is slightly larger than the recent spectroscopic results of Urdahl *et al.* (6.30 eV) [90]. JANAF tables [36] list a value of 6.11 eV based on an earlier spectroscopic measurement which appears to be an underestimate. Very recent mass spectrometric equilibrium measurements performed by Gingerich *et al.* [91a] have also yielded a value (6.27±0.08 eV) similar to that of Urdahl *et al.*

Since C_2 is known to be a highly correlated system where the additivity approximations in G2 theory may not be reliable, we have recently carried out G2(QCI) calculations [18] on C_2 and its ions. At the G2(QCI) level, the calculated binding energy of 6.31 eV is in excellent agreement with the recent experimental values.

C_2 also has a very low-lying $^3\Pi_u$ electronic state, which is only 0.09 eV higher in energy [37]. At the G2 and G2(QCI) levels, the calculated energy differences are 0.28 and 0.23 eV, respectively. The latter value is in slightly better agreement with experiment, though they both underestimate the stability of the triplet state, a known deficiency of such methods.

C_2^- has a $^2\Sigma_g^+$ ground electronic state. Arnold *et al.* [79,80] have determined the adiabatic electron affinity (EA) of C_2 to be 3.273±0.008 eV. At the G2 level, the calculated electron affinity of 3.10 eV is significantly lower, partly due to the overestimation of the stability of neutral C_2. G2(QCI) performs slightly better,

3.14 eV.

C_2^+ has a $^4\Sigma_g^-$ ground state. The G2 and G2(QCI) ionization potentials of 12.05 and 12.00 eV, respectively, are in good agreement with the experimental value of 12.15 eV listed by Huber and Herzberg [37]. However, the experimental value may be uncertain due to complications arising from the presence of low-lying electronic states of C_2, as well as the large bond length difference between neutral C_2 and its cation (1.264 Å vs. 1.403 Å at the MP2/6-31G* level of theory).

4.2 C_3, C_3^+, AND C_3^-

C_3, a molecule that has been observed in astrophysical sources as well as in the laboratory, is well known to have a linear ground state geometry [67]. Part of the interest in this molecule stems from the fact that it exhibits large amplitude vibrational motion due to the presence of a very low bending frequency (63 cm^{-1}).

The structure and harmonic vibrational frequencies of C_3 have been evaluated by a variety of theoretical techniques [98-100]. In particular, accurate levels of theory are necessary to describe the low bending frequency of this molecule. At the QCISD(T)/6-31G* level of theory [99], the optimized bond length (1.310 Å) is slightly larger than the experimental value of 1.297 Å. The calculated QCISD(T)/6-31G* frequencies (66 cm^{-1} (π_u), 1199 cm^{-1} (σ_g), 2063 cm^{-1} (σ_u)) are all in excellent agreement with the corresponding experimental values of 63, 1225 and 2040 cm^{-1}. In fact, this agreement appears to be better than expected considering the moderate size of the 6-31G* basis set.

The G2 binding energy of C_3 is 13.74 eV. Martin *et al.* [100] have also derived a similar value of 13.69 eV by careful scaling techniques. These values are in excellent agreement with the value of 13.71 eV listed in the JANAF tables [36]. However, a significantly smaller value of 13.50±0.13 eV has been derived recently by Gingerich *et al.* from mass spectrometric equilibrium measurements [91a].

The lowest unoccupied orbital of C_3 is a π_g orbital. As shown by Sunil *et al.* [98], the negative ion C_3^- thus has a $^2\Pi_g$ ground state. On going from the neutral to the anion, the C−C bond stretches by only 0.012 Å at the QCISD(T)/6-31G* level [99]. The small change in the bond length is consistent with the fact that the π_g orbital is a non-bonding orbital. QCISD(T) calculations performed by Raghavachari [99] with a 6-311+G(2df) basis set on C_3 and C_3^- yield an adiabatic electron affinity of 1.88 eV. A similar level of theory was used on carbon atom to calibrate these results. The QCISD(T)/6-311+G(2df) level yields an EA of 1.14 eV for the C atom, which is 0.12 eV lower than the experimental value of 1.26 eV. Applying a similar correction to C_3, Raghavachari derived a projected adiabatic electron affinity of 2.00 eV for C_3. At the G2 level, the calculated electron affinity is 1.93 eV. Very recently, Ortiz and Zakrzewski [96] have derived an estimated adiabatic electron affinity of 1.98 eV by electron propagator calculations using large basis sets. The theoretical values are in excellent agreement with the experimental value of 1.98±0.02 eV derived from photoelectron spectroscopy [79,80].

The positive ion C_3^+, however, behaves quite differently. At the QCISD(T)/6-31G* level of theory, it has a cyclic C_{2v} ground state with an apex bond angle of 67° and a $^2B_2$ electronic state [99]. This is quite similar to the structure reported by Grev, Alberts and Schaefer [100], who also concluded that C_3^+ definitely has a cyclic ground state structure. These are consistent with the results of Faibis *et al.* [86] who inferred from Coulomb explosion measurements on C_3^+ that its ground state structure is cyclic. The energy difference between the linear and cyclic forms has been estimated to be in the range 0.1–0.3 eV by different theoretical methods [100].

As in the case of the negative ion, large basis set studies have been used to derive the adiabatic ionization potential (IP) of C_3. At the QCISD(T)/6-311+G(2df) level, the calculated IP is 11.61 eV [99]. Again, the results from carbon atom were used to calibrate these results [99]. At the same level of theory, the calculated IP for C is 11.17 eV, only 0.09 eV smaller than the experimental value of 11.26 eV. Applying a similar correction, the projected adiabatic IP for C_3 is 11.70 eV. This is in perfect agreement with a G2 value of 11.70 eV obtained with the QCISD(T)/6-31G* geometry for C_3^+. Experimental determinations of the ionization potential of C_3 range from 10 to 13 eV. Bracketing experiments from photoionization time-of-flight mass spectrometry gave an ionization potential in the range 9.98–11.61 eV [71]. The theoretical value is near the upper end of this experimental range. Fura *et al.* [92] have recently suggested a value of 11.4 eV based on the observed product abundances in unimolecular dissociation of carbon cluster ions. Charge transfer bracketing reactions of Ramanathan *et al.* [81] have given a value of 12.97 ±0.1 eV, clearly outside the possible error range of G2 theory.

4.3 C_4

C_4 is the smallest neutral carbon cluster to exhibit a competition between a linear triplet isomer and a cyclic D_{2h} rhombus singlet isomer (Figure 2). High levels of theory using large basis sets and accurate electron correlation treatments are important to reach definitive conclusions [101,102]. The largest calculations on this system have been carried out by Watts *et al.* [101] who obtained the rhombus form to be more stable by about 1 kcal/mol with the CCSD(T) method with a large $[5s4p3d2f1g]$ basis set. A similar value was derived earlier by Martin *et al.* [101] using the QCISD(T) method with several large basis sets. At the G1 and G2 levels of theory, the rhombus form has been calculated to be more stable by 3.4 kcal/mol and 5.1 kcal/mol, respectively. It is likely that the G2 value is too large since it tends to underestimate the stability of triplet states by about 2 kcal/mol.

Experimentally, the linear form of C_4 has been clearly observed by ESR as well as by high-resolution laser spectroscopy [84,85]. The ESR hyperfine parameters [85] are consistent with the presence of cumulenic bonding as shown by the theoretical studies. The cyclic isomer has not been detected directly by spectroscopic techniques. Indirect evidence exists for this isomer of C_4 based on Coulomb explosion imaging experiments [87].

Higher temperature effects on the relative stabilities of the two isomers have been considered by Slanina [102] in detail. Entropic factors favor the linear isomer substantially due to the presence of smaller vibrational frequencies. These studies suggest that the linear form should dominate in the gas phase at higher temperatures.

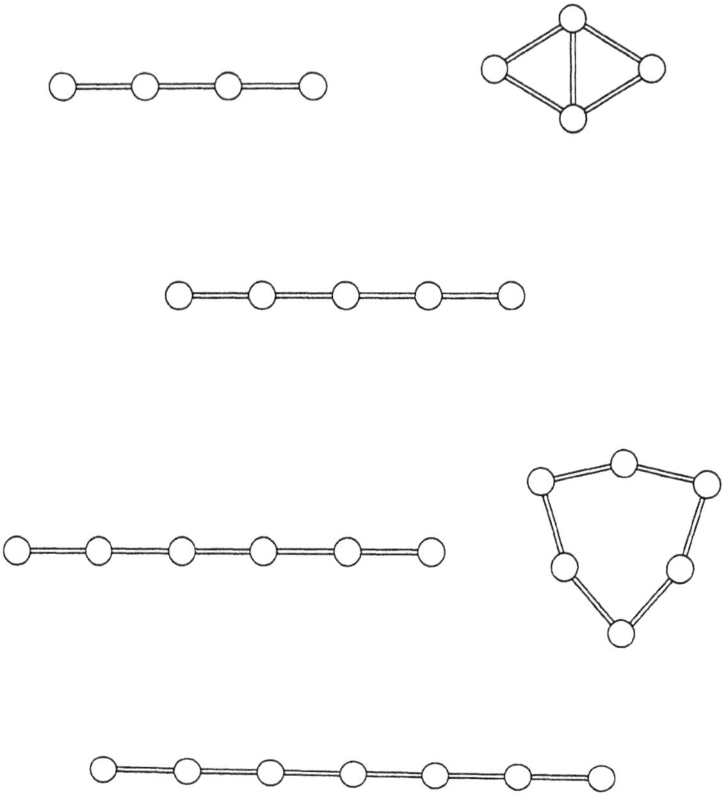

Fig. 2. Structures of carbon clusters. Shown from the top are C_4 ($D_{\infty h}$), C_4 (D_{2h}), C_5 ($D_{\infty h}$), C_6 ($D_{\infty h}$), C_6 (D_{3h}), and C_7 ($D_{\infty h}$) forms.

The binding energy of the linear isomer of C_4 at the G2 level of theory is 18.46 eV. This is in excellent agreement with the recent value of 18.58±0.17 eV derived by Gingerich *et al.* by mass spectrometric equilibrium measurements [91a]. A previous theoretical estimate of Martin *et al.* [101] using a modified G1 method (18.62 eV) is fairly similar. Watts *et al.* [101] using the CCSD(T) method and large basis sets obtain a lower limit of 18.41 eV.

Arnold *et al.* [79,80] have performed photoelectron spectroscopy on C_4^- and have derived an electron affinity of 3.88±0.01 eV. Theoretical studies by Watts *et al.* [101] and others have shown that for the negative ion the linear form is considerably more stable than the rhombus. At the G2 level, the calculated adiabatic electron affinity of the linear form of C_4 is 3.93 eV, in excellent agreement with the experimental value. Watts *et al.* [101] obtain a value of 3.94 eV for the vertical electron detachment energy and an estimated value of 3.78 eV for the adiabatic electron affinity by means of coupled cluster calculations with large basis sets.

4.4 C_5

C_5 has a stable linear singlet ground state at all *ab initio* levels of theory (Figure 2). At the MP2/6-31G* level, the outer and inner C−C bond lengths are found to be 1.300 and 1.291 Å, respectively, clearly indicative of cumulenic bonding [94]. Using a modified version of G1 theory, Martin *et al.* obtained a total binding energy of 25.94 eV [97]. A modified version of G2(MP2) theory without the inclusion of diffuse functions also yields a similar value of 25.91 eV. Diffuse functions were not included in the basis set since they caused linear dependencies. However, as seen below for an isomer of C_6, they contribute very little to the binding energy of neutral carbon clusters. The theoretical values are slightly larger than the recent value of 25.65±0.18 eV derived experimentally by Gingerich *et al.* from mass spectrometric equilibrium measurements [91a].

Arnold *et al.* [79,80] have measured an adiabatic electron affinity for C_5 of 2.839±0.008 eV. Recently, Ortiz and Zakrzewski [96] have used electron propagator techniques and some additivity approximations to derive a theoretical value of 2.82 eV, in excellent agreement with experiment.

4.5 C_6

C_6 is another even-numbered carbon cluster where there is a competition between a linear triplet isomer and a cyclic singlet isomer (Figure 2). Qualitative considerations suggest that a cyclic isomer should be particularly promising for C_6, since such a structure can be expected to have the stabilization resulting from completely filled π-orbitals (aromatic 4n+2 effect). Indeed, MP4 calculations performed by Raghavachari *et al.* [104] suggested that C_6 has a cyclic (D_{3h} symmetry, Figure 2) ground state isomer, which is more stable than a linear form by about 10 kcal/mol. MCSCF calculations by Parasuk and Almlöf [104] appeared to yield a linear ground state, though the results were found to be unreliable due to not correlating all the valence electrons. Using a modified version of G2(MP2) theory without the inclusion of diffuse functions we have obtained a cyclic ground state and a larger energy difference of 15 kcal/mol. Experimentally, the linear form of C_6 has been seen by ESR and by infrared vibrational spectroscopy [84,85]. Only indirect evidence exists for the cyclic isomer [87]. As in the case of C_4, temperature effects on the equilibrium concentrations of the two isomers have been carried out by Slanina and by Van Zee *et al.* [85b,105]. The linear isomer is substantially stabilized by entropic considerations at higher temperatures, such that the equilibrium vapor over graphite is dominated by the linear isomer. This may be responsible for the difficulty of observation of the more stable cyclic isomer experimentally.

The modified G2(MP2) theory without diffuse functions has been used to calculate a total binding energy of 30.84 eV for the linear isomer of C_6. As in the case of C_5, this is very close to the upper limit of the experimental range of 30.63±0.21 eV derived by Gingerich *et al.* from mass spectrometric equilibrium measurements [91b]. For the cyclic isomer where linear dependencies due to dif-

TABLE I
Binding Energies (eV) of Linear Carbon Clusters

Cluster	Experiment[a]	Theory (G2)[b]
C_2	6.27±0.08, 6.30±0.02[c]	6.37
C_3	13.50±0.13, 13.71±0.19[d]	13.74
C_4	18.58±0.17	18.46
C_5	25.65±0.18	25.91
C_6	30.63±0.21	30.84
C_7	37.86±0.21	38.04

[a]Gingerich and coworkers [91] unless otherwise indicated.
[b]G2(MP2) without diffuse functions for C_5 – C_7.
[c]Reference 90.
[d]Reference 36.

fuse functions were not present, the binding energies obtained with the G2(MP2) method with and without diffuse functions were very similar, 31.46 and 31.49 eV, respectively.

4.6 C_7

C_7 has a stable linear singlet ground state similar to that of C_5 (Figure 2). At the MP2/6-31G* level, the outer, middle, and inner C–C bond lengths are found to be 1.301, 1.295, and 1.281 Å, respectively, clearly similar to the values seen in C_5 and indicative of cumulenic bonding [94]. Similar bond lengths have also been obtained by Watts and Bartlett [95]. A modified version of G2(MP2) theory without the inclusion of diffuse functions yields a total binding energy of 38.04 eV in C_7. As in the case of C_5 and the linear form of C_6, the theoretical value is close to the upper limit of the experimental range of 37.86±0.21 eV derived by Gingerich from mass spectrometric equilibrium measurements [91c]. The binding energies of C_2 – C_7 are summarized in Table I.

Ortiz and Zakrzewski [96] have derived an estimated adiabatic electron affinity of C_7 of 3.37 eV by electron propagator calculations. This is in excellent agreement with the experimental value of 3.358±0.014 eV obtained by Arnold *et al.* [79,80].

C_7^+ is the first carbon cluster ion where there is unambiguous experimental evidence for the presence of two different isomers. McElvany *et al.* [74] have observed two different reaction rates for C_7^+ with a number of small molecules. The results were interpreted as resulting from two isomers, a linear isomer (30%) and a more abundant cyclic isomer (70%). Ion chromatography and annealing experiments by von Helden *et al.* [88] later clearly showed that the cyclic isomer is more stable than the linear form. Theoretical calculations using perturbation theory and configuration interaction by von Helden *et al.* [106] are in agreement, yielding the cyclic form as 0.9 eV more stable than the linear isomer.

5 Silicon Clusters

As an important material in the semiconductor industry, the evolution of the properties of silicon clusters has attracted considerable attention. Though it is in the same group in the periodic table as carbon, the behavior of silicon clusters is dramatically different. Silicon shows a much richer variety of structures than carbon, particularly for small clusters. While small carbon clusters are dominated by π-bonding and exhibit chain or ring structures, silicon clusters have complex polyhedral structures, which are three-dimensional starting from Si_5. Excellent experimental information is available for silicon clusters ranging from thermodynamic equilibrium measurements, Raman spectroscopy, photoelectron spectroscopy, photofragmentation, collision induced dissociation, chemical reactivity studies, etc. [109-126]. Silicon clusters thus provide an excellent system for calibration of theoretical techniques. Among the systematic theoretical approaches that have been applied to silicon clusters are accurate quantum chemical calculations including the effects of electron correlation, local density functional calculations as well as semiempirical and empirical schemes. In addition, there have been numerous studies of individual clusters using different theoretical methods with different levels of sophistication [127-147].

For the smallest clusters $Si_2 - Si_7$, it has been possible to search the respective potential energy surfaces carefully in order to identify the ground state geometries (Figure 3) [133]. For $Si_8 - Si_{10}$, the search for minima has been less exhaustive, though excellent agreement regarding the nature of the ground state geometries has been obtained between the quantum chemical studies and local density calculations [133,139]. In this section, we restrict our attention to $Si_2 - Si_7$, since accurate G2(MP2) studies have thus far been performed only on these systems. In addition, experimental binding energies have been recently obtained by Gingerich, Schmude and coworkers [126] on $Si_2 - Si_7$, which provide a calibration of the theoretical work. Vibrational spectra using Raman spectroscopy have been recently measured by Honea *et al.* [122] on Si_4, Si_6, and Si_7, providing additional information on these clusters.

In addition to the neutral clusters, we have also included information on some of the negatively charged silicon clusters. While theoretical investigations of most cluster systems are usually restricted to neutral systems, experimentally they are typically detected as ionic species. In particular, photoelectron spectroscopy on negatively charged ions promises to be a useful tool in determining electronic energy levels in clusters and possibly assigning geometrical structures. Neumark and coworkers [125] have recently reported such spectra on $Si_2^- - Si_4^-$. Thus the correlation between the structures and properties of neutral and ionic clusters can be quite important.

5.1 Si_2 AND Si_2^-

The neutral silicon dimer has a $^3\Sigma_g^-$ ground electronic state ($\sigma_g^2\sigma_u^2\sigma_g^2\pi_u^2$) with a low-lying $^3\Pi_u$ state ($\sigma_g^2\sigma_u^2\sigma_g^1\pi_u^3$). The nature of the triplet ground state for Si_2 is considerably different from that of the first row analogue C_2, which has a singlet ground state. By an analysis of Si_2^- photoelectron and threshold photodetach-

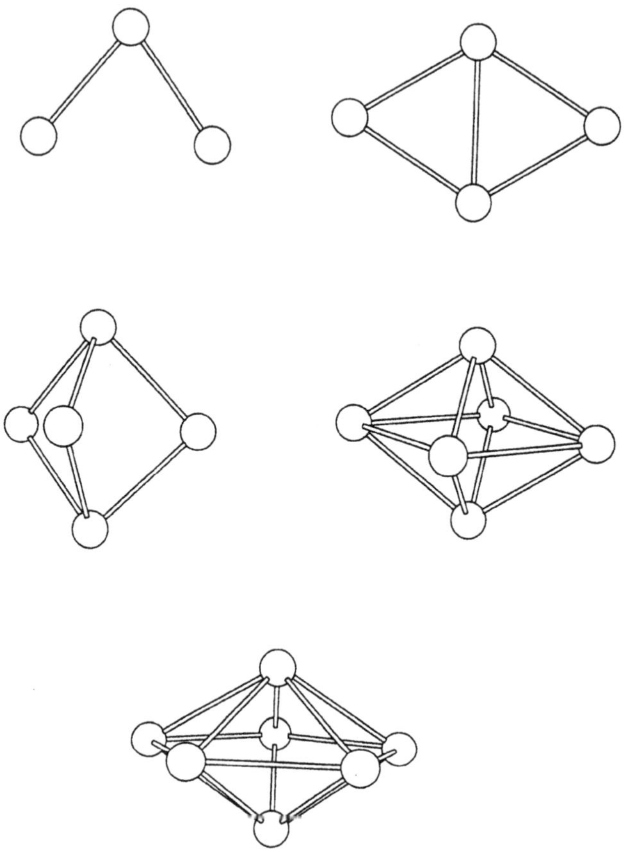

Fig. 3. Ground state structures of silicon clusters $Si_3 - Si_7$.

ment spectra, Arnold *et al.* [125b] have recently determined the energy difference between the two states to be 0.041±0.01 eV. This is very similar to the value of 0.053±0.013 eV derived earlier by Nimlos *et al.* [124]. Theoretically, large basis sets and high levels of electron correlation are necessary to get the correct ordering of the two states. For example, Bauschlicher and Langhoff [129c] have used MRCI techniques with large basis sets to derive an energy difference of 0.05 eV. The calculated energy difference at the G2 level of theory (0.07 eV) [134] is also in excellent agreement with these experimental values. The calculated MP2/6-31G* bond length of 2.260 Å in the $^3\Sigma_g^-$ state is also in excellent agreement with the corresponding experimental value of 2.246 Å [37]. The bond length in the $^3\Pi_u$

state is shorter (2.166 Å at MP2/6-31G*, experiment = 2.155 Å), consistent with the presence of more π bonding electrons.

The G2 binding energy (D_0) of the ground state of Si_2 is 3.19 eV [134]. This is in excellent agreement with the values of 3.21 eV and 3.17 eV listed by Huber and Herzberg [37] and the JANAF tables [36], respectively. Recently, Schmude *et al.* [126] have obtained a slightly larger value of 3.32±0.07 eV from mass spectrometric equilibrium measurements.

The Si_2^- anion also has two electronic states very close in energy, a $^2\Sigma_g^+$ state ($\sigma_g^2\sigma_u^2\sigma_g^1\pi_u^4$) and a $^2\Pi_u$ state ($\sigma_g^2\sigma_u^2\sigma_g^2\pi_u^3$). At the G2 level of theory the two states are close in energy with the $^2\Sigma_g^+$ being more stable by 0.04 eV. This is similar to the values calculated previously by Nimlos *et al.* [124], and by Raghavachari and Rohlfing [133f]. There have been several experimental determinations of this energy difference by means of photoelectron spectroscopy. However, the presence of two low-lying electronic states for *both* the neutral cluster and its negative ion complicates the interpretation of the photoelectron spectra. Earlier experiments by Nimlos *et al.* [124] and by Kitsopoulos *et al.* [125a] concluded that the ground state is $^2\Pi_u$ with the $^2\Sigma_g^+$ slightly higher in energy. Recently, however, Arnold *et al.* [125b] have reassigned the photoelectron spectra of Kitsopoulos *et al.* and determined that the ground state is $^2\Sigma_g^+$, in agreement with the theoretical predictions. Arnold *et al.* obtain an energy difference of 0.025±0.01 eV between the two states, in excellent agreement with the G2 value.

The electron affinity of Si_2 has been determined by Nimlos *et al.* [124] and more recently by Arnold *et al.* [125] to be 2.20±0.01 eV. This is also in excellent agreement with the G2 value of 2.25 eV.

5.2 Si_3 and Si_3^-

The silicon trimer is another cluster system that exhibits fairly complex behavior due to the presence of low-lying electronic states [130]. Neutral Si_3 has a cyclic $^1A_1$ ground state (C_{2v} structure) with an apex angle of 77° at the MP2/6-31G* level (Figure 3) [134]. It results from Jahn-Teller distortion of a $^1E'$ state in D_{3h} symmetry. The bent geometry of Si_3 provides a sharp contrast to the known linear structure of the first row analogue C_3 indicating sharp differences between the two elements. Si_3 also has a low-lying triplet state (D_{3h} structure, $^3A_2'$) which lies 0.15 eV higher in energy at the G2 level of theory. This small energy difference is in excellent agreement with the results from a variety of other theoretical studies [130]. The results for Si_2 and Si_3 which both contain low-lying excited states underscore the importance of high levels of theory including electron correlation effects, which are necessary to provide an accurate description of silicon clusters. Experimental studies using photoelectron and threshold photodetachment have been performed on Si_3^- by Neumark and coworkers [125]. The X band is comprised of at least two different electronic transitions, consistent with the presence of two low-lying electronic states of the neutral cluster. However, definitive experimental assignment of the ground state has not been possible and requires further higher-resolution spectra including both states.

The atomization energy of Si_3 at the G2 level of theory is 7.41 eV [134]. This is in excellent agreement with the most recent experimental determination by Schmude *et al.* of 7.34±0.17 eV [126a]. Earlier experimental values [36] ranged from 7.3 – 7.6 eV with uncertainties of 0.2 – 0.4 eV.

The Si_3^- anion has two states, $^2A_1$ and $^2B_2$ (C_{2v} symmetry), which also result from Jahn-Teller distortion of a $^2E'$ state in D_{3h} symmetry. The two states are nearly isoenergetic and careful studies by Rohlfing and Raghavachari [132] showed that the $^2A_1$ state with an apex bond angle of 66° (MP2/6-31G*) is the ground state. Vibrational frequency analysis confirmed that the $^2A_1$ is a minimum while the $^2B_2$ is a transition state. The energy difference between the two states is only 0.03 eV at the G2 level of theory. The small energy difference is a clear indication of rapid pseudorotation in Si_3^-. This is consistent with the threshold photodetachment studies by Arnold and Neumark [125e], which show features suggesting that the negative ion is a fluxional, Jahn-Teller distorted species.

The G2 electron affinity of 2.24 eV [134] for Si_3 is in excellent agreement with the experimental value of 2.30 eV [125c]. Experimentally, Arnold and Neumark [125e] have reported vibrational frequencies of 501±10 cm^{-1} (a_1) and 337±10 cm^{-1} (e) for the $^3A_2'$ state of neutral Si_3. They are in moderate agreement with the corresponding values of 522 and 285 cm^{-1} reported by Rohlfing and Raghavachari [132] at the QCISD(T)/6-31G* level. Better agreement with experiment is obtained for the e mode by Dixon and Gole [130] using density functional techniques (340 cm^{-1}).

In the photoelectron spectra of Si_3^-, Kitsopoulos *et al.* [125c] have reported the presence of other neutral excited states at 0.45, 0.89, 1.10 and 1.67 eV, respectively, above the neutral ground state. Rohlfing and Raghavachari [132] have assigned the electronic states based on large basis set QCISD(T) calculations on the triplet states, and evaluating singlet-triplet splittings by configuration interaction. The four states have been assigned as $^1B_2$, $^3A_1$, $^3B_1$, and $^1B_1$, respectively. The corresponding theoretical excitation energies are 0.65, 0.92, 1.12, and 1.66 eV. The overall excellent agreement provides additional evidence for the accuracy of the theoretical results.

5.3 Si_4 AND Si_4^-

In contrast to Si_2 and Si_3 which both have low-lying electronic states, Si_4 has a well-defined singlet ground state with a planar rhombus (D_{2h}) geometry (Figure 3) [131,133]. This is similar to the analogous isomer seen for C_4. However, a linear isomer of Si_4 is about 3 eV higher in energy again indicating differences from C_4 where the two forms are almost isoenergetic.

At the MP2/6-31G* level of theory, the Si–Si distance across the short diagonal of Si_4 (2.41 Å) is 0.1 Å larger than the bond length across the perimeter [134]. No direct experimental structural information is available for Si_4. However, recent Raman spectra obtained by Honea *et al.* [122] provide indirect evidence for the theoretical structure. At the MP2/6-31G* level after uniformly scaling the frequencies down by 5% to allow for the systematic overestimations, the rhombus has three allowed Raman lines at 337 (a_g), 440 (b_{3g}), and 463 (a_g) cm^{-1} [122]. The frequencies of the two a_g modes are in excellent agreement with the experimental values of 345 and 470 cm^{-1}. The b_{3g} vibration has not been observed experimentally, probably because it is an order of magnitude weaker than the nearby strong

a_g mode.

The G2 binding energy of Si_4 evaluated by Curtiss *et al.* [134] is 11.95 eV. Recently, Schmude *et al.* [126b] have performed mass spectrometric equilibrium measurements and derived an atomization energy of 12.02±0.23 eV, in excellent agreement with the G2 value.

The ground state of Si_4^- also has a planar rhombus (D_{2h}) structure with a $^2B_{2g}$ state where the singly occupied orbital is principally a π-type nonbonding orbital [132]. The similar geometries of Si_4 and Si_4^- are consistent with the relatively short progression observed in the experimental photoelectron spectra by Kitsopoulos *et al.* [125c]. The G2 electron affinity of 2.06 eV is in good agreement with the corresponding experimental value of 2.17±0.01 eV reported by Arnold and Neumark [125d].

As in the case of Si_3^-, Kitsopoulos *et al.* [125c] have experimentally observed other neutral excited states at 0.81, 1.37, and 2.01 eV. The lowest excited state has been assigned by Rohlfing and Raghavachari [132] as arising from a $^3B_{3u}$ state of the D_{2h} rhombus, with a calculated excitation energy of 0.85 eV, in excellent agreement with experiment. The excited state at 1.37 eV appears to result from $^1B_{3u}$ where the theoretical value (1.7 eV) is significantly larger. Part of the discrepancy is probably due to the difficulty associated with excited singlet states which cannot be described by simple Hartree-Fock wave functions. Finally, the state at 2.01 eV has been assigned to $^3B_{1u}$ where the calculated excitation energy of 1.99 eV is in essentially perfect agreement.

5.4 Si_5 and Si_5^-

Neutral Si_5 has a trigonal bipyramidal geometry (D_{3h}) and a singlet ground state (Figure 3) [133]. The internuclear distances between the atoms in this structure are very sensitive to the level of theory used and change significantly on going from HF/6-31G* to MP2/6-31G* [134]. At the MP2/6-31G* level, the distance between the three atoms in the equatorial plane is 2.96 Å suggesting weak chemical bonding. The distance between the two axial atoms is 3.05 Å, shorter than expected for a trigonal bipyramid, indicating an axially compressed structure. Using a slightly modified version of G2 theory, Curtiss *et al.* [134] obtained a binding energy of 16.15 eV for Si_5. At the simpler G2(MP2) level, the binding energy is similar, 16.20 eV. Very recently, Ran *et al.* [126] have obtained an experimental value of 16.16±0.25 eV based on mass spectroscopic equilibrium measurements, in excellent agreement with G2 and G2(MP2) results.

Si_5 also has a triplet isomer, which has a Jahn-Teller distorted C_{2v} structure. This may be considered as a distorted trigonal bipyramid or more simply as a face-capped tetrahedron with a $^3B_1$ ground state [133d]. At the G2(MP2) level of theory, the triplet is 0.61 eV higher in energy than the ground state of Si_5.

Si_5^- also has a trigonal bipyramidal geometry (D_{3h}) with a $^2A_2''$ electronic state [134]. The axial distance is 3.46 Å indicating significant elongation relative to neutral Si_5. In fact, this distance is closer to the value seen in the neutral triplet state (3.57 Å). This may have implications for photoelectron spectra of Si_5^-, since both the neutral singlet and triplet are accessible from the negative ion. If the negative ion geometry is closer to that of the triplet, it is possible that at the anion geometry the triplet state of the neutral is more stable than the singlet. A similar situation exists for Si_3 where the interpretation of the photoelectron spectra is fairly complicated.

The adiabatic electron affinity of Si_5 has been determined to be 2.36 and 2.34 eV at the modified G2 and G2(MP2) levels of theory, respectively. Accurate experimental values are not yet available. However, the larger electron affinity of Si_5 relative to Si_4 is consistent with the low-resolution photoelectron measurements of Cheshnovsky *et al.* [123].

5.5 Si_6 AND Si_7

Previous experimental and theoretical studies have shown that Si_6 and Si_7 are both relatively stable clusters and can be considered as "magic numbers" [118,133]. The predicted ground state of Si_6 is a distorted octahedron, although the nature of the distortion is very sensitive to the level of theory due to the shallow nature of the potential energy surface. At the HF/6-31G* level of theory, the ground state of Si_6 is an edge capped trigonal bipyramid with C_{2v} symmetry [133d]. At the higher MP2/6-31G* level of theory, which includes electron correlation effects, Si_6 collapses into a tetragonal bipyramidal structure with D_{4h} symmetry (compressed octahedron, Figure 3) [122]. The equatorial bonds in this structure are fairly weak (2.73 Å), though they are somewhat stronger than the corresponding bonds in Si_5. Though no direct structural information is available, Raman spectra on Si_6 reported by Honea *et al.* [122] provide indirect evidence for the theoretically predicted geometry. At the scaled MP2/6-31G* level of theory it has five allowed Raman lines at 209 (b_{2g}), 298 (a_{1g}), 376 (b_{1g}), 425 (e_g), and 457 (a_{1g}) cm^{-1}. They agree quite well with the experimental spectrum, which has lines at 252, 300, 386, 404, and 458 cm^{-1}. The only large discrepancy occurs for the low frequency b_{2g} mode, which may be related to the extent of distortions in the structure.

At the G2(MP2) level of theory Si_6 has a binding energy of 20.64 eV. This is in excellent agreement with the recent experimental determinations of Gingerich, Schmude and coworkers (20.54±0.21 eV) using thermodynamic equilibrium measurements [126d].

The lowest energy isomer of Si_7 is predicted to be a pentagonal bipyramid (D_{5h} symmetry, Figure 3) with a tricapped tetrahedron (C_{3v}) lying about 1 eV higher in energy [133d]. The D_{5h} geometry of this cluster is unexpected, since the two axial silicon atoms have a coordination number larger than 4. In fact, it is axially compressed strongly such that the distance between the two axial atoms is only 2.51 Å, though these two atoms are not connected by a "bond" in Figure 3. At the scaled MP2/6-31G* level of theory, the D_{5h} form has five allowed Raman lines at 300 (e_2'), 339 (e_2'), 346 (e_1''), 352 (a_1'), and 441 (a_1') cm^{-1} [122]. Experimentally, four lines at 289, 340, 358, and 435 cm^{-1} have been observed [122], although the width of the line at 340 cm^{-1} suggests the possibility of two overlapping lines. The agreement between theory and experiment is clearly excellent.

The G2(MP2) binding energy of Si_7 is 24.95 eV. It appears to be within the experimental error limits of the value obtained by Gingerich, Schmude and coworkers (25.19±0.25 eV) [126d]. The excellent performance of the G2(MP2) model for both Si_6 and Si_7 suggests that it should be a very valuable theoretical

TABLE II
Binding Energies (eV) of Silicon Clusters

Cluster	Experiment[a]	Theory (G2)[b]
Si_2	3.32±0.07, 3.17±0.13[c]	3.19
Si_3	7.34±0.17, 7.3±0.4[c]	7.41
Si_4	12.02±0.23	11.95
Si_5	16.16±0.25	16.15
Si_6	20.54±0.21	20.65
Si_7	25.19±0.25	24.95

[a]Gingerich, Schmude, and coworkers [126] unless otherwise indicated.
[b]G2(MP2) for Si_5 – Si_7.
[c]Reference 36.

TABLE III
Vibrational Frequencies (cm^{-1}) for Silicon Clusters

Cluster	Experimental[a] frequency	Theoretical[b] frequency	Absolute deviation	Symmetry assignment
Si_4	345	337	8	a_g
		440		b_{3g}
	470	463	7	a_g
Si_6	252	209	43	b_{2g}
	300	298	2	a_{1g}
	386	376	10	b_{1g}
	404	425	21	e_g
	458	457	1	a_{1g}
Si_7	289	300	11	e_2'
	340	339	1	e_2'
	340	346	6	e_1''
	358	352	6	a_1'
	435	441	6	a_1'

[a]Honea *et al.* [122].
[b]MP2/6-31G* frequencies scaled down uniformly by 5%.

tool to make reliable predictions of the binding energies of clusters in the size range of 5-10 atoms.

In all cases (Si_2 – Si_7), the calculated binding energies at the G2 or G2(MP2) levels are in excellent agreement with the thermodynamic equilibrium measurements (Table II). For the three clusters Si_4, Si_6 and Si_7 where 11 sharp vibrational lines have been seen experimentally, the mean absolute deviation between theory (scaled MP2/6-31G*) and experiment for the 11 lines is only about 10 cm^{-1} (Table III). Overall, the comparison between theory and experiment offers quite convincing evidence for the structural assignments of the calculated ground state isomers of silicon clusters.

6 Phosphorus Clusters

In its elemental form, phosphorus shows a rich structural variety exceeded only by sulfur [148]. As a solid, it exists in a variety of allotropic forms. In the vapor phase, it is dominated by the tetramer (P_4) except at high temperatures when the dimer (P_2) is also prominently seen [149a]. Cationic clusters up to P_{24}^+ have been seen by mass spectroscopy of quenched vapor from red phosphorus [149b].

There is no direct experimental information for neutral species other than P_2 and P_4. An interesting question, which has been addressed in several theoretical studies, is the origin of the stability of the strained tetrahedral P_4 molecule [150-155]. A hypothetical cubic P_8 would have virtually no strain, though this molecule has never been observed.

The most detailed calculations on smaller phosphorus clusters have been performed by Jones and Hohl [154] on P_2 – P_8 and by Jones and Seifert [154] on P_9 – P_{11} by means of local density functional calculations and simulated annealing techniques. Unexpected ground state structures have been found by these authors for clusters such as P_6 and P_8. More recently, Häser *et al.* [155] have performed a detailed investigation of even-numbered clusters from P_4 – P_{28} by *ab initio* Hartree-Fock and MP2 calculations.

6.1 P_2

P_2 has a triple bond similar to that of N_2. At the G2 level of theory, it has a binding energy of 4.97 eV, in good agreement with the experimental value of 5.03 eV.

6.2 P_4

The tetrahedral structure of P_4 (Figure 4) is well described by basis sets containing polarization functions. The MP2/6-31G* bond length [152] of 2.20 Å is in excellent agreement with the corresponding experimental value of 2.21 Å. The MP2/6-31G* vibrational frequencies of P_4 (365 (e), 475 (t_2), 633 (a_1)) are in very good agreement with the corresponding experimental values of 367, 462, and 607 cm^{-1}, respectively. Using a local density functional technique, Jones and Hohl have obtained values of 366, 448, and 597 cm^{-1}, also in very good agreement with experiment. The reaction energy $P_4 \rightarrow 2P_2$ is 2.20 eV at the G2 level of theory, somewhat smaller than the experimental value of 2.33 eV. Large basis sets including multiple sets of polarization functions are important to describe the strained bonding in P_4.

6.3 P_6

Many isomers of P_6 have been studied by several groups. Several initial studies [153] focused on three isomers that are the analogues of benzene, dewar benzene, and prismane forms of $(CH)_6$. Jones and Hohl [154] considered several isomers of P_6 by local density functional calculations. They showed that the most stable isomer of P_6 is a fourth isomer which is analogous to the benzvalene form of $(CH)_6$

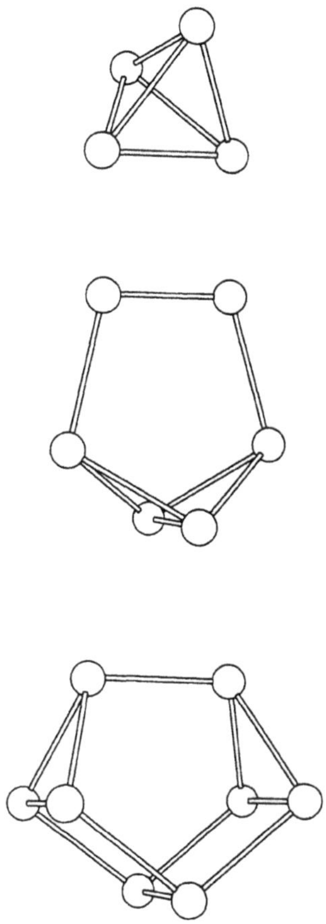

Fig. 4. Ground state structures of phosphorus clusters P_4, P_6, and P_8.

(Figure 4). *Ab initio* MP2 calculations of Häser *et al.* [155] are in agreement. None of the P_6 isomers appears to be energetically more stable than P_4 [155].

6.4 P_8

Many theoretical studies have been performed to consider the stability of the hypothetical cubic P_8 molecule, since it is expected to be relatively strain free. In fact, semiempirical MNDO calculations [151] obtained the cubic form of P_8 to be substantially more stable than two P_4 molecules. However, careful *ab initio* calculations by several groups [152] have shown that the cubic form of P_8 is destabilized by repulsions between parallel bonds, making it considerably less stable than two

P_4 molecules.

Jones and Hohl [154] performed a careful density functional study of many different P_8 isomers and reached the surprising conclusion that several other lower energy isomers exist for P_8. The most stable isomer is found to be a wedge-shaped structure (Figure 4), which is found as a structural unit in violet phosphorus. In fact, Jones and Hohl [154] found that this isomer of P_8 was more stable than two P_4 molecules by 0.47 eV. More recent *ab initio* Hartree-Fock and MP2 calculations by Häser *et al.* [155] agree with the qualitative conclusions of Jones and Hohl that the wedge-shaped isomer of P_8 is its ground state. However, they obtain P_8 to be less stable than two P_4 molecules by 0.44 eV. More definitive studies of P_8 are not yet available.

7 Sulfur Clusters

Elemental sulfur occurs in more allotropic forms than any other element in the periodic table [156]. It is well known that the dimer S_2 has a triplet ground state, similar to that of O_2 [37]. The smaller clusters $S_3 - S_5$ have been studied experimentally [157,158], though their geometric structures have not been determined directly. One of the most interesting aspects in the study of sulfur clusters is that many of the larger clusters (n>5) occur as monocyclic rings. They have been well characterized experimentally [159]: detailed X-ray structures and vibrational spectroscopic information are available for many sulfur rings (S_n, n=6 – 8, 10 – 13), with S_8 being the most stable at room temperature. The interconversions of many of these forms have also been studied. This body of information provides a rigorous means of testing the accuracy of theoretical calculations that attempt to describe the structures and stabilities of sulfur clusters.

There have been several theoretical studies on sulfur clusters, ranging from empirical to *ab initio* studies [160-167]. Most have considered only specific molecular systems though there have also been a few systematic studies on a range of cluster sizes. The most detailed and accurate studies so far have been the density functional calculations performed by Hohl *et al.* [166], and the *ab initio* investigations of Raghavachari *et al.* [167].

7.1 S_2

The sulfur dimer has a $^3\Sigma_g^-$ ground state similar to that of its first row analogue O_2. The calculated MP2/6-31G* bond length (1.920Å) is slightly larger than the experimental value of 1.889 Å. The G2 binding energy of 4.22 eV is somewhat lower than the corresponding experimental value of 4.37 eV.

7.2 S_3

There have been several previous studies [160,161] discussing whether a cyclic D_{3h} form or an open C_{2v} (ozone-like) form is the ground state of S_3. Based on an analysis of the observed Raman spectra, Lenain *et al.* [157] concluded that the ground state geometry corresponds to the bent C_{2v} form (Figure 5). The most accurate *ab initio* calculations [161,166,167] agree with this conclusion though previous semiempirical theories obtained the D_{3h} form as the ground state. The

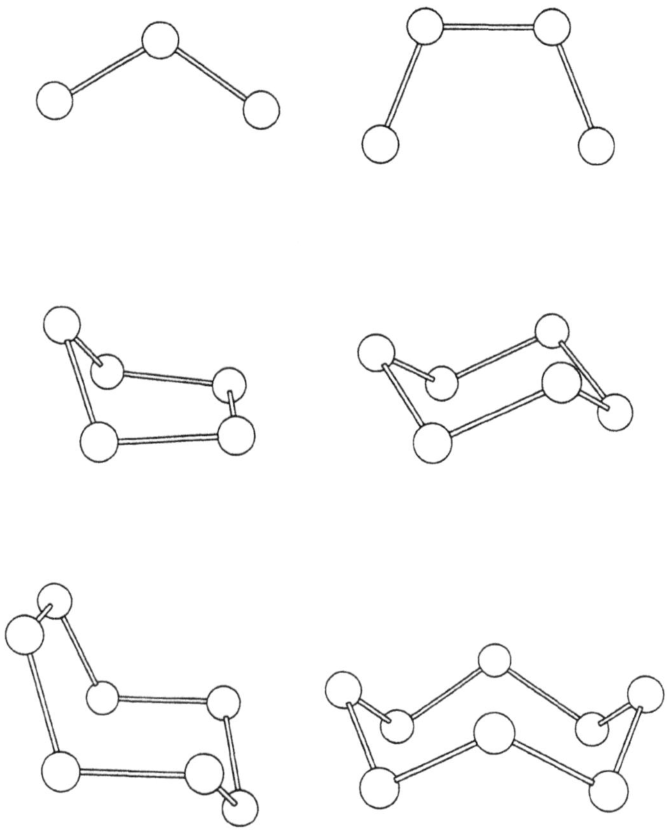

Fig. 5. Ground state structures of sulfur clusters $S_3 - S_8$.

calculated bond length (MP2/6-31G* 1.948 Å) indicates the presence of multiple bonding. The bond angle is not sensitive to the inclusion of correlation effects with different theoretical techniques yielding values of $\approx 117-118°$.

The calculated MP2/6-31G* vibrational frequencies for the C_{2v} form are 263 (a_1), 577 (a_1) and 758 (b_2) cm^{-1} [167]. The symmetric stretching frequencies are in excellent agreement with experiment (256, 575) while the asymmetric stretching frequency still contains considerable error (experiment 656 cm^{-1}). The situation is similar for the first row analogue O_3. The best calculated energy difference favoring the open C_{2v} form is about 9 kcal/mol [161,167].

7.3 S_4

S_4 is another small sulfur cluster where direct experimental information regarding the ground state is not available. Again, the principal question is whether the ground state geometry is cyclic or open. Previous semi-empirical (MNDO or SINDO1) calculations [160] obtained a cyclic puckered (D_{2d}) ground state structure, whereas density functional calculations by Hohl *et al.* [166] predicted an unusual rectangular (D_{2h}) geometry comprised of two weakly bound dimers. Detailed theoretical studies by Quelch *et al.* [162] obtained a cis planar C_{2v} isomer as the ground state (Figure 4). Raghavachari *et al.* [167] found that the potential energy surface connecting the D_{2h} and C_{2v} forms is extremely flat with the nature of the minimum depending on the level of theory. At the QCISD(T)/6-31G* level of theory, the cis planar C_{2v} structure is most stable, in agreement with the conclusions of Quelch *et al.* [162]. The harmonic frequencies at the QCISD(T)/6-31G* level of theory are 99 (a_1), 187 (a_2), 247 (a_1), 303 (b_2), 600 (b_2), and 667 (a_1). The experimentally observed stretching frequencies [158], which have been assigned to S_4 (in the region 640 – 660 cm^{-1}), are consistent with the presence of multiple bonding in the C_{2v} structure.

7.4 $S_5 - S_{12}$

For the larger clusters, it is well known from experimental structure determinations [159] that the ground state geometries are cyclic (Figure 5). This is consistent with the trends in the relative energies of the open and cyclic forms in the smaller clusters. The ring strain involved in forming a cyclic geometry is already minimized sufficiently that even S_5 has a cyclic ground state structure. Since it is reasonable to assume that the strains in the cyclic forms can be even better accommodated in larger rings, the larger sulfur clusters are expected (and indeed observed) to have cyclic geometries. The stereochemical aspects of cyclo–S_5 and cyclo–S_6 exhibit features very similar to that of the valence-isoelectronic hydrocarbons cyclopentane and cyclohexane [167]. The experimental geometries are known for $S_6 - S_8$ and $S_{10} - S_{12}$ [159]. Detailed comparisons between the theoretical (local density functional) and experimental values have been performed by Hohl *et al.* [166]. Raghavachari *et al.* [167] have performed a similar comparison with the HF/3-21G* values. The overall agreement is remarkable with most of the bond lengths being within 0.01 Å of experiment and most of the bond and dihedral angles being within 1° of experiment. In fact, almost all the calculated geometrical parameters are within the reported experimental uncertainties.

Raghavachari *et al.* [167] have also compared the calculated HF/3-21G* frequencies scaled down uniformly by 0.89 (to take into account systematic overestimations) with the corresponding experimental values for S_6, S_8 and S_{12} [159]. Excellent agreement was obtained between the calculated (scaled) frequencies and experiment with a mean deviation (averaged over 35 measured frequencies) of only $\approx$13 cm^{-1}—see Table IV. This gives an estimation of the kind of errors which may be expected for the other clusters where definitive assignments of the measured frequencies have not been made.

Interconversions among different sized sulfur clusters have been studied experimentally [159]. It is known that S_8 and S_{12} are the most stable rings (hav-

TABLE IV
Vibrational Frequencies (cm^{-1}) for S_6, S_8 and S_{12}

S_6		S_8		S_{12}	
Theory[a]	Expt.[b]	Theory[a]	Expt.[b]	Theory[a]	Expt.[b]
457 (e_g)	462	468 (e_2)	475	462 (e_g)	459
457 (a_{1g})	477	465 (e_3)	437	460 (a_{1u})	(464)
447 (e_u)	451	462 (b_1)	415	460 (e_u)	465
439 (a_{1u})	390	454 (a_1)	475	460 (a_{2u})	465
325 (a_{2u})	312	450 (e_1)	471	457 (e_g)	447
259 (a_{1g})	265	256 (b_2)	239	456 (a_{2g})	(410)
214 (e_g)	203	243 (e_3)	248	449 (e_u)	425
159 (e_u)	180	214 (a_1)	218	441 (a_{1g})	459
		192 (e_1)	186	293 (a_{1g})	288
		156 (e_2)	152	277 (a_{2u})	266
		61 (e_2)	86	259 (e_u)	253
				246 (e_g)	243
				182 (e_g)	177
				159 (e_u)	165
				142 (a_{1u})	(158)
				112 (a_{1g})	128
				72 (a_{2u})	72
				66 (a_{2g})	(85)
				43 (e_g)	51
				42 (e_u)	62

[a]HF/3-21G* frequencies scaled down uniformly by a factor of 0.89. Reference 167.
[b]Reference 159. Values in parentheses refer to estimates based on a Urey-Bradley force field.

ing comparable stabilities) and that other cluster sizes are less stable. At the MP3/6-31G* level including zero-point corrections, Raghavachari *et al.* [167] have compared the reaction energies with experiment for the conversion S_8 to the other rings $S_5 - S_{12}$. The energies of conversion of S_8 to S_6 and S_7 are in good agreement with the gas-phase measurements of Berkowitz *et al.* [157]. It is likely that the calculated reaction energies to yield $S_9 - S_{12}$ are reliable estimates as well.

8 Conclusions

The theoretical study of clusters with methods capable of attaining chemical accuracy is still in its infancy. It is clear from this review that the structures and stabilities of small main group clusters are well understood. Theoretical studies of the electronic structures and properties of clusters appear to provide a satisfactory framework for interpreting a variety of experimental measurements. Extensions of these techniques to the study of larger clusters and development of accurate techniques for other systems such as transition metal clusters will be important areas of future research.

9 Acknowledgements

The authors are indebted to Professor John A. Pople for many interesting discussions through the years and to Dr. R. W. Schmude and Prof. K. A. Gingerich for providing information about the heats of formation of silicon clusters prior to publication. This work was supported in part by the U.S. Department of Energy, Office of Basic Energy Sciences, Division of Materials Sciences, under Contract No. W-31-109-ENG-38.

References

1. K. Raghavachari and L. A. Curtiss, in *Modern Electronic Structure Theory*, D. R. Yarkony, Ed. (World Scientific, Singapore, 1994), and references therein.
2. L. A. Curtiss and K. Raghavachari, in *Quantum Mechanical Electronic Structure Calculations with Chemical Accuracy*, S. R. Langhoff, Ed. (Kluwer Academic Publishers, Netherlands, 1994), and references therein.
3. C. W. Bauschlicher and S. R. Langhoff, Science **254**, 394 (1991).
4. Chem. Rev. **86** (1986), Special Issue on Gas Phase Clusters.
5. *Atomic and molecular clusters*, E. R. Bernstein, Ed. (Elsevier, Amsterdam, 1990).
6. *The Chemical Physics of Atomic and Molecular Clusters*, G. Scoles, Ed. (North-Holland, New York, 1990).
7. Proceedings of the 5th International Meeting on Small Particles and Inorganic Clusters, Z. Phys. D **19-20** (1991).
8. V. Bonacic-Koutecky, P. Fantucci and J. Koutecky, Chem. Rev. **91**, 1035 (1991).
9. D. C. Parent and S. L. Anderson, Chem. Rev. **92**, 1541 (1992).
10. *Physics and Chemistry of Finite Systems: From Clusters to Crystals*, Vols. 1-2, P. Jena, S. N. Khanna, and B. K. Rao, Eds., (Kluwer Academic Publishers, Netherlands, 1992).
11. Proceedings of the 6th International Meeting on Small Particles and Inorganic Clusters, Z. Phys. D **26-26S** (1993).
12. *Clusters and Fullerenes*, V. Kumar, T. P. Martin, and E. Tosatti, Eds., (World Scientific, Singapore, 1993).
13. *Advances in Metal and Semiconductor Clusters*, Vol. 2, M. A. Duncan, Ed. (JAI Press, Greenwich, CT, 1994).
14. *Clusters of Atoms and Molecules*, H. Haberland, Ed. (Springer-Verlag, Berlin, 1994).
15. J. A. Pople, M. Head-Gordon, D. J. Fox, K. Raghavachari, and L. A. Curtiss, J. Chem. Phys. **90**, 5622 (1989); L. A. Curtiss, C. Jones, G. W. Trucks, K. Raghavachari, and J. A. Pople J. Chem. Phys. **93**, 2537 (1990).
16. L. A. Curtiss, K. Raghavachari, G. W. Trucks and J. A. Pople, J. Chem. Phys. **94**, 7221 (1991).
17. L. A. Curtiss, K. Raghavachari, and J. A. Pople J. Chem. Phys. **98**, 1293 (1993).
18. L. A. Curtiss, J. E. Carpenter, K. Raghavachari, and J. A. Pople, J. Chem. Phys. **96**, 9030 (1992).
19. L. D. Brown and W. N. Lipscomb, Inorg. Chem. **16**, 2989 (1977); "Boron Hydride Chemistry", E. L. Muetterties, Ed. (Academic, New York, 1975).
20. A. E. Douglas and G. Herzberg, Can. J. Res. A **18**, 165 (1940); W. R. M. Graham and W. Weltner, J. Chem. Phys. **65**, 1516 (1976).
21. J. Berkowitz and W. A. Chupka, J. Chem. Phys. **40**, 2735 (1964).
22. S. Becker and H. J. Dietze, Int. J. Mass Spectrom. Ion. Proc. **73**, 157 (1986).
23. R. Woodward, P. N. Le, M. Temmen, and J. L. Gole, J. Phys. Chem. **91**, 2637 (1987).
24. L. Hanley, J. L. Whitten, and S. L. Anderson, J. Phys. Chem. **92**, 5803 (1988); S. A. Ruatta, L. Hanley, and S. L. Anderson, J. Chem. Phys. **91**, 226 (1989); P. A. Hintz, S. A. Ruatta, and S. L. Anderson, J. Chem. Phys. **92**, 292 (1990).
25. S. J. La Placa, P. A. Roland, and J. J. Wynne, Chem. Phys. Lett. **190**, 163 (1992).
26. (a) M. Dupuis and B. Liu, J. Chem. Phys. **68**, 2902 (1978); (b) P. J. Bruna and J. S. Wright, J. Chem. Phys. **91**, 1126 (1989); J. Phys. Chem. **94**, 1774 (1990). (c) S. R. Langhoff and C. W. Bauschlicher, J. Chem. Phys. **95**, 5882 (1991).
27. P. W. Deutsch, L. A. Curtiss, and J. A. Pople, Chem. Phys. Lett. **174**, 33 (1990).

28. J. M. L. Martin, J. P. Francois, and R. Gijbels, J. Chem. Phys. **90**, 6469 (1989); Chem. Phys. Lett. **189**, 529 (1992).
29. F. Marinelli and A. Pellegatti, Chem. Phys. Lett. **158**, 545 (1989).
30. R. A. Whiteside, "Small Homonuclear Clusters of Beryllium and Boron", Ph.D. Thesis, Carnegie-Mellon University, Pittsburgh, PA, 1981.
31. J. Koutecky, G. Pacchioni, G. H. Jeung, and E. C. Hass, Surface Sci. **156**, 650 (1985).
32. H. Kato, K. Yamashita, and K. Morokuma, Chem. Phys. Lett. **190**, 361 (1992).
33. R. Kawai and J. H. Weare, J. Chem. Phys. **95**, 1151 (1991); R. Kawai and J. H. Weare, Chem. Phys. Lett. **191**, 311 (1992). R. Kawai, M. W. Sung, and J. H. Weare, in "Physics and Chemistry of Finite Systems: From Clusters to Crystals", P. Jena *et al.* Eds., Kluwer Academic Publishers, Vol I, 441.
34. A. C. Tang, Q. S. Li, C. W. Liu, and J. Li, Chem. Phys. Lett. **201**, 465 (1993).
35. I. Boustani (to be published).
36. M. W. Chase, C. A. Davies, J. R. Downey, D. J. Frurip, R. A. McDonald and A. N. Syverud, *JANAF Thermochemical Tables, Third Edition* J. Phys. Chem. Ref. Data, Vol. 14, Supplement 1, 1985.
37. K. P. Huber and G. Herzberg, *Constants of Diatomic Molecules* (Van Nostrand Reinhold, New York, 1979).
38. M. Y. Chou and M. L. Cohen, Phys. Lett. A, **113**, 420 (1986).
39. D. M. Cox, D. J. Trevor, R. L. Whetten, E. A. Rohlfing, and A. Kaldor, J. Chem. Phys. **84**, 4651 (1986); D. M. Cox, D. J. Trevor, R. L. Whetten, and A. Kaldor, J. Phys. Chem. **92**, 421 (1988).
40. W. A. de Heer, P. Milani, and A. Chatelain, Phys. Rev. Lett. **63**, 2834 (1989).
41. M. F. Jarrold, J. E. Bower, and J. S. Kraus, J. Chem. Phys. **86**, 3876 (1987).
42. L. Hanley, S. A. Ruatta, and S. L. Anderson, J. Chem. Phys. **87**, 260 (1987).
43. G. Gantefor, M. Gausa, K. H. Meiwes-Broer, and H. O. Lutz, Z. Phys. D **9**, 253 (1988).
44. K. J. Taylor, C. L. Pettiette, M. J. Craycraft, O. Cheshnovsky, and R. E. Smalley, Chem. Phys. Lett. **152**, 347 (1988).
45. U. Ray, M. F. Jarrold, J. E. Bower, and J. S. Kraus, Chem. Phys. Lett. **159**, 221 (1989); J. Chem. Phys. **91**, 2912 (1989).
46. R. E. Lauchtner, A. C. Harms, and A. W. Castleman, J. Chem. Phys. **94**, 1093 (1991).
47. J. L. Persson, R. L. Whetten, H.-P. Chang, and R. S. Berry, Chem. Phys. Lett. **186**, 215 (1991).
48. K. E. Schriver, J. L. Persson, E. C. Honea, and R. L. Whetten, Phys. Rev. Lett. **64**, 2539 (1990).
49. J. Lerme, M. Pellarin, J. L. Vialle, B. Baguenard, and M. Broyer, Phys. Rev. Lett. **68**, 2818 (1992).
50. T. P. Martin, U. Näher, and H. Schaber, Chem. Phys. Lett. **199**, 470 (1992).
51. M. F. Cai, T. P. Djugan, and V. E. Bondybey, Chem. Phys. Lett. **155**, 430 (1989).
52. J. A. Howard, R. Sutcliffe, J. S. Tse, H. Dahmane, and B. Mile, J. Phys. Chem. **89**, 3595 (1985).
53. Z. Fu, G. W. Lemire, Y. M. Hamrick, S. Taylor, J.-C. Shui, and M. D. Morse, J. Chem. Phys. **88**, 3524 (1988).
54. C. W. Bauschlicher, H. Partridge, S. R. Langhoff, P. R. Taylor, and S. P. Walch, J. Chem. Phys. **86**, 7007 (1987); C. W. Bauschlicher, L. A. Barnes, and P. R. Taylor, J. Phys. Chem. **93**, 2932 (1989); C. W. Bauschlicher and S. R. Langhoff, J. Chem. Phys. **90**, 4627 (1989); S. R. Langhoff and C. W. Bauschlicher, J. Chem. Phys. **92**, 1879 (1990).
55. K. K. Sunil and K. D. Jordan, J. Phys. Chem. **92**, 2774 (1988).
56. U. Meier, S. D. Peyerimhoff, and F. Grein, Z. Phys. D. **17**, 209 (1990).
57. G. Pacchioni and J. Koutecky, Ber. Bunsenges. Phys. Chem. **88**, 242 (1984); J. Koutecky, G. Pacchioni, G. H. Jeung, and E. C. Hass, Surface Sci. **156**, 650 (1985).
58. H. Basch, Chem. Phys. Lett. **136**, 289 (1987); J. S. Tse, J. Mol. Struc. (Theochem) **165**, 21 (1988).
59. T. H. Upton, Phys. Rev. Lett. **56**, 2168 (1986); J. Phys. Chem. **90**, 754 (1986); J. Chem. Phys. **86**, 7054 (1987).
60. C. W. Bauschlicher and L. G. M. Pettersson, J. Chem. Phys. **87**, 2198 (1987).
61. K. Jug, H. P. Schluff, H. Kupka, and R. Iffert, J. Comput. Chem. **9**, 803 (1988).
62. H. P. Cheng, R. S. Berry, and R. L. Whetten, Phys. Rev. B **43**, 10647 (1991); J.-Y. Yi, D. J. Oh, and J. Bernholc, Phys. Rev. Lett. **67**, 1594 (1991).
63. D. R. Salahub and R. P. Messmer, Phys. Rev. B **16**, 2526 (1977); C. W. Bauschlicher and L. G. M. Pettersson, J. Chem. Phys. **84**, 2226 (1986); M. C. McHenry, M. E. Eberhart, R. C. O'Handley, and K. H. Johnson, Phys. Rev. Lett. **56**, 81 (1986).

64. L. G. M. Pettersson, C. W. Bauschlicher, and T. Halicioglu, J. Chem. Phys. **87**, 2205 (1987); T. Halicioglu, Z. Phys. D **12**, 73 (1989); T. Halicioglu and H. O. Pamuk, Surface Sci. **215**, 272 (1989).
65. R. O. Jones, J. Chem. Phys. **99**, 1194 (1993); R. O. Jones, Phys. Rev. Lett. **67**, 224 (1991).
66. K. Raghavachari and L. A. Curtiss (to be published).
67. Review: W. Weltner and R. Van Zee, Chem. Rev. **89**, 1713 (1989).
68. Acc. Chem. Res. **25**, (1992), Special Issue on Fullerenes.
69. *Buckminsterfullerenes*, W. E. Billups and M. A. Ciufolini, Eds., (VCH Publishers, New York, 1993).
70. J. Cioslowski, *Electronic Structures of Fullerenes and Their Derivatives*, (Oxford University Press, Oxford, 1994).
71. E. A. Rohlfing, D. M. Cox, and A. Kaldor, J. Chem. Phys. **81**, 3322 (1984).
72. H. W. Kroto, J. R. Heath, S. C. O'Brien, R. F. Curl, and R. E. Smalley, Nature **318**, 162 (1985).
73. M. E. Geusic, M. F. Jarrold, T. J. McIlrath, R. R. Freeman, and W. L. Brown, J. Chem. Phys. **86**, 3862 (1987).
74. S. W. McElvany, B. I. Dunlap, and A. O'Keefe, J. Chem. Phys. **86**, 715 (1987); S. W. McElvany, J. Chem. Phys. **89**, 2063 (1988); D. C. Parent and S. W. McElvany, J. Am. Chem. Soc. **111**, 2393 (1989).
75. P. P. Radi, T. Bunn, P. R. Kemper, M. Molchan, and M. T. Bowers, J. Chem. Phys. **88**, 2809 (1988); P. Radi, M. E. Rincon, M.-T. Hsu, J. Brodbelt-Lustig, P. Kemper, and M. T. Bowers, J. Phys. Chem. **93**, 6187 (1989).
76. S. Yang, K. J. Taylor, M. J. Craycraft, J. Conceicao, C. L. Pettiette, O. Cheshnovsky and R. E. Smalley, Chem. Phys. Lett. **144**, 431 (1988).
77. W. Krätschmer, L. D. Lamb, K. Fostiropoulos, and D. R. Huffman, Nature **347**, 354 (1990).
78. M. B. Sowa, P. A. Hintz, and S. L. Anderson, J. Chem. Phys. **95**, 4719 (1991).
79. D. W. Arnold, S. E. Bradforth, T. N. Kitsopoulos, and D. M. Neumark, J. Chem. Phys. **95**, 8753 (1991); T. N. Kitsopoulos, C. J. Chick, Y. Zhao, and D. M. Neumark, J. Chem. Phys. **95**, 5479 (1991); C. C. Arnold, Y. Zhao, T. N. Kitsopoulos, and D. M. Neumark, J. Chem. Phys. **97**, 6121 (1992).
80. C. C. Arnold and D. M. Neumark, in *Advances in Metal and Semiconductor Clusters*, Vol. III, M. A. Duncan, Ed. (JAI Press, Greenwich, CT, 1994).
81. R. Ramanathan, J. A. Zimmerman, and J. R. Eyler, J. Chem. Phys. **98**, 7838 (1993).
82. G. von Helden, M. T. Hsu, P. R. Kemper, and M. T. Bowers, J. Chem. Phys. **95**, 3835 (1991); G. von Helden, N. G. Gotts, and M. T. Bowers, Science **259**, 1300 (1993). G. von Helden, M. T. Hsu, N. G. Gotts, and M. T. Bowers, J. Phys. Chem. **97**, 8182 (1993).
83. J. Hunter, J. Fye, and M. F. Jarrold, Science **260**, 784 (1993).
84. J. R. Heath and R. J. Saykally, in *On Clusters and Clustering*, P. J. Reynolds, Ed. (North Holland, New York, 1993), pp 7-21, and references therein; K. H. Hinkle, J. J. Keady, and P. F. Bernath, Science, **145**, 74 (1991); N. Moazzen-Ahmadi, A. R. W. McKellar, J. Chem. Phys. **98**, 7757 (1993); J. R. Heath and R. J. Saykally, J. Chem. Phys. **94**, 3271 (1990); P. F. Bernath, K. H. Hinkle, and J. J. Keady, Science **244**, 562 (1989); N. Moazzen-Ahmadi, S. D. Flatt, and A. R. W. McKellar, Chem. Phys. Lett **186**, 291 (1991); H. J. Hwang, A. Van Orden, K. Tanaka, E. W. Kuo, J. R. Heath, and R. J. Saykally, Mol. Phys. **79**, 769 (1993); J. R. Heath and R. J. Saykally, J. Chem. Phys. **93**, 8392 (1990); A. Van Olden, H. J. Hwang, E. W. Kuo, and R. J. Saykally, J. Chem. Phys. **98**, 6678 (1993); T. F. Giesen, A. Van Orden, H. J. Hwang, R. S. Fellers, R. A. Provencal, and R. J. Saykally, (to be published).
85. (a) W. R. M. Graham, K. I. Dismuke, and W. Weltner, Astrophys. J. **204**, 301 (1976); (b) R. J. Van Zee, R. F. Ferrante, K. J. Zeringue, W. Weltner, and D. W. Ewing, J. Chem. Phys. **88**, 3465 (1988).
86. A. Faibis, E. P. Kanter, L. M. Tack, E. Bakke, and B. J. Zabransky, J. Phys. Chem. **91**, 6445 (1987). Z. Vager and E. P. Kanter, J. Phys. Chem. **93**, 7745 (1989).
87. M. Algranati, H. Feldman, D. Kella, E. Malkin, E. Miklazky, R. Naaman, Z. Vager, and J. Zajfman, J. Chem. Phys. **90**, 4617 (1989); Z. Vager, H. Feldman, D. Kella, E. Malkin, E. Miklazky, J. Zajfman, and R. Naaman, Z. Phys. D **19**, 413 (1991). H. Feldman, D. Kella, E. Malkin, E. Miklazky, Z. Vager, and J. Zajfman, J. Chem. Soc. Faraday Trans. **86**, 2469 (1990).
88. G. von Helden, N. G. Gotts, and M. T. Bowers, Chem. Phys. Lett. **212**, 241 (1993).
89. M. Doverstal, B. Lindgren, U. Sassenberg, and H. Yu, Z. Phys. D **19**, 447 (1991).
90. R. S. Urdahl, Y. Bao, and W. M. Jackson, Chem. Phys. Lett. **178**, 425 (1991).
91. (a) K. A. Gingerich, H. C. Finkbeiner, and R. W. Schmude, (to be published); (b) K. A.

Gingerich, H. C. Finkbeiner, and R. W. Schmude, Chem. Phys. Lett. **207**, 23 (1993); (c) K. A. Gingerich, Chem. Phys. Lett. **196**, 245 (1992).
92. A. Fura, F. Turecek, and F. McLafferty (to be published).
93. K. S. Pitzer and E. Clementi, J. Am. Chem. Soc. **81**, 4477 (1959).
94. K. Raghavachari and J. S. Binkley, J. Chem. Phys. **87**, 2191 (1987); K. Raghavachari, Z. Phys. D **12**, 61 (1989).
95. J. D. Watts and R. J. Bartlett, J. Chem. Phys. **97**, 3445 (1992).
96. J. V. Ortiz and V. G. Zakrzewski, J. Chem. Phys. **100**, 6614 (1994).
97. J. M. L. Martin, J. P. Francois, and R. Gijbels, J. Chem. Phys. **90**, 3403 (1989); **93**, 8850 (1990); **95**, 9420 (1991). J. Comput. Chem. **12**, 52 (1991).
98. K. K. Sunil, A. Orendt, K. D. Jordan, and D. J. DeFrees, Chem. Phys. **89**, 245 (1984).
99. K. Raghavachari, Chem. Phys. Lett. **171**, 249 (1990).
100. R. S. Grev, I. L. Alberts and H. F. Schaefer, J. Phys. Chem. **94**, 3379 (1990); **94**, 8744 (1990); J. M. L. Martin, J. P. Francois, and R. Gijbels, J. Chem. Phys. **93**, 5037 (1990); G. Scuseria, Chem. Phys. Lett. **176**, 27 (1991); J. D. Watts, J. F. Stanton, and R. J. Bartlett, J. Chem. Phys. **94**, 4320 (1991); P. R. Taylor, J. M. L. Martin, J. P. Francois, and R. Gijbels, J. Phys. Chem. **95**, 6530 (1991); J. V. Ortiz, J. Chem. Phys. **97**, 7531 (1992).
101. R. A. Whiteside, R. Krishnan, D. J. DeFrees, J. A. Pople, and P. v. R. Schleyer, Chem. Phys. Lett. **78**, 538 (1981); D. H. Magers, R. J. Harrison, and R. J. Bartlett, J. Chem. Phys. **84**, 3284 (1986); D. E. Bernholdt, D. H. Magers, and R. J. Bartlett, J. Chem. Phys. **89**, 3612 (1988); J. M. L. Martin, J. P. Francois, and R. Gijbels, J. Chem. Phys. **94**, 3753 (1991); J. D. Watts, J. Gauss, J. F. Stanton, and R. J. Bartlett, J. Chem. Phys. **97**, 8372 (1992).
102. Z. Slanina, Chem. Phys. Lett. **173**, 164 (1990).
103. L. Adamowicz, Chem. Phys. Lett. **180**, 466 (1991). J. V. Ortiz, Chem. Phys. Lett. **216**, 319 (1993).
104. K. Raghavachari, R. A. Whiteside, and J. A. Pople, J. Chem. Phys. **85**, 6623 (1986); W. Parasuk and J. Almlöf, J. Chem. Phys. **91**, 1137 (1989). L. Adamowicz, Chem. Phys. **182**, 45 (1991).
105. Z. Slanina, Chem. Phys. Lett. **142**, 512 (1987); Thermochim. Acta **127**, 237 (1988);
106. G. von Helden, W. E. Palke, and M. T. Bowers, Chem. Phys. Lett. **212**, 247 (1993).
107. Z. Slanina, J. Kurtz, and L. Adamowicz, Chem. Phys. Lett. **196**, 208 (1992); Mol. Phys. **76**, 387 (1992).
108. C. Liang and H. F. Schaefer, J. Chem. Phys. **93**, 8844 (1990); J. D. Watts and R. J. Bartlett, Chem. Phys. Lett. **190**, 19 (1992).
109. R. E. Honig, J. Chem. Phys. **22**, 1610 (1954).
110. T. P. Martin and H. Schaber, Z. Phys. B **35**, 61 (1979); T. P. Martin and H. Schaber, J. Chem. Phys. **83**, 855 (1985).
111. T. T. Tsong, Appl. Phys. Lett. **45**, 1149 (1984); T. T. Tsong, Phys. Rev. B **30**, 4946 (1984).
112. L. A. Bloomfield, R. R. Freeman, and W. L. Brown, Phys. Rev. Lett. **54**, 2246 (1985); L. A. Bloomfield, M. E. Geusic, R. R. Freeman, and W. L. Brown, Chem. Phys. Lett. **121**, 33 (1985).
113. W. Begemann, K. H. Meiwes-Broer, and H. O. Lutz, Phys. Rev. Lett. **56**, 2248 (1986); W. Begemann, S. Dreihöfer, G. Ganteför, H. R. Siekmann, K. H. Meiwes-Broer, and H. O. Lutz, "Elemental and Molecular Clusters", T. P. Martin, Ed. Springer, Berlin, 1987.
114. J. R. Heath, Y. Liu, S. C. O'Brien, Q.-L. Zhang, R. F. Curl, F. K. Tittel, and R. E. Smalley, J. Chem. Phys. **83**, 5520 (1985); Y. Liu, Q.-L. Zhang, F. K. Tittel, R. F. Curl, and R. E. Smalley, J. Chem. Phys. **85**, 7434 (1986); Q.-L. Zhang, Y. Liu, R. F. Curl, F. K. Tittel, and R. E. Smalley, J. Chem. Phys. **88**, 1670 (1988).
115. M. L. Mandich, W. D. Reents, and V. E. Bondybey, J. Phys. Chem. **90**, 2315 (1986); W. D. Reents, A. M. Mujsce, V. E. Bondybey and M. L. Mandich, J. Chem. Phys. **86**, 5568 (1987).
116. S. W. McElvany, B. I. Dunlap and A. O'Keefe, J. Chem. Phys. **86**, 715 (1987); W. R. Creasy, A. O'Keefe and J. R. McDonald, J. Phys. Chem. **91**, 2848 (1987).
117. J. M. Alford, F. D. Weiss, R. T. Laaksonen and R. E. Smalley, J. Phys. Chem. **90**, 4480 (1986); J. L. Elkind, J. M. Alford, F. D. Weiss, R. T. Laaksonen and R. E. Smalley, J. Chem. Phys. **87**, 2397 (1987); S. Maruyama, L. R. Anderson, and R. E. Smalley, J. Chem. Phys. **93**, 5349 (1990); J. M. Alford, R. T. Laaksonen, and R. E. Smalley, J. Chem. Phys. **94**, 2618 (1991).
118. M. F. Jarrold and J. E. Bower, J. Phys. Chem. **92**, 5702 (1988). M. F. Jarrold, J. E. Bower, and K. Creegan, J. Chem. Phys. **90**, 3615 (1989). U. Ray and M. F. Jarrold, J. Chem. Phys.

94, 2631 (1991); M. F. Jarrold, Y. Ijiri, and U. Ray, J. Chem. Phys. **94**, 3607 (1991).
119. D. J. Trevor, D. M. Cox, K. C. Reichmann, R. O. Brickman and A. Kaldor, J. Phys. Chem. **91**, 2598 (1987). K. Fuke, K. Tsukamoto, and F. Misaizu, Z. Phys. D **26**, S 204 (1993).
120. M. F. Jarrold and V. Constant, Phys. Rev. Lett. **67**, 2994 (1991).
121. K. D. Rinnen and M. L. Mandich, Phys. Rev. Lett. **69**, 1823 (1992).
122. E. C. Honea, A. Ogura, C. A. Murray, K. Raghavachari, W. O. Sprenger, M. F. Jarrold, and W. L. Brown, Nature **366**, 42 (1993).
123. O. Cheshnovsky, S. H. Yang, C. L. Pettiette, M. J. Craycraft, Y. Liu and R. E. Smalley, Chem. Phys. Lett. **138**, 119 (1987).
124. M. R. Nimlos, L. B. Harding, and G. B. Ellison, J. Chem. Phys. **87**, 5116 (1987).
125. (a) T. N. Kitsopoulos, C. J. Chick, Y. Zhao, and D. M. Neumark, J. Chem. Phys. **95**, 1441 (1991); (b) C. C. Arnold, T. N. Kitsopoulos, and D. M. Neumark, J. Chem. Phys. **99**, 766 (1993); (c) T. N. Kitsopoulos, C. J. Chick, A. Weaver, and D. M. Neumark, J. Chem. Phys. **93**, 6108 (1990); (d) C. C. Arnold and D. M. Neumark, J. Chem. Phys. **99**, 3353 (1993); (e) C. C. Arnold and D. M. Neumark, J. Chem. Phys. (in press).
126. (a) R. W. Schmude, Q. Ran, K. A. Gingerich, and J. E. Kingcade, (to be published); (b) R. W. Schmude, Q. Ran, and K. A. Gingerich, J. Chem. Phys. **99**, 7998 (1993); (c) Q. Ran, M. Miller, R. W. Schmude, and K. A. Gingerich, (to be published); (d) R. W. Schmude and K. Gingerich (private communication).
127. For a review on silicon clusters, see: K. Raghavachari, Phase Trans. **24-26**, 61 (1990).
128. W. L. Brown, R. R. Freeman, K. Raghavachari and M. Schlüter, Science **235**, 860 (1987).
129. (a) K. Kirby and B. Liu, J. Chem. Phys. **70**, 893 (1979); (b) P. J. Bruna, S. D. Peyerimhoff and R. J. Buenker, J. Chem. Phys. **72**, 5437 (1980); (c) C. W. Bauschlicher and S. R. Langhoff, J. Chem. Phys. **87**, 2919 (1987).
130. K. Raghavachari, J. Chem. Phys. **83**, 3520 (1985); G. H. F. Diercksen, N. E. Gruner, J. Oddershede and J. R. Sabin, Chem. Phys. Lett. **117**, 29 (1985); R. S. Grev and H. F. Schaefer, Chem. Phys. Lett. **119**, 111 (1985); R. O. Jones, Phys. Rev. A **32**, 2589 (1985); K. Balasubramanian, Chem. Phys. Lett. **125**, 400 (1986); D. A. Dixon and J. L. Gole, Chem. Phys. Lett. **188**, 560 (1992).
131. K. Balasubramanian, Chem. Phys. Lett. **135**, 283 (1987). L. Adamowicz, Chem. Phys. Lett. **185**, 244 (1991); D. G. Dai and K. Balasubramanian, J. Chem. Phys. **96**, 8345 (1992); L. Adamowicz, Chem. Phys. Lett. **188**, 131 (1992).
132. C. M. Rohlfing and K. Raghavachari, J. Chem. Phys. **96**, 2114 (1992).
133. (a) K. Raghavachari and V. Logovinsky, Phys. Rev. Lett. **55**, 2853 (1985); (b) K. Raghavachari, J. Chem. Phys. **84**, 5672 (1986); (c) K. Raghavachari and C. M. Rohlfing, Chem. Phys. Lett. **143**, 428 (1988); (d) K. Raghavachari and C. M. Rohlfing, J. Chem. Phys. **89**, 2219 (1988); (e) C. M. Rohlfing and K. Raghavachari, Chem. Phys. Lett. **167**, 559 (1990); (f) K. Raghavachari and C. M. Rohlfing, J. Chem. Phys. **94**, 3670 (1991).
134. L. A. Curtiss, P. W. Deutsch, and K. Raghavachari, J. Chem. Phys. **96**, 6868 (1992).
135. G. Pacchioni and J. Koutecky, J. Chem. Phys. **84**, 3301 (1986). G. Pacchioni and J. Koutecky, in *Physics and Chemistry of Small Clusters*, P. Jena, B. K. Rao, and S. N. Khanna, Eds., (Plenum Press: New York, 1987) pp. 439.
136. C. H. Patterson and R. P. Messmer, Phys. Rev. B, **42**, 7530 (1990). R. P. Messmer and C. H. Patterson, Chem. Phys. Lett. **192**, 277 (1992).
137. W. von Niessen and V. G. Zakrzewski, J. Chem. Phys. **98**, 1271 (1993).
138. D. Tomzanek and M. Schluter, Phys. Rev. Lett. **56**, 1055 (1986); D. Tomzanek and M. Schluter, Phys. Rev. B **36**, 1208 (1987).
139. P. Ballone, W. Andreoni, R. Car and M. Parrinello, Phys. Rev. Lett. **60**, 271 (1988). W. Andreoni and G. Pastore, Phys. Rev. B **41**, 10243 (1990). U. Rothlisberger, W. Andreoni, and P. Giannozzi, J. Chem. Phys. **96**, 1248 (1992).
140. R. Fournier, S. B. Sinnott, and A. DePristo, J. Chem. Phys. **97**, 4149 (1992).
141. E. Kaxiras and K. Jackson, Phys. Rev. Lett. **71**, 727 (1993).
142. K. Jug, H.-P. Schluff, H. Kupka, and R. Iffert, J. Comput. Chem. **9**, 803 (1988); H. Kupka and K. Jug, Chem. Phys. **130**, 23 (1989); H. J. Nolte and K. Jug, J. Chem. Phys. **93**, 2584 (1990). K. Jug and M. Krack, Int. J. Quant. Chem. **44**, 517 (1992); K. Jug and M. Krack, Chem. Phys. **173**, 439 (1993).
143. J. C. Phillips, Chem. Rev. **86**, 619 (1986).
144. F. H. Stillinger and T. A. Weber, Phys. Rev. B **31**, 5262 (1985); R. Biswas and D. R. Hamann, Phys. Rev. Lett. **55**, 2001 (1985); Phys. Rev. B **36**, 6434 (1987). J. Tersoff, Phys. Rev. Lett. **56**, 632 (1986); B. W. Dodson, Phys. Rev. B **35**, 2795 (1987); A. D. Mistriotis, N. Flytzanis, and S. C. Farantos, Phys. Rev. B **39**, 1212 (1989). B. C. Bolding and H. C. Andersen, Phys. Rev. B **41**, 10568 (1990). H.-X. Wang and R. P. Messmer, Phys. Rev. B

41, 5306 (1990).

145. S. Saito, S. Ohnishi and S. Sugano, Phys. Rev. **B33**, 7036 (1986); E. Blaisten-Barojas and D. Levesque, Phys. Rev. B **34**, 3910 (1986); B. Feuston, R. Kalia and P. Vashishta, Phys. Rev. B **35**, 6222 (1987); **37**, 6297 (1988); D. A. Jelski, Z. C. Wu, and T. F. George, Chem. Phys. Lett. **150**, 447 (1988); J. R. Chelikowsky and R. Redwing, Solid State Commun. **64**, 843 (1987); J. R. Chelikowsky, Phys. Rev. Lett. **25**, 2669 (1988); J. R. Chelikowsky, J. C. Phillips, M. Kamal and M. Strauss, Phys. Rev. Lett. **62**, 292 (1989).
146. J. C. Phillips, J. Chem. Phys. **88**, 2090 (1988); E. Kaxiras, Chem. Phys. Lett. **163**, 323 (1989); E. Kaxiras, Phys. Rev. Lett. **64**, 551 (1990); D. A. Jelski, B. L. Swift, T. T. Rantala, X. Xia, and T. F. George, J. Chem. Phys. **95**, 8552 (1991); J. Pan and M. V. Ramakrishna (to be published).
147. T. Slee, L. Zhenyang, and D. M. P. Mingos, Inorg. Chem. **28**, 2256, (1989).
148. J. R. Van Wazer, *Phosphorus and its Compounds* (Interscience Publishers, New York, 1958).
149. (a) H. Bock and H. Müller, Inorg. Chem. **23**, 4365 (1984); (b) T. P. Martin, Z. Phys. D **3**, 211 (1986).
150. M. F. Guest, I. H. Hillier, and V. R. Saunders, Trans. Far. Soc. **68**, 2070 (1972); C. R. Brundle, N. A. Kuebler, M. B. Robin, and H. Basch, Inorg. Chem. **11**, 20 (1972); R. Osman, P. Coffey, and J. R. Van Wazer, Inorg. Chem. **15**, 287 (1976); E. Fluck, C. M. E. Pavlidou, and R. Janoschek, Phosphorus Sulfur **6**, 469 (1979); G. Trinquier, J.-P. Malrieu, and J.-P. Daudey, Chem. Phys. Lett. **80**, 552 (1981).
151. E. A. Halevi, H. Bock, and B. Roth, Inorg. Chem. **23**, 4376 (1984). C. Baird, Can. J. Chem. **62**, 341 (1984).
152. K. Raghavachari, R. C. Haddon, and J. S. Binkley, Chem. Phys. Lett. **122**, 219 (1985); M. W. Schmidt and M. S. Gordon, Inorg. Chem. **24**, 4503 (1985); G. Trinquier, J.-P. Daudey, and N. Komiha, J. Am. Chem. Soc. **107**, 7210 (1985); R. Ahlrichs, S. Brode, and C. Ehrhardt, J. Am. Chem. Soc. **107**, 7260 (1985); P. Lazzeretti and J. A. Tossell, J. Phys. Chem. **91**, 800 (1987).
153. S. Nagase and K. Ito, Chem. Phys. Lett. **126**, 43 (1986); M. T. Nguyen and A. F. Hegarty, J. Chem. Soc. Chem. Commun. 383 (1986); R. Janoschek, Chem. Ber. **122**, 2121 (1989).
154. R. O. Jones and D. Hohl, J. Chem. Phys. **92**, 6710 (1990); R. O. Jones and G. Seifert, J. Chem. Phys. **96**, 7564 (1992).
155. M. Häser, U. Schneider, and R. Ahlrichs, J. Am. Chem. Soc. **114**, 9551 (1992).
156. R. Steudel, Topics in Current Chemistry, **102**, 149 (1982); R. Steudel, *Studies in Inorganic Chemistry 5*, edited by A. Muller and B. Krebs (Elsevier, New York, 1974), pp. 3 – 37.
157. J. Berkowitz and J. R. Marquart, J. Chem. Phys. **39**, 275 (1963). W. Rosinger, M. Grade and W. Hirschwald, Ber. Bunsenges. Phys. Chem. **87**, 536 (1983). P. Lenain, E. Picquenard, J. L. Lesne and J. Corset, J. Mol. Struc. **142**, 355 (1986). B. Meyer and T. Stroyer-Hansen, J. Phys. Chem. **76**, 3968 (1972). V. W. Genz and P. W. Schenk, Z. Anorg. Alleg. Chem. **379**, 300 (1970).
158. G. D. Brabson, Z. Mielke, and L. Andrews, J. Phys. Chem. **95**, 79 (1991); P. Hassanzadeh and L. Andrews, J. Phys. Chem. **96**, 6579 (1992).
159. J. Steidel, J. Pickardt and R. Steudel, Z. Naturforsch. **33B**, 1554 (1978). R. Steudel, Spectrochim. Acta **31A**, 1065 (1975). R. Steudel and F. Schuster, J. Mol. Struc. **44**, 143 (1978). S. J. Rettig and J. Trotter, Acta Cryst. **C43**, 2260 (1987). R. Steudel and H.-J. Mausle, Z. Naturforsch. **33A**, 951 (1978). R. Reinhardt, R. Steudel and F. Schuster, Angew. Chem. Int. Ed. Engl. **17**, 57 (1978). J. Steidel and R. Steudel, J. Chem. Soc., Chem. Commun. 1312 (1982). R. Steudel and D. F. Eggers, Spectrochim. Acta **31A**, 879 (1975).
160. N. C. Baird, J. Comput. Chem. **5**, 35 (1984). K. Jug and R. Iffert, J. Mol. Struc. (Theochem) **186**, 347 (1989). R. S. Laitinen, B. Randolph and T. A. Pakkanen, J. Comput. Chem. **8**, 658 (1987).
161. J. Kao, Inorg. Chem. **16**, 2085 (1977). N. R. Carlsen and H. F. Schaefer, Chem. Phys. Lett. **48**, 390 (1977). R. O. Jones, J. Chem. Phys. **84**, 318 (1986). J. E. Rice, R. D. Amos, N. C. Handy, T. J. Lee and H. F. Schaefer, J. Chem. Phys. **85**, 963 (1986). H. Basch, Chem. Phys. Lett. **157**, 129 (1989).
162. D. A. Dixon and E. Wasserman, J. Phys. Chem. **94**, 5772 (1990); G. E. Quelch, H. F. Schaefer, and C. Marsden, J. Am. Chem. Soc. **112**, 8719 (1990); W. von Niessen, J. Chem. Phys. **95**, 8301 (1991); V. G. Zakrzewski and W. von Niessen (to be published).
163. J. Kao, Inorg. Chem. **16**, 3347 (1977).
164. Z. S. Herman and K. Weiss, Inorg. Chem. **14**, 1592 (1975).
165. G. L. Carlson and L. G. Pedersen, J. Chem. Phys. **62**, 4567 (1975).
166. D. Hohl, R. O. Jones, R. Car and M. Parrinello, J. Chem. Phys. **89**, 6823 (1988).
167. K. Raghavachari, C. M. Rohlfing, and J. S. Binkley, J. Chem. Phys. **93**, 5862 (1990);

Electronic spectroscopy of diatomic molecules

Harry Partridge, Stephen R. Langhoff and Charles W. Bauschlicher, Jr.
NASA Ames Research Center, Moffett Field, CA 94035

feb 7

Abstract. This article provides an overview of the principal computational approaches and their accuracy for the study of electronic spectroscopy of diatomic molecules. This is a computationally demanding task, as it requires that the manifold of electronic states be determined at least in the Franck-Condon region for the bands of interest. We include examples from our work that illustrate the obtainable accuracy for a series of applications. First we consider the determination of the vibrational frequencies and transition intensities of the high vibrational levels of the ground states of SiO and NO where a globally accurate dipole moment function is required. We then consider the quintet states of CO and demonstrate that recently observed emission bands in an argon matrix can be explained as intercombination bands involving the $a''^5\Pi$ state. We show the utility of theory in assigning photoelectron detachment spectra by considering NO and the alkali oxide negative ions. We then present several examples illustrating the state-of-the-art in determining radiative lifetimes by considering both valence-valence and valence-Rydberg transitions. For systems involving first- and second-row atoms, theory is capable of yielding radiative lifetimes accurate to within 10%. The next examples illustrate the determination of the potential energy curves and transition moments for a full manifold of states. The last examples involve systems containing transition metal atoms and illustrate the difficulty in describing states with different numbers of d electrons.

Key words: *ab initio* calculations, radiative lifetimes

S. R. Langhoff (Ed.), Quantum Mechanical Electronic Structure Calculations with Chemical Accuracy, 209–260.

Contents

1 Introduction

Improvements in computational methods and computer architecture have greatly enhanced the number of spectroscopic problems that can be treated accurately by *ab initio* methods. Electronic structure calculations are contributing to spectroscopy over a very broad range of wavelengths. For example, theory has contributed in the microwave region by determining accurate geometries and rotational constants [1]. This has led in several cases to the identification of new molecules in interstellar space [2]. Vibrational transition energies and intensities can also be determined very accurately from *ab initio* calculations—see for example the chapter in this volume by Lee and Scuseria [3]. The emphasis in this chapter, however, is on electronic spectroscopy. This is computationally demanding in that generally a manifold of electronic states must be accurately known at least in the Franck-Condon region of the band systems of interest. We include in this category the determination of the frequencies and transition intensities of very high vibrational-rotational levels of the ground state, because this also requires a globally accurate potential and dipole moment function. While progress has been made in applying single reference-based techniques to electronic spectroscopy, multireference-based techniques are generally still required for high accuracy, especially when it is necessary to describe multiple states over a range of internuclear distances.

In this chapter we consider only diatomic systems; the spectroscopy of triatomic molecules is considered in the chapter by Perić *et al.* [4], while the spectroscopy of larger organic systems is considered in the chapter by Roos *et al.* [5]. A key goal of this chapter is to illustrate, for a range of molecular systems, the accuracy that can be currently achieved in calculated spectroscopic constants and band intensities.

In Section 2 we give an overview of the methodology and illustrate how recent benchmark calculations have given insight into the accuracy of approximate treatments of electron correlation. We demonstrate that the transition moment converges much faster in the length gauge than in the velocity representation, which is so sensitive to the level of treatment as to not even offer a reliable measure of convergence. In Section 3 we give several examples that illustrate the accuracy currently achievable in spectroscopic constants and molecular properties. For SiO and NO we report on the determination of a global dipole moment function that yields accurate transition probabilities between ro-vibrational levels with a high degree of vibrational or rotational excitation. We then show how theory has been instrumental in assigning the photoelectron detachment spectra of the NO molecule and the alkali oxide negative ions. Next we consider the quintet states of the CO molecule, where little experimental data exists. We show that emission bands recently observed from excitation of CO in an argon matrix can be explained as intercombination bands involving the $a''^5\Pi$ state. The next two examples illustrate the current state-of-the-art in computing radiative lifetimes. We consider valence transitions in NO and O_2^+ (Ogawa system and the corresponding first negative system, respectively) and valence-Rydberg transitions (β system of NO). Our next example shows how the calculation of potential energy curves and transition moments for a full manifold of states can aid in interpreting experimental spectra. We consider the Al_2, B_2 and AlB systems, which have recently been made in significant quantities using laser vaporization. Our final examples involve molecular systems containing transition metal atoms. We show that the difficulty in describing states with different numbers of *d* electrons significantly complicates an accurate characterization of the electronic states. This is especially well illustrated by calculations on the AlCu molecule. Our conclusions are presented in Section 4.

2 Methods

Although many methods have been developed to accurately solve the correlation problem [6], most are restricted to treating the lowest state of a given symmetry. In the study of electronic spectroscopy it is not only necessary to treat multiple states, often of the same symmetry, it is necessary to describe these over the Franck-Condon region for the transitions of interest. Generally, this requires a multireference configuration-interaction (MRCI) approach, which is in principle completely general, allowing for an arbitrary number of states in a symmetry to be treated. In practice, however, it can be difficult to obtain an equivalent treatment of many states. This is especially true if the orbitals are biased in favor of one state or a group of states, as can occur when the orbitals from an individual state are used. The bias introduced by using the "wrong" orbitals can be so large that prohibitively large CI calculations are required to overcome the bias. To avoid, or at least minimize, the orbital bias in the zeroth-order wave function, we use a state-averaged complete-active-space self-consistent-field (SA-CASSCF) approach [7] to optimize the orbitals. All states of interest of the same spin, regardless of spatial symmetry, are included in the SA-CASSCF calculation. In addition to minimizing the orbital bias, the common orthonormal set of orbitals facilitates the calculation of transition moments.

Prohibitively large active spaces are generally required to obtain a significant fraction of the correlation energy at the CASSCF level. Thus we account for only the most important correlation and near-degeneracy effects in the CASSCF treatment and then account for the external correlation in a subsequent step. Roos *et al.* [5] in their chapter discuss adding external correlation using perturbation theory. In this chapter we use the MRCI approach [8] (and size-extensive modifications [9]) to go beyond the CASSCF zeroth-order reference. Ideally the MRCI reference list includes all configurations in the CASSCF calculation (a so-called CAS-ref). This has become more tractable with the advent of the internal contraction (IC) procedure [10], where the ratios of the coefficients of classes of double excitations are fixed based on the coefficients in the zeroth-order wave function. When reference configuration selection is necessary, the selection is based on the coefficient of the individual configurations at the CASSCF level for each state throughout the Franck-Condon region. We have found that including all occupations that have coefficients with absolute values greater than 0.05 is usually sufficient to obtain accurate results, but there are important exceptions where smaller selection thresholds are required [11]. For more qualitative studies, it is also possible to select (for example, by an energy criteria) the CSF's in the final CI wave function [12]. However, computational advances and the inherent accuracy

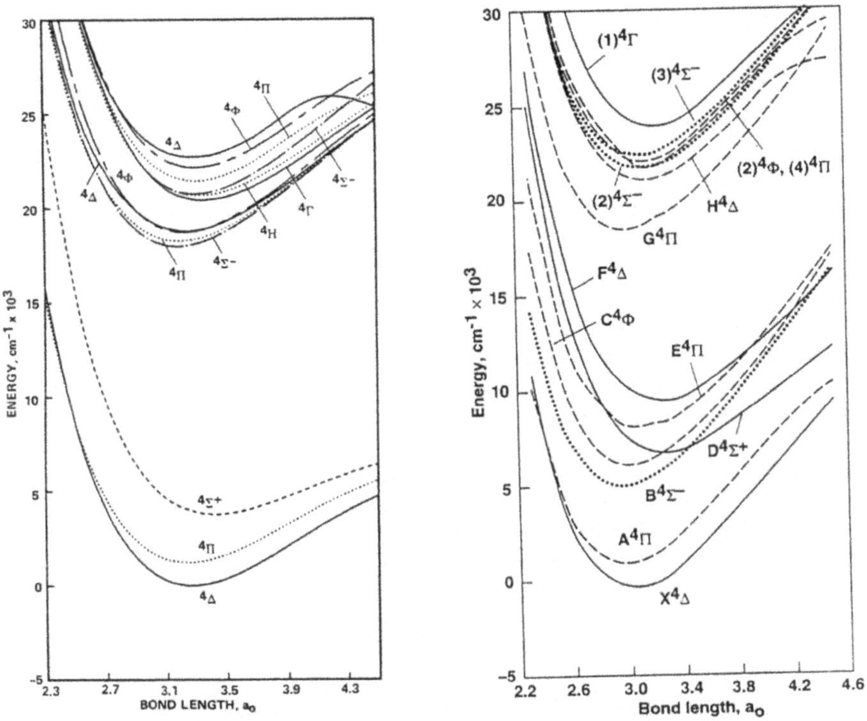

Fig. 1. The SA-CASSCF and MRCI potential energy curves for the quartet states of FeH are shown to the left and right, respectively.

improvements have made such approaches unnecessary for most applications. It is still often employed in schemes which include spin-orbit coupling [4]. The effect of higher excitations can be estimated by using a multireference analog of the Davidson correction [13], denoted +Q, or for the lowest state in each symmetry, more rigorously using the averaged coupled-pair functional (ACPF) approach [9].

Using the SA-CASSCF procedure, the orbital bias is generally only a significant problem in those cases where the states have very different orbital character. For molecules containing first- and second-row atoms, this usually occurs only when valence and Rydberg states are both included in the optimization. The problems are well illustrated by studies of the NO β system considered in Section 3.7. For transition metal systems the orbital bias is a more common problem, because of the different radial extent of the d orbitals in the $d^n s^2$ and $d^{n+1}s^1$ occupations. For example, the error in the carbon $^5S - {}^3P$ separation at the singles and doubles configuration-interaction (SDCI) level is less than 0.01 eV if either the ^5S or ^3P orbitals are used for both states, compared with using separate orbitals for each state. In contrast, an error of about 1 eV is introduced into the Ni $^3F(3d^9 4s^1)$ –

$^3F(3d^84s^2)$ separation if the orbitals from either state are used [14].

The effect of external correlation on the spectra of FeH is demonstrated in Fig. 1 where the quartet states are shown at the SA-CASSCF and MRCI levels of theory [15]. The $^5D(3d^64s^2)$ state of iron bonds to hydrogen by first forming *sp* hybrids. The lowest $^4\Delta$, $^4\Pi$, and $^4\Sigma^+$ states shown in Fig. 1 are obtained, when the electron in the non-bonding *sp* hybrid orbital is low-spin coupled to the $3d$ shell. The $^5F(3d^74s^1)$ state of Fe gives rise to $^4\Phi$, $^4\Delta$, $^4\Pi$, and $^4\Sigma^-$ states by forming a bond between the Fe $4s$ and H $1s$ orbitals. These four states along with the six quartet states derived from the $^3H(3d^64s^2)$ atomic state of Fe are the upper states in the SA-CASSCF potentials in Fig. 1. Since there is a larger correlation contribution to the states derived from $3d^74s^1$, the four states derived from the ^5F state are not as well described as the others at the CASSCF level. Thus these four states drop in energy when more extensive correlation is added in the MRCI calculation. In addition, the mixing of different asymptotes in general stabilizes one state and raises the other. This occurs for the $^4\Pi$ and $^4\Delta$ states, resulting in a stabilization of these two states relative to the others. This coupled with the general lowering of the ^5F derived states results in MRCI potential curves shown to the right in Fig. 1 that differ significantly from the much less accurate CASSCF curves shown to the left. Considering the large correlation contribution to the states, there may still be significant deficiencies in the MRCI potentials, although the limited experimental data [16] suggest that the relative ordering of the states is correct. To significantly improve the calculations would require adding a $3d$ shell to the CASSCF active space, which would dramatically expand the size of the calculation.

Theory can help guide experimental studies by predicting the relative strength of the electronic transitions. The computation of the transition density matrix is greatly facilitated by using a common set of orthonormal orbitals. The computation of the transition moment requires about the same computational effort as the one-electron contribution to the correlation energy. The transition moment can be computed in any one of three representations [17]. The length, velocity, and acceleration formulations are defined as

$$< \Psi_a | \sum_{k=1}^{n} r_k | \Psi_b >, \tag{1}$$

$$(E_a - E_b)^{-1} < \Psi_a | -i \sum_{k=1}^{n} \nabla_k | \Psi_b >, \tag{2}$$

and

$$(E_a - E_b)^{-3} < \Psi_a | - \sum_{k=1}^{n} \nabla_k V | \Psi_b >, \tag{3}$$

respectively, where Ψ_a and E_a are the wave function and energy of state a and V is the potential energy. The acceleration form has been shown to be extremely slowly convergent [18] and is not discussed further. We use only the length representation of the transition moment, denoted $D(r)$, in our work. As we show below, the velocity representation is significantly less reliable, to the point where it gives little insight into the accuracy of the transition moment in the length representation.

The transition moment for a perpendicular transition (such as $\Sigma^+ \rightarrow \Pi$) is

$$D(r) \;= < \Sigma^+|\frac{(x+iy)}{\sqrt{2}}|\Pi > . \tag{4}$$

Note that this definition of the matrix element is equivalent[19] to the Cartesian form $< \Sigma^+|x|\Pi_x >$. However, for all perpendicular transitions not involving a Σ state, the Cartesian form is $\sqrt{2}$ times smaller than that in complex orbitals, e.g.

$$\sqrt{2} < \Pi_x|y|\Delta_{xy} >=< \Pi|\frac{(x+iy)}{\sqrt{2}}|\Delta > . \tag{5}$$

For parallel transitions the transition moment is, for example,

$$D(r) \;= < \Sigma^+|z|\Sigma^+ > . \tag{6}$$

For diatomic molecules it is easy to obtain a numerical representation of the vibrational levels given the potential. We use a spline representation of the potentials and obtain the vibrational wave functions, ψ_v, using a finite difference scheme [20]. Using these vibrational wave functions, a spline representation of the transition moment, and the transition energies, we compute the Einstein A coefficient (s^{-1}) from the relation

$$A_{v'v''} = 2.026 \times 10^{-6}\; g\; \bar{\nu}^3_{v'v''}|D_{v'v''}|^2, \tag{7}$$

where

$$D_{v',v''} = \int_0^\infty \psi_{v'}(r)D(r)\psi_{v''}(r)dr. \tag{8}$$

In Eqs. 7 and 8, $\psi_{v'}$ and $\psi_{v''}$ are the upper and lower state vibrational wave functions, $\bar{\nu}_{v'v''}$ is the transition energy in wavenumbers between levels v' and v'', $D_{v',v''}$ is the transition matrix element between vibrational levels v' and v'' in a.u., and g is a statistical weighting factor, which is given as

$$(2 - \delta_{0,\lambda'+\lambda''})/(2 - \delta_{0,\lambda'}), \tag{9}$$

where λ' and λ'' denote the orbital angular momenta of the upper and lower states.

The radiative lifetime ($\tau_{v'}$) of a vibrational level depends on the sum of the transition probabilities to all lower vibrational levels in all lower electronic states. If rotational effects are neglected, the lifetime can be written as

$$\tau_{v'} = \left(\sum_{v''} A_{v'v''}\right)^{-1} . \tag{10}$$

The lifetimes of vibrational decay within one electronic state can be computed in a similar manner except that there is only one potential and $D(r)$ is replaced by the dipole moment function.

The accuracy of the CASSCF/MRCI approach depends on the completeness of both the one- and n-particle basis sets. Considerable insight into the limitations of various truncated correlation methods was obtained by comparison with full CI (FCI) calculations [21], which include all possible configurations. The FCI represents the exact solution (within the Born-Oppenheimer approximation) for a given one-particle basis and provides an unambiguous standard with which to compare approximate (truncated) correlation methods. The difficulty in applying the FCI approach is that the configuration expansion increases factorially with the number of orbitals and electrons. With improvements in both FCI methodology [22] and computer architecture, it became possible to benchmark a wide range of properties using small but realistic one-particle basis sets. An important conclusion of the FCI studies was that the CASSCF/MRCI approximation consistently agreed well with the FCI result [21]. This showed that much of the error in typical CASSCF/MRCI calculations arose from limitations in the one-particle basis set. This observation led to the development of new one-particle basis sets that are designed for correlated treatments and are systematically expandable. For example, the atomic natural orbital (ANO) basis sets, developed by Almlöf and Taylor [23], employed very large primitive Gaussian basis sets, which were contracted based on the natural orbitals of atomic CI calculations. Widmark *et al.* [24] improved on this approach by averaging sets of natural orbitals derived from the ground state, excited states, ions, solutions including added electric fields, etc. These ANO basis sets perform much better for properties than the original ANOs, which were optimized on the basis of the total energy. More recently, Dunning and co-workers [25] have developed the augmented correlation consistent polarized valence (aug-cc-pV) sets, in which the size of the primitive sets increase with improvements in the basis set quality. The performance of the cc-pV basis sets is demonstrated in Table I, where we summarize the dissociation energy, D_e, of N_2 at the ICMRCI and ICMRCI+Q levels of theory [26]. The convergence of the D_e value with basis set size is logarithmic [27] making it possible to extrapolate to the one-particle limit. If the effect of $1s$ correlation is included, the ICMRCI and ICMRCI+Q results [26] bracket the experimental D_e value. This is consistent with the fact that the FCI value in a smaller basis set also lies between the ICMRCI and ICMRCI+Q D_e values [28]. These calculations demonstrate the phenomenal improvement in the accuracy of *ab initio* calculations over the last few years.

Our FCI benchmark study [29] of the transition moment for the OH ultraviolet band system ($A^2\Sigma^+ - X^2\Pi$) gives some insight into the accuracy that is currently achievable in radiative lifetimes for first-row molecules. We carried out the FCI calculations using a [$4s$ $3p$ $2d/2s$ $1p$] basis set that provides a realistic representation of the transition moment. In Table II we compare several approximate treatments with the FCI for the $A^2\Sigma^+ - X^2\Pi$ energy separation and transition moment at r=1.80 a_0. The difference between the FCI(5) and FCI(7) results shows that the oxygen $2s$-like electrons must be correlated for quantitative results. Thus we correlate seven electrons in all of the approximate treatments tabulated in Table II. The $A - X$ energy separation is insensitive to the level of treatment, with reasonably accurate values being obtained even at the SCF level. However, the SCF level transition moment is much too large. Even though the magnitude de-

TABLE I
Summary of the D_e values (in kcal/mol) for N_2.

Basis	ICMRCI	ICMRCI+Q	Basis set size
cc-pVTZ	217.8	216.6	[4*s* 3*p* 2*d* 1*f*]
cc-pVQZ	224.1	223.1	[5*s* 4*p* 3*d* 2*f* 1*g*]
cc-pV5Z	226.2	225.3	[6*s* 5*p* 4*d* 3*f* 2*g* 1*h*]
cc-pV6Z	227.1	226.3	[7*s* 6*p* 5*d* 4*f* 3*g* 2*h* 1*i*]
cc-pV∞	227.9	227.0	
cc-pV∞+1*s*	228.7	227.8	
Expt	228.4		

TABLE II
Study of the $A-X$ energy separation (Δ) and transition moment of OH with the level of correlation treatment at r=1.80 a_0 in a [4*s* 3*p* 2*d*/2*s* 1*p*] Gaussian basis set.

Calculation	Δ(cm^{-1})	Moment(a.u.)
FCI(5)	34681	0.1010
FCI(7)	33986	0.1195
SCF	34850	0.1731
SDCI	34077(33898)	0.1253
MCPF	33906	0.1214
(222)-CASSCF	35372	0.1618
(222)-CASSCF[a]	36220	0.1693
(222)-MRCI(0.05)	34131(33897)	0.1232
(222)-MRCI(0.0)	34043(33876)	0.1230
(3221)-CASSCF	34609	0.1235
(3221)-MRCI(0.02)	34002(33947)	0.1221
(3221)-MRCI(0.01)	34033(33952)	0.1200
(3221)-MRCI(0.0)	34013(33933)	0.1198

[a] A non-orthogonal transition moment is evaluated based on separate state CASSCF calculations.

creases with the inclusion of correlation at the SDCI and modified CPF (MCPF) levels [30], the value remains too large. Accounting for the near degeneracy effects at the (222)-CASSCF level gives only a marginal improvement over the SCF value. This CASSCF treatment has two active orbitals in the σ and π spaces, which is denoted (222) in C_{2v} symmetry to signify that two a_1, two b_1, and two b_2 orbitals are active. This CASSCF describes the bond breaking as well as the important $p\pi \rightarrow p'\pi$ correlation. Adding more extensive correlation at the (222)-MRCI level gives a result that is only marginally better than the SDCI value. Selecting reference configurations at a threshold of 0.05, denoted MRCI(0.05), does not affect the accuracy of the moment. However, expanding the active space to include a δ

orbital dramatically reduces the CASSCF moment. The (3221)-MRCI moment is now in good agreement with the FCI. The necessity of adding the δ orbital to the active space is unique to the transition moment, as neither the spectroscopic constants nor the dipole moment are sensitive to its inclusion. If reference selection is employed in the subsequent MRCI, a smaller selection threshold of 0.01 must be used to fully account for the effect of the δ orbital, because of the small coefficients of these configurations in the CASSCF wave function. This study clearly shows the value of having FCI calculations to define an accurate truncated n-particle treatment. Having obtained an MRCI approach that reproduces the FCI result in a moderate-sized basis set, we then take this MRCI treatment as close as possible to the one-particle limit.

We believe that the (3221)MRCI(0.01) treatment in a nearly complete one-particle basis gives a transition moment that is accurate to about 2%. This produces a lifetime that is accurate to about 5%, which is sufficient to critically assess the relative accuracy of the experimental values. Also from the pattern of convergence, we believe that the calculated lifetime should be a lower bound. Our best value [29] of 685±20 ns is in very good agreement with two laser excitation fluorescence experiments [31, 32], thus confirming their accuracy and ruling out the shortest[33] and three longest experimental lifetimes [34, 35]. In addition, the theoretical lifetimes agree with the high-frequency deflection experiments [35] for rotational levels above about five, which suggest that cascade effects must be causing an apparent increase in the lifetimes for the lowest few N' values.

Ideally we would like to develop a diagnostic that avoids having to perform active space, threshold, and one-particle basis set saturation studies. Because the transition moment computed in different representations will agree only for a complete CI (*i.e.* a FCI in an infinite one-particle basis set), one possibility is to use the difference between the length and velocity representations as a measure of the accuracy of the computed moment. However, theoretical studies [36] of H_2, BH, and C_2 did not reveal a positive correlation between the accuracy of the transition moment and the difference between the length and velocity representations. For example, consider the transition moments between the ground state of H_2 and the lowest two $^1\Pi_u$ and two $^1\Sigma_u^+$ states given in Table III.

Because SDCI corresponds to a FCI for a two-electron system, this study measures the convergence with respect to the completeness of the one-particle basis set. The basis set completeness is improved from basis set A to G (specific details can be found in the original work). The most striking feature is that for virtually all of the basis sets considered, the differences between the two representations are much less than the difference with the best basis set (Set G). For example, for basis set B, the transition moments in the two representations differ by 0.0182 a.u. for the $(2)^1\Sigma_u^+$ state, but the length value differs from that in basis set G by 0.2001 a.u. Clearly the difference between the length and velocity transition moments is not a good measure of the one-particle basis set completeness.

A study of the C_2 $A^1\Pi_u - X^1\Sigma_g^+$ transition moment in the length and velocity representations is given in Table IV. In the first series of calculations only the $A^1\Pi_u$ and $X^1\Sigma_g^+$ states were included in the CASSCF state averaging (denoted a two-root calculation). The agreement between the length and velocity transition

TABLE III
Summary of H_2 results at 1.4 a_0. The transition moment, given in both the velocity and length representations, is in a.u. and the vertical energy separation, ΔE, is in cm^{-1}.

	Root 1			Root 2		
	ΔE	length	velocity	ΔE	length	velocity
A. [4*s* 2*p* 1*d*] ANO set						
$^1\Sigma_u^+$	116874.6	1.2718	1.2325	239018.3	0.3539	0.3856
$^1\Pi_u$	180758.1	1.1447	0.9915	348453.2	0.0723	0.0895
B. [4*s* 2*p* 1*d*]+(1*s* 1*p*)						
$^1\Sigma_u^+$	103092.7	0.9836	0.9750	124241.6	0.5974	0.5792
$^1\Pi_u$	107611.8	0.7256	0.7143	200930.6	0.9135	0.8184
C. [4*s* 3*p* 2*d*]+(1*s* 1*p*)						
$^1\Sigma_u^+$	103058.2	0.9833	0.9807	124249.5	0.5843	0.5787
$^1\Pi_u$	107390.9	0.7539	0.7493	179409.0	0.8860	0.8388
D. [4*s* 3*p* 2*d*]+(2*s* 2*p*)						
$^1\Sigma_u^+$	102877.9	0.9787	0.9768	119798.0	0.4104	0.4110
$^1\Pi_u$	107026.1	0.7029	0.6948	123149.3	0.4691	0.4679
E. [(4 + 1)*s* (3 + 1)*p* 2*d*]+(2*s* 2*p*)						
$^1\Sigma_u^+$	102768.9	0.9816	0.9809	119778.6	0.4103	0.4110
$^1\Pi_u$	106624.1	0.7394	0.7378	123096.2	0.4788	0.4776
F. [(4 + 1)*s* (3 + 1)*p* (2 + 1)*d*]+(2*s* 2*p*)						
$^1\Sigma_u^+$	102769.9	0.9816	0.9812	119785.2	0.4108	0.4110
$^1\Pi_u$	106571.6	0.7416	0.7407	123073.1	0.4779	0.4771
G. [4*s* 3*p* (2 + 1)*d* 2*f*]+(3*s* 3*p* 1*d*)						
$^1\Sigma_u^+$	102841.7	0.9802	0.9796	119719.7	0.3973	0.3971
$^1\Pi_u$	106787.0	0.7212	0.7170	120921.7	0.3324	0.3302

moments is poor, and the discrepancy increases as the one-particle basis set is improved. In addition, the velocity representation is found to be sensitive to 1*s* correlation while the length representation is not. We have also considered three-root MRCI calculations, where the $(2)^1\Sigma_g^+$ state is added to the SA-CASSCF calculation, due to the avoided crossing that occurs between the lowest two $^1\Sigma_g^+$ states at an r value slightly larger than the r_e of the ground state. Although this should be a superior treatment, the velocity and length transition moments differ by more than in the two-root approach for the smallest basis set considered. Again the velocity transition moment is sensitive to 1*s* correlation. However, unlike the two-root calculations, the length and velocity transition moments agree better with improvements in the one-particle basis set. Because FCI calibration calculations [37] support the accuracy of the three-root MRCI(0.025) treatment, the [(5+1)*s* (4+1)*p* (3+1)*d* 2*f* 1*g*]+(*sp*) MRCI(0.025) length value is probably the most accurate (last row of Table IV). The order of magnitude greater sensitivity of the velocity transition moment to systematic improvements in the one- and

TABLE IV
Summary of $A^1\Pi_u - X^1\Sigma_g^+$ C_2 transition. The transition moment is given in a.u. and the vertical energy separation, ΔE, is in cm^{-1}.

	ΔE	length	velocity
2-root calculations			
[3s 2p 1d] MRCI(0.0)	6003	0.2768	0.2018
[3s 2p 1d] MRCI(0.0)(1s)	6031	0.2766	0.1668
[3s 2p 1d] MRCI(0.05)	5820	0.3040	0.1077
[3s 2p 1d] MRCI(0.025)	6048	0.2833	0.2041
[4s 3p 2d] MRCI(0.025)	5433	0.2966	0.0246
[4s 3p (2 + 1)[a]d] MRCI(0.025)	5437	0.2940	0.0091
[4s 3p (2 + 1)d] MRCI(0.025)[b]	5427	0.2870	0.0204
[(5 + 1)s (4 + 1)p 2d] MRCI(0.025)	5368	0.3007	0.0677
3-root calculations[c]			
[3s 2p 1d] MRCI(0.0)	5974	0.2777	0.1398
[3s 2p 1d] MRCI(0.0)(1s)	6006	0.2777	0.0996
[3s 2p 1d]MRCI(0.025)	5997	0.2798	0.1099
[4s 3p (2 + 1)d] MRCI(0.025)	5366	0.2974	0.1281
[(4 + 1)s (3 + 1)p (2 + 1)d] MRCI(0.025)	5290	0.3013	0.2158
[(4 + 1)s (3 + 1)p (2 + 1)d] MRCI(0.025)(1s)	5375	0.3019	0.2725
[(4 + 1)s (3 + 1)p (2 + 1)d 1f] MRCI(0.025)	5328	0.2999	0.1929
[(4 + 1)s (3 + 1)p 2d] MRCI(0.025)	5303	0.3005	0.1819
[(4 + 1)s (3 + 1)p (2 + 1)d] + (sp)[d] MRCI(0.025)	5282	0.3016	0.2218
[(5 + 1)s (4 + 1)p 3d 2f 1g] + (sp) MRCI(0.025)	5237	0.2924	0.2530
[(5 + 1)s (4 + 1)p (3 + 1)d 2f 1g] + (sp) MRCI(0.025)	5236	0.2912	0.2552

[a] Indicates two ANOs with the outermost primitive uncontracted.

[b] Three-root reference list.

[c] MRCI(0.025) calculations use the 0.025 reference list selected based on r values between 1.8 and 5.0 a_0 from reference [37]

[d] Indicates that an even-tempered s and p function have been added.

n-particle basis sets makes this representation of questionable value even as a diagnostic of the convergence. Thus we conclude that the best method of avoiding a systematic error is still to perform a sensitivity study of the convergence of the length transition moment with respect to expanding the active space, reducing the reference selection threshold, and expanding the one-particle basis set.

3 Applications

In this section we review several applications of state-of-the-art quantum mechanical calculations to molecular systems. The first two illustrate the calculation of transition probabilities between vibrational levels of the ground state. We then demonstrate how calculations can facilitate the interpretation of photoelectron spectra. This requires computing accurate potentials for the n and $n-1$ electron systems and computing Franck-Condon factors between them. We then consider several applications of electronic spectroscopy were the calculation of the spectroscopic constants and radiative lifetimes are the primary purpose of the calculations.

3.1 Dipole moment of SiO

The silicon monoxide molecule is of considerable astrophysical interest due to its prevalence in the atmospheres of many cool stars. Accurate band strengths for the vibrational-rotational bands of SiO are required to determine elemental abundances. Since recent observations and analysis [38, 39] of the cool giant star Alpha Tau suggested that the Einstein coefficients for the fundamental and overtone bands differed significantly from the accepted values, namely those based on the electric dipole moment function (EDMF) of Tipping and Chackerian (TC) [40], we recently redetermined a global EDMF using high-level *ab initio* methods [41]. The EDMF of TC was chosen to reproduce the accurate experimental dipole moments [42] for $v = 0 - 3$ and to have the correct united and long-range limits. In part it was based on the long-range portion of the theoretical dipole moment function of Langhoff and Arnold [43]. While several more recent theoretical studies [44] have suggested that the fundamental band strengths are 15-20% larger than the values reported by TC, no systematic study of the EDMF of SiO was available before our study.

Our calculations [41] showed that while it is straightforward to converge the EDMF with respect to improvements in the one-particle basis set, it is necessary to use a basis set containing both diffuse functions and several sets of polarization functions. Our final contracted Gaussian basis sets were of the form $[8s\,7p\,4d\,2f\,1g]$ for Si and $[7s\ 6p\ 4d\ 2f\ 1g]$ for O. However, it proved to be much more difficult to achieve convergence with respect to the n-particle or correlation treatment. While inner-shell (Si $2s$ and $2p$) correlation did not significantly affect the EDMF, the shape, especially at larger r values, was sensitive to the valence correlation treatment. We considered two CASSCF/MRCI treatments that differed in the choice of the active space. The smaller active space, denoted (433) in C_{2v} symmetry, includes the Si $3s$ and $3p$ electrons and oxygen $2p$ electrons. The larger (633) active space also includes the oxygen $2s$ orbital and a corresponding correlating orbital. The oxygen $2s$ electrons are correlated in all cases in the subsequent ICMRCI calculations.

The magnitude of the dipole moment increases substantially when the oxygen $2s$ electrons are included in the active space. However the difference between the (433) and (633)-based active spaces decreases at higher levels of correlation treatment—see Table V. More specifically, at the CASSCF level the effect of including the oxygen $2s$ electrons in the active space is to increase the dipole moment by about 0.21-0.22 D. At the MRCI level the difference decreases to about 0.07-0.08 D and at the ACPF level the difference is about 0.03-0.04 D. Thus, as the level of correlation increases, the bias from the zeroth-order reference is diminished. The CASSCF(433)/ACPF results are expected to be the best treatment, because the convergence with the level of theory is more rapid for the (433)-based treatment and the calculation can be carried out without reference selection. However, the true EDMF probably lies between the (433) ACPF and (633) MRCI curves shown in Fig. 2. Both theoretical EDMFs differ significantly from that deduced by TC, especially away from r_e where there is no experimental data to guide the

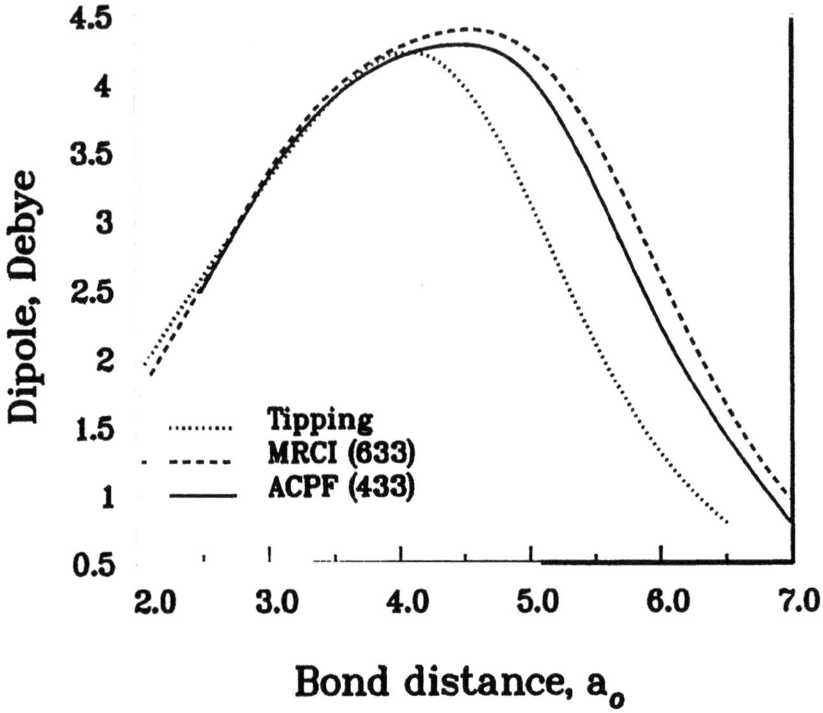

Fig. 2. Three electric dipole moment functions (EDMFs) for the $X^1\Sigma^+$ state of SiO. The solid and dashed lines denotes the (433) ACPF and (633) MRCI EDMFs of Ref. [41]. The EDMF denoted by the dotted line is from the work of Tipping and Chackerian [40].

construction of the empirical EDMF.

It is possible to bring the (433)-based CASSCF/ACPF EDMF into agreement with the accurate experimental dipole moments for v=0-3, by shifting it by 0.010 a_0 to shorter r values and increasing its magnitude by 0.0391 D. The shift in r is consistent with the error in r_e, due primarily to the neglect of inner-shell correlation. Our best transition probabilities were computed using this "shifted" EDMF and the Klein-Dunham potential constructed from the constants given by Huber and Herzberg [45]. To aid in spectroscopic and astronomical studies, we have generated line strengths in the form of gf values for all P and R branch lines for which $J \leq 250$, $v \leq 15$ and $\Delta v \leq 4$. These theoretical line strengths appear[39] to be in better accord with astrophysical observations than those deduced using the results of TC [40].

The SiO study illustrates the limitations of generating a global dipole function using accurate experimental data that is limited to a very small region of r. A better procedure is to correct a globally accurate theoretical EDMF using the

TABLE V

Theoretical dipole moments (Debye), Einstein coefficients (sec^{-1}), and dipole derivatives (a.u./a_0) as a function of the level of correlation treatment.

	(433) MRCI	(633) MRCI	(433) ACPF[a]	Expt. [42]
v=0	3.0541	3.1293	3.0436(3.0983)	3.0982
v=1	3.0746	3.1500	3.0639(3.1184)	3.1178
v=2	3.0951	3.1706	3.0836(3.1380)	3.1372
v=3	3.1153	3.1911	3.1028(3.1572)	3.1574
A_{10}	6.92	6.97	6.73(6.68)	
A_{20}	0.273	0.270	0.264(0.266)	
A_{30}	0.004	0.004	0.005(0.004)	
$(d\mu/dr)_{r_e}$	0.621	0.623	0.613(0.610)	

[a] The values in parentheses correspond to a dipole moment that has been shifted in r by -0.010 a_0 and increased in magnitude by 0.0391 D.

accurate, but limited experimental data. The modified EDMF can then be used to predict accurate transition probabilities for highly rotationally and/or vibrationally excited molecules.

3.2 Dipole moment for NO($X^2\Pi$)

The infrared vibrational transition probabilities for NO($X^2\Pi$) are important for diagnostic applications and for investigations of energy transfer [46, 47]. Because NO is often found highly vibrationally excited, for example, in aurorally disturbed atmospheres [48, 49] and in the shock region in front of vehicles traveling at hypersonic speeds, we have developed [50] a globally accurate EDMF to calculate the Einstein coefficients for high vibrational quantum numbers. There is also some recent experimental data with which to compare our results. Ratios of the first overtone to fundamental vibrational Einstein coefficients for v'=2−13 have been measured by Rawlins *et al.* [51] using spectrally resolved infrared chemiluminescence. These experimental ratios differ from those predicted on the basis of the theoretical EDMF of Billingsley [52]. Rawlins *et al.* argued that this could have significant implications for the shape of the dipole moment function at large internuclear distances (r) and for the behavior of $A_{v',v''}$ for high v' and/or J'.

Calibration calculations were carried out to investigate the sensitivity of the dipole moment to improvements in the one-particle basis sets, size of the CASSCF active space, and the n-particle treatment. We used the (augmented) correlation-consistent polarizedvalence [(aug-)cc-pV] basis sets developed by Dunning and co-workers [25]. The dipole moment was found to be insensitive to basis set saturation. We selected the augmented quadruple-zeta basis (aug-cc-pVQZ), because this gave a dipole moment nearly identical with the larger correlation consistent quintuple-zeta basis as well as an accurate $X^2\Pi$ potential. The dipole moment was found to be much more sensitive to the size of the n-particle treatment, but the dipole moment was less sensitive to expanding the CASSCF active space when it was evaluated using a finite-field approach rather than as an expectation value.

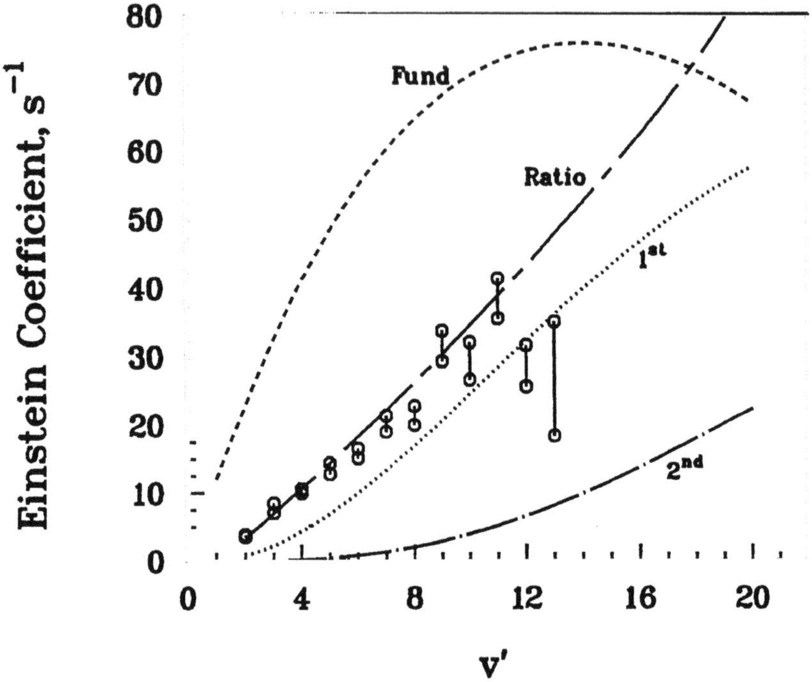

Fig. 3. Einstein coefficients (s^{-1}) for the fundamental, first overtone, and second overtone transitions ($v' \leq 20$) for $NO(X^2\Pi)$. Also shown are the ratios of the first overtone to fundamental transitions ($\times 100$). The experimental ratios ($\times 100$) of Rawlins *et al.* [51] are shown by the connected circles indicating 1σ uncertainties.

We found that a $2p$ CASSCF (a (2220) active space in C_{2v} symmetry) followed by an ACPF treatment of electron correlation gave a dipole moment that was within 0.02 D of the accurate experimental value [53] for v=0. In addition, the difference [54] between the dipole moment for the v=0 and v=1 levels and the Einstein coefficients for the (1-0) and (2-0) bands were also in good agreement with experiment [55, 56], especially the values of Chandraiah and Cho [56]. Furthermore, the ratios of the Einstein coefficients for the first overtone to fundamental bands were found to be insensitive to further refinements of the theoretical treatment. Thus our best Einstein coefficients were obtained using the aug-cc-pVQZ basis set at the (2220)-ACPF level of correlation treatment with the EDMF evaluated using a finite-field approach. The Einstein coefficients were evaluated using a finite difference method with spline representations of the EDMF and the theoretical potential, but were not significantly different when the experimental derived Klein-Dunham potential was used in the vibrational analysis.

The experimental data for the Einstein coefficients involving vibrational levels above v'=2 are rather limited. Recently, Rawlins *et al.* determined the ratios of the overtone to fundamental Einstein coefficients for v'=2-13 by measuring spec-

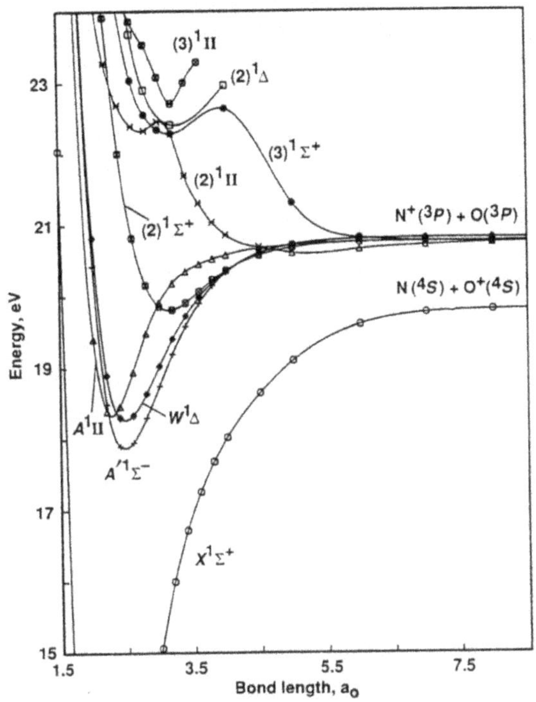

Fig. 4. The MRCI singlet state potentials of NO^+ in the adiabatic representation. The zero of energy is the $X^2\Pi$ state of NO.

trally resolved chemiluminescence in a cryogenic low-pressure reactor. Their data are compared graphically in Fig. 3 with our theoretical values. The experimental data agrees reasonably well with theory for v'=2, 3, 4, 5, 7, 9, and 11, but are significantly different for v'=6, 8, and 10, and differ greatly for v'=12 and 13. The poor agreement between the theoretical and experimental data for v'=12 and 13 may be due to interference from the (2,0) band of the N_2 ($W^3\Delta_u \rightarrow B^3\Pi_g$) electronic transition. The theoretical results plotted in Fig. 3 show conclusively that the ratios increase smoothly with increasing v'. In fact, the increase becomes more rapid for higher v', because the fundamental bands decrease in intensity for v'>14. The theoretical ratios are probably more accurate than the individual Einstein coefficients for the fundamental, first overtone, and second overtone bands plotted in Fig. 3, because they are much less sensitive to refinements in the theoretical calculations. In fact, we believe that the ratios that can be deduced from the Einstein coefficients reported in Ref. [50] are of at least semiquantitative accuracy even for rather high degrees of vibrational excitation (e.g. $v' < 30$). We recommend using these values in partitioning the radiative and collisional quenching of vibrationally excited NO.

TABLE VI
Spectroscopic constants for NO^+. The bond length is in a_0, ω_e, and $T_e(NO^+)$ are in cm^{-1}, and $T_e(NO)$ is in eV.

	MRCI[a]				Expt.[b]			
State	r_e	ω_e	$T_e(NO^+)$	$T_e(NO^c)$	r_e	ω_e	$T_e(NO^+)$	$T_e(NO)$
$X^1\Sigma^+$	2.031	2322			2.009	2378		
$a^3\Sigma^+$	2.457	1209	50979	15.55	2.421	1303	52145	15.70
$b^3\Pi$	2.236	1688	58692	16.51	2.222	1710	59177	16.57
$w^3\Delta$	2.448	1384	61390	16.85	2.417	1317	61947	16.91
$b'^3\Sigma^-$	2.451	1258	67359	17.59	2.411	1283	67776	17.64
$A'^1\Sigma^-$	2.475	1301	69571	17.86	2.432	1280	69543	17.86
$^5\Sigma^+$	3.228	618	71287	18.07				
$W^1\Delta$	2.485	1273	72185	18.18	2.456	1218	71647	18.12
$A^1\Pi$	2.279	1654	73288	18.32	2.256	1601	73472	18.34
$^5\Pi$	2.932	599	82355	19.44				
$(2)^1\Sigma^+$	3.157	711	85289	19.81				
$(2)^5\Pi$	4.056	454	89199	20.29				
$(2)^3\Delta$	3.240	680	89268	20.30				
$(2)^3\Pi$	4.064	407	90980	20.51				
$(2)^3\Sigma^+$	(3.592)	(509)	(91500)					
$(2)^5\Sigma^+$	4.478	192	91143	20.53				
$(3)^3\Delta$	(3.748)	(425)	(97000)					
$(3)^3\Pi$	3.265	485	99116	21.52				
$(2)^3\Pi$	2.200[d]		101100	21.78				
$(3)^1\Sigma^+$	3.235	462	105298	22.28				
$(2)^1\Pi$	2.754	795	105627	22.33				
$(2)^1\Delta$	3.276	584	106273	22.41				
$(3)^1\Pi$	3.199	1440	108723	22.71				

[a] The values in parentheses are the FOCI results. The T_e values are estimates based on relative positions with the MRCI results.

[b] Ref. [60].

[c] Computed using the experimental $X^2\Pi(NO)-{}^1\Sigma^+(NO^+)$ separation.

[d] Inflection in the $(2)^3\Pi$ potential, which is assigned to the 21.7 eV feature in the photoelectron spectrum [59].

3.3 The photoelectron spectra of NO

The most stable diatomic cation NO^+ is important in many ionospheric phenomena such as aurora [57] and is thought to be present in interstellar clouds. The spectroscopy of NO^+ is less well known than the isoelectronic N_2 and CO molecules, largely because of the difficulty in studying its band systems in absorption. Much of what is experimentally known about NO^+ comes from the photofragment spectroscopic studies by Cosby and Helm [58] and from high-resolution photoelectron spectroscopy of NO [59]. A critical assessment of both the experimental and theoretical data though 1978 is given in the paper by Albritton *et al.* [60].

To provide a better understanding of the photoelectron spectra of NO, we stud-

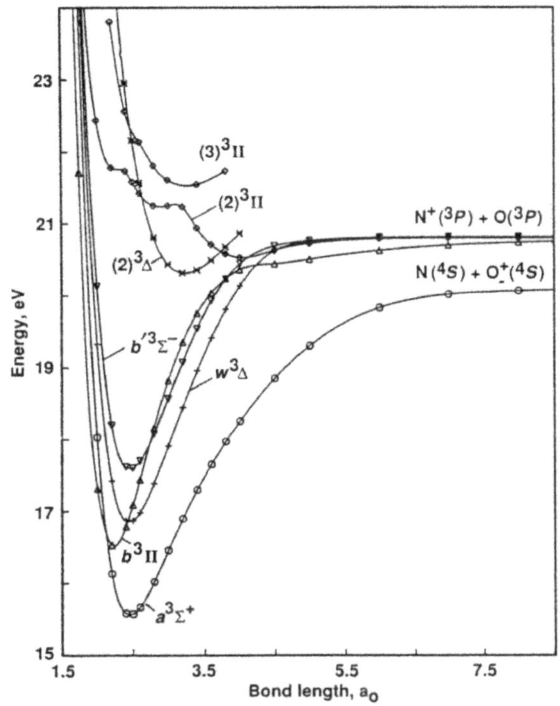

Fig. 5. The MRCI triplet state potentials of NO^+ in the adiabatic representation. The zero of energy is the $X^2\Pi$ state of NO.

ied the singlet, triplet, and quintet states of NO^+ using the first-order configuration-interaction (FOCI) method and the more accurate SA-CASSCF/MRCI procedure [61]. Each multiplicity was studied separately using a common set of molecular orbitals to facilitate the calculation of the radiative lifetimes of selected electronic states of NO^+. The potential energy curves for the singlet and triplet states are plotted in Figs. 4 and 5. The computed spectroscopic constants are compared with the available experimental data in Table VI. Since the MRCI state separations (T_e) are in excellent agreement with the experimental data [60], the theoretical values in Table VI should provide accurate estimates for those states not amenable to photoelectron spectroscopy, such as the quintet states and those states not described by a single electron detachment process from $NO(X^2\Pi)$. The ionization energies were determined by adding the experimental ionization potential (IP) for NO [60] to the computed T_e values for NO^+. The ionization energies computed in this manner agree with experiment to within 0.15 eV for the states below 18.34 eV.

The next features in the photoelectron spectrum [59] are at 21.7 eV, followed by a progression of peaks between about 22.6 and 23.6 eV. Theory strongly suggests assigning the feature at 21.7 eV to the $(2)^3\Pi$ state, as there are no other states with

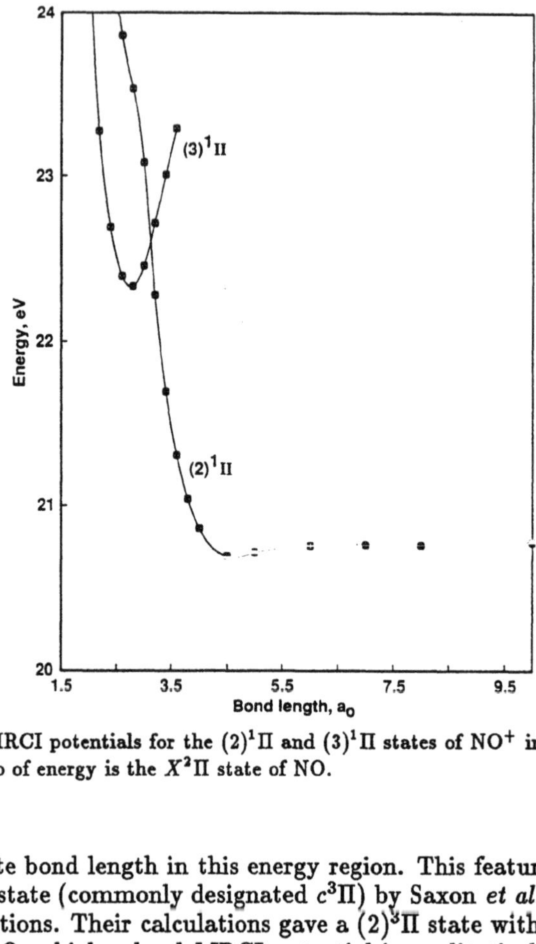

Fig. 6. The MRCI potentials for the $(2)^1\Pi$ and $(3)^1\Pi$ states of NO^+ in the diabatic representation. The zero of energy is the $X^2\Pi$ state of NO.

an appropriate bond length in this energy region. This feature was first assigned to the $(2)^3\Pi$ state (commonly designated $c^3\Pi$) by Saxon *et al.* [62] based on their FOCI calculations. Their calculations gave a $(2)^3\Pi$ state with a quasibound level at 21.84 eV. Our higher level MRCI potential is qualitatively similar to that of Saxon *et al.*, but shows only an inflection point near 2.20 a_0. While still higher levels of theory may produce a shallow well, this is not essential to explain the feature at 21.7 eV. The large curvature of the $(2)^3\Pi$ state in the Franck-Condon region may explain the asymmetry of this peak.

The progression of peaks in the photoelectron spectra between about 22.6 and 23.6 eV has been interpreted in terms of a 729 cm^{-1} vibrational frequency [60]. The peaks were originally assigned by Edqvist *et al.* [59] to the $B'^1\Sigma^+$ state based on analogies with the $b'^1\Sigma_u^+$ state of N_2. However, Lefebvre-Brion [63] suggested that this feature was due to the $B^1\Pi$ state based upon an analysis of the correlation effects, which indicated that the $B^1\Pi$ state should have a larger r_e value and lie about 1.9 eV above the $c^3\Pi$ state. We find the $(3)^3\Pi$, $(2)^1\Pi$, $(3)^1\Pi$, $(3)^1\Sigma^+$, and $(2)^1\Delta$ states to lie in this energy region. The $(2)^1\Pi$ and $(3)^1\Sigma^+$ states, which arise from 4σ and 1π ionizations of NO respectively, both have potential wells that are too shallow to explain the observed vibrational progression. The r_e values of the $(2)^1\Delta$ and $(3)^3\Pi$ states are too large to explain the intensity considering the unfavorable Franck-Condon factors. As the calculations place the $(4)^1\Sigma^+$ state above the $(2)^1\Delta$ state, it cannot explain the observed spectrum. Our FOCI results indicate that the $(2)^3\Sigma^+$ and $(3)^3\Sigma^+$ states are essentially repulsive and that the $(3)^3\Delta$ state has much too large a bond length. However, the explanation for the vibrational progression becomes clear when the $(2)^1\Pi$ and $(3)^1\Pi$ states are drawn in the diabatic representation—see Fig. 6. This is the preferred representation assuming that the two states are weakly interacting in the region of the avoided crossing. Using the diabatic $(3)^1\Pi$ state potential produces Franck-Condon factors with relative intensities similar to that in Fig. 2 of Edqvist *et al.* [59]. Thus the assignment of the spectrum between 22.6 and 23.6 eV to the diabatic $(3)^1\Pi$ (or $B^1\Pi$) state seems secure.

3.4 Photodetachment spectra of the alkali oxides

The alkali oxides are the simplest of the metal oxides, yet their spectroscopy is still not well understood. The bonding in the alkali oxides is known to be ionic, M^+O^-, giving rise to two low-lying states, $^2\Pi$ and $^2\Sigma^+$. Most of the spectroscopic studies have been focused on determining the relative positioning of these two states. It is known [64, 65, 66, 67, 68] that the ground state symmetry changes from $^2\Pi$ for LiO and NaO to $^2\Sigma^+$ for RbO and CsO. For KO there is conflicting experimental data [67, 68] regarding the ground state, but high-level *ab initio* calculations [69] put the $^2\Sigma^+$ state about 200 cm^{-1} below the $^2\Pi$ state. The $^2\Sigma^+-{}^2\Pi$ transition has not been directly observed for any of these systems, which is not surprising considering the extremely small transition moments [70] and energy separations. The low-lying excited states arising from the higher asymptotes are either repulsive or have bond lengths that are so long that transitions to the $v = 0$ levels of the $^2\Pi$ and $^2\Sigma^+$ states carry virtually no intensity [71]–for example, see the potential energy curves for KO shown in Fig. 7. Thus, it is unlikely that these higher valence states will provide a mechanism for determining the $^2\Sigma^+ - {}^2\Pi$ separation. For LiO and NaO estimates of the separation have been deduced from microwave experiments assuming a pure-precession hypothesis [72]. The experimentally derived separations are in excellent agreement with the *ab initio* values [64, 65, 73, 74]. Similar experiments for KO should be possible, but the results have not yet been reported.

Photodetachment from the negative ions offers an attractive alternative for directly determining the $^2\Sigma^+ - {}^2\Pi$ separation. The feasibility, however, depends upon the geometry and occupations of the anion relative to those of the neutral states. If the bond lengths differ appreciably there may be insufficient resolution in the photodetachment spectra to identify the $v = 0$ transition. Furthermore, if the photodetachment corresponds to a two-electron process, the transition may be too weak to be observable. For example, if the ground state of MO^- is $^3\Pi(1\sigma^2 2\sigma^1 1\pi^3)$, then photodetachment experiments might be limited to observing the $^2\Pi(1\sigma^2 1\pi^3)$ state, because the $^2\Sigma^+(1\sigma^1 1\pi^4)$ state corresponds to a two-electron process. Sarkas, Hendricks, and Bowen [75] have recently performed photodetachment experiments for NaO, KO, RbO, and CsO. To help interpret their spectra, we have carried out *ab initio* calculations for the neutrals and negative ions of LiO

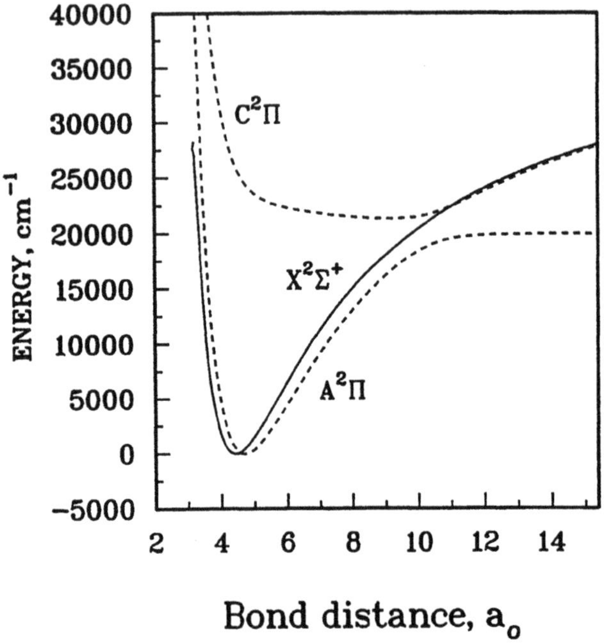

Fig. 7. SA-CASSCF/MRCI potential energy curves for the $X^2\Sigma^+$, $A^2\Pi$, and $C^2\Pi$ states of KO

through KO [76].

The spectroscopic constants for the neutral and anions at the MRCI+Q level are tabulated in Table VII. There is excellent agreement with the limited experimental data for the neutral molecules. The spectroscopic constants were determined using a CASSCF/MRCI procedure where the CASSCF active space consisted of the metal s and oxygen $2p$ and $2p'$ orbitals. The reference configurations were selected using a 0.01 selection threshold and the oxygen $2s$ orbitals were correlated in the MRCI. Core-valence correlation and polarization effects for the alkali atoms were included using the core polarization potential (CPP) approach developed by Müller *et al.* [77]. Previous work has demonstrated that it is necessary to include core effects to get quantitative results [78], and the CPP approach has been shown [79] to account accurately for inner-shell effects.

The $^3\Pi$ and $^3\Sigma^+$ states of MO^- are the analogs of the $^2\Pi$ and $^2\Sigma^+$ states of MO, where the "extra" electron has been added to the metal valence s orbital. The bond lengths for the anions are longer than for the corresponding states of the neutrals, because the extra electron increases the metal-O repulsion. However, the difference in the r_e, ω_e, and T_e values between the Σ^+ and Π states for the anions are similar to those in the neutral systems. The $^1\Pi$ state is nearly degenerate with the $^3\Pi$ state, differing only in the open-shell exchange energy, which is small because the

TABLE VII
Spectroscopic constants[a] for the $^1\Sigma^+$, $^3\Sigma^+$, $^1\Pi$, and $^3\Pi$ states of MO^- and the $^2\Sigma^+$ and $^2\Pi$ states of MO at the MRCI+Q level. The experimental results[b] are given in parentheses.

State	$r_e(a_0)$	$\omega_e(\text{cm}^{-1})$	$T_e(\text{cm}^{-1})$
LiO($^2\Pi$)	3.194(3.190)	808(813)	0
LiO($^2\Sigma^+$)	3.006	862	2496(2516)
LiO^-($^3\Pi$)	3.300	698	−3211
LiO^-($^1\Pi$)	3.300	700	−3104
LiO^-($^1\Sigma^+$)	3.178	750	−2052
LiO^-($^3\Sigma^+$)	3.135	721	−516
NaO($^2\Pi$)	3.866(3.877)	488(492)	0
NaO($^2\Sigma^+$)	3.667	520	2061(2018)
NaO^-($^3\Pi$)	4.011	409	−4217
NaO^-($^1\Pi$)	4.008	410	−4046
NaO^-($^1\Sigma^+$)	3.826	444	−3878
NaO^-($^3\Sigma^+$)	3.835	423	−2402
KO($^2\Pi$)	4.440	404	0
KO($^2\Sigma^+$)	4.147	446	−184
KO^-($^1\Sigma^+$)	4.302	365	−3847
KO^-($^3\Pi$)	4.595	330	−3455
KO^-($^3\Sigma^+$)	4.298	361	−3401

[a] The spectroscopic constants are computed with a quadratic fit in $1/r$.
[b] Ref. [72]. The experimental T_0 values have been converted to T_e values using the theoretical results for ω_e.

open-shell orbitals are on different centers. The $^1\Sigma^+(1\sigma^2 1\pi^4 + 2\sigma^2 1\pi^4)$ state is multireference in character and contains significant M^+O^{2-} character. Based on calculations of the isoelectronic alkaline-earth oxides, the $^1\Sigma^+$ state is expected to require a higher level of computational effort to achieve an equivalent description. Comparable errors are expected for the other states, as they can all be described as MO^- or M^+O^{2-}, and the error in the oxygen electron affinity will effectively cancel.

Theory shows that the ground state changes from $^3\Pi$ for LiO^- to $^1\Sigma^+$ for KO^-. For NaO^-, the $^3\Pi$ and $^1\Sigma^+$ states lie too close together for a definitive prediction of the ground state, considering the greater degree of difficulty in describing the $^1\Sigma^+$ state. The change in the ground state of the negative ion is expected to lead to considerable changes in the photoelectron spectrum with alkali metal. To help analyze the photodissociation experiments, the vibrational levels and the Franck-Condon factors were computed for both negative ion states using the theoretical potential energy curves.

For LiO^-, the Franck-Condon factors were computed between the $^3\Pi$ ground state and the $X^2\Pi$ and $A^2\Sigma^+$ states of the neutral. The $^3\Pi$ state of LiO^- has a

TABLE VIII
Franck-Condon factors for photodetachment from the $^1\Sigma^+$ state of KO^- to form the $^2\Pi$ or $^2\Sigma^+$ states of KO; the MRCI+Q potentials are used.

v(ion)	v(neutral)	$^2\Sigma^+$	$^2\Pi$
0	0	0.62720	0.72030
0	1	0.29861	0.19858
0	2	0.06488	0.05782
0	3	0.00851	0.01649
0	4	0.00076	0.00481
0	5	0.00005	0.00142
0	6	0.00000	0.00042
0	7	0.00000	0.00012
1	0	0.28110	0.26549
1	1	0.16980	0.34633
1	2	0.36376	0.21998
1	3	0.15104	0.10440
1	4	0.03026	0.04096
1	5	0.00372	0.01501
1	6	0.00031	0.00524
1	7	0.00002	0.00176
2	0	0.07368	0.01405
2	1	0.31592	0.41932
2	2	0.01112	0.15538
2	3	0.29732	0.17722
2	4	0.22445	0.12516
2	5	0.06555	0.06308
2	6	0.01072	0.02781
2	7	0.00114	0.01124

longer bond length than either of the neutral states of LiO. From the $v = 0$ level of LiO^-, the $X^2\Pi$ state of LiO is populated primarily in $v = 0$, whereas there is significant probability for populating the $v = 0 - 2$ levels of the $A^2\Sigma^+$ state. However, the transition to the $A^2\Sigma^+$ state may be difficult to observe, because it is a two-electron process. For NaO^-, the Franck-Condon factors were computed for detachment from the $^3\Pi$ and $^1\Sigma^+$ states, as both states could potentially contribute to the photodetachment spectra. A high-resolution study may provide a definitive determination of the ground state of NaO^-.

Photodetachment spectra of KO^- are of particular interest, because they provide the possibility of definitively determining the $^2\Pi - {}^2\Sigma^+$ separation in KO. Detachment to form both states should carry reasonable intensity, as both neutral states are formed by single electron detachment processes from the $^1\Sigma^+$ ground state of KO^-. The bond length for the ground state of KO^- is intermediate between that of the $^2\Sigma^+$ and $^2\Pi$ states of the neutral. As can be seen from the Franck-Condon factors reported in Table VIII, the intensity is spread across a number of transitions, especially for v'(ion) >0. The computed Franck-Condon factors give a simulated photoelectron detachment spectra that agrees reasonably well with the recently determined experimental spectra [75] for electron binding energies less than 0.5 eV. Above 0.5 eV, the computed and measured intensities begin to diverge, probably due to resonances with higher states. The good agreement between the computed and measured spectra [75] for low electron binding energies confirms the theoretical prediction that the ground state of KO is $^2\Sigma^+$ with the $^2\Pi$ state being only about 200 cm^{-1} higher in energy. Thus the availability of theoretical calculations in conjunction with experimental spectra leads to a definitive determination of the state separation in KO. This is an illustration of the utility of having both *ab initio* calculations and experiment for solving difficult spectroscopic problems.

3.5 CO QUINTET STATES

Very little experimental data exists for the high-spin states of first- and second-row diatomics such as N_2 and CO. *Ab initio* calculations are ideally suited to study these states, but experience has shown that these relatively weakly bound states require high levels of theory for a quantitative description. However, with recent improvements in computational techniques, theory becomes the method of choice to characterize these states. An excellent example is our study of the quintet states of N_2 [80]. The $A'^5\Sigma_g^+$ state had been dismissed as being the lower state of the Hermann infrared band system (HIR) [81], and the suggestion [82] that it was a precursor state for the Lewis-Rayleigh afterglow was disputed [83], because earlier calculations (see Refs. [84] and [85] and references therein) found it to be weakly bound. Our more accurate calculations showed that the well depth of the $A'^5\Sigma_g^+$ state was over twice that determined previously. On the basis of these improved calculations, it was possible to assign the HIR system as $C''^5\Pi_u \rightarrow A'^5\Sigma_g^+$ and show that the $A'^5\Sigma_g^+$ state is an important precursor state for the Lewis-Rayleigh afterglow. Our assignment of the HIR system has since been confirmed by experiment [86].

The CO molecule is another system where *ab initio* calculations [87] have been instrumental in explaining the experimental spectra. Bahrdt, Nahme, and Schwentner (BNS) [88] have recently observed two new emission band systems of CO in an Ar matrix upon excitation of the $C^1\Sigma^+(v'=1) \leftarrow X^1\Sigma^+(v''=0)$ transition at 106.4 nm. The progressions observed correspond to emission from a common upper state at 10.3 eV to the $a'^3\Sigma^+$ and $d^3\Delta$ states. Because, the only available *ab initio* calculation [89] had predicted the $^5\Sigma^+$ state, which was bound by only about 600 cm^{-1}, to be the lowest quintet state, BNS dismissed the emission as being due to an intercombination band involving a quintet state. Instead, they suggested that the upper state consisted of a strong mixing of the $a'^3\Sigma^+(v'=31)$ and $b^3\Sigma^+$ states. They hypothesized that after initial excitation of the Rydberg state, there was a crossing to the mixed $a'^3\Sigma^+ - b^3\Sigma^+$ emitting state via an intermediate quintet state. The intermediate quintet state was invoked in order to explain the long (15 ms) lifetime.

On the basis of our earlier study of the quintet states of the isoelectronic N_2 molecule [80], we expected that CO could have significantly bound quintet states

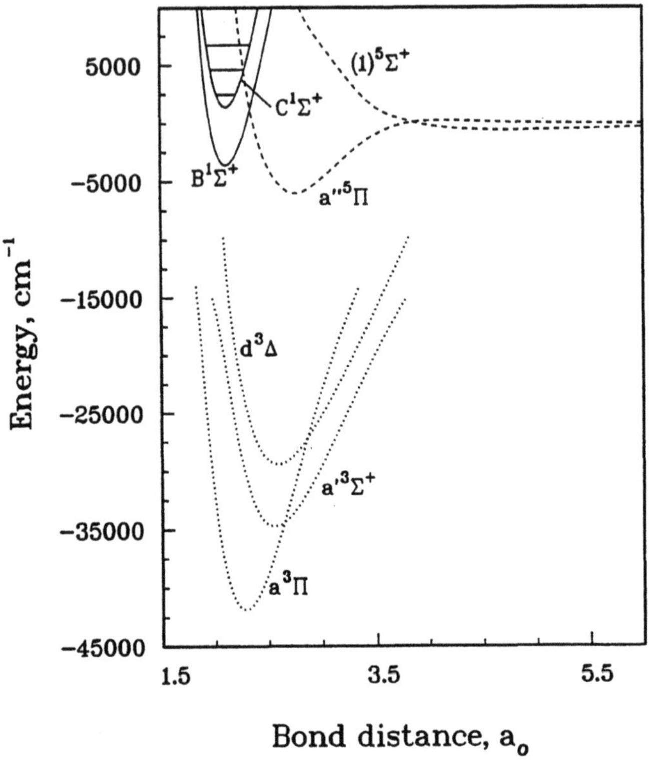

Fig. 8. The computed potential energy curves for the lowest $^5\Pi$ and $^5\Sigma^+$ states and the Klein Dunham curves for the $B^1\Sigma^+$, $C^1\Sigma^+$, $a^3\Pi$, $a'^3\Sigma^+$, and $d^3\Delta$ states of CO. The horizontal lines in the $C^1\Sigma^+$ states denote the positions of the first three vibrational levels.

that played a role in the spectroscopy of this molecule. Our calculations showed that while five of the six states arising from ground state atoms have very shallow wells, the $a''^5\Pi$ state has a significant (6764 cm^{-1}) dissociation energy. Even though the $a''^5\Pi$ state is less strongly bound than the $(1)^5\Sigma^+$ state at large r values, as expected based on electrostatic arguments [90], at shorter r values the formation of sp hybrid orbitals results in a covalent bond with a sizable binding energy.

The $a''^5\Pi$ state is in the correct energy region to be the upper state observed by BNS. Our computed potential for this state along with the Klein Dunham curves for the $B^1\Sigma^+$, $C^1\Sigma^+$, $a^3\Pi$, $a'^3\Sigma^+$, and $d^3\Delta$ states of CO are given in Fig. 8. The positions of the first three vibrational levels of the $C^1\Sigma^+$ state are given by the horizontal lines. The emission spectrum is observed upon excitation of the $C^1\Sigma^+(v' = 1) - X^1\Sigma^+(v''=0)$ band, but is not observed when the $C^1\Sigma^+$ state is excited to the v'=0 or 2 levels. As can be seen from Fig. 8, the intersection of the $C^1\Sigma^+$ and $a''^5\Pi$ states occurs very close to v'=1. Thus, upon excitation of the

$C^1\Sigma^+(v' = 1)$ level there is a rapid intersystem crossing to the $a''^5\Pi$ state followed by rapid vibrational quenching and finally emission from the v'=0 level of the $a''^5\Pi$ state to the $a'^3\Sigma^+$ and $d^3\Delta$ states. Shifting the computed potential energy curve by 0.006 a_0 (which corresponds to the expected error in the calculated r_e value) gives Franck Condon factors that are in excellent agreement with the peak heights in the BNS spectrum. This strongly supports the assignment of the $a''^5\Pi$ state as the upper state in the emission spectrum of BNS. Schwentner [91] has confirmed that our assignment is consistent with their experimental data and that the possibility of a quintet upper state was ruled out only because no significantly bound quintet state had been previously identified. We believe that the existence of significantly bound high-spin states are not unique to N_2 and CO, and that the importance of high-spin states in the spectra of other molecules should not be dismissed on the basis of older less accurate calculations.

3.6 The NO Ogawa system and O_2^+ first negative systems

The Ogawa ($b^4\Sigma^- - a^4\Pi$) and β ($B^2\Pi - X^2\Pi$) band systems are both observed in the NO afterglow arising from the recombination of ground state atoms. It is known [92] that the $b^4\Sigma^-$ state serves as a precursor for the $a^4\Pi$ and $B^2\Pi$ states, which do not correlate with ground state atoms. The radiative lifetimes of these states and the interconversion rate between them are very important in fully understanding the afterglow. This has been part of the motivation for studying the Ogawa band system, which occurs in the infrared region. Unfortunately, there is substantial disagreement between our theoretical radiative lifetimes [93, 94] and the available experimental values [95, 96]. For example, the recent laser-induced fluorescence experiments by Faris, Dyer, and Slanger (FDS) [95] largely agree with the older experimental data by Mohlmann and de Heer [96], when corrections are made for the errors in the earlier vibrational level numbering [97]. FDS contended that while the theoretical lifetimes had approximately the correct dependence on vibrational quantum number, that they overestimated the experimental lifetimes by about 50%. The magnitude of the disagreement was surprising, because calibration calculations suggested that the lifetimes should be accurate to 10-20%.

In addition to calculations for the NO Ogawa system, we carried out analogous calculations for the isoelectronic first negative system ($^4\Sigma_g^- - {}^4\Pi_u$) of O_2^+ for calibration purposes [94]. The excellent agreement between our radiative lifetimes and the experimental results of Erman and Larsson [98] was taken as additional support for our contention that our NO $b^4\Sigma^-$ lifetimes were accurate. However, FDS noted that there were spectral misidentifications [99, 100] in the experiments of Erman and Larsson, and in addition that earlier experiments [101] for O_2^+ had yielded lifetimes significantly shorter than those obtained by theory. Thus they contended that the agreement between the theoretical and experimental O_2^+ lifetimes was no better than for the NO Ogawa band system. However, while there are known problems with the lifetimes measured by Erman and Larsson for $v > 4$, their values for $v \leq 3$ do not appear to be significantly affected by predissociations or spectral misidentifications. Furthermore, Erman and Larsson noted that the shorter lifetimes observed in earlier experimental studies of the $^4\Sigma_g^-$ state of O_2^+ were artificially shortened by Coulomb repulsion effects.

TABLE IX
The O_2^+ $^4\Sigma_g^-$ radiative lifetime, in μs.

	Theory			Expt
v	BPL93[a]	LPB89[b]	Blomberg-Liu[c]	Erman-Larsson[d]
0	1.437	1.560	1.490	1.350±0.010
1	1.455	1.587	1.509	1.400±0.010
2	1.488	1.615	1.544	1.530±0.012
3	1.535	1.675	1.596	1.730±0.013

[a] Ref. [93], [b] Ref. [94], [c] Ref. [102], [d] Ref. [98]

Considering the significant improvements in the accuracy of *ab initio* calculations in the last few years, we decided to revisit [93] the problem of the lifetimes of the O_2^+ first negative and NO Ogawa band systems. We first describe our calibration calculation for O_2^+. For NO additional care was taken to ensure that the results were not biased by underestimating the potential mixing of Rydberg character into the wave functions.

For the O_2^+ $^4\Sigma_g^- - {}^4\Pi_u$ system, we performed a series of CASSCF/ICMRCI calculations. The use of internal contraction was shown to result in a small (2%) increase in the transition moment, but had the very beneficial effect of eliminating the need to select reference configurations. A $(2s2p)$ active space is apparently adequate for accurately determining the transition moment, as expanding the active space by adding the bonding components (the σ_g and π_u orbitals) of a p' correlating orbital increases the transition moment insignificantly at the ICMRCI level. We next considered systematically expanding the one-particle space using the correlation consistent basis sets. Expanding the basis set from cc-pVTZ to cc-pVQZ decreases the moment slightly (by about 1%), while further increasing the basis set to cc-pV5Z or augmenting the basis set with diffuse functions reduced the computed moment very slightly. Thus, internal contraction, expanding the active space, and improving the one-particle basis set all have a small effect on the transition moment. We believe that our best radiative lifetimes are accurate to about 10%, and on the basis of the calibration calculations are likely to be lower bounds to the true values. In the vibrational analysis we employed potential energy curves for the $^4\Sigma_g^-$ and $^4\Pi_u$ states determined using the CASSCF/ICMRCI+Q procedure and the cc-pV5Z basis set. The computed spectroscopic constants for both states are in excellent agreement with experiment [45]; r_e values too long by 0.002 Å, $\Delta G_{1/2}$ values too small by < 5 cm^{-1}, and T_0 too large by 107 cm^{-1}. The errors in the r_e and $\Delta G_{1/2}$ values are due primarily to neglect of core correlation.

The lifetimes for the J=0 levels were computed using spline representations of the theoretical potentials shifted by 107 cm^{-1} to match the experimental T_0 value [45]. The experimental lifetimes of Erman and Larsson are compared with those from theory [93, 94, 102] in Table IX. All three sets of theoretical lifetimes are in mutual accord and agree well with the experimental values for $v \leq 3$, strongly supporting the accuracy of these measurements.

TABLE X
The NO $b^4\Sigma^-$ radiative lifetimes, in μs.

	Theory		Experiment	
v	BPL93[a]	LPB89[b]	MH[c]	FDS[d]
0	18.3	19.9		
1	12.0	13.3		
2	9.5	10.6	6.43±0.20	7.19, 7.47, 5.55
3	8.1	9.2	5.77±0.20	5.84, 5.55
4	7.3	8.4		4.59
5	6.9			4.76, 4.74, 4.24

[a] Ref. [93], [b] Ref. [94], [c] Ref. [96], [d] Ref. [95].

Calculations performed to calibrate the one- and n-particle expansions for the NO Ogawa system indicated that our results were well converged. One possible difference between the band systems of O_2^+ and NO, is the incorporation of Rydberg character into the wave functions of either the $b^4\Sigma^-$ or $a^4\Pi$ states of NO. Although this would be unexpected for states with r_e values this large, it is conceivable at the shorter r values that even a little Rydberg character could substantially change the transition moment. To ensure that there is sufficient flexibility for mixing the correct degree of Rydberg character into the wave function requires using a more diffuse one-particle basis set and possibly a larger n-particle treatment. See, for example, the discussion of the NO β system (next section) where Rydberg character becomes very important.

To facilitate mixing Rydberg character into the CASSCF wave function, we included additional $^4\Sigma^-$ and $^4\Pi$ states in the SA-CASSCF optimization. Rydberg character will be more important for the excited states, and including these states in the CASSCF will enhance the Rydberg character in the valence orbitals, thereby allowing greater flexibility for the $b^4\Sigma^-$ and $a^4\Pi$ wave functions. However, we find that the transition moment is insensitive to the active space and number of roots included in the CASSCF optimization, even when one of the upper roots is clearly of Rydberg character. Expanding the valence basis set and augmenting with diffuse functions has only a small effect on the $b^4\Sigma^- - a^4\Pi$ transition moment. For example, improving the basis set from cc-pVTZ+(spd) to aug-cc-pVQZ+(spd) or cc-pV5Z+(spd) (where the notation +(spd) designates the basis set is augmented with a diffuse s, p, and d function) reduces the moment in the Franck-Condon region by only 0.2%. Thus, we see no indication of an important contribution from Rydberg character. Using an aug-cc-pVQZ+(spd) basis set, the computed spectroscopic constants are in excellent agreement with experiment [45, 97]; the computed r_e values are 0.004 Å longer than experiment and the computed T_0 value is 248 cm^{-1} too large. The $\Delta G_{1/2}$ values for the $^4\Pi$ and $^4\Sigma^-$ states differ from the experimental values by −5 and +10 cm^{-1}, respectively.

The vibrational wave functions and radiative lifetimes were computed using the ICMRCI+Q potentials (shifted down by 248 cm^{-1} to match the experimental T_0 value) and the ICMRCI transition moment. Our most recent lifetimes [93] are compared with our previous theoretical results and with experiment in Table X. The newer theoretical lifetimes are shorter than the older values, but the differences are consistent with the improvements in the calculations. We believe that the newer lifetimes are accurate to 10% based on the convergence with improvements in the level of treatment. In addition, we feel that further refinements in the calculations are much more likely to increase rather than decrease the lifetimes.

The facile energy transfer that occurs between the $b^4\Sigma^-$ state and other electronic states, such as $a^4\Pi$, $A^2\Sigma^+$, and $B^2\Pi$, creates a multi-exponential decay that complicates the extrapolation to zero pressure. Ottinger and Vilesov [103] have recently studied intramolecular collisional transfer in NO. They note that particular $B^2\Pi$ levels are perturbed by specific levels of the $a^4\Pi$ state, serving as gateways to promote the otherwise spin-forbidden $a^4\Pi \rightarrow B^2\Pi$ collision-induced transition. This mechanism can occur at zero pressure for selected ro-vibrational levels of the $a^4\Pi$ state. This facile energy transfer also suggests that some other mechanism other than fluorescence may be contributing to the decay of the $b^4\Sigma^-$ state. For example, one possibility is that there is a direct spin-orbit induced intersystem crossing between the $b^4\Sigma^-$ and $B^2\Pi$ states, followed by fluorescence in the β system. If this is true, the rovibrational levels being investigated by FDS may in actuality be a linear combination of $b^4\Sigma^-$ and $B^2\Pi$ character. In any case, it is clear that the strong coupling between the electronic states of NO makes the unambiguous determination of the radiative lifetime of the $b^4\Sigma^-$ state a difficult experimental problem. Clearly more work is needed to understand the $b^4\Sigma^-$ state decay mechanisms and the NO afterglow.

3.7 NO β SYSTEM

The NO β ($B^2\Pi - X^2\Pi$) system is a prominent feature of shock heated air [104], but it is only recently that accurate Einstein coefficients and radiative lifetimes have become available. Gadd and Slanger [105] have used laser-induced fluorescence techniques to measure zero-pressure radiative lifetimes for v'=0-6. Their lifetimes were considerably shorter than earlier measurements and showed a strong dependence on vibrational quantum number. Using a branching-ratio technique, Piper *et al.* [106] determined the variation of the TMF with internuclear separation over the relatively small region of r sampled by the observed transitions. Their TMF increased sharply with decreasing r in qualitative agreement with two earlier theoretical studies [104, 107]. However, the theoretical TMFs differed substantially from that deduced by Piper *et al.* and neither *ab initio* TMF gave satisfactory agreement with the experimental lifetimes. We investigated [108] this system using SA-CASSCF/MRCI methods to determine the computational requirements for obtaining accurate lifetimes. Previous theoretical studies [109] have shown that high levels of theory are required to describe valence-Rydberg mixing. The *ab initio* calculations for this system are complicated by the fact that the $B^2\Pi$ state undergoes a crossing with the $C^2\Pi$ Rydberg state and the valence-Rydberg mixing occurs for r values in the Franck-Condon region. The $v = 7$ level of the $B^2\Pi$ state is essentially degenerate with the $v = 0$ level of the $C^2\Pi$ state. The $B^2\Pi - X^2\Pi$ system gains intensity by mixing Rydberg character into the $B^2\Pi$ wave function.

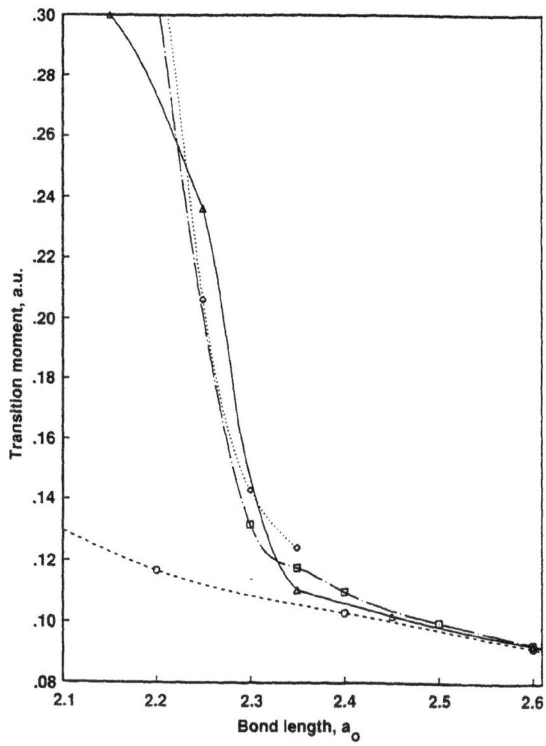

Fig. 9. Shown are various TMFs in the region between r=2.1 and 2.6 a_0, where the degree of Rydberg character in the $B^2\Pi$ state changes significantly. The dashed line corresponds to the 11-electron MRCI calculations in the [5s 4p 2d 1f] ANO basis. The TMF represented by the solid line is the (244) SA-CASSCF/MRCI results (7-electron results shifted to account for 2s correlation-see Ref. [108]) in the larger basis including Rydberg functions. The dash-dot curve with square symbols is the (244)R SA-CASSCF/MRCI results (the (244) active space with one π orbital constrained to be Rydberg). The dotted TMF is the 3-root SA-CASSCF/MRCI results.

The difficulty in treating this system can be seen from Fig. 9, where the $(2)^2\Pi$-$X^2\Pi$ TMF is plotted for several different theoretical treatments. For $r \geq 2.6\ a_0$, both the $X^2\Pi$ and $B^2\Pi$ states are valence states and the transition moment is accurately determined employing a (222) active space (in C_{2v} symmetry) and a good valence basis set. For r values shorter than about $r = 2.25\ a_0$, the $(2)^2\Pi$ state becomes predominantly Rydberg, and thus to describe the lowest two $^2\Pi$ states the basis set must be augmented with diffuse functions and a Rydberg π orbital must be included in the active space. In addition, it is necessary to add a π correlating orbital to the active space, because this orbital is energetically favored over the Rydberg orbital. Employing this (244) active space and the augmented basis set yields a TMF that rises only slightly faster than that given by a valence calculation, until it increases discontinuously near the crossing point between the $B^2\Pi$ and $C^2\Pi$ states. The discontinuity in the TMF occurs from the avoided crossing and the fact that we are treating different upper states on each side of the crossing. For r values inside the crossing point, the CASSCF solution is the average of a valence and a Rydberg state. For r values beyond the crossing point the CASSCF solution is the average of two valence states and the "extra" π orbitals are largely correlating orbitals. The MRCI treatment is not able to overcome the bias in the orbitals and insufficient Rydberg character is built into the excited state. This difficulty is especially apparent at r=2.25 a_o, where the CASSCF solution corresponds to a valence and a Rydberg state, but the MRCI corresponds to two valence states. To obtain a smooth transition moment, it is necessary to provide sufficient flexibility in the reference space to build in the Rydberg character.

We employed two procedures to incorporate Rydberg character into the CASSCF wave functions. The first procedure is to perform a three-root SA-CASSCF calculation for the $X^2\Pi$, $B^2\Pi$, and $C^2\Pi$ states. This provides a consistent mechanism for optimizing the molecular orbitals, because two valence and one Rydberg state are included in the optimization and the character of the orbitals will be consistent across the crossing point. However, this procedure is limited to the region $2.05 \leq r \leq 2.35\ a_0$, because of curve crossings with other states. At shorter r values the $K^2\Pi$ Rydberg state is below the $B^2\Pi$ state and at longer r values several valence states lie below the Rydberg state. The computed transition moment function rises much faster and is smoother than for the two-root determinations. We should note that the TMF between the $X^2\Pi$ and diabatic $B^2\Pi$ potential decreases very rapidly after the crossing point with the $C^2\Pi$ state. We have only shown the adiabatic TMF, because it better illustrates the computational difficulties, and the lifetime calculations described below do not sample the TMF inside the crossing point.

The disadvantage of the 3-root procedure is that it is limited to a restricted r region because of curve crossings with other states. An alternative procedure for building in Rydberg character is to employ the (244) active space, but to require that the fourth π orbital be Rydberg for all r values. This is accomplished by taking the Rydberg orbital from the largest r value that has a Rydberg CASSCF solution, and freezing this orbital (except for orthogonalization) for all larger r values; this solution is denoted (244)R. The resulting TMF also rises significantly faster and is smoother than the two-root (244) TMF, but the rise is slightly slower than for the 3-root calculation. This indicates that the calculation is providing a reasonable description of the Rydberg character in the crossing region. For larger r values, the computed TMF quickly approaches the valence TMF.

The radiative lifetimes were computed using a hybrid TMF where the 3-root TMF is used for the shorter r values and the two-root TMF is used for the longer r values. This TMF is shown by the solid line in Fig. 10. The computed radiative lifetimes are compared with the experimental results of Gadd and Slanger [105] in Table XI. The computed lifetimes fall off more slowly with increasing v' than the experimental values but, overall, they are in good agreement.

A limitation in the calculations is that they may still be underestimating the buildup of Rydberg character. This is supported by the shifted results in Table XI,

TABLE XI
Radiative lifetimes (μs) for the β system of NO.

Level	Theory		Expt.[a]	
v'	unshifted	shifted[b]	actual	linear[c]
0	2.12	1.99	2.00	2.00
1	1.93	1.79	1.82	1.81
2	1.76	1.60	1.52	1.61
3	1.60	1.41	1.46	1.42
4	1.45	1.23	1.19	1.23
5	1.31	1.05	1.07	1.03
6	1.17	0.87	0.85	0.84

[a] Reference [105].
[b] Lifetimes based on the theoretical TMF that has been shifted by 0.045 a_0 to larger r values.
[c] Lifetimes given by the expression 2.00-0.193v'.

which were generated by shifting the TMF by 0.045 a_o to larger r values. As the TMF at long range is nearly linear, this effectively corresponds to building in the Rydberg character at slightly longer r values. The computed lifetimes with the shifted TMF are in excellent agreement with the experimental values. Although the lifetimes can be brought into agreement by shifting the TMF, this procedure is not completely satisfactory. First, shifting the TMF produces Einstein coefficients that do not agree quite as well with the experimental values given for transitions from the v'=6 level to v''=0-16. In addition, the calculations for $r \geq 2.6$ and $r \leq 2.15$ a_o should be more accurate and it is difficult to justify this large of a shift. It is in the intermediate r region where the $B^2\Pi$ state is building in Rydberg character that the results are potentially less accurate. We tested this assumption by constructing an empirical TMF based on the *ab initio* derived values for $r >$ 2.6 and $r \leq 2.15$ a_o and choose the spline function connecting these regions to best reproduce the experimental lifetimes. The resulting TMF is the dotted curve shown in Fig. 10. This TMF accurately reproduces the experimental lifetimes and is consistent with the expected accuracy of the calculations. Furthermore, it does not differ substantially from the "shifted" TMF in the important intermediate region. Hence, both the shifted and empirical TMF results suggest that the *ab initio* calculations are underestimating the degree of Rydbergization with decreasing r values.

Also shown in Fig. 10 is the quadratic TMF derived by Piper *et al.* [106] using a least squares fit to the branching-ratio data. The TMF is defined over the range $2.40 \leq r \leq 3.02$ a_o and is in reasonably good agreement with the theoretical results. Our TMFs rise more rapidly at the shorter r values, which is necessary to reproduce the rapid falloff in the experimental lifetimes.

3.8 Comparison of the spectroscopy of B_2, Al_2, and AlB

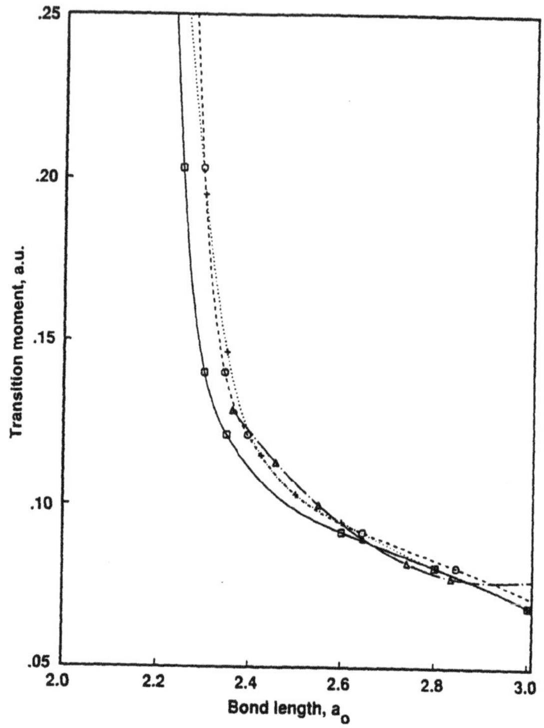

Fig. 10. Transition moment functions (a.u.) for the NO β system. The solid and dashed curves are our directly computed moment and this moment shifted by 0.045 a_0 to larger r values, respectively. The dotted line is the empirical TMF discussed in the text. The dash-dot curve is the quadratic function of Piper *et al.* [106].

In this subsection we compare the bonding and spectroscopy of the isovalent B_2, Al_2, and AlB molecules. The Al_2 molecule has been studied extensively both theoretically [110, 111, 112] and experimentally [113, 114, 115, 116]. Much of the theoretical work has concentrated on determining the identity of the ground state, because while it is now known definitively to be $X^3\Pi_u$, the $A^3\Sigma_g^-$ state lies only about 170 cm^{-1} higher— see the potential energy curves for the triplet states of Al_2 in Fig. 11. Many of the excited states of Al_2 have now been characterized experimentally using resonant two-photon photoionization (R2PI) [116] and laser-induced fluorescence techniques [115]. The theoretical potentials have been very useful in helping to unravel the complex spectra. For example, the theoretical potential curves [111] explain why the $(2)^3\Pi_g$ state is observed in absorption, but not in emission, as well as the anomalous intensity patterns in the $E^3\Sigma_g^-$ –

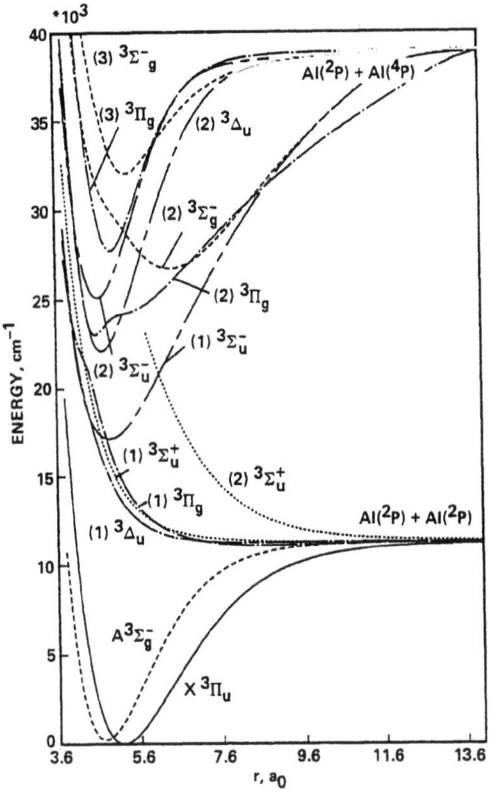

Fig. 11. Potential energy curves for the triplet states of Al_2. The symmetries are differentiated by the various types of chained lines.

$X^3\Pi_u$ band system. More recently, the spectrum of B_2 has been observed by Brazier and Carrick [117] in emission from a Corona excited supersonic expansion source. Again the calculations [118] were useful in the analysis of the spectra. Although the spectra of AlB has not yet been observed, it should be possible to make this molecule in sufficient quantities using co-vaporization techniques to carry out spectroscopic studies. All of these molecules are important as models of simple metal clusters. The B_2 and AlB molecules are potentially important in high-energy density materials such as additives to solid propellants. A comparison of the bonding and spectroscopy of these three systems gives insight into the difference in the bulk metals, as well as between single component and alloy systems.

The triplet states of B_2 are shown in Fig. 12. Unlike Al_2, the ground state of B_2 is $X^3\Sigma_g^-$, with theory predicting that the $A^3\Pi_u$ state is about 3500 cm^{-1} higher in energy. Although both aluminum and boron tend to form delocalized bonds, those for boron are significantly stronger. This is evident in the much larger D_0 value

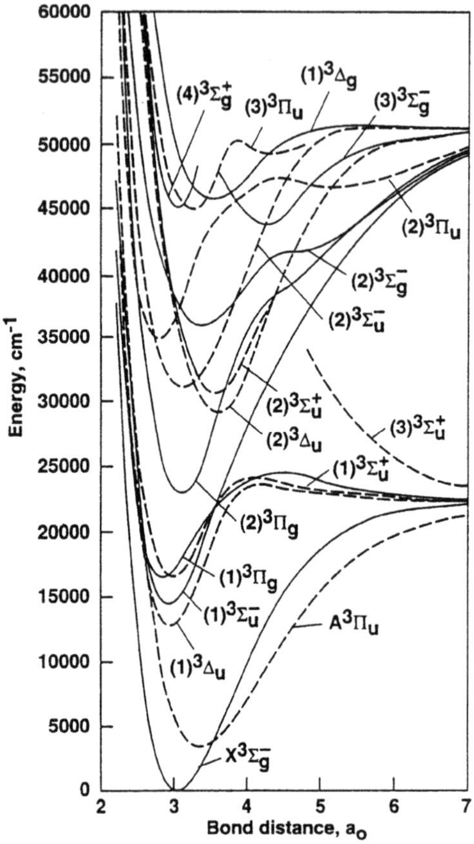

Fig. 12. Potential energy curves for the triplet states of B_2. The states of g and u parity are denoted by solid and dashed lines, respectively.

for B_2(2.85±0.06 eV [118]) compared with Al_2(1.34±0.06 eV [116]), and the much larger cohesive energy of boron metal. The stronger bonding is also apparent in the triplet potential energy curves of these molecules in Figs. 11 and 12. For example, the $(1)^3\Delta$, $(1)^3\Sigma^+$, and $(2)^3\Pi$ states are repulsive for Al_2, but significantly bound for B_2. Another indication is that the strongly bound $^5\Sigma_u^-$ state lies at 1700 cm^{-1} for B_2, but at 12 300 cm^{-1} for Al_2. The $^5\Sigma_u^-$ state lies much lower in B_2 than Al_2, because the smaller radial extent of boron allows for the formation of much stronger π bonds. Thus this state has considerable double bond character in B_2, resulting in a binding energy of 6.2 eV with respect to the $^2P^o + ^4P$ asymptote.

Since the AlB molecule has not been characterized experimentally, we recently carried out SA-CASSCF/ICMRCI calculations [119] comparable to those for Al_2 and B_2 to help aid in future spectroscopic studies and to contrast the bonding in AlB with that in Al_2 and B_2. Our first goal was to unambiguously determine whether the ground state of AlB was the $^3\Pi$ or the $^3\Sigma^-$ state. A summary of

TABLE XII
Calibration calculations for the two lowest states of AlB and AlB^+.

	$r_e(a_0)$	$\omega_e(cm^{-1})$	$D_e(eV)$	$r_e(a_0)$	$\omega_e(cm^{-1})$	$T_e(cm^{-1})$
AlB		$X^3\Sigma^-$			$A^3\Pi$	
ANO ICMRCI	3.895	593	1.84	4.238	483	617
ANO ICMRCI+Q	3.898	592	1.84	4.243	482	392
ANO ICACPF	3.897	592	1.84	4.242	482	453
cc-pVQZ ICMRCI	3.861	599	1.92	4.201	487	757
cc-pVQZ ICMRCI+Q	3.863	598	1.92	4.206	485	539
cc-pV5Z ICMRCI	3.856	610	1.94	4.196	496	772
cc-pV5Z ICMRCI+Q	3.857	608	1.94	4.201	494	552
cc-pV5Z ICACPF	3.857	609	1.94	4.199	494	610
AlB^+		$X^2\Sigma^+$			$A^2\Pi$	
ANO ICMRCI	5.201	245	0.78	4.472	296	3334

these calculations along with our study of the two lowest electronic states of AlB^+ are summarized in Table XII. We use the Dunning correlation consistent basis sets [25] to study the convergence of the $^3\Pi - {}^3\Sigma^-$ separation with improvements in the one-particle basis set. We use both the Davidson correction and ACPF approach to estimate the contribution from higher excitations in the correlation calculations. Our best calculation predicts the $^3\Sigma^-$ state of AlB to lie 610 cm^{-1} below the $^3\Pi$ state. The uncertainty in the T_e value due to the incompleteness in both the one- and $n-$particle expansions is sufficiently small that the $X^3\Sigma^-$ ground state assignment is definitive. If we correct our best D_e value of 1.94 eV in Table XII on the basis of comparable calculations for Al_2 and B_2, we estimate that the D_e value of AlB is 2.00±0.06 eV or a D_0 value of 1.96±0.06 eV. This value is only slightly less than the average of the D_0 values of Al_2 (1.34±0.06 eV) and B_2 (2.85±0.06 eV). The calculations definitively show that the ground state of AlB^+ is the $X^2\Sigma^+$ state. The adiabatic ionization potential is estimated to be 7.05 eV. This should make R2PI experiments on this molecule feasible.

It may be possible to directly measure the $A^3\Pi - X^3\Sigma^-$ energy separation, since this is a dipole-allowed transition. However, the radiative lifetime for the $v' = 0$ level of the $A^3\Pi$ state is rather long (0.21 s) because of the small energy separation and the large difference in the r_e values of the $X^3\Sigma^-$ and $A^3\Pi$ states. Therefore, this band system is probably more amenable to absorption rather than emission studies. The transition moment of about 0.35 a.u. in the Franck-Condon region results in an oscillator strength of about 4×10^{-5} for the $0\leftarrow0$, $1\leftarrow0$, and $2\leftarrow0$ transitions. Thus, the intensity of these transitions are comparable to that of a strong (greater than 200 km/mol) vibrational transition. Hence, it may be possible to directly measure the $^3\Sigma^- - {}^3\Pi$ separation for AlB, whereas the very small separation in Al_2 has to date precluded a direct measurement. However, observing emission to these low-lying states from a common upper state offers another attractive possibility of accurately determining this state separation.

TABLE XIII
Spectroscopic constants for the electronic states of AlB computed using the SA-CASSCF/ICMRCI method and an ANO basis.

State	$r_e(a_0)$	$\Delta G_{1/2}(\mathrm{cm}^{-1})$	$T_e(\mathrm{cm}^{-1})$
$X^3\Sigma^-$	3.893	588	0
$A^3\Pi$	4.234	481	523
$(1)^1\Delta$	3.964	540	4207
$(1)^1\Sigma^+$	4.553	314	4872
$(1)^1\Pi$	4.304	442	5264
$(1)^5\Sigma^-$	3.720	736	8823
$(2)^1\Sigma^+$	4.305	508	8867
$(1)^3\Delta$		repulsive[a]	
$(2)^3\Sigma^-$	3.873	520	17787
$(1)^3\Sigma^+$		repulsive	
$(2)^3\Pi$		repulsive[b]	
$(1)^1\Sigma^-$		repulsive	
$(3)^3\Pi$	3.921	882	23383
$(2)^3\Delta$	4.123	729	23455
$(2)^3\Sigma^+$	4.101	673	25434
$(3)^3\Sigma^-$	3.865	691	25453
$(2)^1\Delta$	4.162	465	25716
$(2)^1\Pi$		repulsive	
$(3)^1\Pi$	3.633	718	26731
$(1)^5\Delta$	4.161	531	27867
$(1)^5\Sigma^+$	4.177	523	28566
$(2)^5\Sigma^-$	4.334	460	29135
$(3)^1\Sigma^+$	4.210	442	30768
$(2)^1\Sigma^-$	3.868	746	30826
$(1)^5\Pi$	3.972	549	31288
$(2)^5\Pi$		repulsive	

[a] At this level of theory there is a very shallow well at $r = 4.08\ a_0$, which is 17010 cm^{-1} above the ground state—see Fig. 13.
[b] At this level of theory there is a very shallow well at $r = 3.69\ a_0$, which is 20241 cm^{-1} above the ground state—see Fig. 13.

Based on the Al_2 and B_2 studies [111, 118], we expect that the low-lying states of AlB will arise from the $^2P^o$+$^2P^o$, $^2P^o$+^4P (28 800 cm^{-1}) and ^4P+$^2P^o$ (29 000 cm^{-1}) asymptotes. The ^2S ($3s^24s^1$) state (25 000 cm^{-1}) of Al does not contribute significantly to the low-lying states of AlB, because the $4s$ orbital is very diffuse and the states derived from Al ^2S + B $^2P^o$ are only weakly bound. In our study of the excited states we included the valence s and p orbitals of Al and B in the active space, but excluded the Al $3d$ orbital, as we did in our studies of Al_2 [111]. This exclusion results in an underestimation of the contribution of

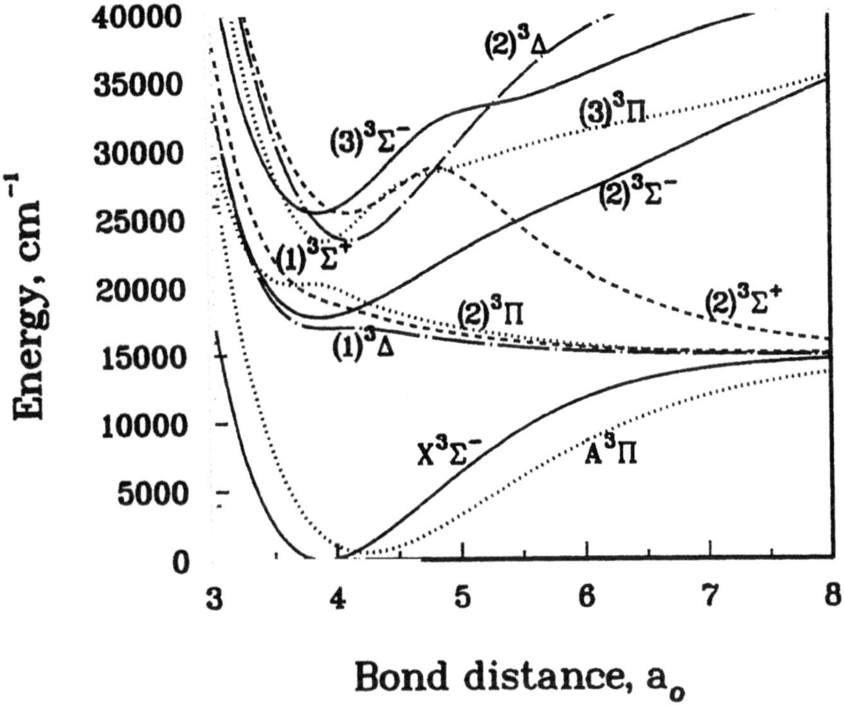

Fig. 13. Potential energy curves for the triplet states of AlB. The $^3\Sigma^-$ states are denoted by solid lines, the $^3\Pi$ states by dotted lines, the $^3\Delta$ states by dot-dashed lines, and the $^3\Sigma^+$ states by the dashed lines.

the Al ^2D state (32 400 cm^{-1}) to the higher-lying states. Thus, in the final SA-CASSCF/MRCI calculations states having energies above about 32 000 were not included in our study.

It is interesting to compare the triplet states of AlB shown in Fig. 13, with the comparable states of Al_2 and B_2 shown in Figs. 11 and 12. Overall, the AlB potential energy curves are more more similar to those of Al_2 than B_2. For example, the $^3\Sigma^- - {}^3\Pi$ separation (computed to be 523 cm^{-1} at this level) is much closer to that in Al_2 (-227 cm^{-1}) than in B_2 (3380 cm^{-1}). The (1)$^3\Delta$, (1)$^3\Sigma^+$, and (2)$^3\Pi$ states of AlB are essentially repulsive, again much more like those of Al_2 than B_2. Finally, the position of the $^5\Sigma^-$ state, derived from the $^2P^o$ ($2s^22p^1$) $+^4P$ ($2s^12p^2$) asymptote, is about 8800 cm^{-1} compared with 12 300 cm^{-1} for the $^5\Sigma_u^-$ state of Al_2, but very different from the value of only 1700 cm^{-1} for B_2.

The spectroscopic constants (r_e, $\Delta G_{1/2}$, and T_e) for the excited states of AlB are summarized in Table XIII. A comparison of the results in Tables XII and XIII shows that the use of the small atomic natural orbital (ANO) set and averaging for the larger number of states in the SA-CASSCF calculations introduces an error of only about 100 cm^{-1} in the T_0 value of the $A^3\Pi$ state at the ICMRCI level. Those electronic states having the $p\sigma$ orbital occupied have significantly longer bond lengths. For example, the bond length of the $A^3\Pi$ ($s^2p\sigma^1p\pi^1$) state of AlB is 0.34 a_o longer than that of the $X^3\Sigma^-$ ($s^2p\pi^2$) state. The occupation of the π^* orbital also results in a lengthening of the bond. Note that the $^5\Sigma^-$ ($s^1p\sigma^1p\pi^2$) state has one of the shortest bond lengths, because it has a larger bond order than any of the other low-lying states.

To facilitate spectroscopic studies of AlB and to identify transitions that might aid in establishing the $A - X$ separation, we determined the radiative lifetimes of the $v' = 0$ and 1 levels of all the bound triplet states lying between about 17 000 and 26 000 cm^{-1}. These results reported in Table XIV were obtained by rigorously treating the bound-bound transitions, but approximating the bound-free contribution using the vertical excitation energies and the transition moments at $< r_v >$. This procedure introduces little error as the bound-free contributions are rather small. Also reported in Table XIV is a decomposition of the lifetimes into their individual components. This clearly identifies several bands that should be readily amenable to experimental study. The transitions from the $(2)^3\Sigma^-$, $(3)^3\Pi$, and $(3)^3\Sigma^-$ states probably offer the best band systems for determining the $A^3\Pi - X^3\Sigma^-$ separation. Both the $(2)^3\Sigma^-$–$X^3\Sigma^-$ and the $(2)^3\Sigma^-$–$A^3\Pi$ transitions are fairly strong with the lifetime of the $(2)^3\Sigma^-$ $v = 0$ level determined almost equally by the two transitions. While neither transition is dominated by the $\Delta v =$ 0 sequence, there are several strong transitions that should be observable.

The Al^-B^+ and Al^+B^- charge transfer asymptotes of AlB lie 7.857 and 5.709 eV above the ground state asymptote. Both of these charge transfer asymptotes give rise to $^3\Sigma^-$ and $^3\Pi$ states. Although the $1/r$ stabilization should place these states in the energy region of the upper states, we see no evidence of them in terms of substantially increased transition strengths, as the radiative lifetimes for the states of AlB and Al_2 are comparable. However, we do see evidence of them at longer r values in the dipole moment functions of some electronic states.

It is hoped that the computed potential energy curves, spectroscopic constants, radiative lifetimes, and Franck-Condon factors for AlB will stimulate spectroscopic studies of this molecule. Clearly, it should be possible to accurately determine the $A^3\Pi - X^3\Sigma^-$ separation by observing decay from a common upper state. An experimental value for this separation would help calibrate the theoretical determinations of the $A - X$ separations for the analogous Al_2 and B_2 systems.

3.9 AlCu and AlAg

In this section we consider AlCu and AlAg, as examples of systems containing a bond between a simple metal and a transition metal. Whereas theory is capable of describing quantitatively the spectra of first- and second-row diatomic systems [120], transition metal containing systems are a greater computational challenge. For AlCu and AlAg much of the difficulty lies in providing an equivalent description for the $d^{10}s^1$ and d^9s^2 occupations of the coinage metals.

TABLE XIV
The lifetime of each excited state and contribution (reported as $1.0/\Sigma A_v$) to this lifetime from each lower state. Decay to the $X^3\Sigma^-$, $A^3\Pi$, and $(2)^2\Sigma^-$ states is computed rigorously, while the contributions from the repulsive $(2)^3\Pi$, $(1)^3\Sigma^+$, and $(1)^3\Delta$ states are computed using the approximate formula (see the text).

State	Lifetime (μs)	Contributions (μs)					
		$X^3\Sigma^-$	$A^3\Pi$	$(2)^3\Sigma^-$	$(2)^3\Pi$	$(1)^3\Sigma^+$	$(1)^3\Delta$
$v'=0$							
$(2)^3\Sigma^-$	14.704	32.820	26.640				
$(2)^3\Sigma^+$	0.335		0.356		15.358	8.843	
$(3)^3\Pi$	0.595	2.328	0.880	14.620	222.398	134.377	29.298
$(2)^3\Delta$	0.297		0.307		38.729		6.316
$(3)^3\Sigma^-$	0.208	0.620	0.330	6.374	108235.		
$v'=1$							
$(2)^3\Sigma^-$	6.392	8.172	29.340				
$(2)^3\Sigma^+$	0.327		0.378		10.102	6.400	
$(3)^3\Pi$	0.382	2.524	0.481	13.250	99.334	69.784	23.324
$(2)^3\Delta$	0.304		0.317		63.643		8.681
$(3)^3\Sigma^-$	0.182	0.447	0.324	6.710	14649.		

The $^2S(d^{10}s^1)$ ground state occupations of Cu and Ag and the $^2P(3s^23p^1)$ ground state occupation of Al give rise to four low-lying states. The two-electron bond for the $X^1\Sigma^+$ ground state arises from a singlet coupling the metal s and the Al $3p\sigma$ electrons. Conversely the repulsive $^3\Sigma^+$ state is obtained when these valence electrons are triplet coupled. If the Al $3p\pi$ orbital is occupied instead of the $3p\sigma$ orbital, two one-electron bonds are formed, giving rise to a $^3\Pi$ and a $^1\Pi$ state. The $^1\Pi$ state naturally lies above the $^3\Pi$ state due to the loss of exchange energy. The higher states of AlCu arise from several possibilities, such as Cu $^2D(3d^94s^2)$+Al $^2P(3s^23p^1)$, Cu $^2S(3d^{10}4s^1)$+Al $^4P(3s^13p^2)$, Cu $^2S(3d^{10}4s^1)$+Al $^2D(3s^23d^1 + 3s^13p^2)$, and Al$^+$+Cu$^-$. For Ag atom, the ^2D occupation lies too high in energy to contribute significantly to the low-lying states.

There is experimental work for both AlCu and AlAg, but not all of the band assignments are definitive. For AlCu, Behm *et al.* [121] observed a band system at 14 892 cm^{-1} with a lifetime of 71 μs. This band system must arise from either the $^3\Pi$ or $^1\Pi$ state, considering that the ground state dissociation energy of AlCu is 2.315 eV. The relatively long lifetime does not preclude an intercombination band involving the $^3\Pi$ state. Thus the assignment of this band system was initially in question.

The next group of excited states, studied by both Behm *et al.* [121] and Cai *et al.* [122], lie in the region between 25 000 to 29 000 cm^{-1}. Cai *et al.* observed two transitions in the 25 000 to 29 000 cm^{-1} region, which they denoted as A and B. More recently, Behm *et al.* [121] observed these transitions as well as two others, which they denoted as C and A', where the prime notation denotes a triplet state that has borrowed intensity from the A state. Tentative assignments were given

TABLE XV
Summary of the ICMRCI+Q spectroscopic constants for the four states of AlCu that correlate with ground state atoms. The experimental results of Behm *et al.* [121] are given in parentheses.

State	$r_0(a_0)$	D_0(eV)/T_0(cm^{-1})	$\Delta G_{1/2}$(cm^{-1})
$^1\Sigma^+$	4.520(4.420)	1.907(2.315)	271(294)
$^3\Pi$	4.434	8874	257
$^1\Pi$	4.606(4.437)	13602(14892)	184(224)
$^3\Sigma^+$		repulsive	

for each of the band systems. The short lifetimes measured for the singlet states led them to suggest that the upper states contained significant charge-transfer character. Clements and Barrows [123] characterized two states of AlAg, which they labelled as $B^1\Sigma^+$ and $C^1\Pi$. More recently Robbins *et al.* [124] have observed several states between the B and C states as well as several states above the C state.

We first studied the sensitivity of the AlCu ground state spectroscopic constants (r_0, D_0, and $\Delta G_{1/2}$) to the level of theory [125]. In our best calculation, which included inner-shell correlation, relativistic effects, very large one-particle basis sets, and high levels of correlation treatment, quantitative agreement with experiment was obtained: the computed r_0, $\Delta G_{1/2}$, and D_0 values, with experiment [121] in parentheses, are: 4.416(4.420) a_0, 295(294) cm^{-1}, and 2.318(2.315) eV. Thus we were able to achieve results of comparable accuracy to those discussed above for first- and second-row systems. Although this level of theory cannot be easily applied to an entire manifold of states, this calibration helps establish the origin and size of the errors in the lower level of theory used to study the excited states.

The four lowest states of AlCu that dissociate to ground state atoms were studied using a SA-CASSCF/ICMRCI approach, with only the valence electrons correlated. The two singlet and two triplet states were treated in separate calculations. The potential energy curves are shown in Fig. 14 and the spectroscopic constants are compared with experiment in Table XV.

The approximations in the computational approach result in errors of about 0.1 a_0 in r_0, 20 cm^{-1} in $\Delta G_{1/2}$, and 0.4 eV in D_0 for the $X^1\Sigma_g^+$ ground state. The errors in the r_0 and $\Delta G_{1/2}$ values are expected to be somewhat larger for the Π states, since they are less strongly bound, as a result of having only one-electron bonds. However, the error for the D_0 value should be less due to the smaller correlation energy contribution. Thus the computed T_0 values should be lower bounds. Nevertheless, there can be no question that the state observed at 14 892 cm^{-1} must be the $^1\Pi$ rather than the $^3\Pi$ state (see Table XV). This contention is further supported by the r_0 and $\Delta G_{1/2}$ values, which must be larger and smaller than experiment, respectively. This assignment results in an error in $T_0(^1\Pi)$ of 1290 cm^{-1}, which implies that we are underestimating the bond strength of one-electron bonds by about half that of two-electron bonds. Using

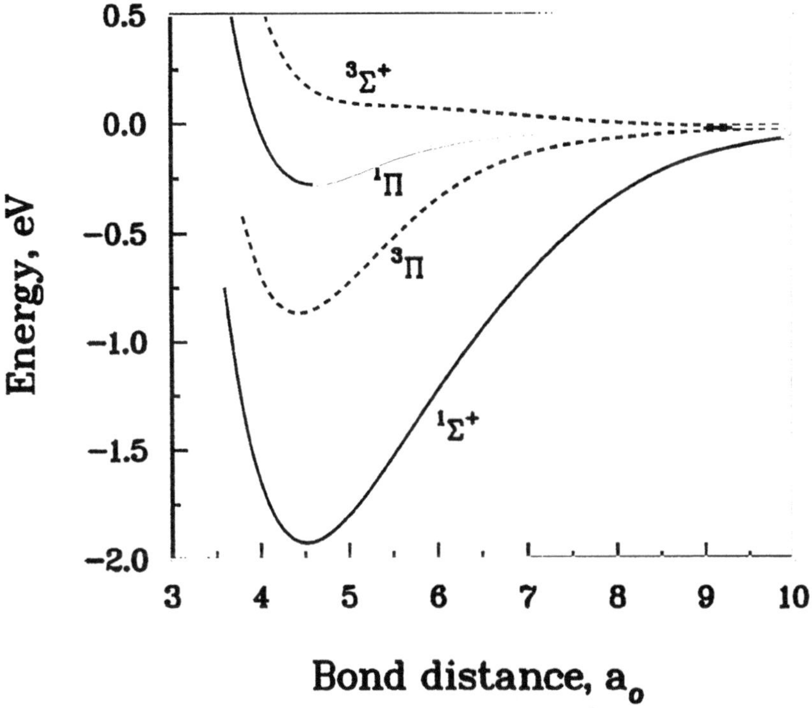

Fig. 14. SA-CASSCF/MRCI potential energy curves for the four electronic states of AlCu dissociating to ground state atoms.

the computed ICMRCI+Q potential, shifted to agree with the experimental T_0 value, and the ICMRCI transition moment yields a lifetime for $v'=0$ of 92 μs. This is in very good agreement with the experimental value of 71 μs, considering the small magnitude of the transition moment. This also suggests that the $^3\Pi - X^1\Sigma^+$ transition, which is expected to borrow its intensity from the $^1\Pi - X^1\Sigma^+$ transition, will give rise to only weak emission near 10 000 cm^{-1}.

Before considering the higher electronic states of AlCu, we first present the results for AlAg, as this molecular system is better described by theory, due to the fact that the d^9s^2 state lies higher for Ag and mixes less into the wave functions. Six singlet states and five triplet states were treated using the SA-CASSCF/ICMRCI approach; the spectroscopic constants (r_e, ω_e, and T_e) for these states are summarized in Table XVI. The lowest four states are analogous to those of AlCu; the computed T_e values and the lifetime of the $^1\Pi$ state are comparable. The spectroscopic constants for the $X^1\Sigma^+$ state are in better agreement with experiment for AlAg than AlCu, because relativistic effects are included for Ag atom using a relativistic effective core potential. The calculated spectroscopic constants for

TABLE XVI
Summary of the AlAg ICMRCI+Q results.

		ICMRCI+Q		
State	$r_e(a_0)$	$\omega_e(\text{cm}^{-1})$	$T_e(\text{cm}^{-1})$	τ
$X^1\Sigma^+$	4.796	240	0	
$(1)^3\Pi$	4.757	221	8679	
$(1)^1\Pi$	4.989	148	13012	5μs[a]
$(1)^3\Sigma^+$		repulsive		
$(1)^3\Sigma^-$	4.569	279	22787	
$(1)^1\Delta$	4.558	279	26543	
$(2)^1\Sigma^+$	4.965	195	27176	3ns
$(2)^3\Pi$	5.085	165	29082	
$(3)^3\Pi$	4.907	205	30160	
$(2)^1\Pi$	4.988	179	30615	21ns
$(3)^1\Sigma^+$	4.632	248	30664	13ns
		Experiment[b]		
State	$r_0(a_0)$	$\omega_e(\text{cm}^{-1})$	$T_0(\text{cm}^{-1})$	
$X^1\Sigma^+$	4.675	256.2	0	
$B^1\Sigma^+$	4.885[c]	206.4	27463	
		180.1	31002	
		247.0	30172	
		226.7	31051	
$C^1\Pi$	4.775[c]	226.1	31743	

[a] On the basis of high-level calculations for AlCu, this value is expected to be only qualitatively accurate.
[b] Taken from Robbins *et al.* [124] unless otherwise noted.
[c] Clements and Barrow [123]

the $(2)^1\Sigma^+$ state are also in good agreement with those measured for the $B^1\Sigma^+$ state, while the r_e and ω_e values for our $(2)^1\Pi$ state differ by somewhat more with the experimental values for the $C^1\Pi$ state. The $(2)^3\Pi$, $(3)^3\Pi$, and $(3)^1\Sigma^+$ states are also found to lie between the $B^1\Sigma^+$ and $C^1\Pi$ states. The r_e and ω_e values for these states agree reasonably well with the experimental values of Robbins *et al.* [124]. At the SA-CASSCF level we find a gap of 9000 cm^{-1} to the next triplet and 15000 cm^{-1} to the next singlet state. This is also consistent with experiment, which shows that the next group of excited states is about 7000 cm^{-1} above the $C^1\Pi$ state. Thus overall the spectroscopic results for AlAg are consistent with the experimental results.

Considering the success of theory in accurately treating the low-lying states of AlCu and the overall accuracy of the results for AlAg, it would seem that theory should be capable of assigning the states of AlCu that have been observed between 25 000 to 29 000 cm^{-1}. The results of SA-CASSCF/ICMRCI calculations for the higher states of AlCu are summarized in Table XVII. We first note that the description of the four lowest states is slightly degraded relative to the four state calculations, due to using a smaller basis set and by averaging for more

TABLE XVII
Comparison of the ICMRCI+Q spectroscopic constants for the singlet and triplet states of AlCu with experiment.

		ICMRCI+Q		
State	$r_e(a_0)$	$\omega_e(\mathrm{cm}^{-1})$	$T_e(\mathrm{cm}^{-1})$	τ
$X^1\Sigma^+$	4.564	260	0	
$(1)^3\Pi$	4.464	253	8526	
$(1)^1\Pi$	4.639	192	13791	$5\mu s^a$
$(1)^3\Sigma^+$		repulsive		
$(1)^3\Sigma^-$	4.280	320	22172	
$(1)^1\Delta$	4.269	298	26705	
$(2)^1\Sigma^+$	4.666	226	27159	5ns
$(2)^3\Pi$	4.810	151	29288	
$(3)^3\Pi$	4.668	253	29460	
$(3)^1\Sigma^+$	4.345	295	30741	17ns
$(2)^1\Pi$	4.675	213	31033	50ns
		Experimental results of Behm *et al.*		
State	$r_0(a_0)$	$\Delta G_{1/2}(\mathrm{cm}^{-1})$	$T_0(\mathrm{cm}^{-1})$	τ
$X^1\Sigma^+$	4.420	294	0	
$(1)^1\Pi$	4.505	224	14892	$71\mu s$
$(2)^1\Sigma^+$	4.864	178	25576	15ns
$(3)^1\Sigma^+$	4.747	210	26371	20ns
$(2)^1\Pi$	5.418	>113	<28400	118ns

[a] The value is very sensitive to the level of correlation treatment—see the text.

states. The lifetime of the $(1)^1\Pi$ state is particularly affected, because the small transition moment is sensitive to the level of theory. Comparing the calculated spectroscopic constants with experiment suggests that the A, B, A', and C states correspond to $(2)^1\Sigma^+$, $(3)^1\Sigma^+$, $(2)^3\Pi$ or $(3)^3\Pi$, and $(2)^1\Pi$, respectively. This is consistent with the assignment made by Behm *et al.* However, the agreement between the theoretical and experimental spectroscopic constants is very poor. The T_e value for the A state is acceptable, but the change in the computed r_e and ω_e values relative to the ground state ($+0.102\ a_0$ and $-34\ \mathrm{cm}^{-1}$), are very different from experiment ($+0.444\ a_0$ and $-114\ \mathrm{cm}^{-1}$). The agreement between theory and experiment is even worse for the B state. The results for the C state are not in good agreement with experiment, but a definitive comparison is difficult, since the lowest vibrational level of this state has not been experimentally observed. We feel the reason that the upper states of AlCu are described so much less accurately than the comparable states in AlAg is that the calculations underestimate the contribution of the $3d^94s^2$ Cu occupation to the excited states. This occupation is unimportant for the lower states of AlCu and does not contribute significantly to the excited states of AlAg, because of the very large $^2D - {}^2S$ separation for Ag. Thus the AlCu calculation serves to illustrate a particularly difficult orbital bias problem, which is often encountered for transition metal systems, but much less frequently so for calculations on systems containing only first- and second-row atoms.

3.10 Spectroscopy of the NbN Molecule

The experimental spectra of molecular systems containing transition metals are generally characterized by the presence of a plethora of lines. Not only are there a large number of low-lying electronic states arising from the different asymptotes of the metal, but large spin-orbit and nuclear hyperfine effects, coupled with the existence in many cases of several abundant isotopes, makes untangling the spectra extremely difficult. For these molecular systems, theory can give much needed guidance by determining the relative ordering of the states and strength of the band systems.

In this section we compare the results of SA-CASSCF/MRCI calculations [126] for NbN with the available experimental data [127, 128, 129, 130, 131]. Recently, new experimental data [130, 131] has become available for the $A^3\Sigma^-$, $a^1\Delta$, $b^1\Sigma^+$, $c^1\Gamma$, $e^1\Pi$ and $f^1\Phi$ states (using our spectroscopic designations). This combined with the previous experimental data [127, 128, 129] for the $X^3\Delta$, $B^3\Phi$, and $C^3\Pi$ states, makes the NbN one of the best studied transition-metal nitrides.

The Nb atom was described using a relativistic effective core potential [132], with the $4s$ and $4p$ orbitals included in the valence space. The nitrogen basis set was supplemented with diffuse functions to provide sufficient flexibility to describe any Nb^+N^- character in the wave function. Both basis sets were supplemented with multiple sets of polarization functions. The SA-CASSCF active space included the Nb $4d$ and $5s$ orbitals as well as the N $2p$ orbitals, except that the $4d\sigma - 2p\sigma$ bonding orbital was not included in the active space for the singlet and triplet calculations. An additional π orbital was added to the active space, due to its importance in describing many of the excited states. Eight electrons were correlated in the MRCI calculations, which includes the Nb-N σ bonding electrons that are not in the CASSCF active space for the singlet and triplet states.

The theoretical spectroscopic constants for the singlet, triplet, and quintet states of NbN are given in Table XVIII. These are in excellent agreement with the available experimental data for the singlet and triplet states. The computed bond lengths are uniformly slightly too long, probably due mostly from the neglect of Nb inner-shell correlation. Note that for the experimental T_e values we have used a weighted average of the fine-structure levels, even though the second-order spin-orbit interaction between the $a^1\Delta_2$ and $X^3\Delta_2$ levels results in shifts of about 50 cm^{-1}. The theoretical T_e values agree with experiment to better than 1000 cm^{-1} for all of the states observed.

The $X^3\Delta$ ground state of NbN arises from the $^6D(4d^45s^1)$ state of Nb and the $^4S^o(2p^3)$ state of N. A triple bond is formed involving the $4d\sigma$ and $4d\pi$ orbitals of Nb and the $2p$ orbitals of N. These orbitals are polarized toward nitrogen resulting in a significant degree of Nb^+N^- character in the wave function (the Mulliken population is $Nb^{+0.6}N^{-0.6}$ at r=3.1 a_o). One open-shell electron is in the nonbonding $5s$ orbital and the other is in the $4d\delta$-like Nb orbital. Note that while we denote the $5s$ orbital as nonbonding and the $4d\sigma$ orbital as bonding, there is significant mixing of these orbitals through the formation of $sd\sigma$ hybrids. The

TABLE XVIII
Spectroscopic constants for the low-lying states of NbN.

State	MRCI+Q			Expt.		
	$r_e(a_0)$	$\omega_e(\text{cm}^{-1})$	$T_e(\text{cm}^{-1})$	$r_0(a_0)^b$	$\omega_e(\text{cm}^{-1})$	$T_e(\text{cm}^{-1})^{b,c}$
$X^3\Delta$	3.165	1008	0	3.141^a	1033.7^c	0
$A^3\Sigma^-$	3.183	999	4925			4932
$B^3\Phi$	3.183	946	16753	3.158^a		16651
$C^3\Pi$	3.183	934	18192	3.157		17430
$a^1\Delta$	3.151	932	3999	3.120		4648
$b^1\Sigma^+$	3.181	976	4904			5307
$c^1\Gamma$	3.178	978	10201			9371
$d^1\Sigma^+$	3.163	1057	12878			
$e^1\Pi$	3.191	911	18939	3.161		18308
$(1)^5\Pi$	3.483	800	17637			
$(1)^5\Sigma^-$	3.362	882	19015			
$(1)^5\Delta$	3.610	769	23148			
$(1)^5\Phi$	3.440	791	28926			

[a] Reference [127].
[b] Reference [131] unless otherwise noted.
[c] The weighted average of fine structure levels.

polarization of the $5s$ orbital away from nitrogen reduces the Nb-N repulsion. This polarization reduces the dipole moment compared with states for which this polarization cannot occur. The dipole moment of 3.43 D indicates a considerable degree of ionic character. The $a^1\Delta$ state has the same occupation as the ground state. It lies about 5000 cm^{-1} higher in energy due to the loss of $d-s$ exchange energy.

The $A^3\Sigma^-$ state is derived from the ground state by promoting the Nb $5s$-like electron to the $4d\delta$ orbital. It has been characterized by the $C^3\Pi - A^3\Sigma^-$ transition. This state has a much larger dipole moment (6.39D) than the ground state, mainly because both of the open-shell electrons are in the $4d\delta$ orbital, which is much less polarizable than the $5s$ orbital. The $4d\delta^2$ occupation also gives rise to the $c^1\Gamma$ state, which lies about 5000 cm^{-1} above the $A^3\Sigma^-$ state. As expected, the dipole moment for this state is comparable to that of the $A^3\Sigma^-$ state.

The unobserved $d^1\Sigma^+$ state arises principally from the $4d\delta^2$ occupation, but it also contains a significant contribution from the $5s^2$ occupation. Conversely, the $b^1\Sigma^+$ state contains some admixture of the $4d\delta^2$ occupation. It is this admixture that results in the strong perturbation of the levels of the $A^3\Sigma^-$ state by the $b^1\Sigma^+$ state.

Exciting the Nb $5s$ electron into the π^* orbital leads to the $B^3\Pi$, $C^3\Pi$, $e^1\Pi$, and $f^1\Phi$ states. The large perturbations in the $C^3\Pi$ state are due to spin-orbit interactions with the nearby $e^1\Pi$ state, both of which are derived from the $4d\pi^1 4d\delta^1$

TABLE XIX
Radiative lifetimes for the $v'=0$ level of selected excited states of NbN.

State	Lifetime	Decay channels[a]
$B^3\Phi$	37 ns	$B^3\Phi - X^3\Delta$
$C^3\Pi$	35 ns	$C^3\Pi - X^3\Delta$, $C^3\Pi - A^3\Sigma^-$
$d^1\Sigma^+$	1010 ns	$d^1\Sigma^+ - b^1\Sigma^+$
$e^1\Pi$	64 ns	$e^1\Pi - a^1\Delta$, $e^1\Pi - b^1\Sigma^+$

[a] The dominant decay channels are listed in order of importance.

occupation. The dipole moments for the $B^3\Pi$, $C^3\Pi$, and $e^1\Pi$ states are all about 3.6 D, comparable to that of the ground state, but significantly less than those states having a $4d\delta^2$ occupation. The smaller dipole moments occur because the π^* orbital mixes in significant Nb $5p$ character to become more non-bonding than anti-bonding. This mixing of $4d\pi$ and $5p\pi$ orbitals leads to a polarization of the open-shell π orbital away from nitrogen. An analogous polarization occurs for the ground state in the σ space.

We have summarized the radiative lifetimes for four states of NbN in Table XIX, along with the dominant decay channels. Theory shows the $B-X$ and $C-X$ transitions to be of comparable strength. Although the lifetimes of these states have not been measured, the $B-X$ transition appears to be several times stronger than the $C-X$ transition in the spectra [130]. This may indicate different populations mechanisms for the two states.

It is interesting to compare the NbN molecule with the isoelectronic ZrO system, for which we have carried out comparable theoretical calculations [133]. The dissociation energy of NbN is computed to be 5.86 eV, which is significantly less than the value of 7.16 eV for ZrO [133]. While higher levels of theory will increase both values, it is clear that even though NbN has formally a triple bond, it is less strongly bound than ZrO. The larger D_e value for ZrO comes in part because back donation from the oxygen lone pair into an empty $4d$ orbital gives ZrO some triple bond character. The spectroscopy of the two molecules is also rather different, since the ground state of ZrO is $^1\Sigma^+$ instead of $^3\Delta$. This is due to the greater stability of the $4d$ orbitals for Nb, which also leads to the states derived from the $4d\delta^2$ occupation being much lower in NbN than in ZrO. This clearly demonstrates the utility of theory for providing insights into the bonding, in addition to accurate spectroscopic constants to guide experimental investigations.

4 Conclusions

In this review we have illustrated the application of computational chemistry methods to a range of problems in electronic spectroscopy of diatomic molecules. We have emphasized the accuracy that is currently achievable and how FCI benchmark calculations have influenced the computational approach. The SA-CASSCF/MRCI method has been shown to be flexible and to be an excellent approach for treating many electronic states in a common molecular orbital basis. For molecules consisting of first- and second-row atoms, radiative lifetimes can be computed with an accuracy of better than 10%. For systems containing transition metal systems, the calculations are more complex but they lead to results that are sufficiently reliable to aid in the analysis of experimental data. Quantitative accuracy for transition metal containing systems will probably require adding the valence *d* shell to the CASSCF active space, which dramatically increases the size of the calculations.

Future methodological developments and improvements in computational hardware are expected to continue to increase the size of the molecular systems amenable to accurate theoretical study. As illustrated by the chapters in the volume by Perić *et al.* [4] and Roos *et al.* [5], considerable insight into the spectroscopy of larger molecular systems is already being obtained.

References

1. P. K. Pearson, G. L. Blackman, H. F. Schaefer, B. Roos, and U. Wahlgren, Astrophys. J. Lett. **184**, L19 (1973).
2. For example see, R. D. Brown and E. H. N. Rice, J. Am. Chem. Soc. **106**, 6475 (1984).
3. T. Lee and G. Scuseria, see the chapter entitled "Achieving Chemical Accuracy with Coupled-Cluster Theory" in this volume.
4. M. Perić, B. Engels, and S. D. Peyerimhoff, see the chapter entitled "Theoretical Spectroscopy on Small Molecules: *Ab initio* Investigations of Vibronic Structure, Spin-Orbit Splittings and Magnetic Hyperfine Effects in the Electronic Spectra of Triatomic Molecules" in this volume.
5. B. O. Roos, M. Fülscher, P.-Å Malmqvist, M. Merchán, and L. Serrano-Andrés, see the chapter entitled "Theoretical Studies of the Electronic Spectra of Organic Molecules" in this volume.
6. "Ab initio methods in quantum chemistry", Adv. Chem. Phys. Vol. 67 and 69, Ed. K. P. Lawley, (Wiley, New York, 1987); "Modern Electronic Structure Theory" Ed. D. R. Yarkony, (World Scientific Publishing Company Co, Singapore 199x).
7. B. O. Roos, Adv. Chem. Phys. **69**, 399 (1987); K. K. Docken and J. Hinze, J. Chem. Phys. **57**, 4928 (1972).
8. V. R. Saunders and J. H. van Lenthe, Mol. Phys. **48**, 923 (1983).
9. R. J. Gdanitz and R. Ahlrichs, Chem. Phys. Lett. **143**, 413 (1988).
10. H.-J. Werner and P. J. Knowles, J. Chem. Phys. **89**, 5803 (1988); P. J. Knowles and H.-J. Werner, Chem. Phys. Lett. **145**, 514 (1988).
11. P. R. Taylor and H. Partridge, J. Phys. Chem. **91**, 6148 (1987).
12. I. Shavitt, p. 189 in "Methods of Electronic Structure Theory", ed. H. F. Schaefer (Plenum, New York 1977).
13. S. R. Langhoff and E. R. Davidson, Int. J. Quantum Chem. **8**, 61 (1974); M. R. A. Blomberg and P. E. M. Siegbahn, J. Chem. Phys. **78** (1983) 5682.
14. C. W. Bauschlicher, S. P. Walch, and S. R. Langhoff, p. 15, in "Quantum Chemistry: the Challenge of Transition Metals and Coordination Chemistry", ed. A. Veillard (Reidel Publishing Company, Dordrecht, Holland 1986).
15. S. R. Langhoff and C. W. Bauschlicher, Ap. J. **375**, 843 (1991).
16. A. E. Stevens, C. S. Feigerle, and W. C. Lineberger, J. Chem. Phys. **78**, 5420 (1983); S. P. Beaton, K. M. Evenson, T. Nelis, and J. M. Brown, J. Chem. Phys. **89**, 4446 (1988); W. J. Balfour, B. Lindgren, and S. O'Connor, Phys. Scr. **28**, 551 (1983); J. G. Phillips, S. P. Davis, B. Lindgren, and W. J. Balfour, Astrophys. J. Suppl. **65**, 721 (1987); J. G. Phillips and S. P. Davis, Astrophys. J. Suppl. **66**, 227 (1988). P. K. Carroll, P. McCormack, and S. O'Connor, Astrophys. J. **208**, 903 (1976); P. McCormack and S. O'Connor, Astron. Astrophys. Suppl. **26**, 373 (1976); A. Dendramis, R. J. van Zee, and W. Weltner, Jr., Astrophys. J. **231**, 632 (1979).

17. H. A. Bethe and E. A. Salpeter, "Quantum Mechanics of One- and Two-electron Atoms" (Springer-Verlag, Berlin 1957).
18. B. Schiff and C. L. Pekeris, Phys. Rev. A **3**, 638 (1964).
19. E. E. Whiting, A. Schadee, J. B. Tatum, J. T. Hougen, and R. W. Nicholls, J. Mol. Spectrosc. **80**, 249 (1980).
20. F. L. Tobin and J. Hinze, J. Chem. Phys. **63**, 1034 (1975).
21. C. W. Bauschlicher, S. R. Langhoff, and P. R. Taylor, Adv. Chem. Phys. **77**, 103 (1990).
22. N. C. Handy, Chem. Phys. Lett. **74**, 280 (1980); P. E. M. Siegbahn, Chem. Phys. Lett. **109**, 417 (1984); P. J. Knowles and N. C. Handy, Chem. Phys. Lett. **111**, 315 (1984).
23. J. Almlöf and P. R. Taylor, J. Chem. Phys. **86**, 4070 (1987).
24. P.-O. Widmark, J. B. Persson, and B. O. Roos, Theor. Chim. Acta **79**, 419 (1991) and P.-O. Widmark, P. A. Malmqvist, and B. O. Roos, Theor. Chim. Acta **77**, 291 (1990).
25. T. H. Dunning, J. Chem. Phys. **90**, 1007 (1989) and D. E. Woon and T. H. Dunning, J. Chem. Phys. **98**, 1358 (1993). R. A. Kendall, T. H. Dunning, and R. J. Harrison, J. Chem. Phys. **96**, 6796 (1992); D. E. Woon, K. A. Peterson, and T. H. Dunning, personal communication and D. E. Woon and T. H. Dunning, personal communication.
26. C. W. Bauschlicher and H. Partridge, J. Chem. Phys. **100**, 4329 (1994).
27. D. E. Woon, Chem. Phys. Lett. **204**, 29 (1993) and J. Chem. Phys. **100**, 2838 (1994).
28. H.-J. Werner and P. J. Knowles, J. Chem. Phys. **94**, 1264 (1991).
29. C. W. Bauschlicher, and S. R. Langhoff, J. Chem. Phys. **87**, 4665 (1987).
30. D. P. Chong and S. R. Langhoff, J. Chem. Phys. **84**, 5606 (1986). Also see R. Ahlrichs, P. Scharf, and C. Ehrhardt, J. Chem. Phys. **82**, 890 (1985).
31. W. L. Dimpfl and J. L. Kinsey, J. Quant. Spectrosc. Radiat. Transfer **21**, 233 (1979).
32. K. R. German, J. Chem. Phys. **64**, 4065 (1976).
33. K. R. German, T. H. Bergeman, E. M. Weinstock, and R. N. Zare, J. Chem. Phys. **58**, 4304 (1973).
34. I. S. McDermid and J. B. Laudenslager, J. Chem. Phys. **76**, 1824 (1982) and T. Bergeman, P. Erman, Z. Haratym, and M. Larsson, Phys. Scr. **23**, 45 (1981).
35. J. Brzozowski, P. Erman, and M. Lyyra, Phys. Scr. **17**, 507 (1978).
36. C. W. Bauschlicher and S. R. Langhoff, Theor. Chim. Acta **79**, 93 (1991).
37. S. R. Langhoff, C. W. Bauschlicher, A. P. Rendell, and A. Komornicki, J. Chem. Phys. **92**, 6599 (1990).
38. M. Cohen, F. C. Witteborn, D. F. Carbon, G. Augason, D. Wooden, J. Bregman, and D. Goorvitch, Astron. J. **104**, 2045 (1992).
39. D. F. Carbon, personal communication.
40. R. H. Tipping and C. Chackerian, J. Mol. Spectrosc. **88**, 352 (1981).
41. S. R. Langhoff and C. W. Bauschlicher, Chem. Phys. Lett. **211**, 305 (1993).
42. J. W. Raymonda, J. S. Muenter, and W. A. Klemperer, J. Chem. Phys. **52**, 3458 (1970).
43. S. R. Langhoff and J. O. Arnold, J. Chem. Phys. **70**, 852 (1979).
44. H.-J. Werner, P. Rosmus, and M. Grimm, Chem. Phys. **73**, 169 (1982); K. A. Peterson and R. C. Woods, J. Chem. Phys. **92**, 6061 (1990); P. Botschwina and P. Rosmus, J. Chem. Phys. **82**, 1420 (1985); V. Kellö and A. J. Sadlej, J. Chem. Phys. **98**, 1345 (1993).
45. K. P. Huber and G. Herzberg, "Constants of Diatomic Molecules", (Van Nostrand Reinhold, New York, 1979).
46. S. Saupe, I. Adamovich, M. J. Grassi, J. W. Rich, and R. C. Bergman, Chem. Phys. **174**, 219 (1993).
47. H. Dünnwald, E. Siegel, W. Urban, J. W. Rich, G. F. Homicz, and M. J. Williams, Chem. Phys. **94**, 195 (1985).
48. B. D. Green, G. E. Caledonia, and R. E. Murphy, J. Quant. Spectrosc. Radiat. Transfer **26**, 215 (1981).
49. G. E. Caledonia and J. P. Kennealy, Planet. Space Sci. **30**, 1043 (1982).
50. S. R. Langhoff, C. W. Bauschlicher, and H. Partridge, Chem. Phys. Lett., in press.
51. W. T. Rawlins, M. E. Fraser, S. M. Miller, and W. A. M. Blumberg, J. Chem. Phys. **96**, 7555 (1992).
52. F. P. Billingsley, J. Chem. Phys. **62**, 864 (1975); **63**, 2267 (1975).
53. R. M. Neumann, Astrophys. J. **161**, 779 (1970).
54. A. R. Hoy, J. W. C. Johns, and A. R. W. McKellar, Can. J. Phys. **53**, 2029 (1975).
55. B. Schurin and R. E. Ellis, J. Chem. Phys. **45**, 2528 (1966).
56. G. Chandraiah and C. W. Cho, J. Mol. Spectrosc. **47**, 134 (1973).
57. E. R. Young, D. G. Torr, and K. C. Clark, J. Geophys. Res. **85**, 4223 (1980).
58. P. C. Cosby and H. Helm, J. Chem. Phys. **75**, 3882 (1981).

59. O. Edqvist, L. Åsbrink, and E. Lindholm, Z. Naturforsch. Teil A **26**, 1407 (1971).
60. D. L. Albritton, A. L. Schmeltekopf, and R. N. Zare, J. Chem. Phys. **71**, 3271 (1979).
61. H. Partridge, S. R. Langhoff, and C. W. Bauschlicher, J. Chem. Phys. **93**, 7179 (1990).
62. R. P. Saxon, D. F.-t. Tuan, H. Helm, and B. Liu, J. Chem. Phys. **80**, 2648 (1984).
63. H. Lefebvre-Brion, Chem. Phys. Lett. **9**, 463 (1971).
64. S. R. Langhoff, C. W. Bauschlicher, and H. Partridge, J. Chem. Phys. **84**, 4474 (1986).
65. J. N. Allison and W. A. Goddard, J. Chem. Phys. **77**, 4259 (1982).
66. M. Freund, E. Herbst, R. P. Mariella, and W. Klemperer, J. Chem. Phys. **56**, 4516 (1972).
67. R. R. Herm and D. R. Herschbach, J. Chem. Phys. **52**, 5783 (1970).
68. D. M. Lindsay, D. R. Herschbach, and A. L. Kwiram, Mol. Phys. **32**, 175 (1974).
69. C. W. Bauschlicher, H. Partridge, and K. G. Dyall, Chem. Phys. Lett. **99**, 225 (1992).
70. K. Yamashita, unpublished, quoted in Ref. [72] and S. R. Langhoff, H. Partridge, and C. W. Bauschlicher, unpublished–taken from the calculations described in Ref. [71].
71. S. R. Langhoff, H. Partridge, and C. W. Bauschlicher, Chem. Phys. **153**, 1 (1991).
72. C. Yamada, M. Fujitake, and E. Hirota, J. Chem. Phys. **90**, 3033 (1989); **91**, 137 (1989).
73. S. P. So and W. G. Richards, Chem. Phys. Lett. **32**, 227 (1975).
74. P. A. G. O'Hare and A. C. Wahl, J. Chem. Phys. **56**, 4516 (1972).
75. H. W. Sarkas, J. Hendricks, and K. H. Bowen, to be published.
76. C. W. Bauschlicher, H. Partridge, and L. G. M. Pettersson, J. Chem. Phys. **99**, 3654 (1993).
77. W. Müller, J. Flesch, and W. Meyer, J. Chem. Phys. **80**, 3297 (1984).
78. S. R. Langhoff, C. W. Bauschlicher, and H. Partridge, p. 357 in "Comparison of *ab initio* Quantum Chemistry with Experiment", ed. R. Bartlett (D. Reidel Publishing Company, Boston 1985).
79. W. Müller and W. Meyer, J. Chem. Phys. **80**, 3311 (1984); L. G. M. Pettersson, H. Åkeby, P. Siegbahn, and U. Wahlgren, J. Chem. Phys. **93**, 4954 (1990); L. G. M. Pettersson and H. Åkeby, J. Chem. Phys. **94**, 2968 (1991).
80. H. Partridge, S. R. Langhoff, C. W. Bauschlicher, and D. W. Schwenke, J. Chem. Phys. **88**, 3174 (1988).
81. R. Hermann, C. R. Acad. Sci (Paris) **233**, 738 (1951).
82. J. Berkowitz, W. A. Chupka, and G. B. Kistiakowsky, J. Chem. Phys. **25**, 457 (1956).
83. I. M. Campbell and B. A. Thrush, Proc. R. Soc. A **296**, 201 (1967).
84. M. Krauss and D. B. Neumann, Mol. Phys. **32**, 101 (1976).
85. P. E. M Siegbahn, Int. J. Quantum Chem. **23**, 1869 (1983).
86. K. P. Huber and M. Vervloet, J. Chem. Phys. **89**, 5957 (1988).
87. C. W. Bauschlicher, S. R. Langhoff, and H. Partridge J. Chem. Phys. **98**, 8785 (1993).
88. J. Bahrdt, H. Nahme, and N. Schwentner, Chem. Phys. **144**, 273 (1990).
89. M. E. Rosenkrantz, J. E. Bohr, and D. D. Konowalow, Theor. Chim. Acta **82**, 153 (1992).
90. B. Bussery, M. E. Rosenkrantz, D. D. Konowalow, and M. Aubert-Frecon, Chem. Phys. **134**, 7 (1989).
91. N. Schwentner, personal communication.
92. I. M. Campbell and R. S. Mason, J. Photochem. **5**, 383 (1976); **8**, 375 (1978); **11**, 53 (1979).
93. C. W. Bauschlicher, H. Partridge, and S. R. Langhoff, Chem. Phys. Lett. **211**, 312 (1993).
94. S. R. Langhoff, H. Partridge, and C. W. Bauschlicher, J. Mol. Spectrosc. **138**, 123 (1989).
95. G. W. Faris, M. J. Dyer, and T. G. Slanger, Chem. Phys. Lett. **199**, 309 (1992).
96. G. R. Möhlmann and F. J. de Heer, Chem. Phys. Lett. **49**, 588 (1977).
97. K. P. Huber and M. Vervloet, J. Mol. Spectrosc. **129**, 1 (1988).
98. P. Erman and M. Larsson, Phys. Scr. **15**, 335 (1977).
99. J. T. Moseley, P. C. Cosby, J.-B. Ozenne, and J. Durup, J. Chem. Phys. **70**, 1474 (1979).
100. F. R. Gilmore, R. R. Laher, and P. J. Espy, J. Phys. Chem. Ref. Data **21**, 1005 (1992).
101. A. R. Fairbairn, J. Chem. Phys. **60**, 521 (1974); M. Jeunehomme, J. Chem. Phys. **44**, 4253 (1966); E. H. Fink and K. H. Welge, Z. Naturforsch. **23a**, 358 (1968).
102. M. R. A. Blomberg and B. Liu, IBM Research Report No. RJ 6080, IBM Research Division, San Jose, CA, February 9, 1988.
103. Ch. Ottinger and A. F. Vilesov, J. Chem. Phys. **100**, 1805 (1994).
104. D. M. Cooper, J. Quant. Spectrosc. Radiat. Transfer **27**, 459 (1982) and references therein.
105. G. E. Gadd and T. G. Slanger, J. Chem. Phys. **92**, 2194 (1990).
106. L. G. Piper, T. R. Tucker, and W. P. Cummings, J. Chem. Phys. **94**, 7667 (1991).
107. R. de Vivie and S. D. Peyerimhoff, J. Chem. Phys. **89**, 3028 (1988).
108. S. R. Langhoff, H. Partridge, C. W. Bauschlicher, and A. Komornicki, J. Chem. Phys. **94**,

6638 (1991).
109. S. R. Langhoff, C. W. Bauschlicher, and H. Partridge, J. Chem. Phys. **89**, 4909 (1988); C. W. Bauschlicher and S. R. Langhoff, J. Chem. Phys. **89**, 2116 (1988).
110. C. W. Bauschlicher, H. Partridge, S. R. Langhoff, P. R. Taylor, and S. P. Walch, J. Chem. Phys. **86**, 7007 (1987).
111. S. R. Langhoff and C. W. Bauschlicher, J. Chem. Phys. **92**, 1879 (1990).
112. H. Basch, W. J. Stevens, and M. Krauss, Chem. Phys. Lett. **109**, 212 (1984); T. H. Upton, J. Phys. Chem. **90**, 754 (1986); C. W. Bauschlicher and S. R. Langhoff, J. Chem. Phys. **90**, 4627 (1989); K. K. Sunil and K. D. Jordan, J. Phys. Chem. **92**, 2774 (1988).
113. D. S. Ginter, M. L. Ginter, and K. K. Innes, Astrophys. J. **139**, 365 (1963).
114. M. A. Douglas, R. H. Hauge, and J. L. Margrave, J. Phys. Chem. **87**, 2945 (1983).
115. M. F. Cai, T. P. Dzugan, and V. E. Bondybey, Chem. Phys. Lett. **155**, 430 (1989).
116. Z. Fu, G. W. Lemire, G. A. Bishea, and M. D. Morse, J. Chem. Phys. **93**, 8420 (1990).
117. C. R. Brazier and P. G. Carrick, J. Chem. Phys. **96**, 8684 (1992).
118. S. R. Langhoff and C. W. Bauschlicher, J. Chem. Phys. **95**, 5889 (1991).
119. C. W. Bauschlicher and S. R. Langhoff, J. Chem. Phys., in press.
120. C. W. Bauschlicher and S. R. Langhoff, Chem. Rev. **91**, 701 (1991).
121. J. M. Behm, C. A. Arrington, J. D. Langenberg, and M. D. Morse, J. Chem. Phys. **99**, 6394 (1993).
122. M. F. Cai, S. J. Tsay, T. P. Dzugan, K. Pak, and V. E. Bondybey, J. Phys. Chem. **94**, 1313 (1990).
123. R. M. Clements and R. F. Barrow, Trans. Faraday Soc. **64**, 2893 (1963).
124. D. L. Robbins, C. S. Yeh, J. S. Pilgrim, G. L. Lang, and M. A. Duncan, J. Chem. Phys. **100**, 4775 (1994).
125. C. W. Bauschlicher, S. R. Langhoff, and H. Partridge, J. Chem. Phys. **100**, 1219 (1994).
126. S. R. Langhoff and C. W. Bauschlicher, J. Mol. Spectrosc. **143**, 169 (1990).
127. T. M. Dunn and K. M. Rao, Nature(London) **222**, 266 (1969).
128. Y. Azuma, J. A. Barry, M. P. J. Lyne, A. J. Merer, J. O. Schröder, and J.-L. Féménias, J. Chem. Phys. **91**, 1 (1989).
129. J.-L. Féménias, C. Athénour, and T. M. Dunn, J. Chem. Phys. **63**, 2861 (1975); E. A. Pazyuk, E. N. Moskvitina, and Yu. Ya. Kuzyakov, Spectrosc. Lett. **19**, 627 (1986); J.-L. Féménias, C. Athénour, K. M. Rao, and T. M. Dunn, J. Mol. Spectrosc. **130**, 269 (1988).
130. A. J. Merer, personal communication.
131. Y. Azuma, G. Huang, M. P. J. Lyne, A. J. Merer, and V. I. Srdanov, preprint communicated to the authors.
132. P. J. Hay and W. R. Wadt, J. Chem. Phys. **82**, 299 (1985).
133. S. R. Langhoff and C. W. Bauschlicher, J. Chem. Phys. **89**, 2160 (1988); Astrophys. J. **349**, 369 (1990).

Theoretical Spectroscopy on Small Molecules : *Ab initio* Investigations of Vibronic Structure, Spin-Orbit Splittings and Magnetic Hyperfine Effects in the Electronic Spectra of Triatomic Molecules

M. Perić*, B. Engels and S.D. Peyerimhoff
Institut für physikalische und theoretische Chemie
Universität Bonn
Wegelerstr. 12
53115 Bonn
Federal Republic of Germany

Feb-1994

S. R. Langhoff (Ed.), Quantum Mechanical Electronic Structure Calculations with Chemical Accuracy, 261–356.

Contents

1 Introduction

In the last decade a number of reviews have been published reporting on progress in *ab initio* quantum mechanical calculations of molecular structure and spectra – let us mention a few of them: 'Third age of quantum chemistry' by G. Richards [1], 'Theoretical chemistry comes alive : Full partner with experiment' by W. A. Goddard III [2], 'Methylene : A paradigm for computational quantum chemistry' by H.F. Schaefer III [3], 'Accurate quantum chemical calculations' by Bauschlicher, Langhoff and Taylor [4] and 'Quantum mechanical calculations to chemical accuracy' by Bauschlicher and Langhoff [5]. All of these document by a number of examples that the numerical accuracy of quantum mechanical calculations on small molecules has reached such a level that it can compete with high–resolution spectroscopy and that the results of these computations allow in many instances a more reliable determination of various molecular parameters than the corresponding experimental findings do. For a pictorial illustration of this matter the reader is referred to the paper by Schaefer [3] in which the struggles, mistakes and successes of both, experimentalists and theoreticians, in trying to resolve the structure of methylene in its two lowest–lying electronic states are described. With the increase in accuracy theoretical chemistry has retained its inherent advantages such as universality (i.e. every molecule can be computed in each of its states independently of the energetic location and regardless of the symmetry of the state, the stability or particular properties which can make its experimental study difficult or even prohibitive) and the ability to give explanation also for mechanisms often hidden by the complicated structure of spectra observed.

At which point the beginning of the 'second', 'third' or 'fourth' age of theoretical (quantum) chemistry is placed depends on the personal experience and the viewpoint of the authors. However, five corner-stones are central to the progress in quantum chemical calculations over the past decades:

1. Implementation of the LCAO idea into the molecular structure computations [6] – [8];

2. Construction of appropriate atomic orbital (AO) bases making actual computations according to this scheme possible [9]–[11];

3. Use of fast computers which makes it possible to closely approach the solution of the molecular Schrödinger equation;

4. Accounting for the electronic correlation effects by the configuration interaction (CI) method or equivalent procedures (for an overview see e.g. [12]);

5. Possibility of benchmark calculations in which large atomic orbital bases are employed and the full CI limit is effectively reached (see, for example [4, 5]).

The above quoted major steps are related mainly to the solution of the electronic Schrödinger equation, which is a consequence of the application of the Born–Oppenheimer approximation [13] introduced shortly after discovering the laws of quantum mechanics. This approximation introduces the concept of potential surfaces and in this manner makes the handling of the otherwise hopelessly complicated molecular Schrödinger equation possible. To the full appreciation of quantum chemistry have, however, also contributed struggles of both experimentalists and theoreticians to join each other: In the early phase of quantum chemical calculations a number of quantities have been computed which generally cannot be directly compared to measurements, for example equilibrium geometry parameters, vertical electronic transition energies, and potential surfaces. The bridge between these quantities and the results of experimental observations (positions of spectral lines and their intensities) has been built from both sides. Fitting their data to various kinds of effective Hamiltonians the experimentalists have succeeded in extracting from their measurements the above mentioned quantities resulting from quantum chemical structure calculations. On the other hand, theoreticians have developed approaches for calculating rovibrational structure of spectra employing the computed potential surfaces. With the improvement in the accuracy of these surfaces it has become realistic to include constantly smaller effects (vibronic and spin–orbit couplings, hyperfine interactions) into the theoretical treatment, which then allows the construction of the molecular spectrum with all its details.

Parallel with the progress in quantum chemical calculations the attitude of the experimentalists towards this field has also changed: From large scepticism in the early phase of computational chemistry, to the recognition that such calculations can be of help in the interpretation of observed spectra, and finally to a situation, in which theoretical chemistry is generally accepted as a 'full partner to experiment'. In 1974 at the NATO Conference in Breukelen (Holland) the renowned spectroscopist D.A. Ramsay presented a table to show what the experimentalists expected from the theoreticians twenty years ago (Table I): Not necessarily data of spectroscopic accuracy but rather results with definite error limits that would help answer some important questions, which could not be answered on the basis of experiments alone. So, for example, the theoretical information about the occurrence of various electronic states in a certain energy range and their interactions was very valuable for the experimentalists; the help of the theoreticians was sought if it was not possible to assign unambiguously the spectral lines observed. It is clear that the error limits quoted in Table I as 'acceptable' possess hardly any predictive value – they could only be of some help in the interpretation of general features of the spectra. The situation is different, however, with the theoretical results referred to as 'good': in a number of cases it has been shown that results of such accuracy, corresponding to the state-of-the-art of that time, have lead to reliable predictions of spectra not observed thus far.

In this review we present the results of *ab initio* calculations for the structure and spectra of a number of triatomic molecules performed in our laboratory in the

TABLE I
Requirements for accuracy of quantum chemistry calculations twenty years ago

	good	acceptable
r,Δr	0.01 Å	0.05 Å
θ	1 degree	5 degrees
ΔE,D,I	0.1 eV	0.5 eV
ω	50 cm^{-1}	250 cm^{-1}
μ	0.1 D	0.5 D
f	50 %	250 %

r $\equiv$ equilibrium bond length
$\theta \equiv$ equilibrium bond angle
ΔE $\equiv$ vertical electronic transition energy
D $\equiv$ dissociation energy
I $\equiv$ ionization potential
$\omega \equiv$ vibrational frequency
$\mu \equiv$ dipole moment
f $\equiv$ oscillator strength

last decade. We will concentrate in particular on the interplay between vibronic interaction, spin–orbit coupling and hyperfine effects in the low–lying states. Our aim is to demonstrate the possibilities of modern quantum chemical computations to explain and predict details of spectra of small molecules, but we will also comment on the difficulties of such a task and on the error limits to be expected for the results.

2 *Ab initio* Calculation of Molecular Energies and Wave Functions

2.1 How to solve the molecular Schrödinger equation?

If we consider an isolated molecule consisting of N nuclei and n electrons, then the postulates of quantum mechanics say that all observable information on it can be obtained by solving the corresponding Schrödinger equation. In this review we consider only the time–independent Schrödinger equation

$$H_{tot}\Psi_{tot}(\mathbf{r},\mathbf{s},\mathbf{R},\mathbf{I}) = E_{tot}\Psi_{tot}(\mathbf{r},\mathbf{s},\mathbf{R},\mathbf{I}) \tag{1}$$

in which H_{tot} represents the exact molecular Hamiltonian, E_{tot} its eigenvalues and Ψ_{tot} the total wave function depending on 3n electronic spatial coordinates (denoted collectively by **r**), n electronic spin coordinates (s), 3N nuclear spatial (**R**) and N spin coordinates (**I**). This Schrödinger equation, involving 4(n+N) coordinates can be solved analytically only for the most simple systems and one is therefore forced to seek alternative ways to solve it for the problems of chemical interest. Three main directions are conceivable:

- Solutions of Eq. 1 can in principle be obtained 'brute force' by direct numerical integration. Because of the immense computational requirements this is only possible for very small molecules, but will probably never be feasible for larger molecules in spite of the rapid progress in developing powerful computers.

- The terms in the Schrödinger equation (1) which cause the difficulties for the solutions are replaced by simpler expressions, for example, by introducing suitable parameters (fitted to experimental data or based on intuition) or by omitting them entirely. A serious drawback of such semiempirical approaches is loss of generality and often lower accuracy of the results so that they are not adequate for many purposes, particularly to predict details of molecular spectra.

- One can simplify the Schrödinger equation by separating it into relatively weakly coupled parts, so that each part depends directly only on a smaller number of variables. The combined solutions of such simpler problems then represent an approximate solution to the general problem. The results can be improved, if necessary, by introducing the coupling into these zeroth–order wave functions by perturbation or variational methods. Typical examples are the separation of nuclear and electronic coordinates or of spin and electronic coordinates; spin–orbit interaction or coupling between electronic and nuclear (vibrational, rotational) degrees of freedom is generally introduced after the zeroth–order electronic wave functions and the vibrational or spinfunctions have been generated. The first and at the same time the most important step in this scheme is to find the convenient 'coordinate systems' which allows the separation of variables such that the products of the zeroth–order wave functions are as close as possible to the exact solution.

The last approach is the basic philosophy of an *ab initio* treatment of the molecular problem. Although the *ab initio* method is commonly defined as an exact approach, which does not build on any *a priori* knowledge about the system to be considered (except for basic data such as the number of electrons, the masses and charges of the nuclei building the molecule), its actual application requires always some specific approximations whose role is to reduce the computational requirements. In contrast to the situation with semiempirical approaches, however, these approximations are physically well justified, and can be easily controlled. If necessary, they can be removed whenever their reliability is questionable. In spite of these simplifications, the computational requirements for calculations remain relatively high and therefore calculations that treat all the different coordinates shown in Eq. 1 have thus far been restricted to a large extent to relatively small molecular systems.

2.2 Born–Oppenheimer approximation

In the most commonly applied simplification of the Schrödinger equation, the terms that couple the spatial and spin coordinates in the Hamiltonian are neglected (for a survey of *ab initio* calculations beyond this approximation performed in our laboratory the reader is referred to [14]). In this case Eq. 1 is reduced to

$$H\Psi(\mathbf{r},\mathbf{R}) = E\Psi(\mathbf{r},\mathbf{R}) \tag{2}$$

where H is the nonrelativistic Hamiltonian:

$$\begin{aligned} H &= -\frac{1}{2}\sum_{a}^{N}\frac{1}{m_a}\Delta_a - \frac{1}{2}\sum_{i}^{n}\Delta_i - \sum_{a}^{N}\sum_{i}^{n}\frac{Z_a}{r_{ia}} + \sum_{i<j}^{n}\frac{1}{r_{ij}} + \sum_{a<b}^{N}\frac{Z_aZ_b}{R_{ab}} \\ &\equiv T_N + T_e + V_{eN} + V_{ee} + V_{NN} \equiv T_N + H_e + V_{NN} \end{aligned} \tag{3}$$

(atomic units are used throughout in the present paper) and Ψ represents the wave function depending only on spatial coordinates of the electrons and nuclei. Equation 2 is still much too complicated to be solved directly and a number of additional approximations are commonly introduced. The most important of them is the Born–Oppenheimer approximation, which separates the electronic and nuclear problems. This approximation is motivated by the experimental finding that the structure of molecular spectra can in most cases be interpreted in terms of nearly complete separation of electronic and nuclear motions. The physical background is that the 'electronic cloud' follows adiabatically the motion of the nuclei and adapts itself to any new nuclear arrangement. This idea is expressed mathematically through representation of the wave function $\Psi(\mathbf{r},\mathbf{R})$ in a product form

$$\Psi(\mathbf{r},\mathbf{R}) = \psi_{\mathbf{R}}(\mathbf{r})f_N(\mathbf{R}) \tag{4}$$

where $\psi_{\mathbf{R}}(\mathbf{r})$ represents the solution of the electronic Schrödinger equation at fixed nuclear positions $\mathbf{R}$

$$H_e\psi_{\mathbf{R}}(\mathbf{r}) = E_e\psi_{\mathbf{R}}(\mathbf{r}) \tag{5}$$

Due to the fact that the form of the electronic Hamiltonian H_e depends on the actual positions of the nuclei, the corresponding wave function $\psi_{\mathbf{R}}(\mathbf{r})$ depends also on the nuclear positions; this dependence is, however, weak ('parametrical'), so that in many cases it is justified to neglect first and second derivatives of the electronic wave function with respect to nuclear coordinates. The nuclear function $f_N(\mathbf{R})$ satisfies then the Schrödinger equation

$$[T_N + E_e(\mathbf{R}) + V_{NN}]f_N(\mathbf{R}) = Ef_N(\mathbf{R}) \tag{6}$$

in which the sum of the electronic energy, E_e, and the nuclear repulsion term, V_{NN}, plays the role of the potential, calculated at a number of various nuclear arrangements.

The Born–Oppenheimer approximation is very successful in many cases, but there are a number of well–specified instances in which it breaks down. An important example is the Renner–Teller effect being the main topic of the present study.

2.3 SOLUTION OF THE ELECTRONIC SCHRÖDINGER EQUATION

The electronic (nonrelativistic) Hamiltonian consists of two terms:

$$H_e = \sum_i^n h_i + \sum_{i<j}^n \frac{1}{r_{ij}} \tag{7}$$

the first one being a sum of one–electron operators taking account of the kinetic energy of the electrons and the electron–nuclear attraction

$$h_i = -\frac{1}{2}\Delta_i - \sum_a^N \frac{Z_a}{r_{ia}} \tag{8}$$

and the second term representing a two–electron operator that describes the electron–electron repulsion. If the two–electron operator did not exist, H_e would describe a system of mutually independent electrons moving in the field produced by N fixed nuclei. The solution of the corresponding Schrödinger equation would have the form of a product of n one–electron wave functions generally written as spin orbitals $u_i(j) = \phi_k(j)\cdot\omega(j)$, $\phi_k(j)$ being a spatial molecular orbital (MO) and $\omega(j)$ one of the eigenfunctions (α,β) of the spin operator s_z. According to the Pauli principle this product has to be antisymmetrized (Slater determinant). In the framework of the Hartree–Fock (HF) approach [15, 16], this simple antisymmetrized product form (or more generally a linear combination of a few Slater determinants for the systems containing also partially populated shells) is assumed and the optimal orbitals $\phi_k(j)$ are obtained by minimizing the expectation value of H_e and integrating over the spin coordinates, in short by solving the Hartree–Fock (HF) equations

$$F_k\phi_k \equiv [h_k + \sum_j (J_j - K_j)]\phi_k = \varepsilon_k\phi_k \tag{9}$$

where F_k is the Fock operator involving besides the one–electron operator h_k (see Eq. 8) a sum of Coulomb (J_j) and exchange operators (K_j) representing a mean-field for electron k caused by all other electrons j. Since J_j and K_j are defined in terms of ϕ_k, Eq. 9 has to be solved iteratively (self-consistent-field (SCF) approach).

In practical molecular calculations Eqs.9 are solved according to the scheme proposed by Roothaan, by expanding the molecular orbitals into linear combinations of atomic orbitals (AO) χ_p [6], whereby the expansion coefficients are treated as variational parameters. In this way the integro–differential equations (9) are converted into a system of algebraic equations. Roothaan's idea introduces new problems however, because the form of the atomic orbitals χ_p is needed which is known exactly only for the hydrogen atom. Furthermore there is no guarantee that the optimal orbitals for isolated atoms would be appropriate when these atoms are part of a molecule. An efficient application of this approach has become possible thanks to the work of a number of authors [9, 10, 11] who have published the approximate orbitals for the atoms in the form of linear combinations of certain elementary functions (Slater, Gaussian). The deformed electron density distributions in molecules relative to those of isolated atoms is generally accounted for by adding 'polarization functions', namely AOs in the form of higher spherical harmonics or in some instances functions simply placed in the bonding area.

The HF approach is essentially a one–electron approximation, which considers only the interaction of each electron with the meanfield produced by the others, but does not account for the actual correlation in the movement of the electrons, and this simplification leads to an inherent correlation error being of different magnitude for various electronic states. Furthermore the HF approach generally gives unreliable results when the electronic energy is computed over a large range of nuclear coordinates up to dissociation. In such cases an electronic state can only rarely be described by a single Slater determinant ('configuration'). A much more reliable representation of an electronic state can be achieved if the corresponding wave function assumes the form of a linear combination of many configurations:

$$\psi_k = \sum_i^M c_{ki}\Phi_i \tag{10}$$

where the basis functions Φ_i (Slater determinants) are commonly constructed from the molecular orbitals obtained in the HF procedure or modifications thereof, which include parts of the correlation effect {multi-configuration SCF (MCSCF) [17] or complete active space SCF (CASSCF) [18]}. This approach is referred to as the configuration-interaction (CI) procedure. The expansion coefficients c_{ki} and the energy values E_k are obtained as solutions of a homogeneous system of linear equations whose size is determined by the number of expansion terms M in Eq. 10:

$$\sum_i^M c_{ki}(H_{ki} - E_k\delta_{ki}) = 0 \tag{11}$$

$$k = 1, 2, \ldots.., M$$

where

$$H_{ki} = < \Phi_i|H_e|\Phi_k > \tag{12}$$

are the matrix elements of the electronic Hamiltonian (Eq. 7). (It is assumed that the basis functions Φ_i form an orthonormal set.)

The most difficult technical problem in applying the CI approach is the choice of the configurations Φ_i entering into the expansion in Eq. 10. The number of configurations, which can in principle be constructed from the MOs generated in a realistic AO basis, is very large so that a full CI can approach $10^{10} - 10^{30}$ configurations. (A basis of P atomic orbitals in a molecule with n electrons can generate $\binom{2P}{n}$ determinants and a basis of 100 AOs is quite usual). Clearly, secular equations of such size cannot be solved routinely. Therefore a number of approaches that approximate the full CI treatment has been elaborated. The scope includes perturbation–like treatments [19] and CI procedures that truncate the full–CI expansion [12]. The CI approach employed in this work is to carry out a rigorous selection, which extracts those configurations whose contribution to the total wave function is significant [20].

The first step in our MRD–CI (multi–reference, single and double–excitation CI) calculation consists in the choice of a moderate number of the most important configurations (between one and hundred), which will be abbreviated as 'references' or 'mains'. This set includes commonly all species that appear in the expansion of Eq. 10 with coefficients $|c_{ki}|^2 \geq 0.005$; they can be extracted quickly from a few test calculations. In the second step all configurations are generated that can be derived form these reference species by single and double excitations. (This restriction is motivated by the fact that H_e has at most two–particle interactions, so that the matrix elements of H_e between configurations differing by more than two orbitals are zero). If spin– and symmetry considerations are taken into account, the size of the MRD–CI space is on the order of 10^5 to 10^{10}. An efficient way to obtain the eigenvalues and eigenvectors is to carry out a selection among the generated configurations according to their individual contribution to the states studied, i.e. all those configurations which cause a lowering of the energy for these states relative to the calculation with the reference species alone, by more than a chosen threshold value $\mathrm{T_{CI}}$ (typically 10^{-6} hartree) are included into the final secular equation. The effect of the unselected configurations is taken into account indirectly by repeating the diagonalization of the Hamiltonian matrix for several values of $\mathrm{T_{CI}}$ and by extrapolating the $\mathrm{E(T_{CI})}$ curve to $\mathrm{T_{CI}} = 0$ [20]. The contribution of configurations which differ from the main set by more than two MOs is estimated by means of a generalized Davidson perturbative formula [21].

With the help of the CI approach in the variant described above, the most important part of the electron correlation ($\approx 90\%$) is accounted for; experience has furthermore shown that this procedure can produce a highly balanced description of correlation effects in various electronic states and thus a reliable computation of the energy differences between states even if the absolute percentage of the correlation errors accounted for is not extremely high. Thus the quality of the final results is predominantly determined by the AO basis set employed. It is clear that it is not sensible to save computer time by employing a small AO basis in order to reduce in this way the number of MOs and consequently the number of possible configurations. However, it is often justified to build a 'core' of the lowest MOs (for example the 1s orbitals), which are kept doubly occupied in all configurations; similarly it is possible without loss of accuracy to discard some of the MOs characterized by very high energies.

The procedure described above is designed primarily to obtain reliable electronic transition energies. The expansion length of the wave function corresponds to the size of the secular equation actually solved (determined by the chosen value of T_{CI}), however. For certain purposes (particularly for computing the hyperfine coupling constants) the accuracy of this wave function is not sufficient.

The problems arising from the truncation of the wave function are easily explained in terms of the partitioning perturbation theory [22, 23]. Suppose the Hamiltonian matrix $\mathbf{H}$ of the MRD-CI space is partitioned as

$$\begin{pmatrix} \mathbf{H_0} & \mathbf{h^T} \\ \mathbf{h} & \mathbf{H_1} \end{pmatrix} \begin{pmatrix} \mathbf{c_0} \\ \mathbf{c_1} \end{pmatrix} = \mathbf{E} \begin{pmatrix} \mathbf{c_0} \\ \mathbf{c_1} \end{pmatrix} \tag{13}$$

where H_0 is an KxK submatrix of H containing all important configurations. Then H_1 is a (M-K)x(M-K) matrix formed from configurations of lesser importance and h contains the connecting matrix elements between the two sets of configurations. Within selective MRD-CI techniques, $\mathbf{h}$ and $\mathbf{H_1}$ are neglected and only

$$\mathbf{H_0 c_0''} = \mathbf{E c_0''} \tag{14}$$

is solved. If the truncated wave function $\mathbf{c_0''}$ is used to calculate a property as an expectation value, the result differs from that calculated with the correct wave function ($\mathbf{c_0}$,$\mathbf{c_1}$) due to two reasons: The first effect arises since the correct wave function contains more expansion terms ($\mathbf{c_1} \neq 0$). The second reason for the disparity results from the differences between $\mathbf{c_0''}$ and $\mathbf{c_0}$ which have different values of the expansion coefficients for the same configurations. This originates from neglecting less important configurations, if Eq. 14 instead of Eq. 13 is solved. It contains not only normalization effects, but also changes in the ratio among the individual coefficients. In a previous study [29] it was shown that for hyperfine coupling constants the latter effect is much more important than the first one. A perturbation–like estimation of the latter effect can be obtained by the B_K method. Reformulating Eq. 13 by using the partitioning techniques, it can be written as

$$\left[\mathbf{H_0} + \mathbf{h^T}(\mathbf{E} - \mathbf{H_1})^{-1}\mathbf{h}\right] \mathbf{c_0} = \mathbf{E c_0} \tag{15}$$

and if $\mathbf{H_1}$ is replaced by its diagonal part, the formula of the B_K method according to Davidson [25]–[28] is obtained:

$$\left[\mathbf{H_0} + \mathbf{h^T}(\mathbf{E'} - \mathbf{D})^{-1}\mathbf{h}\right] \mathbf{c_0'} = \mathbf{E c_0'} \tag{16}$$

Depending on how E' in Eq. 16 is chosen, one arrives at Brillouin-Wigner (E' is equal to E and the equation is solved iteratively) or Rayleigh-Schrödinger (E' is equal to the energy in Eq. 14) perturbation theory [25, 26]. In the present work the Rayleigh-Schrödinger perturbation theory is adopted. Using $\mathbf{c_0'}$ to calculate a property the second effect described above is accounted for because the new vector $\mathbf{c}_0'$ contains the relaxation of $\mathbf{c}_0''$ due to the neglected configurations in second-order perturbation theory. Furthermore it could be shown [29] that it is sufficient to correct only the expansion coefficients of the single–excitations. This is convenient because the computation of all terms in $\mathbf{h}^\mathbf{T}(\mathbf{E}' - \mathbf{D})^{-1}\mathbf{h}$ is very time consuming. A calculation consists of two steps. After choosing the reference space, all single and double excitations with respect to the reference configurations are generated. In the first step, the $\mathbf{H}_0$ matrix is diagonalized. It contains all configurations that within the selection procedure of the MRD-CI program [20] lower the energy by more than a given threshold $\mathrm{T_{CI}}$. In the second step, the B_K method is applied. Corrections are calculated for all single–excitations and for the most important configurations possessing a coefficient greater than a given threshold $\mathrm{T_{BK}}$. More detailed studies of the modified B_K treatment are given in [30]. They show that the modified B_K method described above is very effective in obtaining e.g. hyperfine coupling constants (see section 2.5.3) close to those calculated with the correct MRD–CI ($\mathrm{T}_{CI} = 0$) wave function.

Various molecular properties (dipole and electronic transition moments, electronically averaged values for the spin–orbit operator and hyperfine coupling constants) are generally computed using the same one–electron basis for the electronic states involved. This is justified by the fact that the form of the wave functions is insensitive to the choice of the MO basis if highly sophisticated correlation treatments are applied. This facilitates the computation of the properties considerably.

2.4 Solution of the nuclear Schrödinger equation

The Schrödinger equation, Eq. 6, involves all 3N (Cartesian) coordinates of the nuclei and thus includes more degrees of freedom than necessary to describe the vibrational and rotational motion of the molecule. It is therefore much more convenient to consider the nuclear Schrödinger equation expressed in terms of the coordinates more closely related to the three types of nuclear motions: translations, rotations and vibrations. This goal can be achieved by introducing a moving frame bound to the molecule in such a way that it takes over, as completely as possible, the translations and rotations of the whole molecule. The only remaining motion observable in this coordinate system is the relative displacement of the nuclei with respect to each other (vibrations). The conditions which satisfy the axes of such a frame were given by Eckart [31]

$$\sum_i^N m_i \mathbf{r}_i = 0 \tag{17}$$

$$\sum_i^N m_i (\mathbf{r}_i^e \times \mathbf{r}_i) = 0 \tag{18}$$

where $\mathbf{r}_i$ is the position vector, $\mathbf{r}_i^e$ the equilibrium position vector of nucleus i in the moving system. The first condition places the origin of the system into the nuclear center of mass and the second determines the orientation of its axes. While fulfillment of the condition of Eq. 17 ensures a complete separation of the translations, it should be noted that the molecular rotations are not entirely removed from the vibrations by the condition of Eq. 18, except for the case of infinitesimal vibrations when $\mathbf{r}_i^e \approx \mathbf{r}_i$ and the time derivative of the expression on the left-hand side of Eq. 18 becomes equal to the angular momentum of the nuclei.

The general expression for the rotation-vibration Hamiltonian corresponding to the conditions of Eqs. 17 and 18 was derived by Wilson [32] and simplified by Watson [33]. It has the form

$$H = \frac{1}{2}\sum_{\alpha\beta} \mu_{\alpha\beta}(J_\alpha - p_\alpha - L_\alpha)(J_\beta - p_\beta - L_\beta) \tag{19}$$

$$+\frac{1}{2}\sum_{r}^{3N-6} P_r^2 - \frac{1}{8}\sum_{\alpha}\mu_{\alpha\alpha} + V \tag{20}$$

where $\mu_{\alpha\beta}^{-1}$ $(\alpha, \beta \equiv x, y, z)$ represent the elements of a matrix which is nearly equal to the instantaneous moment of inertia matrix, J_α and J_β are the components of the rovibronic angular momentum operator (i.e. of the total angular momentum excluding spin), p_α and p_β are the components of the vibrational and L_α and L_β of the electronic angular momentum; P_r are the momenta conjugate with the normal coordinates and V is the potential for the vibrational motion expressed in terms of the normal coordinates Q_r. According to the method of Wilson and Watson, $\mu_{\alpha\beta}$ and V are expanded into Taylor series about the equilibrium geometry of the molecule. The zeroth-order Hamiltonian is obtained by neglecting p_α, L_α and all but the leading terms in the expansions for $\mu_{\alpha\beta}$ and V. The corresponding Schrödinger equation is then separated into an equation describing the rotational motion of the molecule and 3N–6 mutually independent ones-dimensional harmonic oscillator equations whose eigenvalues and eigenfunctions are well known. If the higher terms in the Hamiltonian are taken into account, the interaction between the vibrational and rotational motion (Coriolis interaction) as well as between various vibrational modes occur and therefore it is not possible to reduce the dimensionality of the problem to less than 3N–3. This has the consequence that a 3N–6 dimensional potential surface has to be calculated and thereupon the complete rotation-vibration Schrödinger equation involving all 3N–3 coordinates has to be solved.

The traditional Eckart-Wilson-Watson (EWW) formalism outlined above reflects in the most natural way the experience from spectroscopic observations that the main features of molecular spectra can be understood on the basis of the three relatively weakly coupled types of motion: electronic, vibrational and rotational. Furthermore, this formalism is predestined for a perturbative treatment of the vibration-rotation problem, an approach inevitable in the precomputer era. Use of the Hamiltonian in which the vibrational and rotational motions, as well as different vibrational modes (through the use of normal coordinates) are maximally decoupled from each other should be expected to enable also the most consequent realization of the main idea followed in actual *ab initio* calculations, namely to separate the variables in the Schrödinger equation as much as possible from each other before the actual computations are started. However, while the separation of the nuclear from electronic coordinates in the framework of the Born–Oppenheimer approximation is practically always carried out (at least as the starting point) the EWW scheme is not the only possible and often even not the most convenient for the treatment of the vibration–rotation problem. Let us stress first that Eckart's approach assumes the existence of one minimum on the potential surface being deep enough to allow small amplitude vibrations. Secondly, the Wilson–Watson Hamiltonian has the serious drawback that it has different forms for molecules with nonlinear and linear equilibrium geometries. Both of these facts cause difficulties for the description of quasilinear molecules. Furthermore, the nature of the interaction between certain vibrational and rotational modes is sometimes such that they have to be treated simultaneously, as for example by the Renner–Teller effect in triatomics. It should also be mentioned that the form of the normal coordinates is not known in advance, but can be determined only if the equilibrium geometry and the force field around it is calculated. Furthermore, the transformation of the potential, normally computed as a function of some internal coordinates into the series in normal coordinates is complicated by the fact that the transformation from internal into normal coordinates is nonlinear for noninfinitesimal vibrations [34].

Application of the Eckart condition in Eq. 18 looses much of its attractiveness if one intends to solve the complete vibration–rotation problem by a variational approach. Therefore, it is often advantageous to abandon the Eckart constraints and the use of normal coordinants in order to gain additional flexibility that can be utilized for designing more convenient forms for particular parts of the Hamiltonian {for example for its rotational part by employing the instantaneous principal inertia axes frame; or an appropriate choice of the internal coordinates can help to avoid some singularities in the vibrational Hamiltonian (see. e.g. [35])}; furthermore this enables derivation of explicit analytical expressions for the vibrational–rotational Hamiltonian (at least for smaller, i.e. triatomic or tetraatomic molecules [35]–[44]), and is much easier to handle in a variational calculation than the Taylor expansion within the EWW scheme. For a more detailed discussion of this matter the reader is referred to a series of excellent articles by Sutcliffe [45].

The general approaches for treating the vibration–rotation problem described above lead to considerable computational requirements if some of the vibrational modes are characterized by large amplitudes. The EWW scheme becomes impractical because of the slow convergence of the Taylor series of $\mu_{\alpha\beta}$ and V. Therefore other, more pragmatic strategies, adapted to concrete problems in question, are adopted. They generally divide various vibrational modes into two classes: those which are accompanied by relatively small displacements of the nuclei and can be handled perturbationally (or even in the harmonic approximation) and those for which a large–amplitude handling is necessary so that they require a much more sophisticated treatment.

In triatomic molecules the bending vibration plays a special role. Quite frequently in spectra one observes relatively long progressions in the bending vibration, because the equilibrium angles can have very different values in various electronic states, in contrast to bond lengths, which generally vary less from state to state. This makes a large–amplitude treatment for the bending motion imperative, while the stretching motions can usually be assumed to occur with small amplitudes. Another peculiarity of the bending vibrations is their connection with the rotations: when a molecule during the bending vibrations approaches linear geometry, a gradual transformation of one of its rotational degrees of freedom (corresponding to the rotation around the smallest moment of inertia axis) into the degenerate component of the bending vibration takes place. Particularly interesting phenomena occur when two electronic states correlate at the linear geometry to the same orbitally degenerate state (Renner–Teller effect). Such situations have motivated a number of authors to develop various methods in which the emphasis is placed on an accurate handling of the effective bending problem while the other nuclear modes are treated more or less conventionally [37, 46, 47]; the most widely used is that elaborated by Bunker and coworkers [48]–[52]

In the formalism of Bunker *et al.* the 'reference configuration' plays the role of the equilibrium configuration in the traditional EWW scheme. It is characterized by fixed bond lengths and by a variable valence angle and is chosen such that the bending motion at each value of the bending angle ρ (measured as π minus the bond angle) is maximally decoupled from the other two modes. The Eckart condition, Eq. 18, is thus replaced by

$$\sum_i m_i(\mathbf{r}_i^{ref} \times \mathbf{r}_i) = 0 \qquad (21)$$

$$\sum_{\alpha,i} m_i(\alpha_i - \alpha_i^{ref})\frac{\partial \alpha_i}{\partial \rho} = 0 \qquad (22)$$

where $\alpha = x, y$ or z, and superscript ref stands for reference. Equation 22 corresponds to the Sayvetz condition [53]. The $\mu_{\alpha\beta}$ and V terms in the Hamiltonian analogous to that of Eq. 19 are expanded into Taylor series in the two normal coordinates corresponding to the stretching motions, with coefficients depending on ρ:

$$\mu_{\alpha\beta} = \mu_{\alpha\beta}^{ref} - \sum_{\gamma\delta r} \mu_{\alpha\beta}^{ref} a_r^{\gamma\delta} \mu_{\alpha\beta}^{ref} Q_r + \ldots. \qquad (23)$$

$$V = V_0(\rho) + \sum_r V_r Q_r + \frac{1}{2}\sum_r \lambda_r Q_r^2 + \frac{1}{6}\sum_{rst} V_{rst} Q_r Q_s Q_t + \ldots. \qquad (24)$$

$$(r = 1, 2)$$

with $\alpha, \beta, \gamma = x$, y, or z. Because the stretching vibrational displacements with respect to the reference configuration always have small amplitudes, the series in Eqs. 23 and 24 should converge quickly. The zeroth-order Hamiltonian is obtained by neglecting all but the leading terms in these expansions, $\mu_{\alpha\beta}^{ref}$ and $V_0(\rho) + 1/2\sum_r \lambda_r Q_r^2$ and has the form

$$H = \frac{1}{2}\sum_{\alpha\beta} \mu_{\alpha\beta}^{ref} J_\alpha J_\beta + \frac{1}{2}\mu_{\rho\rho} J_\rho^2 + V_0(\rho) \tag{25}$$

where $J = -i\,\partial/\partial\rho$ and $\mu_{\rho\rho}$ is the inverse of the 'reduced mass' for the large amplitude vibrations. The Hamiltonian in Eq. 25 is an essentially one–dimensional operator with respect to the coordinate ρ. The coupling between ρ and the other degrees of freedom, as well as contributions from higher terms in the small–amplitude vibrations can be taken into account perturbationally [50], or indirectly as in the framework of the semirigid bender model by Bunker and Landsberg [51] by allowing the bond length to vary smoothly with changes in ρ and by correcting correspondingly the potential $V_0(\rho)$.

The approach of Bunker and coworkers does not require knowledge of the entire three–dimensional potential surfaces for the motions of the nuclei. Instead only the bending potential curve and depending on the variant applied (rigid [48], semirigid [51] or nonrigid bender [50] Hamiltonian), one or a few one–dimensional stretching curves, or in the most sophisticated model several two–dimensional sections of the three–dimensional potential surface have to be known, but always only for a small range of bond length variations. In the simpler variants, the computation of vibronic energies and wave functions is reduced to an effectively one–dimensional (bending) problem. Application of this philosophy allows a drastic reduction of computer time requirements in an *ab initio* treatment, which is of crucial importance when highly excited vibronic states of strongly interacting electronic species have to be calculated. All the results we present in Section 3 of this chapter are obtained by solving the rovibronic problem in this manner.

2.5 Coupling effects

The effects of the couplings between various types of variables can be accounted for either perturbationally or by diagonalizing the Hamiltonian matrix (including the coupling terms) in the basis built by solutions of the zeroth–order (neglected couplings) problem. In both cases the matrix elements

$$< \Psi_\alpha | O | \Psi_\beta > \tag{26}$$

have to be computed where $\Psi_{\alpha/\beta}$ are the products of electronic, vibrational, rotational and spin species and O is the coupling operator whose form depends on the type of interaction considered. For example, for the spin–orbit coupling $O = AL_zS_z$ (phenomenological form); if the nonadiabatic effects neglected by introducing the

Born–Oppenheimer approximation are taken into account, O involves the derivatives of the electronic wave functions with respect to the nuclear coordinates. In all applications, group theoretical considerations are very useful for reducing the computational requirements, since many of the matrix elements of Eq. 26 vanish by symmetry. In the present paper we concentrate on three types of coupling : vibronic (Renner–Teller effect), spin–orbit, and hyperfine coupling. The next three subchapters are devoted to a more detailed description of the methods we employ in the study of these phenomena.

2.5.1 Renner–Teller effect

The Renner–Teller effect in triatomic molecules has been one of the best studied examples for the break–down of the adiabatic (Born–Oppenheimer) approximation. The consequences of the coupling between electronic and bending motions in linear triatomic molecules onto their spectra were foreseen [54] and the corresponding theory was elaborated by R. Renner [55] almost 25 years before the first example of its experimental evidence was reported [56]. Experimentally it manifested itself in a complicated irregular vibrational structure of the electronic spectra; this structure reflects the existence and mutual coupling of two electronic states in the bent nuclear conformations, which correlate with the same degenerate ($\Pi, \Delta, ..$) species at the linear molecular geometry. The Renner–Teller effect can be considered in the neighborhood of the linear geometry as a consequence of the electrostatic interaction between two components of an electronic state with a non–zero angular momentum. At strongly bent geometries it is more plausible to think about two different electronic states coupled with one another through the electronic–rotational Coriolis interaction [58, 59]. From the group–theoretical standpoint, the splitting of the components of the $\Pi, \Delta, ..$ states is a consequence of reducing the axial symmetry $C_{\infty v}$, $D_{\infty h}$ upon bending to C_s, C_{2v} possessing only one–dimensional irreducible representations.

The original perturbative method of Renner [55], suitable for a treatment of the weak Renner–Teller effect (both components of the electronic state have linear equilibrium geometry) has been extended and generalized by other authors [47, 57],[59]–[61] who have developed more refined computational techniques, with the help of which it has become possible to achieve a very accurate fitting of experimentally measured spectra. For a very comprehensive review of various ideas, the reader is referred to excellent articles by Jungen and Merer [58], Duxbury [62], Brown and Jorgensen [63], and Köppel *et al.* [64].

For a long time all studies have been based on the potential surfaces (and other relevant parameters) derived from experimental data. For this reason it has not been possible to make quantitative predictions concerning the spectra prior to the actual measurements. The situation has changed in the last decade when a series of full *ab initio* treatments of the Renner–Teller effect appeared in the literature (see e.g. [65] and references cited therein and [44, 67, 68, 69]).

According to the considerations at the end of subsection 2.4, we employ in the *ab initio* treatment of the Renner–Teller effect the Hamiltonian (for the sake of simplicity we omit for the moment the spin–orbit coupling term) :

$$H = H_e + V_{NN} - \frac{1}{2}T_{S_1}\frac{\partial^2}{\partial S_1^2} - \frac{1}{2}T_{S_2}\frac{\partial^2}{\partial S_2^2} - \frac{1}{2}[T_1(\rho)\frac{\partial^2}{\partial \rho^2} + T_2(\rho)\frac{\partial}{\partial \rho} + T_3(\rho)\frac{\partial^2}{\partial \phi^2} + T_4(\rho)] \tag{27}$$

In Eq. 27, T_{S_1} and T_{S_2} are the inverse reduced masses for the stretching vibrations. The stretching coordinates S_1 and S_2 are either exact or approximate normal coordinates so that the cross term involving $\partial^2/\partial S_1 \partial S_2$ vanishes. ρ is the bending coordinate defined as $\rho \equiv \pi - \textit{bond angle}$, ϕ is the angle between the molecular plane (xy) and the space fixed XY plane, describing the rotation of the molecule around the axis corresponding to the smallest inertial moment ($a \equiv z$–axis). The T_i (i=1,...,4) functions, whose form depends on the choice of the moving coordinate system attached to the molecule and on the internal coordinates employed, are assumed to be dependent on ρ to enable a treatment of large amplitude vibrations. In the linear limit the term in the square brackets on the right–hand side of Eq. 27 reduces to the kinetic energy operator of a two–dimensional harmonic oscillator.

The terms describing the rotations around the $b(\equiv y)$ and $c(\equiv x)$ axis are omitted in expression 27, because we do not consider these motions in the present study. It is assumed that these rotations are effectively decoupled from the other degrees of freedom. The validity of this approximation is discussed elsewhere [65, 77]. Thus the Hamiltonian of Eq. 27 is less sophisticated than that employed in the references [67]–[70], but it includes (when the spin–orbit coupling term is added) all the terms needed for a reliable description of the phenomena to be treated.

The wave function Ψ corresponding to the Hamiltonian of Eq. 27 can be expanded in a basis consisting of products of conveniently chosen electronic wave functions and the corresponding rovibrational functions. As the electronic basis functions, we can employ the solutions of the electronic Schrödinger equation obtained in the framework of the Born–Oppenheimer approximation (adiabatic basis), but also, if more convenient, certain transformations thereof (diabatic basis). Thus we assume Ψ has the form

$$\Psi = \sum_i \psi^i \xi^i(S_1)\eta^i(S_2)\Phi^i(\rho,\phi) \tag{28}$$

where ψ^i is the electronic wave function for the state i, ξ^i and η^i the corresponding stretching functions and Φ^i the function describing the bending and a–axis rotations. In the framework of the Born–Oppenheimer approximation the dependence of the electronic wave functions ψ^i on the nuclear coordinates is assumed to be so weak that the corresponding derivatives can be neglected. In our treatment we assume ψ^i to be generally dependent on all coordinates S_1, S_2, ρ *and* ϕ.

The model Hamiltonian in Eq. 27 commutes with the operator J_z corresponding to the projection of the total angular momentum along the z–axis – the validity of this approximation is discussed by Carter and Handy [44] – thus we write the function $\Phi^i(\rho,\phi)$ in the form

$$\Phi^i(\rho,\phi) = \frac{1}{\sqrt{2\pi}} e^{iK\phi} f^i(\rho) \tag{29}$$

assuming the quantum number K, corresponding to J_z, to be conserved. Thus the Hamiltonian matrix will be divided into mutually noncoupled K–blocks. After integrating over the electronic coordinates and ϕ, the matrix elements of the Hamiltonian of Eq. 27 in the basis given by Eq. 28 become

$$\begin{aligned}
&< f^i\eta^i\xi^i\psi^i|H|\psi^j\xi^j\eta^j f^j >= \\
&< f^i\eta^i\xi^i|[V^{ij} - \frac{1}{2}T^{ij}_{S_1}\frac{\partial^2}{\partial S_1^2} - \frac{1}{2}T^{ij}_{S_2}\frac{\partial^2}{\partial S_2^2} - \frac{1}{2}(T^{ij}_1\frac{\partial^2}{\partial\rho^2} + T^{ij}_2\frac{\partial}{\partial\rho} - K^2T^{ij}_3 + T^{ij}_4)]\delta_{ij} \\
&-T^{ij}_{S_1}B^{ij}_{S_1}\frac{\partial}{\partial S_1} - \frac{1}{2}T^{ij}_{S_1}C^{ij}_{S_1} - T^{ij}_{S_2}B^{ij}_{S_2}\frac{\partial}{\partial S_2} - \frac{1}{2}T^{ij}_{S_2}C^{ij}_{S_2} \\
&-T^{ij}_1B^{ij}_{\rho}\frac{\partial}{\partial\rho} - \frac{1}{2}T^{ij}_1C^{ij}_{\rho} - \frac{1}{2}T_2B^{ij}_{\rho} \\
&-iKT_3B^{ij}_{\phi} - \frac{1}{2}T_3C^{ij}_{\phi}|\psi^j\xi^j\eta^j f^j >
\end{aligned} \tag{30}$$

where

$$\begin{aligned}
V^{ij} &\equiv < \psi^i|H_e + V_{NN}|\psi^j > \\
B^{ij}_q &\equiv < \psi^i|\frac{\partial\psi^j}{\partial q} > \\
C^{ij}_q &\equiv < \psi^i|\frac{\partial^2\psi^j}{\partial q^2} > \\
T^{ij} &\equiv \frac{1}{2}(T^i + T^j)
\end{aligned} \tag{31}$$

and $q = S_1, S_2,\ \rho\ or\ \phi$. Superscript in the expression T^i indicates that different forms of the kinetic energy operator are generally assumed in various electronic states. The terms B_q and C_q are responsible for nonadiabatic effects.

The actual form of the quantities in Eq. 31 depends on the choice of the electronic basis. If the solutions of the electronic Schrödinger equation obtained in the framework of the Born–Oppenheimer approximation are employed, i.e. the functions satisfying the eigenvalue equations

$$(H_e + V_{NN})\psi^i = V^i\psi^i \tag{32}$$

the terms V^{ii} reduce to the eigenvalues of Eq. 32, V^i (potentials) and V^{ij} become equal to zero for $i \neq j$.

In the framework of the usual Born–Oppenheimer approximation, the rovibronic wave function is represented by only one term in the expression 28. In this case each electronic state of the molecule is described by its own wave function and the corresponding energy is determined by a single diagonal term. Note that even in the case that the presence of other electronic states can safely be neglected, there exists the nonadiabatic term C^{ii} (which can be understood as a correction to the potential V^i) in the Schrödinger equation describing nuclear motions [1].

The Born–Oppenheimer electronic basis is, however, not always the most convenient one. This is particularly the case when the electronic states of the same spatial and spin symmetry undergo sharp avoided crossings resulting in strong change of the character of the adiabatic electronic wave functions, and consequently in large values for the B^{ij} and C^{ij} matrix elements whose calculation is computationally intensive. In such situations it is sensible to employ another electronic bases, for example those in which the kinetic energy operator is diagonal (i.e. in which $B^{ij} \simeq 0$, $C^{ij} \simeq 0$, 'diabatic basis') rather than H_e. The key step in applying this concept consists in finding a simple way to perform the corresponding transformation of the electronic basis (see for example [78]).

A full variational treatment according to the scheme outlined above is still connected with great computational requirements if two or more interacting electronic states are involved and if the coupling between stretching and bending modes is not negligible. For this reason certain additional simplifications are made in concrete computations [79, 72]. In order to avoid the need for the calculation of full three-dimensional potential surfaces, in accord with the semirigid bender model by Bunker and Landsberg [51], the bond lengths are normally optimized for a number of values of the bending coordinate in each of the adiabatic or diabatic electronic states in question, and in this way the effective potentials for the bending vibrations are obtained. The stretching parts of the potentials are then added to them, whereby the stretch-bend coupling can be taken into account through the coefficients in the polynomial expansions of the stretching potentials, assumed to be generally ρ–dependent. When the stretching frequencies are much larger than the bending frequency (as usually the case for triatomic molecules containing hydrogen atoms) another reasonable simplification of the problem can be made by performing first an integration over the stretching coordinates in the Hamiltonian matrix elements of Eq. 30, after which the full rovibronic problem is essentially reduced to a treatment of effective bending (involving also a-type rotations) coupled with electronic motions.

In the treatment of the usual Renner-Teller effect, two terms corresponding to the electronic states connected at the linear molecular geometry have to be picked up from the expansion of Eq. 28. In this case an additional approximation is also

[1] Kolos [73] distinguishes between the 'adiabatic' approximation in which off–diagonal coupling elements (for $i \neq j$) in equation 30 are neglected and still cruder 'Born–Oppenheimer' approximations when the C^{ii} terms are also set equal to zero. This distinction is, however, not accepted by all authors (see e.g. [64])

justified, namely that the form of the stretching functions is the same for both the electronic states in question, so that the stretching problem can be effectively separated. Thus we assume vibronic wave functions of the form

$$\Psi = \frac{1}{\sqrt{2\pi}}[\psi^{\alpha}\sum_i c_i^{\alpha}\chi_i^{\alpha}(\rho) + \psi^{\beta}\sum_i c_i^{\beta}\chi_i^{\beta}(\rho)]e^{iK\phi} \tag{33}$$

where it is supposed that the bending functions $f_{\alpha/\beta}(\rho)$ (α/β means α or β) are expanded in a basis $\{\chi_i^{\alpha}(\rho)\}$ with coefficients $c_i^{\alpha/\beta}$ to be determined variationally. We neglect the derivatives of the electronic wave functions $\psi^{\alpha/\beta}$ with respect to all nuclear coordinates except ϕ, so that the effective Hamiltonian, acting still only on the bending basis functions, becomes

$$H^{\alpha\beta} = [V^{\alpha\beta}(\rho) - \frac{1}{2}(T_1^{\alpha\beta}\frac{\partial^2}{\partial\rho^2} + T_2^{\alpha\beta}\frac{\partial}{\partial\rho} - K^2T_3^{\alpha\beta} + T_4^{\alpha\beta})]\delta_{\alpha\beta} - (iKB_{\phi}^{\alpha\beta} + \frac{1}{2}C_{\phi}^{\alpha\beta})T_3^{\alpha\beta} \tag{34}$$

Two sets of the electronic functions are used most frequently in the treatment of the common (two-state) Renner-Teller effect:

(i) The electronic wave functions ψ^+ and ψ^- for the component electronic states calculated in the Born-Oppenheimer approximation. The signs + and − denote the electronic wave function being either invariant or changing sign upon reflection in the molecular plane, respectively. In the linear limit ($\rho \to 0$), the angular parts of these functions are usually assumed to have the form $\Psi^+ \sim cos\Lambda(\vartheta-\phi)$ and $\Psi^- \sim i\ sin\Lambda(\vartheta-\phi)$ where Λ represents the eigenvalue of the operator L_z (equal 1 for a Π state, 2 for Δ etc.) and ϑ is the coordinate conjugate to L_z. In most of the treatments of the Renner-Teller effect the matrix elements $B_{\phi}^{-+} = B_{\phi}^{+-}$, C_{ϕ}^{++} and C_{ϕ}^{--} (the elements B_{ϕ}^{++}, B_{ϕ}^{--}, C_{ϕ}^{+-} and C_{ϕ}^{-+} vanish for symmetry reasons) have been computed employing these asymptotic forms of the electronic wave functions, leading to the values $B_{\phi}^{-+} = -i\,\Lambda$, $C_{\phi}^{++} = C_{\phi}^{--} = -\Lambda^2$. This is motivated by the fact that they appear in the secular equation weighted by the factor T_3 being proportional to $1/\rho^2$ as $\rho \to 0$. Use of the asymptotic forms for ψ^+ and ψ^- as given above has been critically analyzed by Brown and Jørgensen [63]. In the *ab initio* calculations presented here there is no need to make any *a priori* assumptions concerning the form of the electronic wave functions, because the matrix elements B_{ϕ}, C_{ϕ} being actually functions of the bending coordinate (ρ), can be calculated explicitly [74]; for an example of such computations see [75, 76]. If, however, the asymptotic values for $B_{\phi}^{-+}, C_{\phi}^{++}$ and C_{ϕ}^{--} are set into the expression 34, one obtains the usual form employed by most authors (see, e.g. [59])

$$H^{\alpha\beta} = \{V^{\alpha/\beta} - \frac{1}{2}[T_1^{\alpha/\beta}\frac{\partial^2}{\partial\rho^2} + T_2^{\alpha/\beta}\frac{\partial}{\partial\rho} - (K^2+\Lambda^2)T_3^{\alpha/\beta} + T_4^{\alpha/\beta}]\}\delta_{\alpha\beta} - K\Lambda T_3^{\alpha\beta}(1-\delta_{\alpha\beta}) \tag{35}$$

where $\alpha, \beta = +$ or $-$. Note that the vibronic secular equation has electronic off-diagonal elements (electronic-rotational Coriolis terms [58, 59]) being proportional to $K\Lambda$. This reflects the fact the Σ electronic states (Λ=0) do not exhibit the Renner–Teller effect, but also that there is no coupling between K=0 vibronic levels belonging to different electronic species correlating with a spatially degenerate electronic state at the linear geometry.

(ii) A new electronic basis ψ_1, ψ_2 can be derived by a unitary transformation

$$\psi_1 = \frac{1}{\sqrt{2}}(\psi^+ + \psi^-) \qquad \psi_2 = \frac{1}{\sqrt{2}}(\psi^+ - \psi^-) \tag{36}$$

In the linear limit ψ_1 and ψ_2 become eigenfunctions of the spatially degenerate electronic state $\lim_{\rho\to 0}\psi_1 \sim exp[i\Lambda(\vartheta - \phi)]$, $\lim_{\rho\to 0}\psi_2 \sim exp[-i\Lambda(\vartheta - \phi)]$. Replacing again the corresponding B_ϕ and C_ϕ elements with their $\rho \to 0$ asymptotic values we obtain another commonly used form for the effective Hamiltonian:

$$\begin{aligned} H^{\alpha\beta} = \{ & \frac{V^+ + V^-}{2} - \frac{1}{2}[\frac{T_1^+ + T_1^-}{2}\frac{\partial^2}{\partial\rho^2} + \frac{T_2^+ + T_2^-}{2}\frac{\partial}{\partial\rho} \\ & -(K \mp \Lambda)^2\frac{T_3^+ + T_3^-}{2} + \frac{T_4^+ + T_4^-}{2}]\}\delta_{\alpha\beta} \\ + \{ & \frac{V^+ - V^-}{2} - \frac{1}{2}[\frac{T_1^+ - T_1^-}{2}\frac{\partial^2}{\partial\rho^2} + \frac{T_2^+ - T_2^-}{2}\frac{\partial}{\partial\rho} \\ & -(K^2 + \Lambda^2)\frac{T_3^+ - T_3^-}{2} + \frac{T_4^+ - T_4^-}{2}]\}(1 - \delta_{\alpha\beta}) \end{aligned} \tag{37}$$

$\alpha, \beta = 1$ or 2; $(K - \Lambda)^2$ and $(K + \Lambda)^2$ in the first expression on the right–hand side correspond to H^{11} and H^{22}, respectively. The convenience of using this electronic basis consists in the fact that the off–diagonal element of Eq. 35, being singular in the linear limit ($\lim_{\rho\to 0} T_3 \sim 1/\rho^2$), is now absorbed in the diagonal elements, where it is exactly cancelled by other singular terms when $\rho \to 0$. Note that in this 'linear' basis the off–diagonal element is predominantly determined by the electrostatic term $(V^+ - V^-)/2$.

In variational computations of the vibronic energies and the corresponding wave functions according to the scheme outlined above, two sets of the basis functions $\chi_i(\rho)$ have been employed: i) The eigenfunctions of the radial part of the Hamiltonian for a two–dimensional harmonic oscillator; the potentials and the kinetic energy terms $(T_i - \lim_{\rho\to 0} T_i)$ are expanded into polynomial series in ρ and their contribution to the secular equation is computed by using recursive formulae published in refs [65, 76, 79]. ii) A basis consisting of trigonometric functions $sin(nx)$, $cos(nx)$, having certain advantages for handling of the potentials, which cannot be fitted well by polynomial series, and in the calculation of the transition moments [80].

It has been demonstrated [76, 80] that the numerical results obtained employing this approach are practically identical (with the largest differences of only a few

cm^{-1} in the energy range between 0 and 30 000 cm^{-1}) with those generated in the calculations in which the same potentials and other relevant parameters, but entirely different computational procedures (particularly those based on a numerical integration of the Schrödinger equation [81, 82]), have been used.

The stretching energy levels and wave functions are obtained by representing the potentials by polynomial series and expanding the wave functions in the basis consisting of eigenfunctions of the corresponding one–dimensional harmonic oscillator [84].

In calculations of vibronically averaged molecular properties the electronic matrix elements of the corresponding operators are expanded into polynomial or trigonometric series in the nuclear coordinate in question and the vibronic matrix elements are computed as integrals over them, as described in detail in refs. [65, 72, 76, 79, 80, 84]

2.5.2 Spin–Orbit coupling

Computation of the matrix elements of the spin–orbit operator is described in detail in a recent review [14] and only a very brief summary will be given here. The spin–orbit operator is taken in the Breit–Pauli form [85]

$$H_{SO} = \frac{1}{2c^2}\{\sum_\alpha \sum_i \frac{Z_\alpha}{r_{i\alpha}^3}(\mathbf{r}_{i\alpha} \times \mathbf{p}_i) \cdot \mathbf{s}_i - \sum_{i \neq j} \frac{1}{r_{ij}^3}(\mathbf{r}_{ij} \times \mathbf{p}_i)(\mathbf{s}_i + 2\mathbf{s}_j)\} \tag{38}$$

where the summation index α denotes the nuclei and the indices i and j refer to the electrons. Z_α is the charge of the nucleus α, $\mathbf{r}_{i\alpha}$ is the radius vector for the i-th electron with respect to the nucleus α, $\mathbf{p}_i$ is the spatial momentum and $\mathbf{s}_i$ the spin operator for the electron i. To facilitate comparison with experimental findings it is convenient to represent the spin–orbit operators in the phenomenological form

$$H_{SO}^z = A_{SO} L_z S_z \tag{39}$$

where we consider only the z–component of the spin–orbit Hamiltonian[2], A_{SO} is the 'spin–orbit coupling constant'. In order to include the effects of the spin–orbit coupling we generalize the basis functions for the vibronic treatment multiplying them by ω, representing the eigenfunctions of the spin–operator.

The effect of the spin–orbit coupling on the vibronic structure of spectra (spin–orbit splitting of vibronic levels) can be accounted for either perturbationally or variationally. In the first case the mean values of the operator of Eq. 39 have to be calculated employing the vibronic wave functions obtained by diagonalizing the matrix for the Hamiltonian without the spin–orbit term. In the second case the spin–orbit operator is added directly to the vibronic Hamiltonian and the

[2] For a more rigorous treatment of the interplay between vibronic and spin–orbit effects see [63].

corresponding matrix is thereupon diagonalized. Because of the normally small contribution of the spin–orbit operator to the total vibronic energy (at least for the systems with relatively light atoms considered here) both approaches give practically the same numerical results. The spin–orbit operator couples the basis functions with the same K value and the same spin part but different spatial symmetries of the electronic functions; the product of the irreducible representations for the electronic species must be the same as that of the operator L_z. Thus the nonvanishing vibronic matrix elements of the operator H^z_{SO} in the case of the classical Renner–Teller effect and for the electronic basis consisting of the eigenfunctions ψ^+, ψ^- of the electronic operator are:

$$< \omega\chi_i\psi^{\pm}|H^z_{SO}|\psi^{\mp}\chi_j\omega >= \Sigma < \chi_i| < \psi^{\pm}|A_{SO}L_z|\psi^{\mp} > |\chi_j > \tag{40}$$
$$\simeq \Sigma A^o_{SO} < \chi_i| < L_z > |\chi_j >$$

where

$$< L_z >=< \psi^{\pm}|L_z|\psi^{\mp} >= -iB^{\pm\mp}_{\phi} \tag{41}$$

Σ is the quantum number for the projection of the electronic spin onto the z–axis and A^o_{SO} is the value for the spin–orbit constant computed at the linear molecular geometry.

2.5.3 Magnetic hyperfine interaction

The magnetic hyperfine interaction results from the interaction of the nuclear spin I and the total spin S of the electrons (for more details and references see the review given by D.M. Chipman, also presented in this book). It can be divided into the dipole–dipole (anisotropic) part and the Fermi contact (isotropic) term; the latter possesses no classical analogue and arises from the interaction of **I** with the magnetic field produced at the nucleus by the spin of the unpaired electron. The isotropic hyperfine coupling constant, A_{iso}, being the proportionality factor between the scalar product $\mathbf{I} \cdot \mathbf{S}$ and the corresponding part of the Hamiltonian, is given for a nucleus N by [86]

$$A_{iso}(N) = \frac{8\pi}{3} g_N g \beta_N \beta_e \frac{1}{\Sigma} < \Psi \mid \sum_k^n \delta(r_k - r_N) s_z(k) \mid \Psi > \tag{42}$$

where β_N, g_N are the nuclear magneton and nuclear g factors, respectively. The term g is the g value for the electron in the free radical while β_e is the Bohr magneton. In the present paper a value of 2.0 is assumed for g. Ψ represents in principle the total molecular wave function. If A_{iso} is assumed to represent only the electronically averaged matrix element ('electronic hfccs'), it is proportional to the spin density at the nucleus N given by the expression $< \psi \mid \sum_k^n \delta(r_k - r_N) s_z(k) \mid \psi >$. The delta function $\delta(r_k - r_N)$ picks out the density at the nucleus N, s_z indicates that the spin density and not the electron density is taken; $1/\Sigma$ is a normalization factor [87]

The anisotropic part of the hf operator for a nucleus N is represented by a traceless tensor whose cartesian components with respect to a molecule fixed frame are given by

$$A_{ij}(N) = g_N g_e \beta_N \beta_e \frac{1}{\Sigma} < \Psi \mid \sum_k^n (\frac{3ij - r^2\delta_{ij}}{r^5})_{kN} 2s_{zk} \mid \Psi > \tag{43}$$

with $i, j = x, y, z$.

Energetic effects (splittings) caused by the hf interactions are generally several orders of magnitude smaller than even those resulting from the spin–orbit coupling, and can thus be accounted for quite reliably by averaging over the vibronic wave function for a particular vibronic level. The nonvanishing vibronic matrix elements involving the components of the hf–tensor correspond to the following selection rules

$$\begin{aligned} &< K| < A_{iso} > |K > \\ < K| < A_{aa} > |K > = - &< K| < A_{bb} + A_{cc} > |K > \\ &< K| < A_{ab} \pm iA_{ac} > |K \pm 1 > \\ &< K| < A_{bb} - A_{cc} \pm 2iA_{bc} > |K \pm 2 > \end{aligned} \tag{44}$$

where $a \equiv z$, $b \equiv y$, $c \equiv x$ and $< A_{iso} >, \ldots$ denote electronic hfccs whose vanishing or non–vanishing is governed by the symmetry properties of the operator and the electronic wave functions in question. The elements with $\Delta K \neq 0$ arise because A_{ij} is defined within the molecule fixed frame while the nuclear spin $\mathbf{I}$ (and depending on the coupling scheme also the electron spin $\mathbf{S}$) is coupled to the space fixed frame, so that transformation between both frames is necessary. A description can be found in [88]. Because of the weakness of the hyperfine interaction only the effects within a particular vibronic state can generally be measured. This has as a consequence that normally only experimental data concerning diagonal matrix elements of Eq. 44 are available. An exception is the matrix element between the K=–1 and K=1 components of a $|K| = 1$ vibronic level. The value for A_{ab} can be indirectly extracted from experimental findings provided the measurements have been carried out for various isotopomers of the same molecule (see Subsection 3.3.3).

For linear molecules the hyperfine coupling constants (hfccs) are in the literature usually given in the Frosch–Foley notation [89]. If the hf interaction is taken into account by first–order perturbation theory the Hamiltonian is usually represented in the phenomenological form, which for linear molecules can be reduced to

$$H_{hf}^N = a^N L_z I_z + b^N \mathbf{I} \cdot \mathbf{S} + c^N I_z S_z \tag{45}$$

The quantities b^N and c^N are related to A_{iso} and A_{zz} from 42 and 43 (see e.g. [86])

$$b^N = A_{iso} - \frac{A_{zz}^N}{2} \tag{46}$$

$$c^N = \frac{3A^N_{zz}}{2} \tag{47}$$

3 Results and discussion of *ab initio* calculations of molecular spectra

3.1 Overall accuracy for spectroscopic constants

A summary of the results from calculations of the various spectroscopic constants for the molecules considered in this paper is presented in Table II. Computed spectroscopic constants are compared with their experimentally derived counterparts whenever the latter were available at the time when the calculation was finished. Before discussing the overall agreement between theoretical and experimental results we want to address some critical points.

What are spectroscopic constants? This term generally refers to molecular parameters such as equilibrium internuclear distances and bond angles, force constants and anharmonicities, vibrational frequencies, electronic transition energies, i.e. those quantities which are, for example, collected in Appendices of Herzberg's first and third book [96, 143]. A widely spread misinterpretation is that these quantities follow directly from experimentally observed spectra and that the disagreement between the experimental and theoretical values is a direct measure of the error in the calculations. In fact, only positions of spectral lines/bands and their intensities are directly measured, the latter quantities being often only of semiquantitative reliability. In order to extract the spectroscopic constants, the experimentalist has to fit the measured data to certain more or less idealized models. The values for the spectroscopic constants depend not only on the choice of that model (i.e. its proximity to reality) but also on various technical details (e.g. on the number of parameters entering into the model, on the quantity and quality of available experimental data) required to apply the model.

The extent of the model–dependence of the various quantities presented in Table II is very different. The experimentally derived energy difference between the lowest vibrational levels of two electronic states, T_{00}, should be exact to within the uncertainty caused by the measuring device, and the discrepancy between this number and the corresponding theoretical result should be ascribed to the inaccuracy of the latter. However, this energy difference is very often not directly measured but instead it is obtained with the help of various extrapolation procedures, which are not always unambiguous. Examples of such situations are BH_2 and H_2O^+, in which the disagreement between the theoretical and experimental values for T_{00} is almost certainly caused by an erroneous assignment of the observed band progressions leading to the wrong extrapolated estimate of T_{00} [91, 68, 104, 69].

Other quantities that can also be extracted rather directly from the experimental findings are the vibrational frequencies, corresponding to nearly constant energy

TABLE II

Summary of theoretically and experimentally derived data for the spectroscopic constants of various triatomic molecules in the lowest-lying electronic states exhibiting the Renner–Teller effect. The values for the equilibrium bond lengths and angles are rounded off to 0.01 Å and 1 degree, respectively. Stretching vibrational frequencies are computed as energy differences between $v=1$ and $v=0$ vibrational levels. Bending frequencies in the lower electronic states are generally the differences between $v_2^{\prime\prime bent}=1$ and $v_2^{\prime\prime bent}=0$ ($K=0$) vibronic levels and those for the upper electronic states are computed as the energy difference between the lowest two $K=0$ vibronic levels ($v_2^{\prime\prime lin}=1$, $v_2^{\prime\prime lin}=3$ for Π electronic species, $v_2^{\prime\prime lin}=2$, $v_2^{\prime\prime lin}=4$ for Δ electronic states) divided by two. The exception are the A^2A_1 states of PH_2 and SH_2^+ and the A^2A' state of HNF, possessing strongly bent equilibrium geometry, for which ω_2 is the energy difference between $v_2^{\prime\prime bent}=1$ and $v_2^{\prime\prime bent}=0$ ($K=0$) levels. For the last three species considered, C_3, NCO and BO_2, having linear equilibrium geometry in both components of the Renner–Teller pair of states, the bending frequency given represents the mean value connected by Renners formula $\omega_2 = \sqrt{(\omega_2^-)^2 + (\omega_2^+)^2}/\sqrt{2}$ with those of the lower (-) and upper (+) component of the Π electronic states. The quantity quoted as T_{00} represents the term value for the lowest $K=0$ vibronic level ($v_2^{\prime\prime lin}=1$ for Π, and $v_2^{\prime\prime lin}=2$ for Δ states).

System	State	$r(A^{\prime\prime})$ theor.	$r(A^{\prime\prime})$ exp.	α(degree) theor.	α(degree) exp.	$\omega_1(cm^{-1})$ theor.	$\omega_1(cm^{-1})$ exp.	$\omega_2(cm^{-1})$ theor.	$\omega_2(cm^{-1})$ exp.	$\omega_3(cm^{-1})$ theor.	$\omega_3(cm^{-1})$ exp.	T_{00} (cm^{-1}) theor.	T_{00} (cm^{-1}) exp.
BH_2	$^2A_1$	1.18	1.18 [92]	129	131 [92]	2585		1057	1083 [92]	2863		4145	4194[a] [92]
$^2\Pi_u$ [91]	$^2B_1$	1.17	1.17 [92]	180	180 [92]	2671		978	954 [92]	2991			
CH_2^+	$^2A_1$	1.09		141		2934		1034		3324		3058	
$^2\Pi_u$ [93, 94]	$^2B_1$	1.09		180		2964		1228		3416			
CH_2	$^1A_1$	1.11	1.11 [97]	103	102 [97]			1520	1352 [97]			8970	8210[c] [97]
$^1\Delta_g$[95, 98]	$^1B_1$	1.07	1.05 [97]	137	140 [97]			606 706[d]	560[b] [97] 698[d] [97]				
NH_2^+	$^1A_1$	1.04		108				1450				6300	
$^1\Delta_g$ [100, 101]	$^1B_1$	1.02		155				727					
NH_2	$^2B_1$	1.02	1.02 [96] 1.03 [81]	103	103 [96] 102 [81]	3307	3221 [102]	1540	1497 [81]	3627	3300 [103]	11150	11126 [81]
$^2\Pi_u$ [75]	$^2A_1$	0.99	1.00 [96] 1.01 [81]	143	144 [81]	3467	3325 [96]	586	633 [96] 584 [81]	3504			

Continuation of Table II

System	State	$r(Å)$ theor.	$r(Å)$ exp.	α(degree) theor.	α(degree) exp.	$\omega_1(cm^{-1})$ theor.	$\omega_1(cm^{-1})$ exp.	$\omega_2(cm^{-1})$ theor.	$\omega_2(cm^{-1})$ exp.	$\omega_3(cm^{-1})$ theor.	$\omega_3(cm^{-1})$ exp.	T_{00} (cm^{-1}) theor.	T_{00} (cm^{-1}) exp.
	$^2B_1$	1.01	1.00 [105]	109	111 [105]	3350	3271 [108]	1435	1408 [105]	3400	3253 [106]		
OH_2^+			1.01 [81]		109 [81]		3213 [106]		1434 [108]				
									1408 [81]			10196	9783[g] [105]
$^2\Pi_u$ [104]	$^2A_1$	1.00	1.00 [81]	180	180 [81]	3350	3547 [108]	764	900[e] [107]	3500			9835[g] [108]
									800[f] [81]				9912[g] [81]
	$^2A_1$	1.61	1.59 [96]	118	119 [96]	1760		750		2240			
AlH_2												9000	
$^2\Pi_u$ [109]	$^2B_1$	1.56	1.53 [96]	180	180 [96]	1910		530		2250			
SiH_2^+	$^2A_1$			119				895 [110]					
												9200	
$^2\Pi_u$ [110]	$^2B_1$			180				618 [110]					
	$^2B_1$	1.42	1.42 [115]	91	91 [115]	2330	2270[h] [113]	1110	1101 [99]	2495			
PH_2			1.43 [99]		92 [112]				1103 [114]				
												18820	18277 [99]
$^2\Pi_u$ [111]	$^2A_1$	1.40	1.40 [99]	122	123 [99]	2430		973	951 [96]	2660			18272 [115]
			1.39 [115]		124 [112]				949 [112]				
	$^2B_1$	1.36	1.36 [117]	93	93 [117]	2710	2516[h] [120]	1240	1000 [120]	2780			
							2485 [119]		1165 [117]				
SH_2^+												18620	18520 [117]
$^2\Pi_u$ [116]	$^2A_1$	1.37	1.37 [117]	128	127 [117]	2620	2040[h] [118]	1020	940[h] [118]	2580			
									855 [120]				
									1088 [121]				
									960 [117]				
									1080 [122]				
	$^2A'$	1.07[i]		125		2880		890		2180			
HNO^+		1.16[j]										7350[k]	
$^2\Pi$ [83, 123]	$^2A''$							790					

Continuation of Table II

System	State	$r(A'')$ theor.	$r(A'')$ exp.	α(degree) theor.	α(degree) exp.	$\omega_1(cm^{-1})$ theor.	$\omega_1(cm^{-1})$ exp.	$\omega_2(cm^{-1})$ theor.	$\omega_2(cm^{-1})$ exp.	$\omega_3(cm^{-1})$ theor.	$\omega_3(cm^{-1})$ exp.	T_{00} (cm^{-1}) theor.	T_{00} (cm^{-1}) exp.
HNF $^2\Pi$ [125]	$^2A''$	1.03 [l] 1.39 [m]	1.06 [l] [126] 1.37 [m] [126]	101	105 [126]	3200		1430	1432 [127] 1441 [128]	1000	1000 [127]	20600	20141 [1[illegible]
	$^2A'$	1.01 [l] 1.34 [m]	1.03 [l] [126] 1.34 [m] [126]	123	125 [126]	3600		1176	1081 [127] 1074 [126]	1262			
C_3 $^1\Pi_u$ [129, 132]	$^1B_1$, $^1A_1$	1.32	1.31 [130]	180	180 [130]	1185	1125 [131] 1086 [96]	364	305 [131] 308 [96]				
NCO $^2\Pi$ [133]	$^2A'$,$^2A'$	1.25 [n] 1.18 [p]	$\leq$ 2.41 [q] [134]	180	180 [96]	1250	1275 [184]	550	539 [134] 487 [184] 539 [136]	2000	1933 [184]		
BO_2 $^2\Pi_g$ [137]	$^2B_2$,$^2A_2$	1.27	1.27 [138]	180	180 [96]	1065	1070 [138] 1056 [141] 1071 [139] 1067 [140]	481	464 [138] 454 [141] 447 [139] 451 [140]	1270	1322 [138] 1332 [141] 1336 [139] 1337 [140]		

[a] Term value for the $v_2^{lin}=1(K=0)$ level of the A^2B_1 state not actually observed but extrapolated after the numbering of the progression detected in [92] has been reassigned from v_2' to $v_2'+2$

[b] Extrapolated value resulting from fitting higher members of the bending progression observed.

[c] Extrapolated value for the $v_2^{\prime lin}=2(K=0)$ term, not actually observed.

[d] Energy difference between $v_2^{\prime lin}=8$ and $v_2^{\prime lin}=6(K=0)$ vibronic levels divided by two.

[e] Extrapolated from experimentally observed higher-term values ($v_2^{\prime lin}=7-15$).

[f] Extrapolated after reassigning the original vibrational numbering from $v_2^{\prime lin}$ to $v_2^{\prime lin}+2$.

[g] Not observed but extrapolated after reassigning the bands observed from $v_2^{\prime lin}$ to $v_2^{\prime lin}+2$.

[h] Values with high uncertainties [i] H–N bond length [k] N–O bond length

[k] Experimentally observed intensity maximum in the spectrum ascribed to HNO^+ lies at 13890 cm^{-1} [124]. Maximum in the computed spectrum is at 13833 cm^{-1} and corresponds to the transition into $v_2^{\prime lin}=9(K=0)$ vibronic state.

[l] H–N bond length [m] N–F bond length [n] N–C bond length [p] C–O bond length

[q] Sum of the N–C and C–O bond lengths

differences between the members of a band progression. However, in many cases the corresponding potential surfaces are highly anharmonic even in the geometry range close to the equilibrium geometry (for example in the lowest lying electronic states of C_3 and C_2H, and generally in all quasilinear molecules) and then it has to be first defined precisely what is really meant by a vibrational frequency. In order to avoid ambiguities, we indicate in Table II the way in which the values for the computed vibrational frequencies are derived. Let us note that in this paper we employ, depending on the concrete case, two types of labelling of the bending vibronic levels: Bending levels of electronic states possessing bent equilibrium geometries are labelled by conventional 'bent' quantum numbers $v_2^{bent} = 0, 1, 2, \ldots$. For the states with linear or nearly linear equilibrium geometry it is more appropriate to employ the 'linear' notation: $v_2^{lin} = 2 \times v_2^{bent} + \Lambda + K$.

Experimentally derived values for the parameters describing equilibrium geometry (bond lengths and angles) should be very accurate if they are extracted from the rotational structure of spectra. If the rotational structure is not resolved the changes of the equilibrium bond lengths and angles upon an electronic transition can be estimated from the intensity distribution within the corresponding band progression. In such cases not only the lower accuracy, but also the possibility of an erroneous assignment of the quantum numbers for the members in the progression should be taken into account. As an example for this let us mention the estimate of the change of 22 degrees in the H–N–N bond angle of N_2H_2 upon $^1B_g \leftarrow X^1A_g$ electronic transition in [145]; the corresponding theoretical result was 13 degrees [146] and it turned out that the experimental number was in error, because of an incorrect assignment of vibrational quantum numbers in the observed bands (see also [147]).

One has to be very cautious of judging the reliability of the calculated potentials (particularly the force constants, not presented in Table II) on the basis of their agreement with those derived from experimental findings. While the situation for diatomic molecules is relatively simple, and very reliable RKR potential curves can be constructed, provided a sufficient number of experimental lines are available, determination of the potential surfaces for polyatomics from measured spectra is a very difficult task (see, for example [59, 81]). In this case any incorrect assumption concerning the form of the kinetic energy operator (generally allowing for large amplitude vibrations) affects the reliability of the potential derived by fitting the experimental data on the basis of the model Hamiltonian. The inaccuracy of the potential derived in such a procedure can be hidden because of possible error cancellation between the kinetic and potential energy parts of the model Hamiltonian. An example is discussed in more detail in chapter 3.2.3.

In accord with the general pragmatic philosophy taken in this paper, namely to obtain as much data as possible, which can serve for a reliable interpretation of measurements and prediction of results by spending not more computer (and also generally working–) time than necessary, relatively little attention was paid to the determination of the spectroscopic parameters. The main emphasis was placed on obtaining sections of the entire potential surface for various electronic states in order to predict and characterize pertinent features of the electronic spectra. In this sense the equilibrium constants are only a by–product and not as accurate as they would be if emphasis were placed on determining details around the minimum of the surface. If the main goal is to obtain spectroscopic parameters, higher accuracy can routinely be obtained as illustrated in the works of Schaefer III and Botschwina (see, for example their recent papers [148, 149]).

The values for the spectroscopic constants presented in Table II are derived from the one–dimensional potential curves generally calculated varying the bond angle in intervals of typically 20 degrees between 180 degrees (linear geometry) and 40–80 degrees, while the interval of 0.2 bohr has usually been chosen in treating the stretching coordinates. The calculations are performed employing the AO bases involving s and p functions for hydrogen and s, p and d species for heavier atoms. In addition, s– and p–type polarisation functions are commonly located at the middle of the bonds. Only in two cases (BH_2 [91], NH_2 [75]) are the AO bases augmented with f–functions. Dimensions of the MRD–CI spaces vary between a few thousand (in old calculations) to a few million (in the recent studies), those of the secular equations (involving selected configurations) actually solved vary between several hundred and 18 000. For more technical details the reader is referred to Table 2b of [65]. Furthermore this table together with Table II of the present paper illustrates how the overall accuracy of the results has changed over the years with increased efforts in the technical treatment.

In spite of the rather simple computational procedure applied, the overall agreement between the theoretical and experimental results presented in Table II can be considered as satisfactory, particularly if the year of the computation is also taken into account. Maximal discrepancies between computed and experimentally derived values for the equilibrium bond lengths do not exceed 0.01 Å and discrepancies in the bond angles are only 1 degree. It should be noted that in some cases similar disagreement exists between the results obtained in various experimental studies. Thus the theoretical results for these quantities represent in practically all instances ‘finger–prints’ of their experimental counterparts and could serve by themselves already as a tool for identification of the corresponding molecules. In this connection one should also remember that the calculated equilibrium data refer to the bottom of the potential surface while experimentally the results are derived for the lowest vibrational states.

Computed values for the stretching vibrational frequencies disagree with (generally overestimate) those derived on the basis of observed spectra – the largest discrepancies are about 10 %. This is a logical consequence of the rather crude approximation in which they have been calculated (neglect of coupling between vibrational modes, zeroth-order kinetic energy operator, relatively small number of computed points). Systematic overestimation of the stretching frequencies indicates that the computed potential curves are somewhat too steep; this can be attributed to the fact that they are calculated at constant values for the other two

TABLE III

Error limits for calculated data presented in Table II.

Quantity	Calculating error
r	0.01 Å
θ	1–2 degrees
ΔE	0.05 eV
ω (stretching)	100–200 cm^{-1}
ω (bending)	50 cm^{-1}
quantum number of most intense band in progression	0 (± 1)

Abbreviations are given in Table I

vibrational coordinates (particularly the bond angle) being generally optimal only at the minimum of the curve in question.

The agreement between computed and experimentally derived values for the bending frequency is considerably better than for the stretching frequencies. This should be attributed to a more careful determination of the bending potential curves, often including optimization of the stretching coordinates, as well as to the use of the kinetic energy operator suitable for the treatment of the large–amplitude vibrations and enabling (at least approximate) incorporation of the bend stretch coupling. Few striking exceptions from the uniformly good agreement (e.g. CH_2) correspond to older calculations and should be ascribed to deficiences in the theoretical treatment (small AO basis, limited CI, zeroth–order kinetic energy operator).

The agreement between theoretical and experimental results for T_{00} is in all cases (in which reliable experimental results are available) better than 0.05 eV ($\simeq$ 1 kcal/mole) and has been found to become continuously better with increasing level of sophistication in the calculations (see e.g. NH_2 [75]) The theoretical results presented in Table II correspond to the term values for the lowest Σ (K'=0) vibronic level of the upper electronic state. A summary of calculation error limits is presented in Table III. It is the counterpart to Ramsay's table (Table I) discussed above.

Is it possible to improve the accuracy of the calculations markedly? Which part in the chain of approximations should be changed in order to do that? What is the price? We discuss these question in this and the next subsection comparing the results of our investigation of the spectrum of the BH_2 radical, performed in 1980 [91], with the recently published benchmark study by Brommer *et al.* [68]. Comparison of relevant technical parameters and the most important computation results are given in Table IV.

While the computations in the old study were performed employing an AO basis consisting of only 38 Gaussian groups (s, p and d species for boron, s and p

TABLE IV

Comparison of technical details and results for molecular structure parameters in the X^2A_1, A^2B_1 system of BH_2 of *ab initio* studies [68] and [91].

	Year	1981	1992	1967
	References	[91]	[68]	exp [92]
	number of AO basis functions	38	140	
	CI–treatment	MRD–CI (see text)	CASSCF/MR–CI	
	number of ref.–configurations	$7(A_1)$ $5(B_1)$	$115(A')$ $95(A'')$	
Technical	dimension of the CI space	46252 (A_1) 31988 (B_1)		
Details	dimension of secular equation	2200		
	number of points calculated	10–15	85	
	vibrational treatment	1D	3D	
	number of basis functions in the vibr. treatment	10 (stretch.) 2×30 (bend.)	18 (stretch.)× 18 (stretch.) × 40(bend.)=12960	
X^2A_1	H–B (Å)	1.183	1.184	1.181
	∠ (degree)	129	128.68	135
	ω_1 (cm^{-1})	2585	2517	
	ω_2 (cm^{-1})	1057 (992$^{\oplus}$)	986	
	ω_3 (cm^{-1})	2863	2667	
	barrier (cm^{-1})	2675	2693 †	
A^2B_1	H–B (Å)	1.172	1.167	1.17
	∠ (degree)	180	180	180
	ω_1 (cm^{-1})	2671	2597	
	ω_2 (cm^{-1})	978	950	955‡
	ω_3 (cm^{-1})	2991	2830	
$A^2B_1 \leftarrow X^2A_1$	T_{01-0} (cm^{-1})♣	4145	4216	
	T_{09-0} (cm^{-1})	12060	11768	11767 *
	T_{011-0} (cm^{-1})	14064	13640	13641 *
	T_{013-0} (cm^{-1})	16084	15507	15506 *

$^{\oplus}$ Obtained employing the potential curves calculated with the AO basis involving f–functions

† After correcting the upper potential by 97 cm^{-1}

‡ Result of a long extrapolation

♣ Term value of the lowest K=0 level (v_1=0, $v_2^{lin} = 1^-$, $v_3 = 0$) of the upper electronic state.

* After revising the assignment as proposed in [91]

for hydrogen) for the electronic part of the Schrödinger equation, Brommer *et al.* used as many as 140 basis functions, involving s, p, d, f and g species on boron and s, p, d and f on hydrogen. The MRD–CI space used in the old study was generated by single and double excitations with respect to 7 reference configurations for A_1 and 5 for B_1 symmetry leading to dimensions of 46 252 (A_1) and 31988 (B_1). The electronic Hamiltonian was diagonalized in a basis consisting of roughly 2200 selected configurations. The recent treatment employs a CASSCF calculation with 115 reference species for A' and 95 for A'' symmetry. In calculations of the vibronic energy levels and wave functions the old work employs three one–dimensional potential curves, each of them computed while the other two vibrational coordinates are kept fixed. The recent work computes three–dimensional potential surfaces and applies a fully–coupled vibrational treatment.

In spite of the huge difference in the computational efforts, the equilibrium geometry parameters determined in the two studies are essentially the same. The agreement for the computed barriers to linearity in the X^2A_1 state is also excellent. Discrepancies in computed vibrational frequencies are 120–160 cm^{-1} for ω_3, about 70 cm^{-1} for ω_1 and as small as 6 cm^{-1} (if the better value for the X^2A_1 state, obtained employing the bending potential curve calculated with the AO basis involving also the f boron function [91] is taken) for ω_2. The T_{00} term values obtained in these two studies differ from one another by only 70 cm^{-1}. [3] Finally, maximal discrepancies between computed positions of the experimentally observed bands in the energy region 11 800 – 15 500 cm^{-1} are roughly 400 cm^{-1}.

There is no doubt that in all instances in which the numerical results generated in these two studies differ from each other, those obtained in the recent study are more reliable. However, all key problems in the analysis of the BH_2 spectrum where solved already in the more modest treatment, namely i) the equilibrium bond angle in the ground state is significantly smaller than the value derived on the basis of experimental findings [92], ii) the energy difference between the minima of the A^2B_1 and X^2A_1 states is roughly 2700 cm^{-1} and not about 4300 cm^{-1} as concluded by Herzberg and Johns [92]; the first Σ vibronic level of the A^2B_1 state lies at ~4200 cm^{-1} above the lowest level of the ground state, X^2A_1, and not at 6100 cm^{-1} as derived on the basis of a long extrapolation in the forementioned experimental study [92]; iii) the quantum numbers of the observed bands should be reassigned from 7,9,11 to 9,11,13.

Concluding this subsection we want to stress that even the very extensive calculations of Brommer *et al.* [68] were not able to determine the relative location of the BH_2 Born–Oppenheimer surfaces to within an accuracy of a few wavenumbers by a purely *ab initio* approach. To obtain accuracy to within a few wavenumbers is even less feasible for larger systems with more complex electronic structure (i.e. more electrons and more electronic states that can interact) by a simple straight–

[3] It should be noted that Brommer *et al.* had to correct their directly calculated energy difference between the minima of the A^2B_1 and X^2A_1 states by 97 cm^{-1} in order to get better agreement with the experimentally measured band positions.

forward application of present-day's techniques. On the other hand, a skillful design of an *ab initio* computation, possibly with an adjustment to one measured energy separation may be at present the best choice for reliable work on larger systems. Furthermore, as Brommer *et al.* [68] state in accordance with our own experience [76], the discrepancies between results of very sophisticated studies and the experimental observations cannot be removed without an explicit consideration of the effects beyond the adiabatic approximation. For BH_2 this means considering the non-adiabatic correction terms $< A_1|L_z^2|A_1 >$ and $< B_1|L_z^2|B_1 >$ to the potential surface and the geometrical dependence of the coupling term $< A_1|L_z|B_1 >$.

3.2 COMPUTATION OF VIBRONIC ENERGIES AND WAVE FUNCTIONS

The results of *ab initio* calculations of the vibronic structure of a number of triatomic molecules carried out in our laboratory up to 1984 are summarized elsewhere [65]. They concern mainly the lowest-lying electronic states, exhibiting the classical two-state Renner-Teller effect. Since that time a number of studies on other molecules has been published [71, 72, 78, 79, 80, 90, 93, 94, 104, 133, 137],[150]-[163] in which the Renner-Teller and other fine structure effects (spin-orbit, hyperfine coupling) are also treated in electronically excited electronic states of both valence and Rydberg character [152, 153, 161]; furthermore tetraatomic molecules have been considered [151]. A series of papers has been devoted to a more general problem involving three strongly interacting electronic states of C_2H [72, 78, 79, 90], [154]-[157] For the present context we want to discuss two examples, the $3p^2\Pi$ (Rydberg) state of HCO and the $X^2\Sigma^+$, $A^2\Pi$ system of C_2H; these studies demonstrate the possibilities and advantages, but also the shortcomings of the theoretical approach described in Section 2 of the present paper.

3.2.1 $3p\ ^2\Pi$ (Rydberg) state of HCO

The HCO radical is an important intermediate in many thermal and photochemical reactions occurring under various experimental conditions, as well as in different parts of the atmosphere. For this reason it has been the subject of extensive experimental and theoretical studies. For a literature overview the reader is referred to [160]. Here we present the results of our recent *ab initio* studies [160, 161, 162]. In these calculations an AO basis consisting of 69 contracted Gaussian orbitals is used. For carbon and oxygen the $(9s5p) \rightarrow [4s3p]$ GTO bases suggested by Dunning [11] are augmented by four uncontracted *d* polarization functions for carbon and for oxygen. The hydrogen atom is described by the $(8s2p)/[5s2p]$ basis of Lie and Clementi [164]. For the representation of the lowest Rydberg species two *s*-(α = 0.025 and 0.007) and two *p*-type functions ($\alpha = 0.023$ and 0.006) located at the midpoint between the C and O centers are used [165]. The SCF calculations are performed for the ground electronic state

$$...6a'^2 1a''^2 7a' \qquad X^2A' \tag{48}$$

and the corresponding MO's are employed as the one–electron basis in the CI computations of the doublet electronic species of both A' and A'' symmetries. In the MRD–CI calculations the K shells of the C and O atoms, building $1a'$ and $2a'$ MO's are kept doubly occupied while the two virtual MO's with the the highest orbital energies are completely excluded from consideration. All other orbitals are in principle allowed to be variably occupied by the 11 valence electrons. Up to nine of the lowest doublet electronic states of both irreducible representations are computed employing up to 40 reference species. The dimension of the CI space generated by single and double excitations with respect to the reference configurations is roughly 4 000 000 for A' and 2 000 000 for A'' symmetry. The secular equations of dimensions between 12 000 and 18 000 are actually diagonalized, the contribution of the unselected configurations is estimated by the extrapolation procedure [20] and by employing the Davidson correction [21]. Instead of determining the full three-dimensional potential surfaces, only important one–dimensional sections (potential curves) are computed. The bending curves are calculated with the bond lengths kept fixed at their equilibrium values for the ground state ($R_{HC} = 2.098$ bohr, $R_{CO} = 2.213$ bohr). The H–C and C–O stretching curves are computed for linear nuclear arrangement and at the H–C–O angle of 120 degrees, corresponding nearly to the equilibrium bond angle in the ground electronic state. Three of these one–dimensional sections are displayed in Figures 1–3.

The electronic spectrum of HCO can be looked upon as typical for a triatomic molecule. The lowest lying species of both A' and A'' symmetry are of (predominantly) valence character in the entire spectroscopically relevant range of geometry variations. Excited doublet species are generally of mixed valence–Rydberg character, resulting from a number of (avoided) crossings between the potential surfaces corresponding to the pure Rydberg species (preferring linear geometry and short C–O bond lengths) and those of the valence–type states (characterized with elongated C–O bonds and many of them with strongly bent equilibrium structure). These and other non–adiabatic effects (Renner–Teller coupling, conical intersections) make both the analysis of the experimentally observed spectra and the theoretical investigation of the spectral structure very difficult.

At the same time as the potential curves shown in Figs. 1–3 were computed, Song and Cool [166] published the results of their experimental investigations of the $3p^2\Pi(Rydberg) \leftarrow X^2A'$ transition in HCO and DCO, followed by an analysis differing significantly from those based on earlier experimental studies [167, 168, 169]. In order to resolve this controversy, we performed an *ab initio* treatment of the Renner–Teller effect in the $3p^2\Pi$ state, which we found to correlate to $4^2A'$, $2^2A''$ species at bent molecular geometries (Fig. 1) applying the approach described in Subsection 2.6 of the present paper. The coupling with other electronic species and the coupling between the bending and stretching modes was completely neglected, except for the fact that the bending curves for the $3p^2\Pi$ state are corrected to account for the shortening of the C–O bond length upon electronic transition from

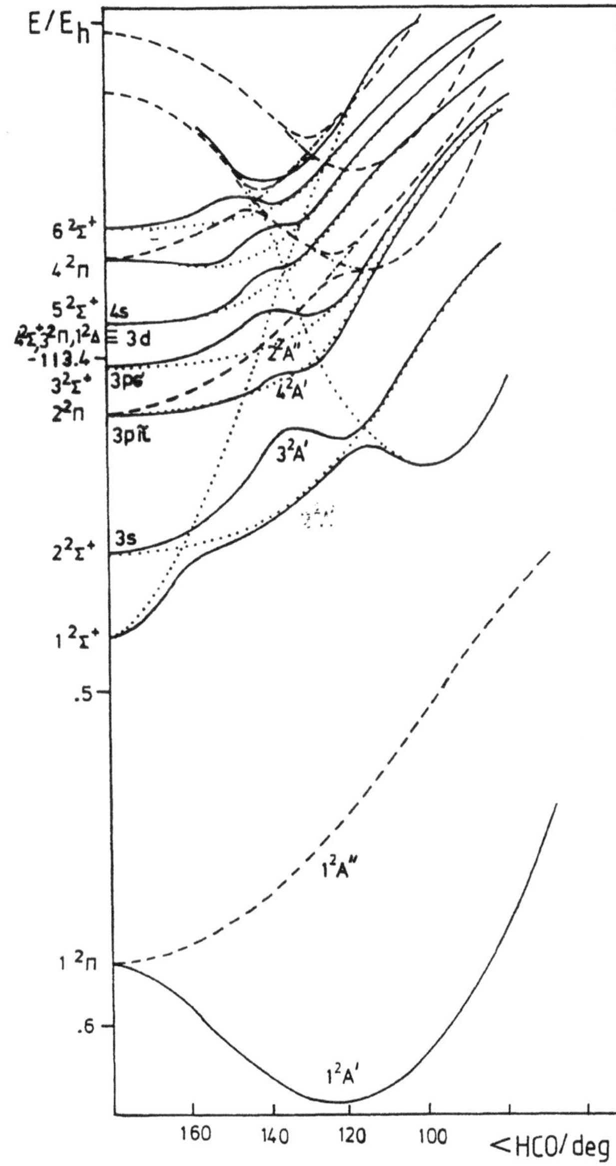

Fig. 1. Bending potential curves for the doublet electronic states of HCO taken from the *ab initio* study [160]. Full lines: calculated adiabatic curves for the $^2A'$ species; dashed lines: adiabatic curves for the $^2A''$ states. Dotted lines: estimated diabatic curves. Species of predominantly Rydberg character are denoted by $3s$, $3p\pi$,... Estimated positions of the $3d$–Rydberg species (not actually calculated) at the linear geometry is indicated. The H–C and C–O bond lengths are kept fixed at 2.098 and 2.213 bohr, respectively.

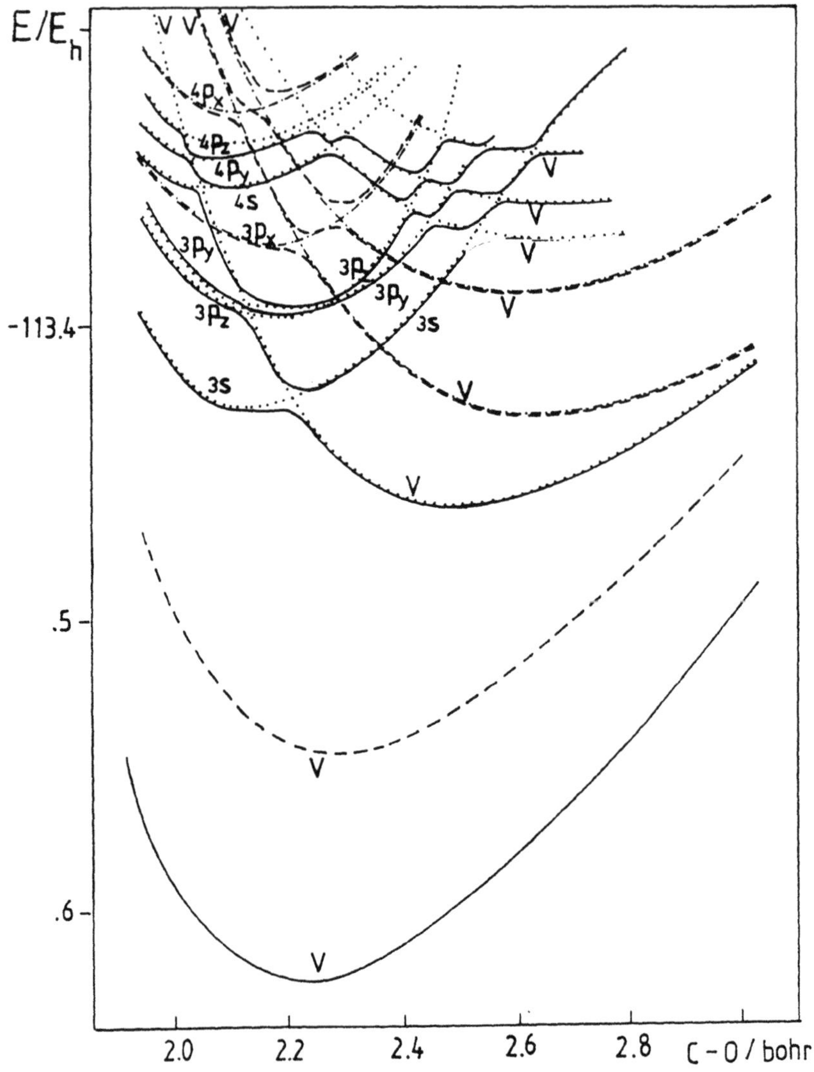

Fig. 2. C–O stretching potential curves for HCO calculated at the bond angle value of 120 degree and the H–C bond kept fixed at 2.098 bohr [160]. Full lines: computed ${}^2A'$ adiabatic states; dotted lines: estimated ${}^2A'$ diabatic species; dashed lines: computed ${}^2A'$ adiabatic states; dashed–dotted lines: estimated ${}^2A''$ adiabatic states, prevailing character of electronic states is indicated.

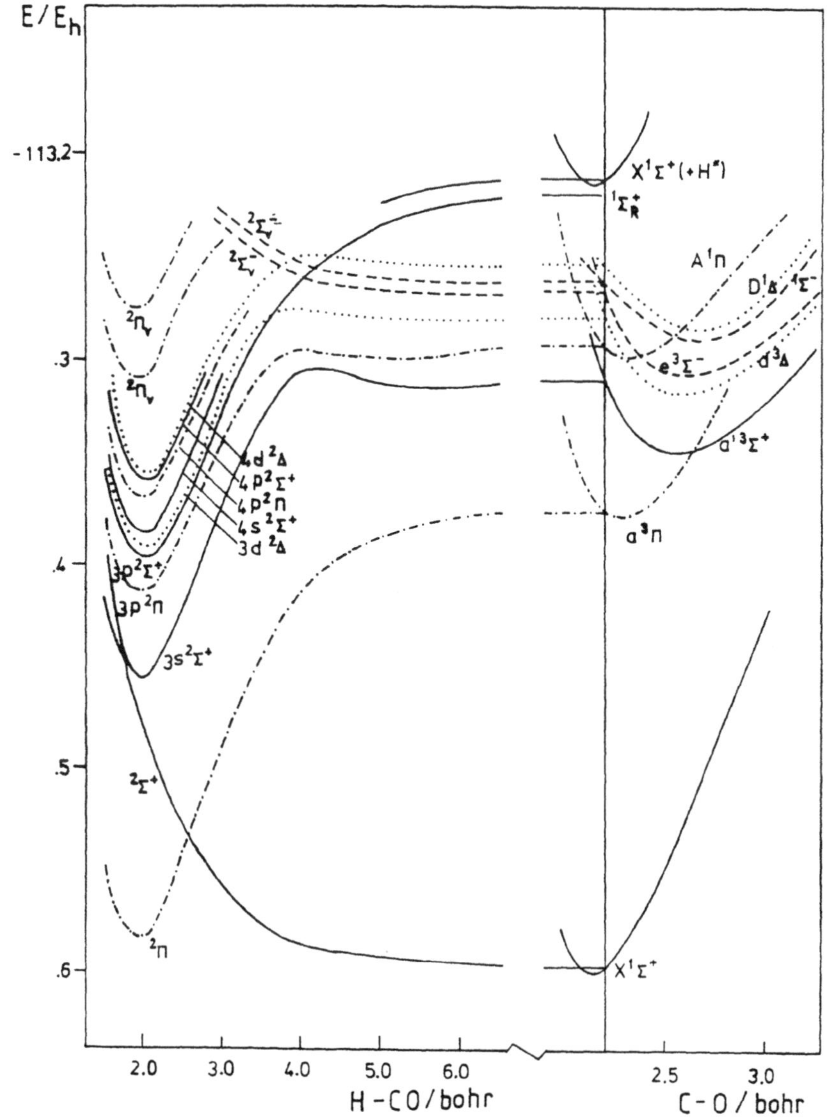

Fig. 3. Left hand side: *Ab initio* H–CO potential curves for the doublet electronic states of HCO calculated at the linear nuclear arrangement and the C–O bond length kept fixed at 2.213 bohr [162]. Full lines: $^2\Sigma^+$ states; dashed lines: $^2\Sigma^-$ species; dashed–dotted lines $^2\Pi$ states; dotted lines $^2\Delta$ states. Right hand side: Potential curves for the lowest-lying singlet and triplet species of CO computed as doublet states of HCO at the linear molecular arrangement and the H center fixed far away (12.098 bohr) from the CO fragment. The highest lying curve corresponds to the $X^1\Sigma^+$ ground state of CO combined with the ^2S excited state of the hydrogen atom.

TABLE V

Comparison of *ab initio* computed K = 0,1 and 2 vibronic term values for the $3p^2\Pi$ state of HCO with the corresponding experimental findings (in parentheses), derived from Ref. [166] ($v_1 = v_3' = 0$).

v_2^{lin}		A'			A''	
	K=0	K=1	K=2	K=0	K=1	K=2
0		45450 (45534)			45450 (45534)	
1	46138 (46271)		46227 (46325)	46328 (46388)		46227
2		46861 (47039)			47117 (47199)	
3	47524 (47787)		47581 (47817)	47898 (48005)		47905 (48008)
4		48214 (48521)			48667 (48806)	
5	48875		48918 (49290)	49429 (49595)		49439 (49604)
6		49569			50190 (50383)	
7	50252		50286	50948 (51159)		50959 (51169)
8		50967			51710 (51934)	
9	51678		51707	52471 (52698)		52483 (52709)
10		52418			53239 (53461)	
11	53154		53180	54006		54017

the ground state. The results of the computations of the vibronic energy levels in HCO are presented together with the corresponding experimental findings in Table V and Figure 4.

In spite of the relatively simple effective one–dimensional theoretical treatment, the agreement between the two sets of results is very reasonable; moreover, the results of the *ab initio* calculations strongly support the analysis of the observed spectra carried out by Song and Cool [166].

At this place we want to stress two other points. It is striking that the term values for the vibronic levels attributed to the A' component of the $3p^2\Pi$ state, published by Song and Cool, lie in the energy range between zero and 3 000 cm^{-1}, while those for the A'' component extend to roughly 8 000 cm^{-1}. An inspection of Fig 1 offers a simple explanation of this fact: The bending potential curve for the $4^2A'$ state undergoes at the bond angle value of roughly 120 degrees and about 3 000 cm^{-1} above its minimum (at linear geometry) an avoided crossing with a valence–type species of the same symmetry, and it is expected that the vibronic structure in this energy is very irregular, with the consequence that its

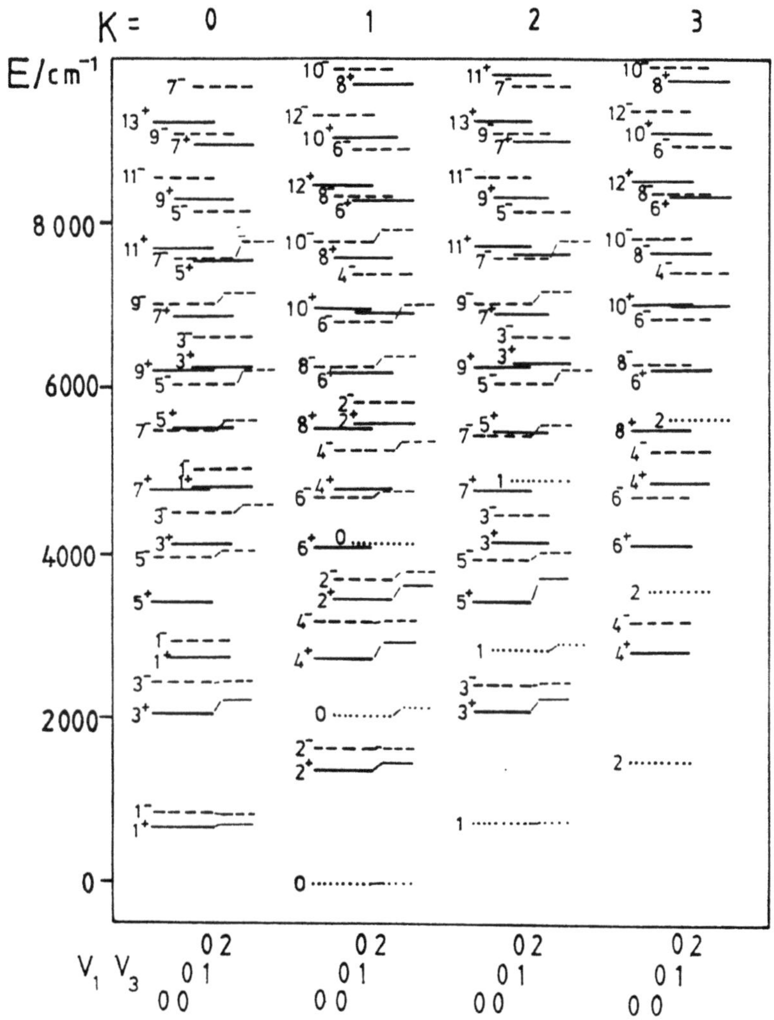

Fig. 4. K=0–3 vibronic levels in the $3p^2\Pi$ (Rydberg) state of HCO computed in Ref. [161]. Full lines: levels belonging predominantly to the A' component electronic state; dashed lines: levels of the A'' component; dotted lines: 'unique' levels. Bending series corresponding to the C–O stretching quantum number $v_3' = 0,1$ and 2 (displaced slightly with respect to each other) are presented. Shorter lines (on the right): experimentally derived values [166]. Vibronic levels are denoted by v_2^+ (A' state) and v_2^- (A'' state). Zero energy corresponds to the position of the lowest vibronic level [0,0(K=1),0].

pattern cannot be fitted to the simple Renner–Teller model assumed by Song and Cool. The avoided crossing of the A'' component of the $3p^2\Pi$ species with the corresponding valence–type state occurs at approximately 9 000 cm^{-1}. The simple one–state model for A'' symmetry breaks down at appreciably higher energies. Another point of interest is the systematic overestimation of the computed vibronic spacings with increasing energy in accord with our general experience [65]. The analysis of possible reasons for this error will be discussed in Subsection 3.2.3.

3.2.2 $X^2\Sigma^+, A^2\Pi$ *system of* C_2H

Elucidation of the structure of the long wavelength spectrum of the C_2H radical has been challenging for both the experimentalists and theoreticians. The ethynyl radical is of considerable astrophysical importance – it is assumed to be one of the most abundant interstellar polyatomic molecules. It also plays an important role as an intermediate in many chemical reactions, particularly in various combustion processes in connection with the photolysis of acetylene. The very complicated structure of the infrared and visible spectra of C_2H is caused by the presence of three close lying electronic species, $1^2A'$, $2^2A'$ and $1^2A''$ in a narrow energy region and their strong interactions via various coupling mechanisms. For a literature survey the reader is referred to [72]. In this paper we give a brief review of our *ab initio* investigation [72, 78, 79, 90], [154]–[157].

In the lowest two electronic states C_2H possesses a linear equilibrium geometry; these states are predominantly characterized by the electronic configurations

$$...4\sigma^2 1\pi^4 5\sigma^1 \qquad X^2\Sigma^+$$
$$...4\sigma^2 1\pi^3 5\sigma^2 \qquad A^2\Pi$$

The second of these species ($A^2\Pi$) arises from a $\pi - \sigma$ excitation with respect to the ground state ($X^2\Sigma^+$). The energy separation between these two states is small because the $5\sigma(3\sigma_g$ character) and $1\pi(1\pi_u$ character) orbitals are nearly isoenergetic. The equilibrium C–H bond lengths are very similar in these two species, but the equilibrium C–C distance in the excited state is significantly larger due to the weaker bonding character of the σ_g type orbital compared to that of π_u type.

For bent molecular geometries, and at the equilibrium bond lengths of the ground state, the $X^2\Sigma^+$ state correlates with the first state of $^2A'$ symmetry having as a dominant configuration $...4a'^2 5a'^1 6a'^2 1a''^2$; the $A^2\Pi$ state splits into $1^2A''$ and $2^2A'$ components with the leading configurations $...4a'^2 5a'^2 6a'^1 1a''^2$ $2^2A'$ and $...4a'^2 5a'^2 6a'^2 1a''$ $1^2A''$, respectively. On the orbital level this assignment corresponds to a correlation of $5a'$ to 5σ and of the 1π MO to $6a'$ and $1a''$. At larger C–C distances (C–C $\equiv$ S $\geq$ 2.5 bohr) the energetic ordering of the Σ^+ and Π species is inverted and the Π state becomes the lower of the two (Fig. 5). Other electronic species, particularly those of Rydberg character, lie considerably higher in energy and can thus be safely ignored in a theoretical treatment of the $X^2\Sigma^+$, $1^2\Pi$ system.

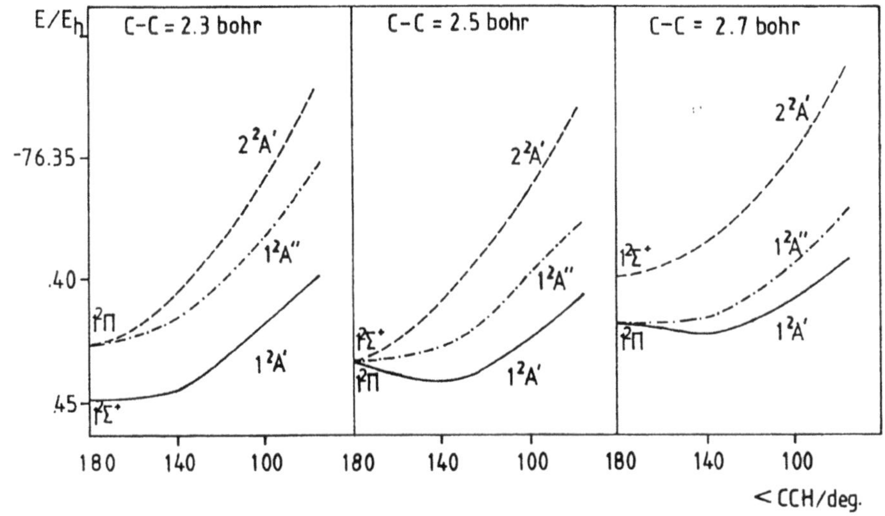

Fig. 5. Bending potential curves for the three lowest-lying electronic states of C_2H calculated at the C–C bond length of 2.3, 2.5 and 2.7 bohr [154, 78]. In all cases the C–H distance is kept fixed at 2.0 bohr.

The most important effects contributing to the complexity of the $X^2\Sigma^+ – ^2\Pi$ spectral system of C_2H are thus:

- Vibronic interaction of all three electronic states in question, $1^2A', 2^2A'$ and $1^2A''$ including a generalized Renner-Teller effect (coupling via operators $\partial/\partial\phi$, $\partial^2/\partial\phi^2$), but also strong coupling of the two species of A' symmetry through the bending and C–C stretching operators $\partial/\partial\rho, \partial^2/\partial\rho^2, \partial/\partial S, \partial^2/\partial S^2$ in the regions of the avoided crossing of the corresponding potential surfaces.
- Spin–orbit coupling
- Interaction of the nuclear spin with the total electron spin, i.e. hyperfine coupling.

In this subsection we concentrate on the vibronic aspects. Spin–orbit and hyperfine interactions will be discussed in Subsection 3.3.

The potential energy surfaces for the lowest three lying electronic states of C_2H are generated in SCF/CI calculations with an AO basis consisting of 72 contracted Gaussian functions. It involves s, p and d species on the carbon atoms and s and

p functions for the hydrogen. In addition, one s and one p polarization function is placed in the middle of the C–C bond and one s function in the midpoint of the C–H bond. All calculations, including those for linear nuclear arrangements, are performed assuming C_s symmetry of the molecule. The only exception is the computation of the spin–orbit coupling constant, which is carried out for linear molecular geometries in the C_{2v} subgroup of the $C_{\infty v}$ point group. The SCF computations are performed for the $(...5a'^2 1a''^2 6a')\ ^2A'$ state. It was found that due to significant mixing on both the MO and CI level, the description of the two species of A' symmetry was sensitive to the choice of the MO basis and thus approximate natural orbitals (averaged for the $1^2A'$ and $2^2A'$ species), ensuring a balanced treatment for both species in question, were employed as the one-electron basis in the CI calculations. In all CI calculations, except for those in which the electronic matrix elements of the hyperfine coupling constants are computed, a core corresponding to the K-shells of the carbon atoms is kept doubly occupied and the two highest MOs are excluded from consideration. Twenty reference configurations are employed for both (A' and A'') symmetries. The MRD–CI space consists of approximately 10^6 configurations ($\sim 3 \times 10^6$ in computations of the hfccs in which all electrons are correlated) and the dimensions of the secular equations actually solved (involving the selected configurations according to a threshold criterion $T_{CI} = 5\mu$hartree) vary between 5000 and 13000 depending on the symmetry and geometry.

Results of computations of the H–C stretching potential curves (Fig. 6) show that their shape is very similar for all three electronic species in question and does not depend significantly on the value of the other two vibrational coordinates (ρ, C–C). This represents the basis for the assumption that the C–H stretching can to a good approximation be decoupled from the other degrees of freedom, enabling the reduction of the vibrational problem to an effectively two dimensional one. The two–dimensional bending/C–C stretching potential surfaces computed in the framework of the Born–Oppenheimer approximation (adiabatic surfaces) are displayed in Fig. 7. The front plane represents the C–C stretching potential curves for the linear molecular geometry where a crossing between $^2\Sigma^+$ and $^2\Pi$ potential curves in the $C_{\infty v}$ classification would be allowed. Upon bending the $^2\Pi$ state splits into A'' and A' components and the two adiabatic potential surfaces $1^2A'$ and $2^2A'$ (one resulting from the $^2\Sigma^+$ state) exhibit an avoided crossing. Figure 7 also shows how the $1^2A''$ potential surface is connected in the linear limit with the $2^2A'$ and $1^2A'$ surfaces at smaller and larger C–C distances, respectively.

The dramatic change in the (adiabatic) wave function character of the two states of A' symmetry due to their avoided crossing are also clearly reflected in the geometry dependence of various molecular properties. As an example Fig 8 presents the adiabatic electric dipole moments for the lowest three electronic states of C_2H as functions of the C–C bond length at four values of the bond angle (full lines). These quantities show a complicated dependence on geometry, characterized by abrupt changes in their magnitude in the area of avoided crossings. It can

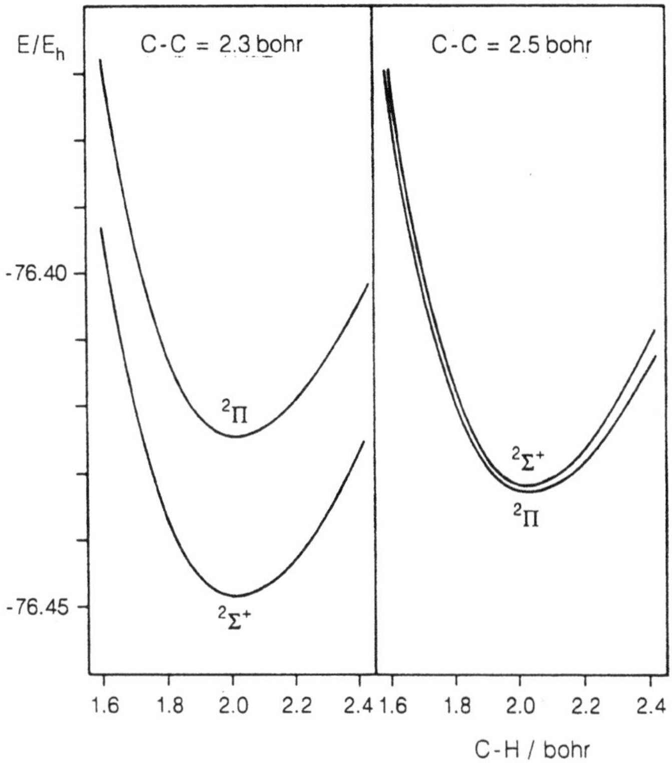

Fig. 6. C–H stretching potential curves for the two lowest-lying electronic states of C_2H calculated at the linear molecular geometry and the C–C bond length kept fixed at 2.3 (left) and 2.5 bohr (right) [154, 72, 79].

also be noted that the bending dependence of these (as well as of other related) quantities shows a significantly different behaviour at different C–C distances.

Direct calculation of vibronic energies and wave functions for the $X^2\Sigma^+ - A^2\Pi$ system of C_2H by employing the adiabatic potentials presented in Fig 7, would lead to considerable technical difficulties. In the regions where the two ${}^2A'$ potential energy surfaces undergo an avoided crossing and the corresponding electronic wave functions strongly change their character, the non–adiabatic matrix elements $< 1A'|\partial 2A'/\partial q > \equiv B_q^{1A'2A'}$ and $< 1A'|\partial^2 2A'/\partial q^2 > \equiv C_q^{12}$, $q = S$ or ρ (and also diagonal C_q terms C_q^{11}, C_q^{22}) that appear in the treatment of the vibronic problem may become very large. Thus their contribution cannot be neglected as is done in more conventional cases in which the electronic states are well separated from each other. Furthermore, the complicated form of the adiabatic potentials and

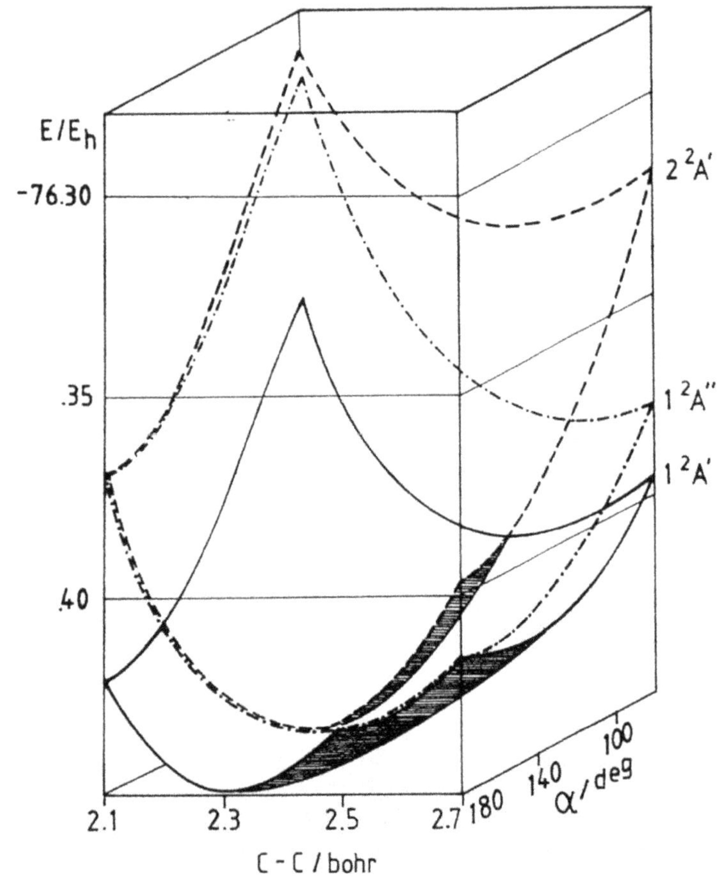

Fig. 7. Two–dimensional C–C, (C–C–H) angle adiabatic potential surface for the three lowest–lying electronic states of C_2H calculated with the C–H distance kept fixed at 2.0 bohr [72, 154]. Full lines: $1^2A'$ states; dashed lines: $2^2A'$; dashed–dotted lines: $1^2A''$.

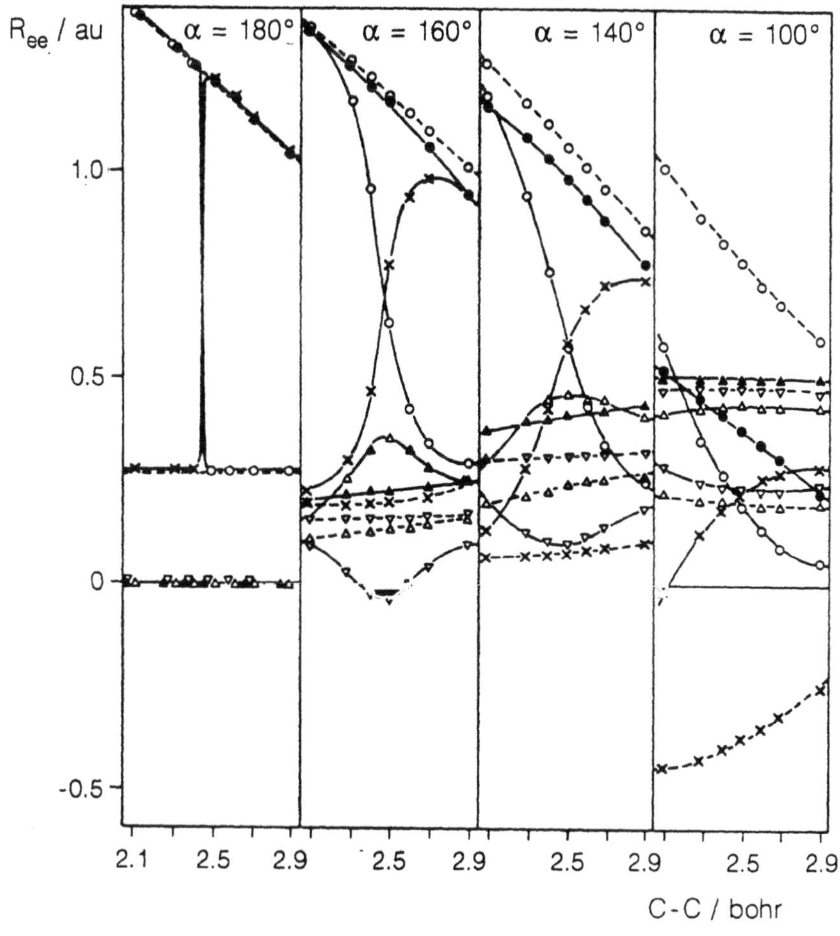

Fig. 8. Dependence of the electric dipole (diagonal transition moment) components for the three lowest-lying electronic states of C_2H on the C–C bond length at bond angle values of 180, 160, 140 and 100 degrees [72, 78]. Full lines: adiabatic electronic basis; dashed lines: diabatic basis. The molecule is assumed to lie in the (y, z) plane with the z–axis along the C–C bond:
$\bigtriangledown$: $\langle 1A'|y|1A' \rangle$, $\langle \Sigma|y|\Sigma \rangle$; $\times$: $\langle 1A'|z|1A' \rangle$, $\langle \Sigma|z|\Sigma \rangle$;
$\triangle$: $\langle 2A'|y|2A' \rangle$, $\langle \Pi^+|y|\Pi^+ \rangle$; $\circ$: $\langle 2A'|z|2A' \rangle$, $\langle \Pi^+|z|\Pi^+ \rangle$;
$\blacktriangle$: $\langle A''|y|A'' \rangle \equiv \langle \Pi^-|y|\Pi^- \rangle$; $\bullet$: $\langle A''|z|A'' \rangle \equiv \langle \Pi^-|z|\Pi^- \rangle$.

the strong dependences of the electronic transition moments (as well as of other quantities needed for the calculation of various features of the spectra) prevents the use of typical simplifications.

An elegant way to avoid the above mentioned difficulties is based on introducing a more convenient (diabatic) electronic basis ($|\Sigma>, |\Pi^+>, |\Pi^->$) by simply rotating the adiabatic functions $|1A'>, |2A'>$ [78, 72]

$$
\begin{aligned}
|\Sigma> &= |1A'> \cos\gamma - |2A'> \sin\gamma \\
|\Pi^+> &= |1A'> \sin\gamma + |2A'> \cos\gamma \\
|\Pi^-> &= |1A'>
\end{aligned}
\tag{49}
$$

The rotation angle γ is generally a function of all nuclear coordinates. The most convenient choice for γ is one that minimizes the non–adiabatic matrix elements $B_q^{\Sigma,\Pi^+}, C_q^{\Sigma,\Pi^+}$, i.e. fulfills the condition (see, e.g. [64, 170, 171, 172])

$$
<\Sigma|\frac{\partial}{\partial q}|\Pi^+> = 0 \tag{50}
$$

i.e. by taking into account relations of Eq. 49.

$$
<1A'|\frac{\partial}{\partial q}|2A'> + \frac{\partial\gamma}{\partial q} \equiv B_q^{12} + \frac{\partial\gamma}{\partial q} = 0 \tag{51}
$$

and consequently

$$
\gamma = -\int B_q^{12} dq \tag{52}
$$

where the integration has to be carried out over the entire region of q for which the coupling of the electronic states in question is significant.

In the case when there are two (or more) nuclear coordinates involved, and/or in which the potential surfaces interact over a wide range of the geometrical variables, the straightforward evaluation of γ on the basis of Eq. 52 is cumbersome and requires considerable expenditure of computing power. For this reason it is very convenient to find a reliable indirect way to determine the transformation angle γ. The success of an alternative way of determining the transformation, Eq. 49, depends on finding a simple quantity that characterizes the diabatic states. This idea seems to have been applied first by Werner and Meyer [173] and its validity has been checked explicitly on the systems similar to that considered here [174]–[177]. Several variants for determining the transformation angle γ for the present case were proposed and explicitly checked in Ref. [78]. The most convenient one consists of diagonalizing the 2×2 matrix of the z–components of the electronic transition moments $<nA'|z|mA'>$ with $n, m = 1$ or 2, leading to the condition

$$
tan2\gamma = \frac{2<1A'|z|2A'>}{<2A'|z|2A'> - <1A'|z|1A'>} \tag{53}
$$

Fulfillment of Eq. 53 ensures that the matrix element $< \Sigma|z|\Pi^+ >$ vanishes at all geometries, i.e. it forces the diabatic states $|\Sigma >$ and $|\Pi^+ >$ to behave at all geometries as the real Σ^+ and Π state do at the equilibrium geometry.

It was shown [78] that various approaches proposed give essentially the same values for the transformation angle γ when the molecular geometry is not far from the linear one, the agreement becoming significantly worse at strongly bent geometries. It should be noted, however, that the diabatic transformation represents only a rotation of the electronic basis functions and thus the final numerical results must be identical regardless of which transformation angle is employed, provided the corresponding Hamiltonian matrix is properly constructed. In the present case it is only of importance that the angle γ is chosen so as to minimize the matrix elements involving the derivatives of the electronic wave functions with respect to the nuclear coordinates in the vicinity of the linear geometry and in this geometry region the above described approach for determining the transformation angle γ works very well.

Introduction of the diabatic electronic basis changes the structure of the secular equation for the vibronic problem (for details see refs. [79, 72]). Diagonal elements (with respect to the electronic basis functions) contain the diabatic potentials

$$V^{\Sigma} \equiv V^{1A'} cos^2\gamma + V^{2A'} sin^2\gamma \tag{54}$$

$$V^{\Pi^+} \equiv V^{1A'} sin^2\gamma + V^{2A'} cos^2\gamma \tag{55}$$

$$V^{\Pi^-} \equiv V^{A''} \tag{56}$$

A crossing of the so constructed species is 'allowed' and the forms of these potential surfaces are much simpler than those for their adiabatic counterparts (Fig. 9). The electronic off–diagonal element involving the diabatic species $|\Sigma >$ and $|\Pi^+ >$ is of the form

$$V^{\Sigma\Pi^+} \equiv \frac{1}{2}(V^{1A'} - V^{2A'})\ sin2\gamma \tag{57}$$

i.e. the Σ and Π^+ blocks are coupled through the difference of the adiabatic potentials [multiplied by $1/2\ sin(2\gamma)$] being much simpler to handle than the nonadiabatic $B_q^{1A',2A'}$, $C_q^{1A',2A'}$ terms coupling the adiabatic species $1A'$ and $2A'$. Concrete computations showed that the quantity $V^{\Sigma\Pi^+}$ (Eq. 57) hardly depends on the stretching coordinates and its dependence on the bond angle is very nearly linear (Fig. 10). This is in accordance with the prediction by Köppel *et al.* (see Sect. IVc of [64]) based on a simple three–state model. Finally, the geometry dependence of the cartesian components of the electronic transition moments (Fig. 8, dashed lines, as well as of other properties (see Section 3.3 of the present paper) is much simpler than that of their adiabatic counterparts. For the variations of these quantities with the change of the C–C distance one obtains practically straight lines, and the majority of the moments show a very weak C–C dependence. The

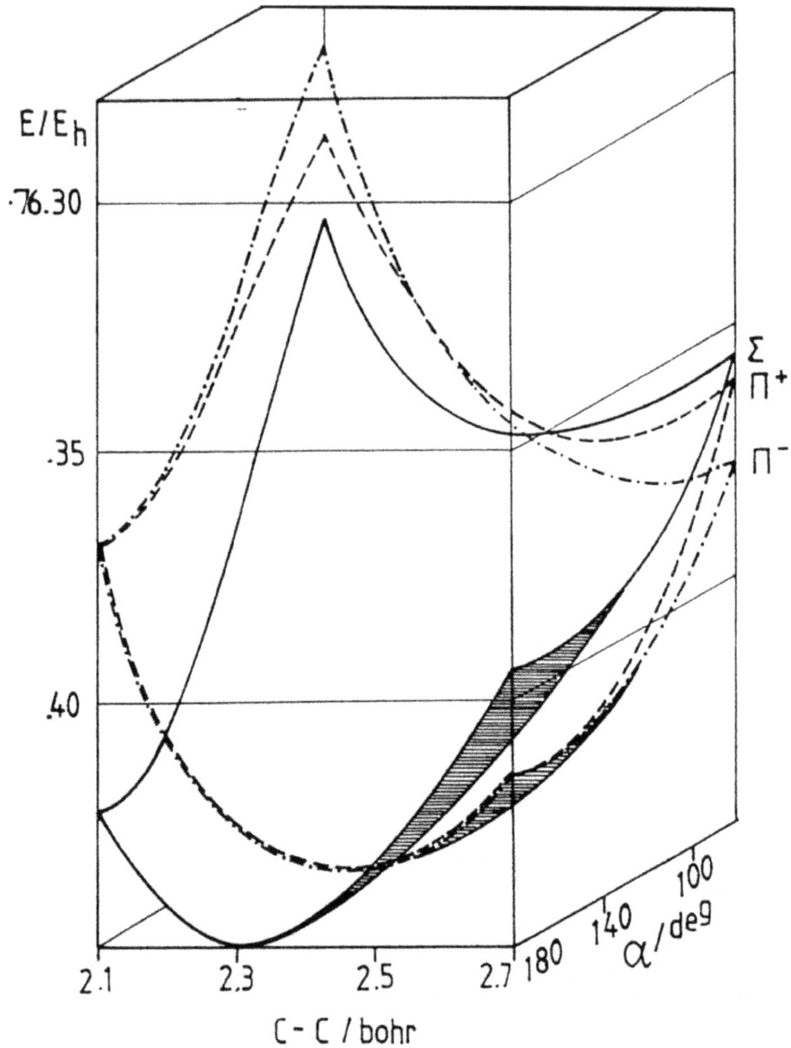

Fig. 9. system of C_2H calculated with the C–H bond length fixed at 2.0 bohr [72, 78]. Full lines: Σ state; dashed lines: Π^+, dashed–dotted lines: Π^- state.

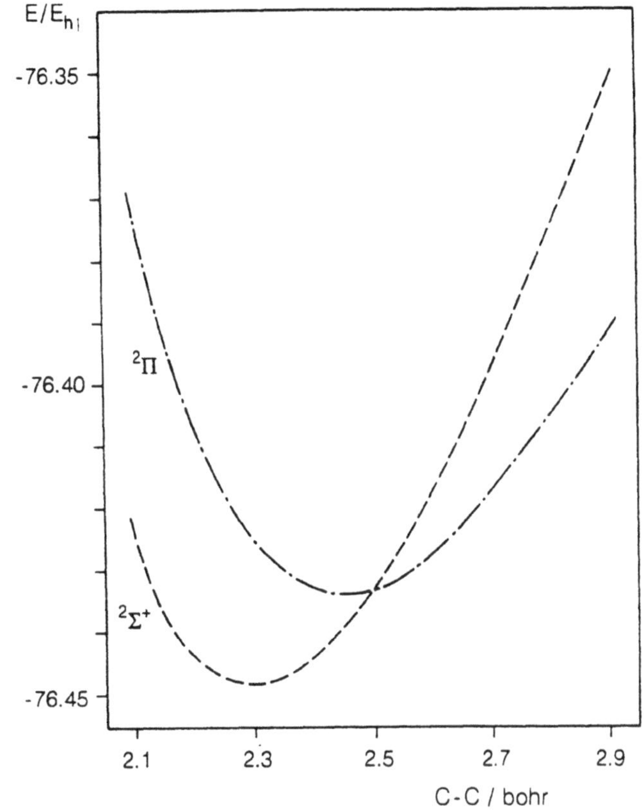

Fig. 10. Diabatic C–C stretching potential curves for the two lowest lying states of C_2H calculated at the linear nuclear arrangement and C–H = 2.0 bohr [72, 79].

most striking difference in the behavior of diabatic and adiabatic matrix elements consists in the very similar bond–angle dependence of the diabatic quantities for different values of the C–C bond length.

Simplification of geometry dependences of all relevant quantities (potential surfaces, electronic transition moments) in the diabatic representation makes a number of approximations possible that reduces the computational requirements and gives insight into the various coupling mechanisms. The relatively weak dependence of the form of the diabatic C–C stretching curves on the bond angle (see [79]) allows for a quite reliable representation of the two–dimensional potential surfaces by two one–dimensional sections (Figs. 10 and 11). The interaction between the bending and stretching modes can indirectly be incorporated into the vibronic treatment through the use of effective bending potentials (corresponding to the optimized C–C bond lengths) and the overlap of the stretching functions

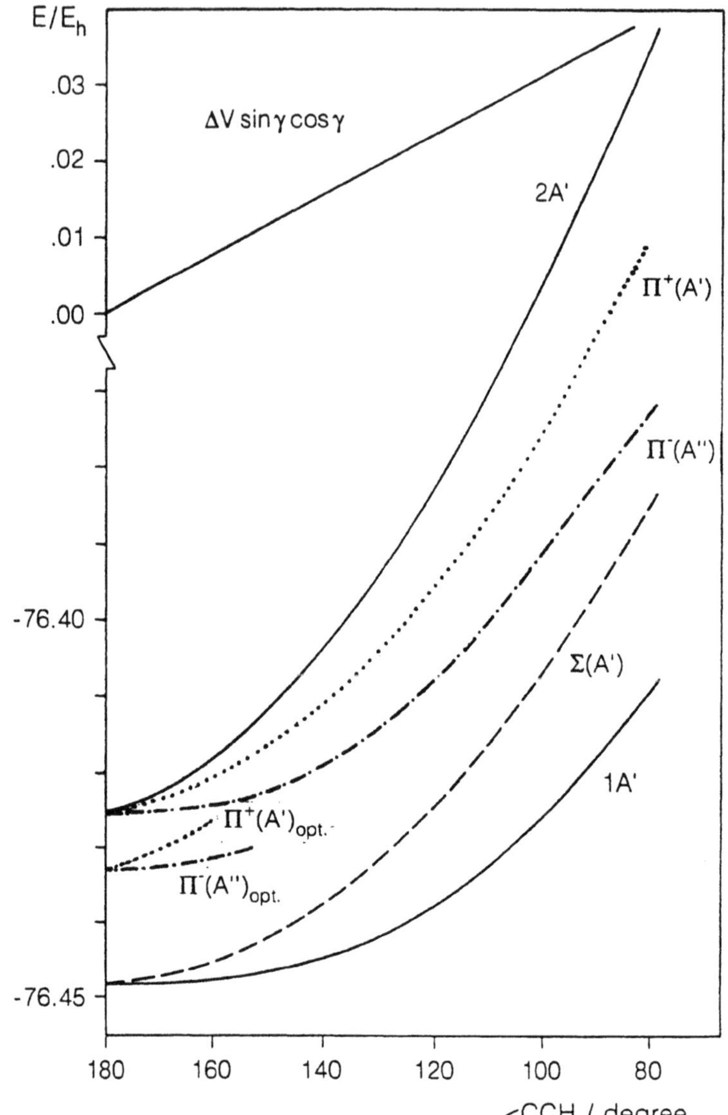

Fig. 11. Effective bending potential curves for adiabatic (full lines) and diabatic A' electronic species and the $1^2A''$ state in C_2H obtained by optimizing the C–C distance for each value of the C–C–H bond angle [72, 79]. The curves for the Π^+ and Π^- states are parallel to the minima of the corresponding two-dimensional potential valleys, but are shifted upwards so that they cross the E-axis at the energy value that the Π state possesses at C–C = 2.3 bohr. The absolute energy minimum of the Π state is also indicated ($\Pi^{\pm}$ opt). At the top of the diagram the angular dependence of $\frac{1}{2} \triangle V sin2\gamma$ is presented. The C–H distance is kept fixed at 2.0 bohr.

corresponding to the different diabatic species, leading to the coupling of the vibrational basis functions for particular combinations of v_2 and v_3 (C–C) quantum numbers. Thanks to the above mentioned simplification of the treatment, the vibronic energy levels and wave functions for a particular K-quantum number in the energy range from 0–7000 cm^{-1} are obtained by solving the secular equations of dimensions not exceeding 300.

The result of computations of the vibronic levels for all possible isotopomers of C_2H involving ^{12}C,^{13}C, H and D in the energy range from 0 to 7000 cm^{-1}, as well as of vibronic transition probabilities in the absorption spectra of these species are published in refs. [79] and [155]. As illustration of the reliability of these results, we compare in Fig. 12 the computed vibronic energies for $^{12}C_2H$ with available experimental data. An inspection of Fig. 12 leads to the following conclusions

- All observed bands can be assigned[4], including those for which an assignment has been lacking so far.
- In some instances previous assignments of measured data are corrected.
- The disagreement between the computed band positions and their experimentally observed counterparts in the majority of cases does not exceeded 100 cm^{-1}. Exceptions are the vibronic levels belonging predominantly to the Π electronic state for which the discrepancies in vibronic term values are as much as 300 cm^{-1}. This indicates strongly that the energy difference between the Σ and Π potential surfaces is underestimated in the present calculations by about 270 cm^{-1} [5]. As a consequence there are difficulties in providing a reliable description of the vibronic wave functions with appreciable contributions from both electronic species (see discussion in Section 3.3)
- The location of many yet unobserved spectral lines is predicted.

Examples of the usefulness of the present results for the explanation of certain features in the C_2H spectrum will be given in Section 3.3.

3.2.3 Analysis of calculation errors in computations of vibronic levels

In this section we want to analyze possible sources for discrepancies between the vibronic energies computed according to the scheme described in section 2 and their experimental counterparts. The central question is if the discrepancies result from inaccuracy in the potential surfaces or from approximations in solving the nuclear Schrödinger equation employing these potentials. Unfortunately, an unambiguous

[4] The assignment of the vibronic levels in Fig. 12 is not unambiguous in some instances, but depends on the criterion applied. In this procedure the following considerations are made: composition of the corresponding wave functions, comparison of the positions of the levels with those obtained in calculations in which various kinds of coupling are not considered, and also some qualitative requirements concerning the vibronic scheme of a combined $\Sigma - \Pi$ electronic system; for details see [72, 79]

[5] This statement is in agreement with the results of the very extensive *ab initio* study of Kraemer *et al.* [190] who obtained an electronic transition energy that was 350 cm^{-1} larger

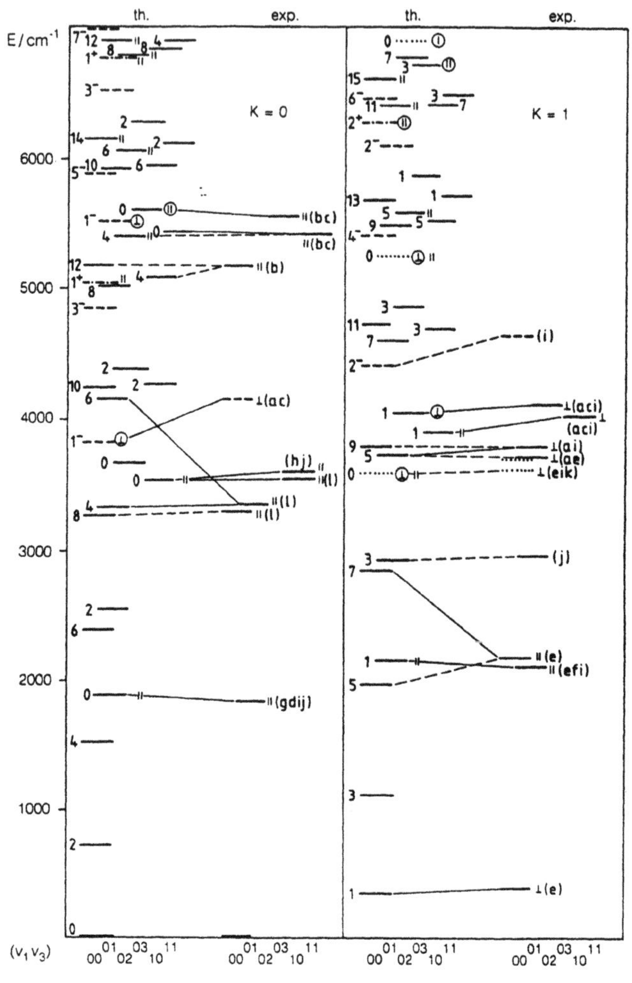

Fig. 12. Comparison of calculated energy values for K=0 and 1 vibronic levels in the $X^2\Sigma^+, A^2\Pi$ system of $^{12}C_2H$ [79, 72] with the available experimental findings. The numbers left to the calculated vibronic levels denote the bending quantum numbers ('linear' notation). Full line levels: levels belonging predominantly to the Σ electronic state; dotted line levels: unique levels; dashed lines: levels of the Π^- electronic species; dashed–dotted lines: levels belonging predominantly to the Π^+ state. The stretching quantum numbers are given at the bottom of the picture. The types of the observed bonds are indicated; $\|$: parallel; $\perp$: perpendicular. Calculated levels predicted to be involved in strong transitions from the (0,0,0) K=0 and (0,1,0) K=1 lower states are denoted by $\|$ or $\perp$ depending on the transition type. (‖) and (⊥) denote very strong transitions predicted. Full lines correlating a calculated term value with its counterpart derived from experimental data correspond to the assignment proposed by the authors of the experimental study, dashed correlation lines represent most probable assignment according to the present study.
References: a: [178]; b: [179]; c: [180]; d: [181]; e: [182]; f: [183]; g: [184]; h: [185]; i: [186]; j: [187]; k: [188]; l: [189].

answer to this question cannot be obtained directly, at least not for polyatomic molecules, because the experimental findings, i.e. vibronic term values, are related to the eigenvalues of the total molecular Hamiltonian and generally cannot give unambiguous information on its constituent parts, i.e. the potentials and the kinetic energy operator. Except for a few examples for which the number of available experimental data has been sufficient to determine the complete multidimensional potential surfaces (e.g. for the ground state of HCN [191]), the information about the molecular potentials derived by fitting the measured band positions is more or less dependent on the model Hamiltonian applied. As already stressed in the subsection 3.1, even excellent agreement between the original experimental results and those obtained by solving the nuclear Schrödinger equation employing a model Hamiltonian and the potentials extracted from the measured data does not prove that these potentials are the corresponding Born–Oppenheimer energies, because of the possibility of cancellation of the errors in the potential and kinetic energy part of the model Hamiltonian can generally not be excluded. One such example is the $X^2B_1 - A^2A_1$ system of H_2O^+: Jungen *et al.* [81] succeeded in fitting all the experimentally observed data by employing potentials derived on the basis of almost certainly the wrong numbering of the vibronic levels observed. The last conclusion follows from our *ab initio* study [104] as well as from that carried out by Brommer *et al.* [69] at a higher level of sophistication.

A very illustrative example are the potential curves obtained on the basis of experimentally measured band positions for the two lowest lying electronic states of NH_2 [81] by employing the semirigid bender Hamiltonian by Bunker and Landsberg [51]. They are presented in Fig. 13. In the same figure the adiabatic *ab initio* bending potential curves computed by Buenker *et al.* [75] are displayed. In these calculations a relatively large AO basis (72 contracted Gaussian functions) including *d*– and *f*–functions for the nitrogen atom and *s*– and *p*–type bond functions was employed. Comparison with the results obtained with the smaller AO basis involving 52 Gaussians (without *f*–functions) showed that a further enlargement of the basis would not improve the results noticeably. The *ab initio* curves are adjusted so that the energy minimum of the lower one coincides with that of its counterpart derived by Jungen *et al.* Although the overall agreement between two sets of potentials is quite satisfactory, some significant differences can be noted. While the T_e values agree very well with one another (*ab initio* 11350 cm^{-1}, Jungen et al 11293 cm^{-1}) the *ab initio* computed barrier towards linearity in the upper, $^2A_1$, state is significantly larger (980 cm^{-1}) than that derived by fitting the experimental findings (729 cm^{-1}). The most striking difference between the two sets of results is the steeper form of the *ab initio* potential curves for both electronic states at strongly bent geometries. As a consequence the *ab initio* computed term values for larger vibrational quantum numbers lie systematically above the values published by Jungen *et al.* as well as above the available experimentally derived counterparts. This feature does not seem typical for the NH_2 molecule, but seems quite general [65].

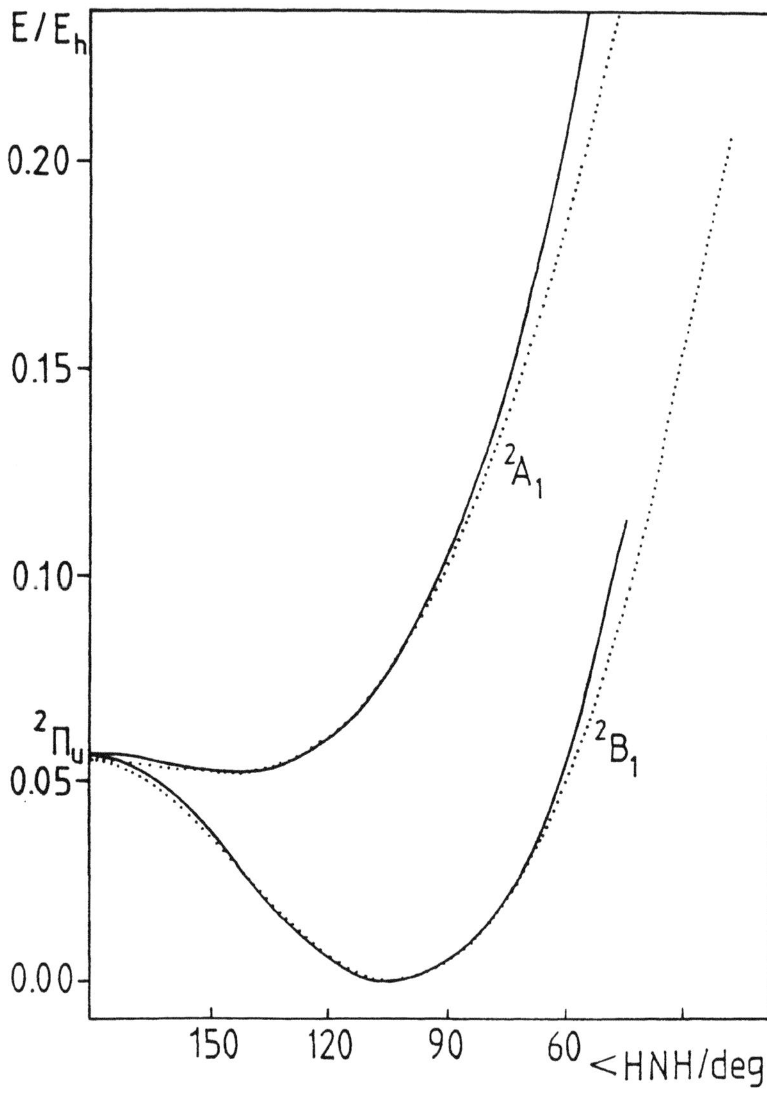

Fig. 13. Comparison of *ab initio* calculated effective bending potential curves for the X^2B_1 and A^2A_1 states of NH_2 (full lines) [75] with the counterparts derived by fitting experimentally observed vibronic band positions (dotted lines) [81].

Although the possibility cannot be excluded that the *ab initio* potentials presented in Fig. 12 are simply not accurate enough (particularly at smaller bond angle values), there are certain arguments against such an interpretation of the discrepancies between two sets of results. For example, Jungen *et al.* assumed a simple bond length dependence on the bending coordinates (see Fig. 14 of [75]), which is in distinct contrast to what is found in *ab initio* calculations and must influence the reliability of the potentials derived. Another indication that the potentials by Jungen *et al.* are not the exact counterpart of the Born–Oppenheimer potentials obtained in *ab initio* calculations follows from the results of the study in Ref. [192], where these NH_2 potentials were employed to compute the vibronic energies in the other two isotopomers NHD and ND_2. The agreement between the results of these calculations and the corresponding experimental findings (deviations on the order of 100 cm^{-1}) was found to be significantly worse than for the NH_2 isotopomer. Of course the experimentally derived potentials need not necessarily be isotopically invariant, because the optimized bond length involved in the kinetic energy expression of the model Hamiltonian represent averages over stretching vibrational states. Nevertheless, the discrepancies are much larger than expected.

On the other hand if the *ab initio* computed potential curves were in error, they should be improved using a better AO basis or by an enlargement of the configuration space in the CI calculations. Additional calculations and experience with a number of similar systems indicates that shortcomings in the *ab initio* treatment are not likely to be responsible for rather pronounced discrepancies between the *ab initio* curves and those derived by Jungen *et al.* Consistent with this is the fact that the calculated T_{00} energy is in excellent agreement with the corresponding experimental value (the difference is only 25 cm^{-1}, Table II). Thus it seems logical to seek other reasons for the disagreement. First, it has been shown [193] that the A^2A_1 state of NH_2 undergoes at a bond angle of about 70 degrees a crossing (forbidden in the C_s symmetry, which is the point group of the molecule at antisymmetric stretching distortions) with a third electronic species (B^2B_2) so that a two-state (X^2B_1, A^2A_1) treatment at strongly bent geometries is not adequate anymore. Secondly, the disagreement in the shape of the *ab initio* and experimentally derived bending curves could partly be caused by the fact that the former (obtained by optimization of the N–H bond lengths for various values of the bending coordinate) correspond to the minima of the potential curves for the stretching vibrations and the latter to the zeroth vibrational levels of both stretching modes. Finally, the three-dimensional vibrational problem is reduced to an effectively one-dimensional one in the computation of vibronic energies. It is true that part of the coupling between the stretching and bending vibrations can be incorporated into the bending problem by use of suitable forms of the Hamiltonian as, for example, the semirigid bender Hamiltonian by Bunker and Landsberg [51], but the question still remains if the solution of any one-dimensional (effective) Schrödinger equation can reliably describe real three-mode vibrations. Thus it is not surprising that the potential curves derived from experimental data differ from their *ab initio* counterparts.

To summarize: The 'experimental' potentials correspond to expressions with a certain number of free parameters, which are chosen such that the eigenvalues of the one-dimensional model Hamiltonian (into which these potentials enter) match experimentally observed vibronic levels. In this way they incorporate effectively all kinds of coupling with the other degrees of freedom and the other electronic states. The *ab initio* curves, on the other hand are well defined one-dimensional sections of the three-dimensional potential surfaces computed in the framework of the Born-Oppenheimer approximation. A part of the coupling of the mode considered with the other degrees of freedom can indirectly be incorporated, but this represents always an approximation to the coupling between various degrees of freedom. Consequently, the differencies between 'experimental' and theoretical potential curves for polyatomic molecules do not necessarily imply that the latter are inaccurate, because both sets of potential curves do not represent the same quantity; they are similar if a one-dimensional approach is a realistic description and less agreement is expected when this approach breaks down, as it does when interactions between various Born-Oppenheimer states occur and the coupling of modes is strong.

The reliability of the potentials employed in the computation of the vibronic energies presented in this study can be judged much more directly by comparing them with those generated in more sophisticated *ab initio* studies. Fig. 14 displays the bending potential curves for the X^2A_1 and A^2B_1 states of BH_2 computed long ago [91] with the bond lengths kept fixed at 2.21 bohr together with the one-dimensional section (corresponding to the same bond lengths) of the full three-dimensional surfaces published recently [68]. The curves are adjusted to have the same energy at the linear nuclear arrangement. It is obvious that the differences in the two sets of bending potentials are not significant. This fact (accompanied with the finding that other relevant parameters such as equilibrium bond lengths are also in excellent agreement, see Table IV) is also reflected in the good agreement between the vibronic energies generated for these two sets of potential curves —see Table VI and Fig. 15. Maximal discrepancies do not exceed 150 cm^{-1} in the energy range from 0 to 16 000 cm^{-1}. Replacement of the bending potentials computed at fixed bond lengths by the effective potentials corresponding to optimized bond lengths at each angle (also given in Fig. 14) leads in this case to a lowering of the vibronic energies by at most 200 cm^{-1} for the lower and 40 cm^{-1} for the upper electronic state. Additional energy lowerings of about 300 cm^{-1} and 30 cm^{-1} for highest vibronic levels in the $^2A_1$ and $^2B_1$ states, respectively, is achieved if the bond angle dependence of the bond lengths is taken into account in the construction of the kinetic energy operator. Thus the total effect of the replacement of the rigid bender treatment by Hougen *et al.* [48] by the semirigid bender model by Bunker and Landsberg [51] amounts to roughly 500 cm^{-1} for the highest levels of the X^2A_1 state and 50 cm^{-1} for those of A^2B_1. The differences between these

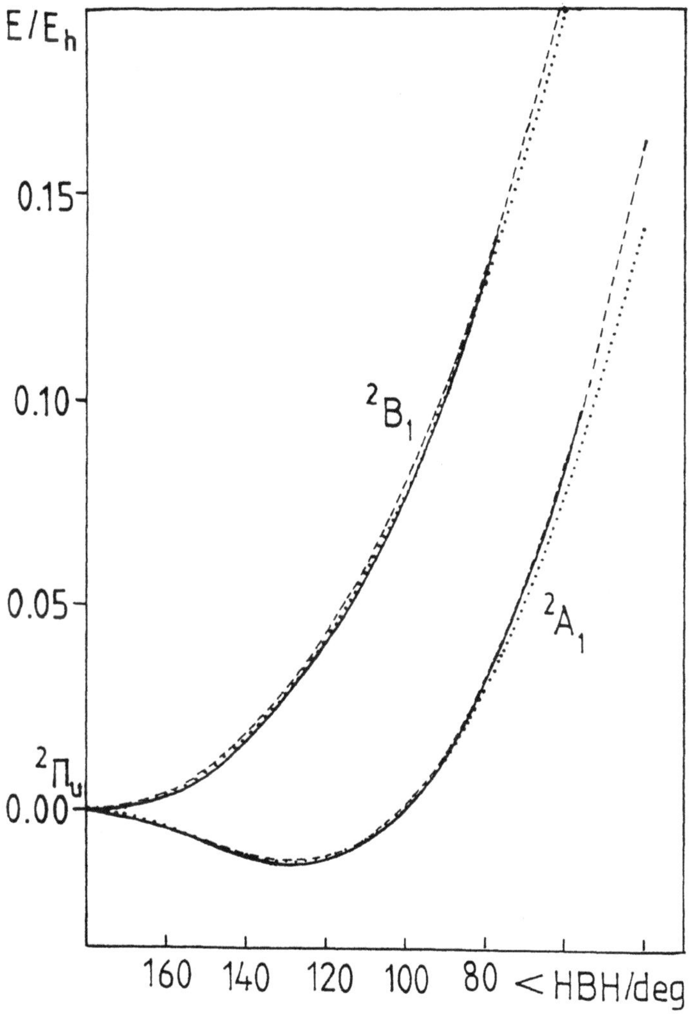

Fig. 14. Comparison of the bending potential curves for the X^2A_1 and A^2B_1 electronic states of BH_2 calculated in the *ab initio* study [91] with the bond lengths kept fixed at 2.211 bohr (full lines) with those from the work of Brommer *et al.* [68], corresponding to the same bond lengths (dashed lines) and optimized values of the bond distances (dotted lines).

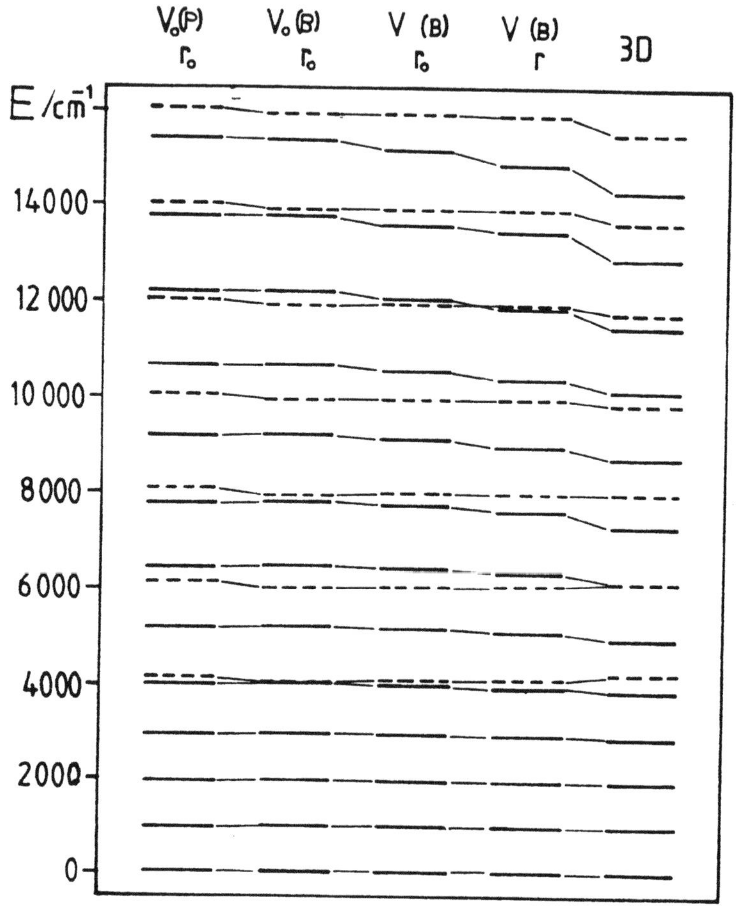

Fig. 15. Analysis of the effect of various approximations in the vibronic treatment to the positions of the K=0 vibronic levels calculated for the X^2A_1, A^2B_1 system of $^{11}BH_2$.
First column: results obtained employing the bending potentials [$V_0(P)$] from [91] computed with the B–H bond lengths kept fixed at r_0 = 2.211 bohr;
second column: potentials from [68], B–H = 2.211 bohr;
third column: effective bending potentials [V(B)] corresponding to the optimized bond lengths from [68], rigid bender kinetic energy operator [48];
fourth column: effective bending potentials [68], semirigid bender kinetic energy operator [51];
fifth column: full three-dimensional vibrational treatment [68].
Full lines: vibronic levels of the X^2A_1 electronic state; dashed lines: levels of the A^2B_1 state.

TABLE VI

Influence of various approximations in the vibronic treatment on the calculated vibronic energies (K=0 levels) in the X^2A_1, A^2B_1 system of BH_2. The vibronic levels are labeled by the bending quantum number v_2^{bent}, those belonging to the A^2B_1 state also by v_2^{lin} (given in parentheses). The superscripts + and − correspond to the X^2A_1 and A^2B_1 components of the electronic state, respectively (see text).

v_2	V_0 (*ab initio*) r=2.21 [91]	V_0(R) r=2.21	V_{opt} r=2.21	V_{opt} r=opt [150]	3D [68]
0^+	0	0	0	0	0
1^+	992	1003	1001	991	986
2^+	1955	1968	1962	1942	1930
3^+	2945	2956	2941	2910	2879
4^+	4018	4030	4003	3957	3888
5^+	5189	5206	5164	5098	4975
6^+	6451	6471	6412	6324	6124
7^+	7793	7814	7735	7622	7302
8^+	9206	9222	9122	8981	8749
9^+	10683	10690	10564	10392	10165
10^+	12218	12209	12057	11849	11490
11^+	13802	13775	13590	13345	12900
12^+	15428	15383	15161	14876	14299
$0^-(1)$	4141	4056	4098	4108	4216
$1^-(3)$	6125	6003	6044	6054	6115
$2^-(5)$	8097	7961	8000	8007	8006
$3^-(7)$	10072	9932	9966	9968	9890
$4^-(9)$	12060	11916	11945	11937	11768
$5^-(11)$	14064	13915	13935	13917	13640
$6^-(13)$	16084	15929	15939	15906	15507

results and those obtained in the full three–dimensional treatment remain however still as large as 500 cm^{-1} for the highest levels (v_2 = 12^+, Table VI) of the lower and 400 cm^{-1} for those (v_2=6^-) of the upper electronic state. This analysis shows that the main source for discrepancies between the *ab initio* computed vibronic energies presented in this paper and their experimental counterparts does not seem to lie in the inaccuracy of the potential curves calculated, but in the fact that the effect of the stretch–bend interaction is underestimated in the effectively one–dimensional treatments such as that introduced by Bunker and Landsberg.

The restricted accuracy for the vibronic energies computed in the framework of an essentially one–dimensional vibronic treatment causes serious difficulties in describing 'local perturbations', which occur when two vibronic states of the same symmetry belonging in zeroth–order approximation (neglected coupling between them) to different electronic states lie close to one another. This concerns particularly all K≠0 vibronic levels of the electronic states exhibiting a Renner–Teller effect. The strength of the coupling is very sensitive to the magnitude of the energy difference between such levels and the inaccuracy in determining their zeroth–order positions can dramatically affect the final results for the vibronic energies and especially the composition of the corresponding wave functions. This topic will be taken up again in Subsections 3.3.2 and 3.3.3.

Let us stress at the end of this analysis that it should not be expected that these 'local effects' can be described absolutely reliably even if highly accurate *ab initio* potential surfaces obtained in the framework of the Born–Oppenheimer approximation were employed in a full three–dimensional vibronic treatment. As has been shown explicitly [76] for the X^2B_1, A^2A_1 system of NH_2, the neglect of the terms going beyond the Born–Oppenheimer approximation, i.e. derivatives of the electronic wave functions with respect to the nuclear coordinates, (or replacement of those by their asymptotic values) may lead to an error in the computed vibronic energies on the order of ten wavenumbers. Furthermore, as mentioned above, the accuracy of computed vibronic energies for high–lying levels is very often obscured by the presence of other electronic states in the same energy region (as for example the 1^2B_2 state in NH_2, PH_2, SH_2^+ and similar systems [65]) leading to the breakdown of the two–state Renner–Teller model.

3.3 *Ab initio* CALCULATION OF VIBRONICALLY AVERAGED MOLECULAR PROPERTIES

3.3.1 Reliability of ab initio computed properties

In this section we will discuss the computation of various matrix elements involving vibronic wave functions. We are particularly interested in vibronic transition moments, determining the intensity distribution within a band progression, in the mean values of the spin–orbit operator being connected with the fine structure of the spectrum, and in vibronically averaged values for the isotropic hyperfine coupling constants as well as the elements of the anisotropic hyperfine tensor. In accord with the model we have assumed throughout this study (effective separation of bending motion from other vibrational degrees of freedom, conservation of the projection of the total angular momentum onto the a–axis) we suppose the corresponding operators in the form $A(\rho)e^{in\phi}$ with n=0,1,2. Reliable *ab initio* computation of the vibronic matrix elements for these operators is generally a difficult task, because they are very sensitive to both the quality of the electronic wave functions and the composition of the vibronic wave functions. We begin with an analysis of the expected reliability of the computed values for the various operators mentioned above.

Let us restrict ourselves to the case of the classical two–state Renner–Teller problem. We write the vibronic wave functions of Eq. 33 for the state characterized by the bending quantum numbers v and K in the form

$$\begin{aligned}\Psi_{Kv} &= \frac{1}{\sqrt{2\pi}}[\psi^+ \sum_i c^+_{Kvi}\chi_{Ki}(\rho) + \psi^- \sum_i c^-_{Kvi}\chi_{Ki}(\rho)]e^{iK\phi} \\ &\equiv \frac{1}{\sqrt{2\pi}}[\psi^+ f^+_{Kv} + \psi^- f^-_{Kv}]e^{iK\phi} \qquad (58)\end{aligned}$$

where ψ^+ and ψ^- represent the electronic wave functions computed in the framework of the Born–Oppenheimer approximation. It is assumed that the bending basis functions $\chi_{ki}(\rho)$ are the same for both electronic states. Assuming that the total vibronic wave function as well as the electronic functions are normalized, one obtains

$$< f^+_{Kv}|f^+_{Kv} > + < f^-_{Kv}|f^-_{Kv} >\equiv |f^+_{Kv}|^2 + |f^-_{Kv}|^2 = 1 \qquad (59)$$

Thus the quantities $|f^+_{Kv}|^2$ *and* $|f^-_{Kv}|^2$ represent the percentage of the electronic species + or −, respectively, in the vibronic wave function.

The matrix element of the operator $A(\rho)e^{in\phi}$ between the vibronic wave functions $\Psi_{Kv'}$ and Ψ_{Kv} is given by

$$\begin{aligned}&< \Psi_{K'v'}|Ae^{in\phi}|\Psi_{Kv} > \\ &= \delta_{K',K+n} < \psi^+ f^+_{K'v'} + \psi^- f^-_{K'v'}|A|\psi^+ f^+_{Kv} + \psi^- f^-_{Kv} > \\ &= \delta_{K',K+n}[< f^+_{K'v'}|A^{++}|f^+_{Kv} > + < f^-_{K'v'}|A^{--}|f^-_{Kv} > \\ &\quad + < f^-_{K'v'}|A^{-+}|f^+_{Kv} > + < f^+_{K'v'}|A^{+-}|f^-_{Kv} >]\end{aligned} \qquad (60)$$

where

$$A^{\alpha\beta} \equiv < \psi^\alpha|A|\psi^\beta > \qquad (61)$$

($\alpha = +\text{or}-$) is the electronic matrix element for the operator in question.

Let us consider first the case when only diagonal matrix elements (Eq. 60) are needed, i.e. when $K' = K$ and $v' = v$. This situation arises for the computation of the mean values of the hfccs or spin–orbit splittings of particular vibronic levels. The expression of Eq. 60 is then reduced to

$$\begin{aligned}&< \Psi_{Kv}|A|\Psi_{Kv} >= \\ &< f^+_{Kv}|A^{++}|f^+_{Kv} > + < f^-_{Kv}|A^{--}|f^-_{Kv} > \\ &+ < f^-_{Kv}|A^{-+}|f^+_{Kv} > + < f^+_{Kv}|A^{+-}|f^-_{Kv} >\end{aligned} \qquad (62)$$

the last two terms on the right hand side of Eq. 62 become identical if A is a hermitian operator. If A is totally symmetric as A_{iso} and A_{aa}, $A^{-+} = A^{+-} = 0$ and Eq. 62 reduces to

$$< \Psi_{Kv}|A|\Psi_{Kv} >=< f^+_{Kv}|A^{++}|f^+_{Kv} > + < f^-_{Kv}|A^{--}|f^-_{Kv} > \qquad (63)$$

If the ρ–dependence of the electronic matrix elements A^{++} and A^{--} is weak in the region where the vibrational wave functions f_{Kv} does not vanish, the right hand side of Eq. 63 becomes $A^{++}|f_K^+|^2 + A^{--}|f_K^-|^2$. The reliable computation of the matrix elements is relatively straightforward. For vibronic states belonging predominantly to one of the electronic species (i.e. if f_{Kv}^- or f_{Kv}^+ are $\simeq 0$) the accuracy depends only on the quality of the Born–Oppenheimer electronic wave functions and the quality of the vibrational functions corresponding to the uncoupled electronic states. In the critical area around the linear geometry in which vibronic coupling is strong, the computed matrix elements of Eq. 63 should also be accurate because $A^{++} \simeq A^{--} \neq 0$. Serious problems can arise only for local interactions of close–lying vibronic levels and then only if A^{++} is substantially different from A^{--} (which can happen at bent geometries).

If A is not a totally symmetric but hermitian operator, as for example L_z which transforms as B_1 *or* A'' irreducible representations, the expression of Eq. 62 reduces to

$$< \Psi_{Kv}|A|\Psi_{Kv} >= 2 < f_{Kv}^+|A^{+-}|f_{Kv}^- > \tag{64}$$

The *ab initio* calculation of the matrix elements of Eq. 64 is required, for example, to compute the spin–orbit splitting in vibronic levels. This can be very difficult, because the results are very sensitive to the amount of mixing of the two electronic species in the vibronic state in question. In vibronic states belonging predominantly to one electronic species, which means f_{Kv}^+ or $f_{Kv}^- \simeq 0$ the spin–orbit splitting is effectively quenched and the magnitude of the vibronic mean values for the operator $H_{SO}^z = A_{SO}L_zS_z$ becomes much less than the value of $A_{SO}(\times 2\Lambda\Sigma)$. The spin–orbit splitting is comparable in magnitude to A_{SO} only in the 'unique levels' [59], the lowest lying $K \neq 0$ levels of molecules exhibiting weak Renner–Teller effects, characterized by bending curves with linear equilibrium geometry for both electronic states. These levels possess roughly equal contributions of both Born–Oppenheimer electronic states, i.e. $f_{Kv}^+ \simeq f_{Kv}^-$ and the matrix element (64) becomes nearly equal to the value of A_{SO} $(\times 2\Lambda\Sigma)$ at the linear geometry.

Ab initio computation of off–diagonal matrix–elements (60) is generally less reliable than for their diagonal counterparts, because the accuracy of the results depends on the reliability of the wave functions for both vibronic states involved. This concerns particularly the matrix elements of the operator x, y and A_{ab} (see subsection 2.5.3), belonging to the type $Ae^{i\phi}$, because of the inherently strong dependence of the corresponding electronic matrix elements on the bending coordinate (they are zero at the linear molecular geometry). Only one operator of the type $Ae^{2i\phi}$ is considered in the present study, namely the $A_{bb} - A_{cc}$ combination of the elements of the anisotropic hf tensor and only between degenerate vibronic species K=+1 and K=-1 of Π electronic states. The corresponding wave functions can be written in the form

$$\Psi_{K=+1,v} = \frac{1}{\sqrt{2\pi}}[\psi^+ f_{Kv}^+ + \psi^- f_{Kv}^+]e^{i\phi} \tag{65}$$

$$\Psi_{K=-1,v} = \frac{1}{\sqrt{2\pi}}[\psi^+ f^+_{Kv} - \psi^- f^+_{Kv}]e^{-i\phi} \tag{66}$$

and the matrix elements of the operator $A_{bb} - A_{cc}$ are given by

$$< \Psi_{K=+1,v}|(A_{bb} - A_{cc})|\Psi_{K=-1,v} >= \tag{67}$$
$$< f^+_{Kv}|(A_{bb} - A_{cc})^{++}|f^+_{Kv} > - < f^-_{Kv}|(A_{bb} - A_{cc})^{--}|f^-_{Kv} >$$

Due to the fact that at the linear molecular geometry $A^{++}_{bb} = A^{--}_{cc}$ and $A^{--}_{bb} = A^{++}_{cc}$ and thus $(A_{bb} - A_{cc})^{++} = -(A_{bb} - A_{cc})^{--}$ the computation of the off–diagonal (in K) matrix element of Eq. 67 is equivalent to that for the diagonal elements $< K = +1||K = +1 >$ (or $< K = -1||K = -1 >$) in which the sign of the electronic matrix element $A_{bb} - A_{cc}$ for one of the electronic states is changed.

In the present paper we will not discuss in detail the computation of vibronic transition moments. The reader is referred to original references (see, e.g. [75, 192] for NH_2, [104] for H_2O^+, [93] for CH_2^+, [152, 153] for HCN, [79] for C_2H as well as the reviews [65, 80]. Instead we will concentrate on the *ab initio* investigation of the fine and hyperfine structure of vibronic spectra in triatomic molecules.

The effects of the spin–orbit and nuclear–electronic spin couplings can normally be treated independently from one another, because of clearly pronounced energetical hierarchy of the effects. The spin–orbit splitting of vibronic levels can nevertheless affect the magnetic hyperfine coupling constants in quasilinear molecules, in which the magnitude of this splitting in the 'unique' and neighboring vibronic levels is not negligibly small compared to the energy difference between various vibronic levels. This can lead to different composition of the vibronic wave functions corresponding to different spin–components of the same vibronic state. As a consequence noticeable effects on the mean values of the hfccs result. One such example is the low–lying vibronic levels of the $1^2\Pi_u$ state of CH_2^+ [94].

3.3.2 Ab initio calculation of spin-orbit splittings of vibronic levels

In this Subsection we present the results of *ab initio* computations of spin–orbit splittings of vibronic levels in the electronic states X^2B_1, A^2A_1 of NH_2, $X^2\Pi_g$ of BO_2 and $X^1\Sigma^+, A^2\Pi$ of C_2H, all of which exhibit a Renner–Teller effect.

The electronic matrix element $< {}^2B_1|H^z_{SO}|^2A_1 >$ of the spin–orbit operator H^z_{SO} in its phenomenological form, Eq. 39, is computed as a function of the bond angle for the first two electronic states of NH_2 with the N–H bond lengths kept fixed at the value of 1.935 bohr [194]. In these calculations a relatively small AO basis consisting of 32 cartesian Gaussian functions is employed. The nitrogen atom is described by the $(10s6p)$ set of Huzinaga [10] in the $[5s3p]$ contraction recommended by Dunning [11], augmented with a single d–function with the exponent $\alpha_d = 0.95$ proposed by Roos and Siegbahn [195]. An $(8s1p)$ basis in the contraction scheme $[3s1p]$ is used for the hydrogen atoms. The s–Gaussians of Lie and Clementi [164] are employed and the exponent of the p function is taken to be 0.75.

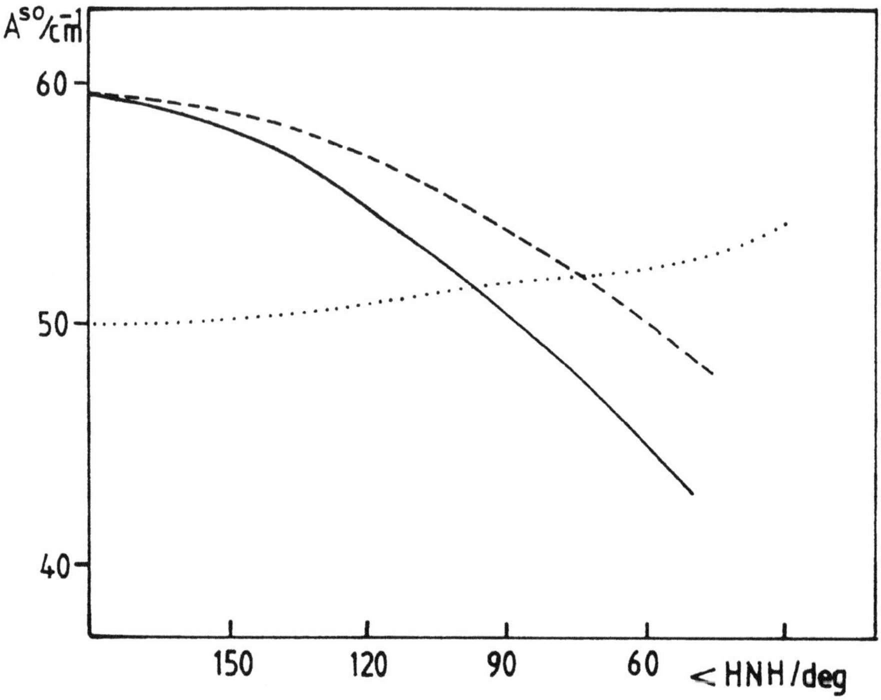

Fig. 16. *Ab initio* calculated dependence of the matrix element $| <^2 B_1|H_{SO}|^2A_1 > |(\times 2)$ for the NH_2 on the bending coordinate (full line) [194]. Dashed line: angular dependence of the quantity $|A_{SO} < L_z > |$. Dotted line: function assumed by Jungen *et al.* [196].

No bond functions are used. The dependence of the value of the matrix element $< {}^2B_1|H^z_{SO}|^2A_1 >$ (multiplied by two) on the bending coordinate is shown in Fig. 16. The value of this function at the linear geometry (59.5 cm^{-1}), overestimates by about 20 % the value of 50 cm $^{-1}$ derived by Jungen *et al.* [196] employing the available experimental data. The shape of the curve is similar to that obtained by multiplying the value of the spin–orbit constant at the linear geometry with the function $< {}^2B_1|L_z|^2A_1 >$ along the bending coordinate, it differs significantly, however, from the angular dependence of the function $A_{SO} < L_z >$ assumed by Jungen *et al.*, also shown in Fig. 16.

The results of our computations of the spin–orbit splittings of the vibronic levels for the X^2B_1, A^2A_1 system of NH_2 are presented in Fig. 17, together with the experimentally observed splittings as well as with those calculated by Jungen *et al.* who employed the potential curves and other relevant parameters derived on the basis of experimental data.

The agreement between the results of both calculations is good for the lowest vibronic levels belonging (predominantly) to the X^2B_1 state. For the $v_2 = 0$

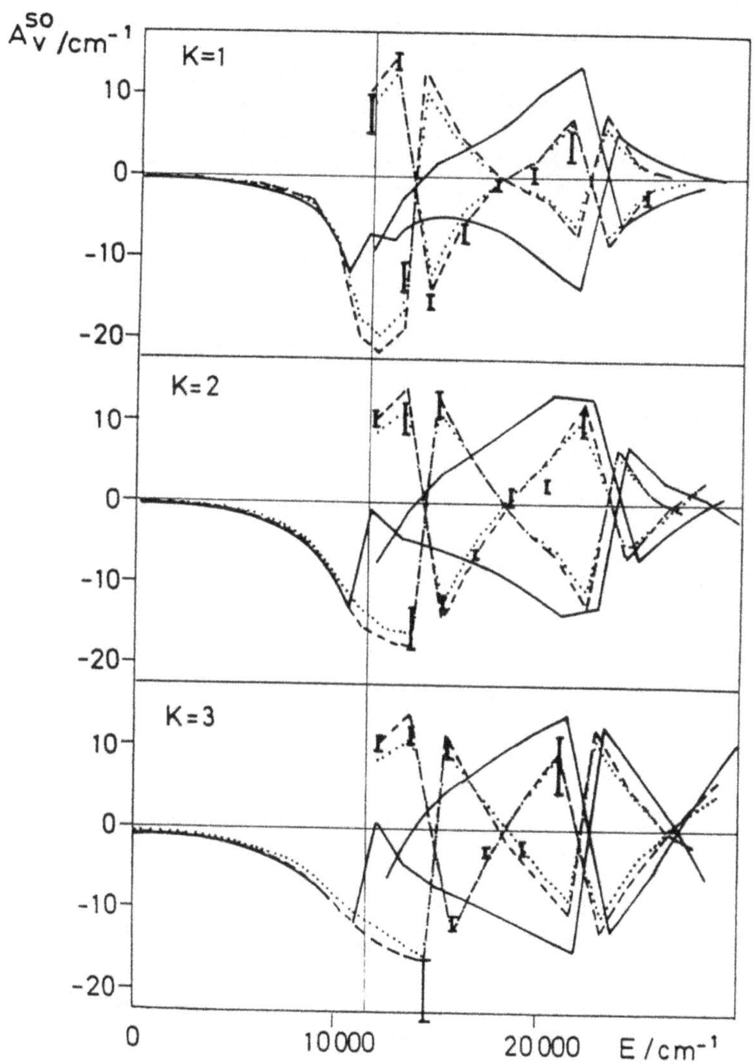

Fig. 17. Spin–orbit splittings for the vibronic levels of the X^2B_1, A^2A_1 electronic states of NH_2. Bars are observed values with error limits taken from [196]. Broken lines connect the values computed (i) by Jungen *et al.* [196] (dotted lines); (ii) in the pure *ab initio* study (full lines); (iii) employing the potential curves of Jungen *et al.* [196] and the *ab initio* computed $< H^{SO} >$ (dashed lines). Vertical thin line denotes the energetic position where both electronic states touch one another at the linear molecular geometry.

level the *ab initio* computed spin–orbit splittings are -0.33 cm^{-1} for K=1 vibronic level, -0.65 cm^{-1} for K=2, -0.97 cm^{-1} for K=3 and -1.26 cm^{-1} for K=4. The corresponding results of Jungen *et al.* are -0.308, -0.612, -0.906 and -1.185 cm^{-1}. For $v_2 = 1$ the *ab initio* calculated splittings are -0.45, -0.87, -1.28 and -1.64 cm^{-1} for $K = 1, 2, 3$, and 4, respectively, while the corresponding results of Jungen *et al.* are -0.387, -0.768, -1.130 and -1.462 cm^{-1}. For the low levels of an electronic state with strongly bent equilibrium geometry, lying far below the potential surface of the other electronic species, the magnitude of the spin–orbit splitting is nearly proportional to K. This follows directly from the structure of the secular equation (34) (completed by addition of the spin–orbit operator to its off–diagonal terms): Assuming for the sake of simplicity that the form of the kinetic energy operator is the same for both Born–Oppenheimer electronic states and taking into account that $|V^+ - V^-| \gg |iKB_\phi^{+-}T_3| \gg H^{SO} \simeq -iA_{SO}^0 B_\phi^{+-}\Sigma$, one finds that the spin–orbit splitting for the low–lying vibronic levels of the lower electronic state is nearly equal to $4K\Sigma A_{SO}^0 T_3 {B_\phi^{+-}}^2/(V^+ - V^-)$. Reliable *ab initio* computation of the spin–orbit splittings of these vibronic levels does not represent a serious problem, because it is primarily determined by the values of the electronic matrix element and by the difference of the potential energies which are computed with relatively high accuracy. For higher vibrational levels of the X^2B_1 state the agreement between the *ab initio* results and those of Jungen *et al.* becomes continuously worse, the former overestimating systematically the latter. This is primarily caused by a slight overestimation of the bending vibrational frequency for the X^2B_1 state in *ab initio* computations (1540 cm^{-1}, compared with the experimental value of 1497 cm^{-1}) pushing the *ab initio* calculated vibronic levels somewhat closer to the upper state potential.

The agreement between the *ab initio* calculated spin–orbit splittings and those computed by Jungen *et al.* becomes very poor in the vicinity of the linear geometry and for levels above the barrier. The *ab initio* results show the same saw–tooth pattern at the places where the upper state levels overtake the lower state ones as those of Jungen *et al.*, but because the crossing positions are calculated at somewhat different energies, the quantitative agreement is lacking. Some of the vibronic levels in the vicinity of the barrier to linearity show relatively large spin–orbit splittings (up to 20 cm^{-1}), but still appreciably smaller than the value of A_{SO} (roughly 50 cm^{-1}). This results from the fact that both the X^2B_1 and A^2A_1 states have nonlinear equilibrium geometries (the upper state is actually 'quasilinear' – it is characterized by a relatively small barrier to linearity so that only the lowest vibrational level lies below it) so that the 'unique' character is spread over a number of vibronic levels around the barrier. The magnitude of the spin–orbit splitting for these levels shows no systematic correlation to the value of the quantum number K. High above the barrier to linearity the quenching of the spin–orbit splitting becomes continuously more efficient.

The above analysis shows that errors in the relative location of the vibronic levels on the order of 100 cm^{-1}, does not permit consistently quantitative agree-

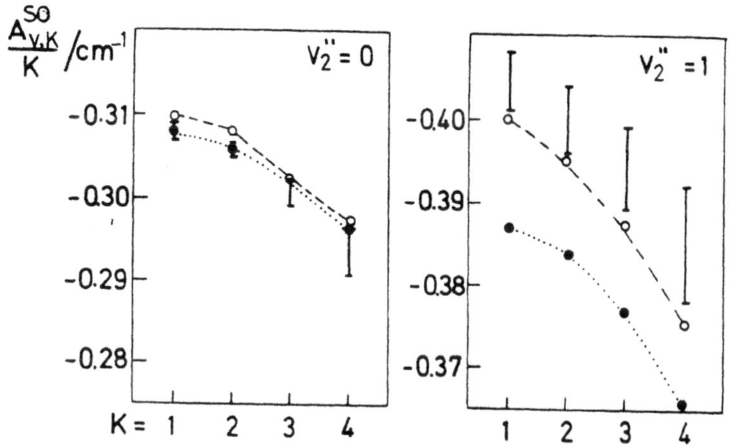

Fig. 18. Spin splittings divided by K for the $v_2 = 0$ and 1 levels of the X^2B_1 state of NH_2. Bars are experimental values with the error limits taken from [196]. Broken lines connect calculated values; dotted lines – results obtained by Jungen *et al.* [196] employing the parameters (potentials, intermolecular distances, spin–orbit constant) derived by fitting experimental data; dashed lines: results obtained using the same potential and kinetic operators, but *ab initio* calculated $< H^{SO} >$.

ment of the computed vibronic spin–orbit splittings with experiment. A possible way to overcome this problem is to modify the calculated potential curves to fit the measured vibronic levels or to use the experimentally derived curve and combine these data with the *ab initio* computed electronic matrix elements for the spin–orbit operator. The results of such computations are also shown in Fig. 17. It is seen that the spin–orbit splittings computed with the empirical potential function match perfectly those published by Jungen *et al.*, as well as the corresponding experimental findings, in spite of significant differences in the angular dependence of the *ab initio* computed $< H^{SO} >$ and that assumed by Jungen *et al.* (see Fig. 16). Moreover, the agreement between the results of this combined semiempirical – *ab initio* computations are for the lowest two bending levels of the $^2B_1$ state in even better agreement with measurements than the results of Jungen *et al.* (Fig. 18). On the basis of these observations it can be concluded that the disagreement between the calculated value for A_{SO} at the linear geometry and that derived from experimental data, as well as the difference in the angular dependence of the electronic matrix element of H^z_{SO} (Fig. 16), does not mean that the *ab initio* results are in error, but rather demonstrates that various sets of parameters can be derived from the same experimental data and that solution of the Schrödinger equation employing different sets can nevertheless satisfactorily reproduce the original experimental findings.

The BO_2 molecule possesses a linear equilibrium geometry in its $^2\Pi_g$ ground state. Upon bending this state splits into B_2 and A_2 components and represents the classical example of a weak Renner–Teller coupling [55]. Because both electronic states, $^2B_2$ and $^2A_2$, have linear equilibrium geometries, the 'unique' character of the lowest-lying $K \neq 0$ vibronic levels is very pronounced. The spin–orbit splittings for the lowest K=1 level is computed to be 134.1 cm^{-1} and that for the lowest K=2 state 146.8 cm^{-1} [137]. Their magnitude is very similar to the value for the spin–orbit constant A_{SO} computed at the linear geometry and the *ab initio* results are in very reasonable agreement with the corresponding experimental findings 149.8 cm^{-1} [141], 149.7 cm^{-1} [140] for K=1 and 146.8 cm^{-1} [141] for K=2. For the lowest vibronic levels above the unique ones the values obtained in the *ab initio* study are 15.6 cm^{-1} (K = 1, v_2 = 2^+), 29.6 cm^{-1} (K = 2, v_2 = 3^+) and 42.1 cm^{-1} (K = 3, v_2 = 4^+) and their experimental counterparts are 15.3 cm^{-1} [141], 15.8 [140], 25.8 cm^{-1} [141] and 43.04 cm^{-1} [139]. Such good agreement between the theoretical and experimental results is due to the fact that the lower-lying vibronic levels in the case of weak Renner–Teller effect contain a significant contribution from the 'unique' level so that the spin–orbit splitting is not as efficiently quenched as in the higher-lying vibronic species.

The last example for an *ab initio* treatment of the interplay between the vibronic and spin–orbit couplings concerns the $X^2\Sigma^+$, $A^2\Pi$ system of C_2H [156]. The electronic matrix elements of the spin–orbit operator for this three-state system computed as a function of the C–C stretching at different values of the bending coordinate are displayed in Fig. 19. At linear molecular geometry the matrix elements $< B_2|H^z_{SO}|B_1 >=< \Pi^-|H^z_{SO}|\Pi^+ >$ and $< A_1|H^x_{SO}|B_2 >$ were explicitly computed in the C_{2v} subgroup of the $C_{\infty v}$ molecular point group. The value of the $< B_2|H^z_{SO}|B_1 >$ (×2) element obtained at the equilibrium geometry of the Π electronic state ($\simeq$ 25 cm^{-1}) confirmed quantitatively the prediction of Curl and coworkers [188, 189]. The values for the matrix elements $< 1A'|H^z_{SO}|A'' >$ and $< 2A'|H^z_{SO}|A'' >$ at nonlinear molecular arrangements are approximated by the product of the element $< B_2|H^z_{SO}|B_1 >$ with the electronic mean value of the operator L_z. As noted in the case of electronic transition moments (see Fig. 8), the geometry dependence of the adiabatic electronic matrix elements of the spin–orbit operators is rather complicated, reflecting strong changes in the composition of the electronic wave functions in the regions where two species of A' symmetry undergo avoided crossings. In the diabatic electronic basis the geometry dependence of the matrix elements of the same operator is much simpler; this is particularly true for the dependence on the C–C bond length, which is nearly linear.

The results of *ab initio* calculated splittings of the K=1 vibronic levels for the $X^2\Sigma^+, A^2\Pi$ system of C_2H are compared in Fig. 20 with the corresponding experimental findings. As discussed above it is very difficult to achieve quantitative accuracy in such computations unless the *ab initio* treatment furnishes the vibronic levels with an accuracy of a few wavenumbers. Having an error limit of a few hun-

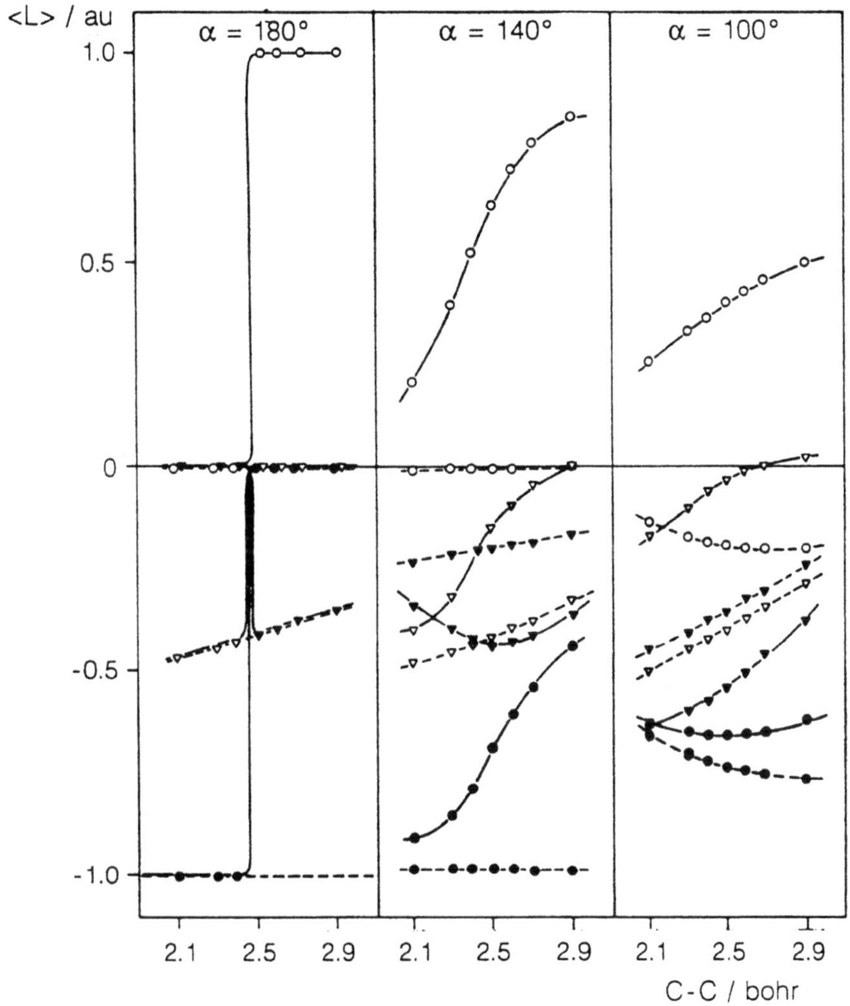

Fig. 19. Electronic matrix elements of the $y-$ and $z-$ components of the electronic angular momentum operator **L** in the $X^2\Sigma^+$, $A^2\Pi$ system of C_2H as functions of the C–C bond length at the bond angle values of 180, 140 and 100 degrees and the C–H bond length kept fixed at 2.0 bohr [72, 156].

∇ : $< 1A'|L_y|A'' >$, $< \Sigma|L_y|\Pi^- >$; $\circ$: $< 1A'|L_z|A'' >$, $< \Sigma|L_z|\Pi^- >$;

$\blacktriangledown$: $< 2A'|L_y|A'' >$, $< \Pi^+|L_y|\Pi^- >$; $\bullet$: $< 2A'|L_z|A'' >$, $< \Pi^+|L_z|\Pi^- >$.

Full lines: adiabatic electronic basis; dashed lines: diabatic basis.

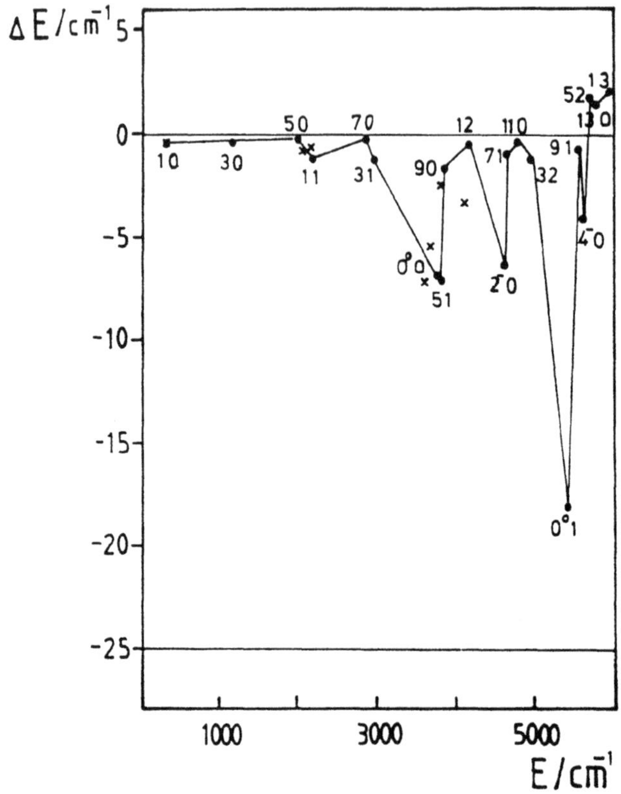

Fig. 20. Comparison of theoretical results for the spin–orbit splittings of the $K=1$ vibronic levels in the $X^2\Sigma^+$, $A^2\Pi$ system of C_2H (•) [156] with the available experimental findings (×) [182, 188]. Vibronic levels are labelled by (v_2, v_3); $v_1=0$ in all cases.

dred wavenumbers in mind, the agreement between both sets of results presented in Fig. 20 is surprisingly good. The *ab initio* calculations also give insight into the mechanisms causing certain interesting irregularities in the spin–orbit splittings scheme (see 'example 3' in chapter 7 of paper [72]). For example, the spin–orbit splitting of a vibronic level belonging to a Σ^+ electronic state should be zero. The computed spin–orbit splitting in the lowest-lying $K=1$ vibronic level assigned to the $X^1\Sigma^+$ electronic state is –0.35 cm^{-1} in excellent agreement with the measured splitting of –0.3468 cm^{-1} [182]. This finding can only be explained by an admixture of the higher lying Π species to the Σ state. The quantitative agreement achieved in this case indicates that the coupling in question is not of the 'local character'; it represents also an illustration of the high reliability with which the couplings of this kind between various electronic states can be accounted for in *ab initio* calculations.

An example of a much more complicated situation when such 'local interaction' takes place is given by the relatively close lying $K=1$ vibronic levels assigned to $(v_1, v_2, v_3)=(0,5,0)$ and $(0,1,0)$. Computed values for the spin–orbit splitting in these vibronic states, -0.23 and -1.50 cm^{-1} [156], differ significantly from the measured values of -0.65644 and -1.0175 cm^{-1}, respectively [172]. However, in the computation in which the relative energetic position of these levels is slightly changed to match the experimentally observed energy difference of 75 cm^{-1}, the results of -0.56 and -1.11 cm^{-1} are obtained. This illustrates the sensitivity of such calculations involving 'locally interacting' vibronic states. It shows also that this difficulty can easily be overcome by combining the results of pure ab initio calculations with certain experimental findings.

3.3.3 Ab initio investigation of the hyperfine structure in electronic spectra

We conclude the present review by considering the interactions between vibronic and hyperfine couplings in triatomic molecules. We will focus our interest on three molecules, NH_2, BH_2 and C_2H.

It has been shown (see, for example the studies [90, 94, 150, 157, 158, 159, 163, 197, 198]) that a reliable reproduction and prediction of the values for the hyperfine coupling constants (hfccs) derived from experiments can in general be achieved only if proper averaging of the corresponding *ab initio* calculated quantities over the vibronic states in question is performed. Only for the lowest lying vibrational states are the averaged and the non–averaged values normally very similar.

The change of the vibronically averaged hfccs (in what follows we shall use the term 'vibronic hfccs' for these quantities and 'electronic hfccs' for the electronic matrix elements of the hfccs) from one vibronic level to another reflects generally three interrelated effects: i) geometry dependence of the electronic hfccs; ii) composition of the vibronic wave function, particularly the amount of mixing of the electronic states; iii) strong local coupling between (accidentally) close–lying vibronic species belonging in the zeroth–order approximation (neglected coupling) to different electronic states. The last two effects are obviously intimately connected with one another because near degeneracy of interacting vibronic states has dramatic consequences on the composition of the corresponding wave functions. In triatomics exhibiting a Renner–Teller effect, the vibronic coupling is especially large around the point where both electronic surfaces touch each other, i.e. in the neighborhood of the linear molecular geometry, so that vibronic averaging of hfccs is particularly important in this area. For levels far away from this point, vibronic coupling is significant only in the case of near degeneracies. However, it has been shown [90] that experimentally detectable effects can be expected even if the interacting states are separated from one another by as much as a thousand wave numbers (see also the discussion at the end of this section). Thus generally only those hfccs can be compared with experimental data, which are calculated by taking into account the influence of both the electronic and nuclear motions.

Experimental determination of the values for the hfccs in various vibronic levels is a difficult and expensive task. First, numerous measurements are necessary and secondly, in order to extract the values for hfccs a knowledge of the relevant structural parameters for each of the vibronic levels of interest is required. Furthermore, not all anisotropic hfccs can be obtained by the current experimental techniques [200] and finally very often it is not possible to understand the reasons for variation of hfccs from one vibronic level to another solely on the basis of the measured data.

The above listed disadvantages are not present in the theoretical determination of the hfccs. All components of the hf tensor can be computed, and a single solution of the nuclear Schrödinger equation gives the values of the vibronic hfccs for all states of interest. Not only are the actual values of the hfccs obtained for various vibronic states, but also an explanation for the origins of their variation can be extracted. A serious drawback of theoretical investigations has been for a long time the lower accuracy compared to experiment. However, recent studies have shown that it is possible to overcome to a large extent the well-known difficulties of the *ab initio* computations of the electronic (particularly isotropic) hfccs by improving the electronic wave functions employing approaches like B_K, described in paragraph 2.3 of the present study.

The electronic hfccs for the lowest two states of NH_2, X^2B_1 and A^2A_1 were computed [158] employing for the description of the nitrogen atom the $(13s8p)$ basis given by van Duijneveldt [201], in the $[8s5p]$ contraction [202], augmented by two d-functions with exponents $\alpha_d = 1.9$ and 0.5 and an f-function with exponent $\alpha_f = 1.0$ [203]. For the hydrogen atoms the $(8s)$ to $[5s]$ [201, 202], $2p$ ($\alpha_p = 1.4$ and 0.25) AO basis was used. All computations were performed within the C_{2v} point group. In contrast to the CI treatment applied for the computation of the potential curves [75], in which the $1s$ orbital of the carbon is kept doubly occupied, in calculations of the electronic hfccs all electrons are correlated and no virtual orbitals are discarded, because correlation effects involving the inner-shell electrons are very important for a reliable description of the isotropic hfccs. Diagonal electronic matrix elements for the isotropic hfccs and all the elements of the anisotropic hf tensor were computed for ^1H and ^{14}N in both electronic states. The term A_{bc} (which transforms according to the B_1 irreducible representation of the C_{2v} point group, ϵB_1) having a nonvanishing value between the $^1B_1$ and $^2A_1$ electronic wave functions for both the N and H atoms, as well as $A_{ac}(\epsilon A'')$ for H atoms connecting the states $^2A''$ and $^2A'$, which correlate in the lower C_s symmetry to $^2B_1$ and $^2A_1$, respectively, were not computed; these neglected terms connecting vibronic states with energy differences of generally at least several hundreds of wavenumbers are expected to have insignificant effects on the hyperfine structure of the spectra. The terms $A_{ac}(\epsilon A_2)$ and $A_{ab}(\epsilon B_2)$ for the nitrogen atom vanish for symmetry reasons.

Computed bond angle dependences of the electronic hfccs at fixed N–H bond

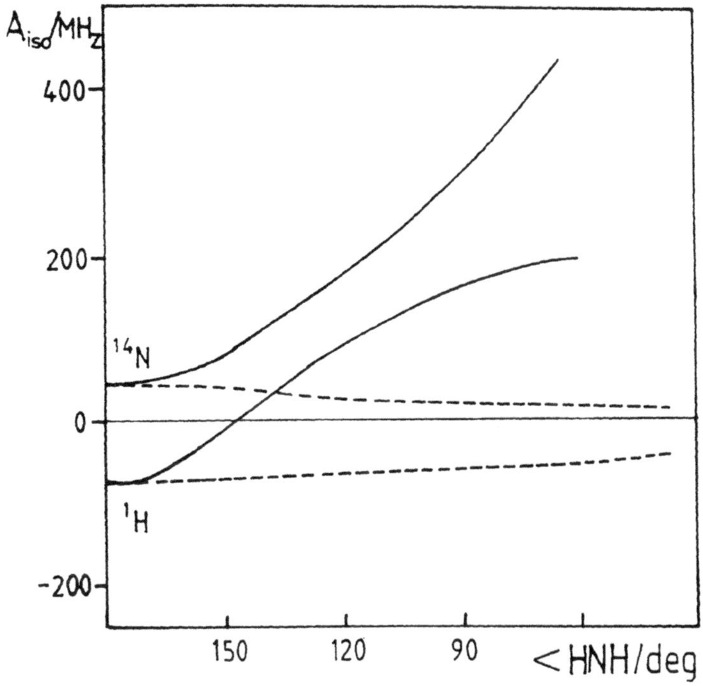

Fig. 21. Bond angle dependence of the electronic mean values for the isotropic hfccs of ^{14}N and ^{1}H in the X^2B_1 (dashed lines) and A^2A_1 (full lines) states of NH_2 [158, 159].

lengths are displayed in Figs. 21–25. They were generated in CI computations in which the threshold value of T_{CI}=0.3 μhartree was employed as a criterion for configuration selection. The influence of electronic wave function quality on the values of the hfccs was examined in Refs. [198, 158]. It was found that the values for the isotropic hfccs for both ^{14}N and ^{1}H increase by improving the electronic wave functions, the effect becoming less pronounced for decreasing bond angles. For example, the value for A_{iso} at the nitrogen center obtained in the computation with T=0.3 μhartree (15 000–20 000 configurations selected) was higher by 10 MHz than in the more restricted calculation with T_{CI}=10 μhartree (2 000–3 000 configurations) at the linear molecular geometry; the corresponding difference was 5 MHz for the bond angle value of 103 degrees and the increase due to the better wave function was only 2 MHz for the angle of 40 degrees (X^2B_1 electronic state). Use of still better wave functions (T_{CI}= 0.1 μhartree, 27755 configurations selected) lead to a further increase of A_{iso} at 103 ° by 2 MHz. Because the improvement of the electronic wave functions pushes each value for A_{iso} towards slightly larger values, a fairly reliable qualitative behavior of the functions A_{iso} with the bending coordinate is achieved already at small CI spaces. The dependence of the isotropic

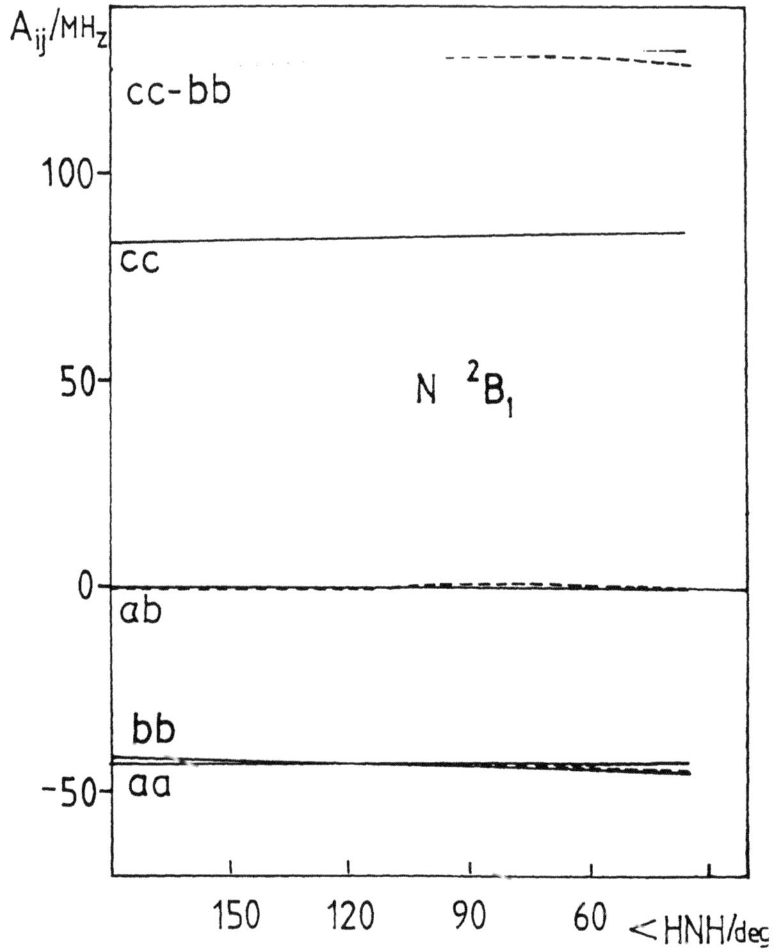

Fig. 22. Bond angle dependence of the electronic mean values of the components of the anisotropic hf tensor for ^{14}N in the X^2B_1 state of NH_2 [158, 159]. Full lines: along the principal inertia axes (PIA) of NH_2; dashed lines : along PIA of NHD. For the notation of the various components see section 2.5.3.

hfccs on the quality of the electronic wave function was found to be weaker for the $^2A_1$ state. As a result the hfcc values for nitrogen obtained in the calculations with T=0.3 are between 3 and 5 MHz higher than in the simpler calculation with T_{CI}=10 μhartree.

In contrast to the isotropic hfccs the electronic mean values of the components

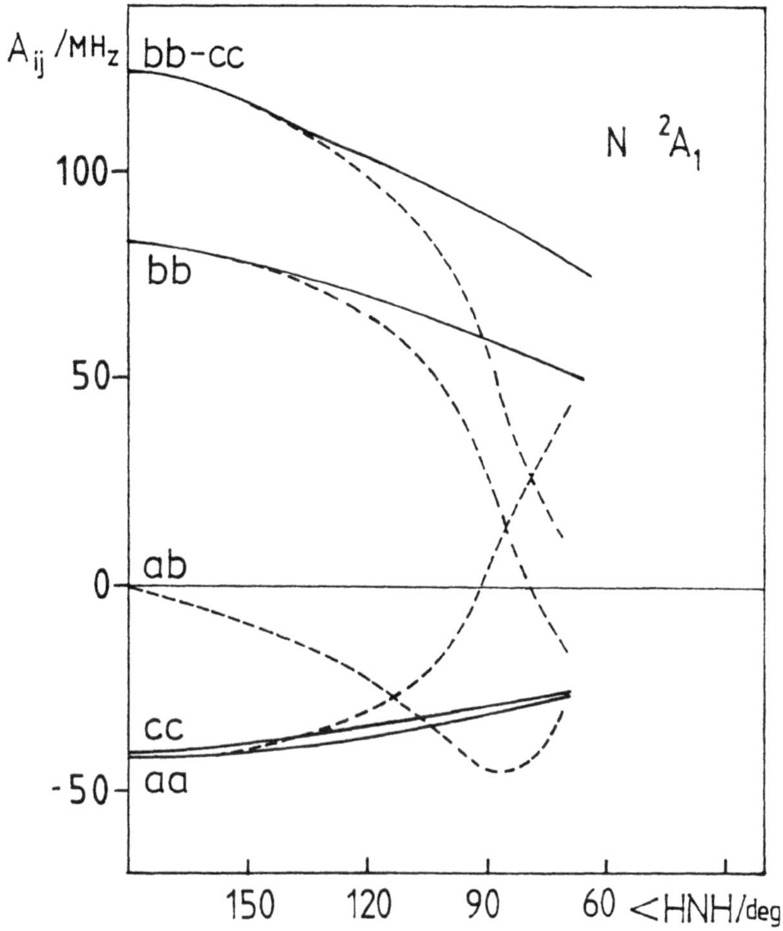

Fig. 23. Bond angle dependence of the anisotropic hfccs for ^{14}N of the A^2A_1 electronic state of NH_2 [158, 159]. Full lines: in PIA system of NH_2; dashed lines: in PIA system of NHD.

of the anisotropic hf tensor are computed rather reliably already at the HF level of sophistication. In this case an improvement of the CI functions showed no significant effect on the computed hfccs.

Figure 21 shows that A_{iso} exhibits in the X^2B_1 state a very different dependence on the bond angle than in A^2A_1 electronic state. This is easily understandable if the character of the singly occupied orbitals in the leading electronic configuration

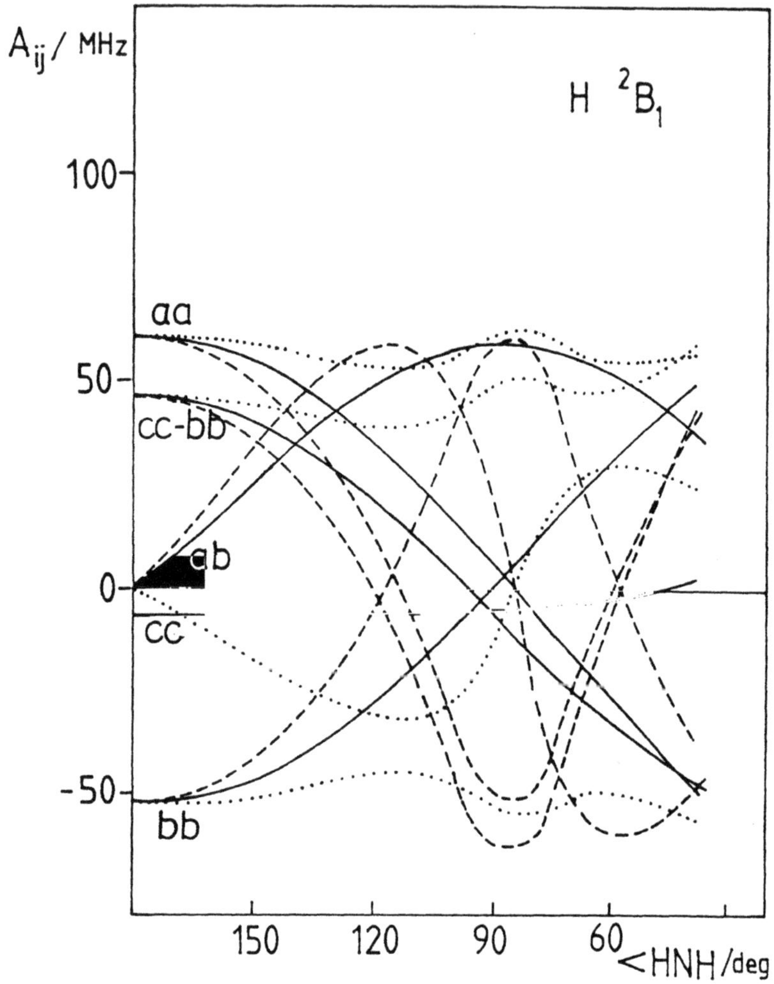

Fig. 24. Anisotropic hfccs for the hydrogen atom in the X^2B_1 electronic state of NH_2 along the PIA of NH_2 (full lines) and NHD (dashed and dotted lines) [158, 159]. Dashed lines correspond to the ^1H isotope, dotted line functions (when multiplied by the factor $g_N(D)$=0.153506) to deuterium.

of each state is considered. The $1b_1$ orbital, partially populated in the X^2B_1 ground state, constitutes at all bond angles essentially the nitrogen $2p_x$ orbital with the node in the molecular plane and its contribution to the isotropic hfccs is determined solely by spin–polarization effects. This causes relatively small values for A_{iso} in the entire range of bond angle variations and a relatively weak dependence on the

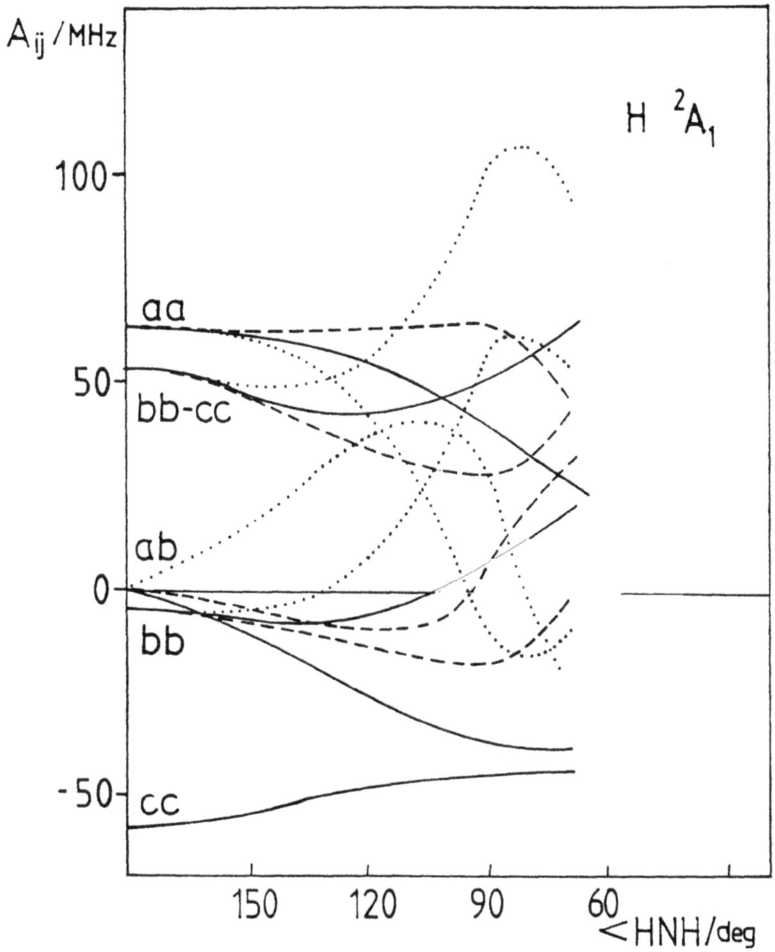

Fig. 25. Anisotropic hfccs for the hydrogen atom in the A^2A_1 electronic state of NH_2 along the PIA of NH_2 (full lines) and NHD (dashed and dotted lines) [158, 159]. The notation is equivalent to Fig. 24.

bond angle. In particular, A_{iso} is negative for the protons. On the other hand, the $3a_1$ orbital, singly occupied in the A^2A_1 state, possesses at the linear geometry pure p character, but transforms upon bending into a σ–type MO; Because singly occupied σ orbitals lead to large A_{iso} values, a much stronger dependence on the bond angle value is found. Note that the function A_{iso} (^{1}H) for the A^2A_1 state

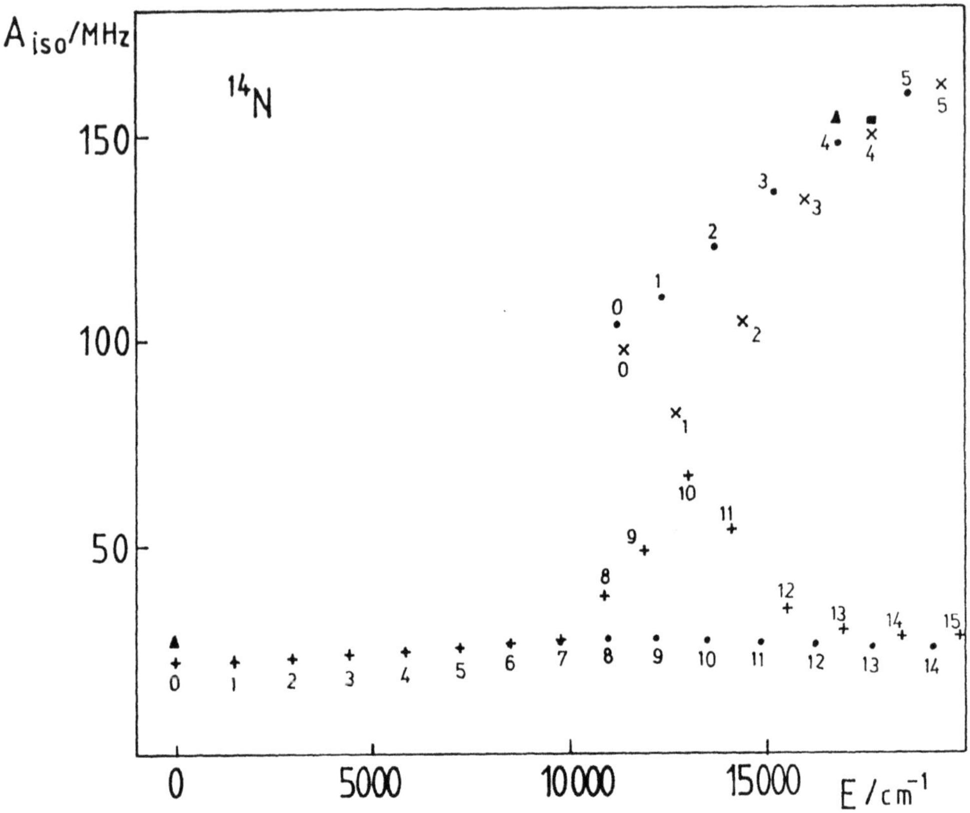

Fig. 26. Vibronic mean values for the isotropic hfcc of ^{14}N computed employing the potential curves derived by Jungen *et al.* [81] and ab initio calculated electronic hfcc [158];
•: $K=0$ vibronic states of the X^2B_1 electronic species;
o: $K=0$ vibronic states of A^2A_1;
+ : $K=1$ vibronic states belonging predominantly to the X^2B_1 electronic species;
×: $K=1$ vibronic levels belonging predominantly to the A^2A_1 state;
△, △: experimental results for $K=0$ levels of the X^2B_1 and A^2A_1 electronic states, respectively;
: experimental results for the $v_2'=4$, $K=1$ level (see Table XIX).

becomes equal to zero at the bond angle corresponding roughly to the equilibrium value for this species.

While correlation effects are very important to describe the isotropic hfcc, they are nearly negligible for the anisotropic tensor. A description of the signs and the geometry dependence of the components of the anisotropic hfccs can already be obtained applying very simple models [207, 199].

The results of purely *ab initio* computations (*ab initio* potential curves and ab initio electronic hfccs) of vibronically averaged isotropic hfccs for ^{14}N in the $K=0$

TABLE VII

Vibronically averaged values for the isotropic hfcc of ^{14}N in the $^2\Pi_u$ electronic state of NH_2 computed by employing the *ab initio* potential curves (BPPM) [75] and those derived by fitting of experimental data (JHM) [81]. For notation see table VI.

	K=0			K=1	
v_2	BPPM	JHM	v_2	BPPM	JHM
0^-	22.5	22.3	0^-	22.5	22.4
1^-	23.1	22.9	1^-	23.1	22.8
2^-	23.7	23.4	2^-	23.7	22.4
3^-	24.3	23.9	3^-	24.2	24.0
4^-	25.0	24.7	4^-	24.9	24.6
5^-	25.8	25.4	5^-	25.8	25.4
6^-	26.7	26.3	6^-	26.9	26.4
7^-	27.7	27.2	7^-	29.4	28.1
8^-	28.4	27.7	8^-	75.1	38.0
$0^+(1)$	106.7	103.8	$0^+(2)$	71.9	97.6
9^-	28.4	27.8	9^-	51.5	48.9
$1^+(3)$	111.7	110.4	$1^+(4)$	103.9	82.1
10^-	27.9	27.3	10^-	40.3	66.9
$2^+(5)$	123.5	123.1	11^-	35.4	53.8
11^-	27.3	26.7	$2^+(6)$	122.1	104.7
$3^+(7)$	135.8	135.8	12^-	33.7	34.9
12^-	26.8	26.1	$3^+(8)$	135.5	134.3
$4^+(9)$	147.6	148.0	13^-	34.4	29.7
13^-	26.4	25.6	$4^+(10)$	145.8	150.4
$5^+(11)$	158.9	159.9	14^-	40.5	28.2
14^-	26.0	25.1	$5^+(12)$	150.4	162.0
$6^+(11)$	169.9	171.5	15^-	81.0	28.5

and $K=1$ vibronic states of NH_2 in the energy range 0–20 000 cm^{-1} are presented in Table VII, together with those obtained by using the potential curves given by Jungen *et al.* [81] in combination with the *ab initio* electronic hfccs. The latter results are also displayed in Fig. 26. The values for the other hfccs, as well as the results for NHD and ND_2 are published elsewhere [159, 158, 204]. 'Bent' labeling of the vibronic levels is employed for both the electronic states. Averaging over the stretching coordinates is not performed because it is found that in this case it causes only a small change ($\leq 1\%$) for the hfccs if the molecule is assumed to be in the zeroth stretching state.

The $K=0$ vibronic levels can unambiguously be attributed to either one of the two electronic states in question and in this case the variation of the vibronically averaged hfccs is caused predominantly by the geometry variation of the corresponding electronic mean values. The values of A_{iso} for $K=0$ levels of the X^2B_1 state show relatively weak dependence on the vibrational quantum number. For the $K=0$ vibronic states of the A^2A_1 electronic species strong dependence of the electronic mean values of A_{iso} on the bending coordinate for all centers is reflected in the sharp increase of the values for the vibronic hfccs with increasing v_2. The situation is different for the $K=1$ vibronic levels. While in the energy range 0–10 000 cm^{-1} ($v_2 \leq 7^-$) in which the Renner–Teller coupling is weak, the vibronic levels belong almost exclusively to the lower electronic state and the corresponding vibronic hfccs have very similar values as those for their $K=0$ counterparts, in the region around the barrier to linearity, where the vibronic coupling is strong, the values for vibronic hfccs show rather irregular behavior with v_2. Far above the barrier the behavior of the vibronically averaged values for hfccs becomes again regular, except for the cases of possible local interactions between accidentally close lying levels of the same symmetry.

In order to check the reliability of the vibronic hfccs obtained by pure ab initio calculations, the computations have been repeated by replacing the *ab initio* computed potentials by the 'experimental' potentials derived by Jungen *et al.* [81] The results for the isotropic hfccs are presented in Table VII and Fig. 26.

As expected, on the basis of the analysis performed at the beginning of subsection 3.3, the general agreement between the results generated by employing these two sets of potentials is quite satisfactory. Small systematic differences are caused by the above discussed overestimation of the vibronic energies in the pure *ab initio* computations. Significant discrepancies are noted solely at strong local interactions of $K=1$ levels, taking place in this concrete case only in the vicinity of the barrier to linearity. For example, values of 75 and 72 MHz, respectively, are obtained for the mean values of the isotropic hfcc for nitrogen in the $v_2 = 8^-$ level belonging predominantly to the X^2B_1 electronic state and the lowest $K=1$ vibronic level of the A^2A_1 vibronic species ($v_2 = 0^+$) when the *ab initio* potentials are employed; the corresponding results generated using the experimentally derived potentials are 38 and 97 MHz. This discrepancy is explained by the fact that these levels are computed to lie much closer to one another (300 cm^{-1}) in the pure *ab initio* calculations than in the case when the potentials of Jungen *et al.* are employed (420 cm^{-1}), as a consequence the mixing of the electronic character is much stronger in the former case.

The results of this *ab initio* study are compared in Tables VIII and XIX with the available experimental findings. The theoretical results for A_{bb} and A_{cc} components of the anisotropic hf tensor are derived from the vibronic mean values for $A_{aa} = A_{bb} + A_{cc}$ and $A_{bb} - A_{cc}$ actually computed i.e. via the same procedure as their experimental counterparts. The theoretical results, particularly for the isotropic hfccs, generally underestimate slightly their experimental counterparts, in accordance with our experience with similar systems such as BH_2 [150], and C_2H [90] to be discussed below. A relatively large difference in A_{bb} for the hydrogen centers in the $v_2 = 4$ ($K=1$) level of the A^2A_1 electronic state reflects the sensitivity of the computation of this quantity in the energy region in which the magnitude of the corresponding electronic matrix element in both the $^2B_1$ and

TABLE VIII

Comparison of the results of the *ab initio* study [159] for vibronically averaged hfccs in the lowest vibronic level of the X^2B_1 state of NH_2 with the corresponding experimental data.

X^2B_1		NH_2		NHD			ND_2	
(000)		N	H	N	H	D	N	D
	th. I	22.4	-62.2	22.4	-62.2	-9.55	22.3	-9.55
	th. II	22.5	-62.3					
A_{iso}		27.9 (0.6)[a]		28.0[c]	-67.2[c]	-10.3[c]		-10.1(3)[f]
(MHz)	exp	28.2±0.4[b]	-67.2±0.4[b]				28.055(33)[g]	-10.241(28)[g]
		27.88(20)[d]	-67.59(28)[d]					
		28.11(34)[d]	-67.22(43)[d]					
	th. I	-42.4	19.6	-42.5	-23.7	8.45	-42.4	3.01
	th. II	-42.4	20.2					
A_{aa}		-43.3 (1.5)[a]		-43.0[c]	-25.5[c]	8.2[c]		3.3(7)[f]
(MHz)	exp	-42.8±1.3[b]	18.6±1.3[b]				-43.136(48)[g]	2.874(45)[g]
		-43.38(30)[d]	17.54(63)[d]					
		-43.17(88)[d]	18.61(125)[d]					
	th. I	-42.9	-12.9	-42.8	30.3	-7.43	-42.9	-1.99
	th. II	-42.9	-13.4					
A_{bb}		-44.2 (1.6)[a]		-44.4[c]	30.5[c]	-7.4[c]		-2.3[f*]
(MHz)	exp	-44.7±1.0[b]	-13.4±0.9[b]				-44.272(63)[g]	-2.108(69)[g]
		-44.75[d*]	-12.75[d*]					
		-44.53[d*]	-13.41[d*]					
	th. I	85.4	-6.67	85.3	-6.63	-1.02	85.4	-1.03
	th. II	85.4	-6.68					
A_{cc}		87.4 (0.9)[a]		85.3[c]	-5.2[c]	-0.8[c]		-1.0(4)[f]
(MHz)	exp	87.5±0.5[b]	-5.2±0.6[b]				87.413[g*]	-0.766[g*]
		88.13(18)[d]	-4.79(47)[d]					
		87.70(47)[d]	-5.20(58)[d]					
A_{ab}	th. I		55.3					
(MHz)	exp.		58.5(2.5)[c+]					

thI: Theoretical results obtained by using the potentials of Jungen et al. [81]
th II : Pure *ab initio* results.
a = [205]; b= [206]; c= [207]; d= [208]; e= [209]; f= [210]; g= [211]
Values in parentheses denote third standard deviations and apply to the last digits of the constants. Exceptions are data from Refs. f (1 standard deviation) and c (95 % confidence limit).
* denotes the values obtained from the published data using the relation $A_{bb} + A_{cc} = -A_{aa}$.
\+ Obtained indirectly by comparing the values for A_{aa} and A_{bb} in NH_2 and NHD.

TABLE IX

Comparison of the results of the *ab initio* study [159] for vibronically averaged hfccs with the corresponding experimental data for the (0,4,0) K=0 and K=1 vibronic levels ('bent' notation is used) of the A^2A_1 state of NH_2.

		A^2A_1 (040) K=0		A^2A_1 (040) K=1	
		N	H	N	H
	th. I	148.0	48.3	140.4	49.9
A_{iso}	th. II	147.6	47.8	145.8	45.7
(MHz)	exp.[a]	154.6 (26)	49.4 (55)	153.0 (13)	52.2 (14)
	th. I	-37.7	56.3	-37.6	54.2
A_{aa}	th. II	-37.7	56.3	-37.8	53.1
(MHz)	exp.[a]	-48.7 (49)	48.5 (109)	-39.5 (33)	59.5 (34)
	th. I			73.3	-4.92
A_{bb}	th. II			73.7	-4.97
(MHz)	exp.[a]			76.9 (14)	-9.5 (16)
	th. I			-35.7	-49.2
A_{cc}	th. II			-36.0	-48.1
(MHz)	exp.[a]			-37.4 (19)	-50.0 (20)

thI: Theoretical results obtained by using the potentials of Jungen *et al.* [81]
th II : Pure *ab initio* results.
a = Ref. [209].
Values in parentheses denote third standard deviations and apply to the last digits of the constants.

$^2A_1$ states is small and their geometry dependence is very strong (see Figs. 24 and 25). However, the overall agreement is good, especially when taking into account the relatively large experimental uncertainties.

The next example is the BH_2 radical in its $^2\Pi_u(X^2A_1, A^2B_1)$ state. In contrast to NH_2 in its $^2\Pi_u(X^2B_1, A^2A_1)$ state just considered, BH_2 is only slightly bent (equilibrium bond angle of about 130°) in its ground state and the A^2A_1 state possesses a linear equilibrium geometry. Hence BH_2 has much more pronounced quasilinear characteristics, which is also seen in the hyperfine structure of the BH_2 spectrum. Besides, the ordering of the electronic species is inverted, the $^2A_1$ being the ground state of BH_2.

The AO basis employed in computations of the electronic matrix elements of the hfccs [150] consists of 90 Gaussian functions. The boron atom is described by the ($13s8p$) basis in the [$8s5p$] contraction proposed by van Duijneveldt [201], augmented by three d--functions with exponents α_d=1.10, 0.405 and 0.145 [212] for a better description of the polarization effects, as well as by two additional d species (α_d = 8.0, 0.1) which are known to improve the computation of isotropic hfccs [213]. For the hydrogen centers the ($9s$) set of van Duijneveldt [201] in a [$7s$] contraction is augmented by one s function (α_s = 0.01) and three p functions (α_p=1.848, 0.649 and 0.228) [212]. To both basis sets an s function with a very high exponent ($\alpha_B = 166801, \alpha_H = 8510.53$) is added, because it is found that such functions improve the description of isotropic hfccs of electronic states with the singly occupied orbital of σ character. All calculations are performed within the C_{2v} point group. The electronic mean values for hfccs are computed employing the MRD–CI method in connection with a modified B_K correction of the wave functions [30]. Approximate natural orbitals for both electronic states are used as the one–electron basis. Twenty six reference species for X^2A_1 and sixteen for the A^2B_1 state are employed. Their contribution (sum of squared coefficients) to the final CI wave function varies between 97 and 98 %. The dimension of the CI spaces generated by single and double excitations with respect to them are of roughly one million and the number of the configurations selected according to the energy lowering criterion (T=0.1μhartree) is between 17 000 and 23 000 depending on the molecular geometry. In the framework of the B_K treatment all coefficients in the CI expansion with a magnitude greater than 0.01 (roughly 400) as well as those configurations generated by single excitations from the leading reference configuration are corrected. Computations are carried out for eight bond angle values between 180 and 140 degrees (with the bond length kept fixed at 2.211 bohr) and for six values of the symmetric stretching coordinate (at the bond angle value of 131 degrees).

As expected, the influence of the compact basis functions (with exponents α_B=166801, α_H=8510.53) on the values for the isotropic hfccs in the X^2A_1 state of BH_2 is found to be relatively large. The value for $A_{iso}(^{11}B)$ at the equilibrium geometry obtained in the CI calculation (without B_K correction) is 345.7 MHz when these functions are present and 320.3 in their absence. The corresponding numbers for ^1H are 35.6 and 28.7, respectively. A very similar trend is found in the computations with the B_K corrected wave functions . The B_K correction by itself causes at the linear molecular geometry an increase in the absolute value for $A_{iso}(^{11}B)$ of 2.7 MHz and for $A_{iso}(^1H)$ of 1.3 MHz. At bent geometries the effect of the B_K improvement on the A_{iso} values in the X^2A_1 state becomes smaller, and those for the A^2B_1 electronic state somewhat larger. The values for the components of the anisotropic hf tensor are found, in accordance with experience on similar systems, to depend insignificantly on the improvement of the electronic wave functions.

In the case of the X^2A_1, A^2B_1 system of BH_2, vibronic averaging plays an important role already in the lowest vibronic state. Let us analyze the situation for the isotropic hfccs at the boron center in the (v_1, v_2, v_3) = (0,0,0), K=0 and 1 levels of the X^2A_1 electronic state. For the boron center the electronic isotropic hfcc calculated at the experimentally estimated equilibrium bond angle value of 131 degrees [92] and a bond length of 2.211 bohr is 325 MHz. At the same angle and the experimentally determined equilibrium bond length of 2.230 bohr [92], a value of 330 MHz is obtained. It compares fairly well with the computations by Fernandez *et al.* [214] performed at the same geometry. However, at the equilibrium geometry determined in the very extensive *ab initio* calculations by

Brommer *et al.* [68], being essentially the same as published in our previous work [91] ($\angle$ HBH=126.68 degree, R_{BH} = 2.2373 bohr) we obtain for A_{iso} (^{11}B) the value of 347 MHz. This value rises slightly (by about 1 %) upon averaging over the symmetric stretching coordinate in the zeroth ($v_1 = 0$) vibrational state becoming 351 MHz. Finally, integration over the bending coordinate results in the value of 341.3 MHz (i.e. in a lowering of about 3 %). This result is in good agreement with the experimental finding by Knight *et al.* [215] of 358 $\pm$ 2 MHz. In the lowest K=1 level, besides the above effects, a small admixture of the upper electronic species contributes non–negligibly to the vibronic mean value.

The reasons for the importance of vibronic averaging in the lowest–lying levels of BH_2 can be summarized as follows i) The ground state of BH_2 has a slightly bent equilibrium geometry – it belongs to the geometry range where the singly occupied orbital in the leading electronic configuration of this state, $3a_1$, changes its character from π (at linear geometry) to σ (at bent geometries), resulting in a strong dependence of the electronic hfccs on the molecular geometry, particularly on the bond angle; ii) the quasilinearity of the X^2A_1 electronic state has as a consequence high anharmonicity of the corresponding bending potential curve – thus the form of the bending wave functions are even in the lowest vibronic levels significantly asymmetric with respect to the equilibrium bond angle value; iii) the upper electronic state lies only 2700 cm^{-1} above the minimum of the lower one thus contributing appreciably to all $K \neq 0$ vibronic levels, also to those lying below the potential barrier.

Our last example discusses part of the *ab initio* results on the hyperfine structure of the $X^2\Sigma^+, A^2\Pi$ system of C_2H [90, 157]. In this case, besides the effects discussed above, the existence of nonvanishing off–diagonal electronic matrix elements connecting both species of the A' symmetry (derived from $^2\Sigma^+$ and $^2\Pi$ in bent geometry) plays an important role.

The electronic hfccs are computed employing the same basis that was used in calculations of the potential surfaces and the electronic transition moments (see paragraph 3.3.2). The only difference in the CI treatment is that now all electrons are correlated. Electronic matrix elements of the hfccs are computed employing the CI truncated wave functions, i.e. the B_K correction is not carried out. The angular dependence of the electronic isotropic hfccs at three values of the C–C bond length for the hydrogen atom are presented in Fig. 27. The corresponding functions for both carbon atoms, as well as the results for the components of the anisotropic hf tensor are published in the original reference [157]. In Fig. 27 the geometry dependences of both the adiabatic and diabatic electronic matrix elements of $A_{iso}(^1H)$ are shown; the diabatic representation is again much simpler as was pointed out for other molecular properties (Figs. 8 and 19). The electronic hfccs computed at the equilibrium geometry of the ground state generally overestimate slightly the experimental findings for the lowest–lying vibronic species, apparently in contrast to the situation in similar systems. However, after performing vibronic averaging the results are in accord with the usual trend (see below).

TABLE X

Comparison of mean values for the isotropic hfcc (A_{iso}) and for $\frac{1}{2}A_{aa}$ in the lowest K=0 and K=1 vibronic levels of various isotopomers of C_2H calculated in the *ab initio* study [90] with the corresponding experimentally derived data. a=Ref. [216], b=Ref. [217], c=Ref. [218], d=Ref. [219], e=Ref. [220], f=Ref. [221], g=Ref. [222], h=Ref. [223], and i=Ref. [224]. The values denoted by an asterisk are recalculated from the originally published experimental results for b (see formula 45). Theoretical results are given with five significant figures to show the calculated isotopic shifts.

	$v_1\ v_2\ v_3$	$< A_{iso} >$ / MHz C_1 th	C_1 exp	C_2 th	C_2 exp	H th	H exp	H exp
	000					39.260	43.30^{b*}	44.51^{c*}
$^{12}C^{12}CH$							44.91^{d*}	44.511^{h}
	010					31.838	35.37^{g}	
$^{12}C^{13}CH$	000			153.01	177.9^{i}	39.271	55.5^{i}	
	010			118.10		31.879		
$^{13}C^{12}CH$	000	881.47	939.0^{i}			39.224	49.4^{i}	
	010	790.50				31.800		
$^{13}C^{13}CH$	000	881.53	902.0^{a}	152.78	156^{a}	39.231	44^{a}	
	010	790.87		117.88		31.835		
$^{12}C^{12}CD$	000					6.2029	7.159^{e}	6.88^{f*}
	010					5.2085		
$^{12}C^{13}CD$	000			159.02		6.2075		
	010			127.33		5.2193		
$^{13}C^{12}CD$	000	895.47				6.1960		
	010	814.88				5.2009		
$^{13}C^{13}CD$	000	895.91		158.80		6.2011		
	010	815.81		127.10		5.2123		

	$v_1\ v_2\ v_3$	$1/2 < A_{aa} >$ / MHz C_1 th	C_1 exp	C_2 th	C_2 exp	H th	H exp	H exp
	000					4.4264	4.077^{b}	4.110^{c}
$^{12}C^{12}CH$							3.977^{d}	4.0847^{h}
	010					4.6437	4.413^{g}	
$^{12}C^{13}CH$	000			19.351	16.3^{i}	4.4278	4.0^{i}	
	010			19.288		4.6454		
$^{13}C^{12}CH$	000	42.600	44.0^{i}			4.4288	4.0^{i}	
	010	41.044				4.6463		
$^{13}C^{13}CH$	000	42.591	39^{a}	19.346	17^{a}	4.4305	4^{a}	
	010	41.037		19.282		4.6482		
$^{12}C^{12}CD$	000					0.6776	0.712^{e}	0.530^{f}
	010					0.71108		
$^{12}C^{13}CD$	000			19.259		0.67773		
	010			19.129		0.71125		
$^{13}C^{12}CD$	000	42.717				0.67808		
	010	41.268				0.71163		
$^{13}C^{13}CD$	000	42.715		19.252		0.67818		
	010	41.269		19.121		0.71177		

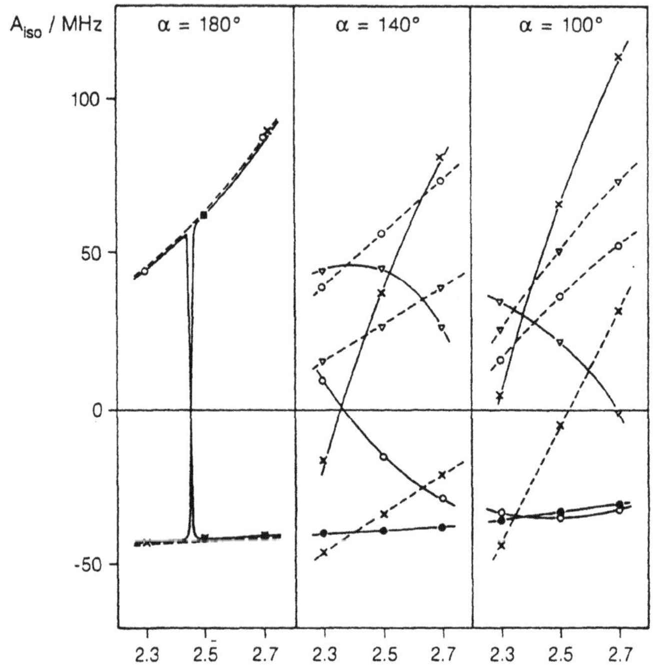

Fig. 27. C–C bond length dependence of the isotropic hfcc for the hydrogen atom in the $X^2\Sigma$, $A^2\Pi$ system of C_2H calculated at the bond angle values of 180, 140 and 100 degrees and the C–H bond length kept fixed at 2.0 bohr [90, 157]. Full lines: adiabatic electronic matrix elements; dashed lines: diabatic matrix elements.
o : $< 1A'|A_{iso}|1A' >$, $< \Sigma|A_{iso}|\Sigma >$; × : $< 2A'|A_{iso}|2A' >$, $< \Pi^+|A_{iso}|\Pi^+ >$;
• : $< A''|A_{iso}|A'' > \equiv < \Pi^-|A_{iso}|\Pi^- >$; ▽ : $< 1A'|A_{iso}|2A' >$, $< \Sigma|A_{iso}|\Pi^+ >$.

The results are given in Table X. The influence of the vibronic couplings can be seen if the value of $A_{iso}(^{13}C_1)$ obtained at the equilibrium geometry of the $X^2\Sigma^+$ electronic state (993 MHz) [157] is compared to the value calculated for the lowest vibronic level $(v_1, v_2, v_3) = (0,0,0)$ $K=0$ of $^{13}C^{12}CH$ [$A_{iso}(^{13}C_1)$= 881 MHz] [90]. This result is very different from that in which the admixture of the upper electronic species is neglected in the vibronic wave functions and differ also from that in which the presence of the off–diagonal electronic matrix element $< 1A'|A_{iso}|2A' >$ is ignored. However, the best illustration of the reliability of the *ab initio* results is obtained by comparing them with some hitherto puzzling experimental findings. Cochran *et al.* [225] stated that the ratio of the splittings of

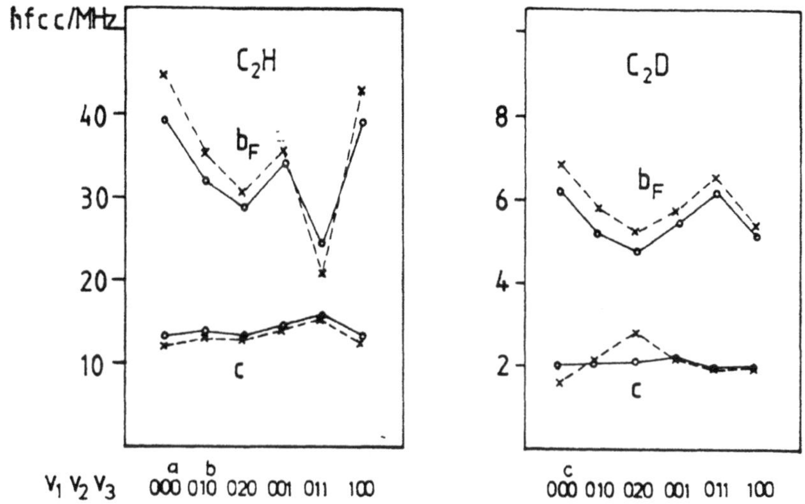

Fig. 28. Comparison of theoretical results for hfccs b_F and c in the low-lying vibronic levels of the $X^2\Sigma^+$, $A^2\Pi$ system of C_2H and C_2D (o) [90] with their experimentally derived counterparts (×). Full lines connect theoretical, dashed lines experimental results. The vibronic levels for which the hfccs were determined prior to the theoretical study are denoted by a=[222], b=[223], c=[221]. The other experimental results are from yet unpublished work of Killian and Gottlieb [226].

the lines in the ESR spectra of C_2H and C_2D attributed to the hyperfine structure of the hydrogen isotopes is 6.33, differing significantly from the theoretically expected value of 6.51 (g_H/g_D). The authors concluded that 'the mechanism which produces the observed splitting in ethyne is rather complicated, and at the present we cannot explain this isotope effect'. The result of the *ab initio* study for this ratio, namely 6.32, is in quantitative agreement with the experimental finding and moreover, offers a simple explanation. In C_2D the ground vibronic level lies somewhat lower than in C_2H and consequently possesses a slightly smaller 'percentage' of Π character than the corresponding C_2H level (6 vs. 7 %). Since according to our calculations, the electronic matrix element of the isotropic hfcc for the Σ electronic state has the value of $\simeq$ 45 MHz at the linear molecular geometry, and since that for the Π state is $\simeq$ -43 MHz (see Fig. 27), it follows that the vibronically averaged value for the isotropic hfcc in C_2D should slightly increase relative to that in C_2H.

Woodward *et al.* [223] found that in the (010) K=1 vibronic state of C_2H 'the dipole–dipole parameter c is 7% larger than in the ground state, but the Fermi contact parameter b_F is 21 % smaller, possibly reflecting the effect of vibronic mixing on the unpaired electron'. The *ab initio* study predicts almost exactly the same decrease of the b_F value in going from the (000) to (010) vibronic level (39.26 and 31.84 MHz, respectively) and a very similar ($\simeq$ 5 %) increase of c [4.426 and 4.664 for the (000) and (010) levels, respectively]. Both of these effects are easily explainable in terms of a significantly stronger admixture of the Π electronic state in the (010) vibronic species and of the very different values of the isotropic hfcc and the A_{zz} component of the anisotropic hf tensor (see Figs 6 and 8 of [157]) in both electronic states.

Finally we present in Fig. 28 an illustration of the predictive value of the above discussed theoretical approach. It contains the results of the most recent experimental study by Killian and Gottlieb [226], together with the corresponding theoretical values computed three years ago [90]. It is obvious that not only the numerical values, but also erratic variations of the values for hfccs from one vibronic level to another were reliably predicted in the *ab initio* study.

4 Conclusion

In this study we have attempted to show that it is possible to predict and reproduce many details of molecular spectra of small polyatomic molecules with chemical accuracy by carrying out *ab initio* calculations at an economical level. The main philosophy adopted is to separate the entire problem into a number of not strongly interacting problems of smaller dimension which are solved separately and coupled in a subsequent step depending on which effects will be studied. Prime examples are the separation of electronic, spin–orbit and hyperfine effects in the electronic Schrödinger equation and the separation of weakly interacting vibrational and rotational modes in the nuclear Schrödinger–type equation. The interaction of nuclear and electronic motion cannot be neglected if details of molecular spectra are sought, but this coupling can be achieved quite satisfactorily for a few important degrees of freedom as shown explicitly for a number of molecules exhibiting vibronic coupling via the Renner–Teller effect. The main advantage of this approach is that it gives a quick overall picture of many details that characterize the structure of molecular spectra, and enables the interpretation of the prime spectral features; moreover, it gives insight into the mechanisms determining the structure of the spectra. This approach can easily be extended to larger polyatomic molecules for which the extension of more sophisticated treatments considering simultaneously high number of degrees of freedom is likely to break down because of large computational requirements.

Our goal in this study was not to give an extensive survey of the literature, but rather to report on results and experiences of our own laboratory. In selecting the examples we hope that we were able to show that quantum chemical investigations of electronic and nuclear motions including spin properties have become a full partner to experiment; not necessarily to compete with the accuracy of high–resolution spectroscopy, but rather to work complementary and to address themselves to those questions which are difficult to answer by direct experimental observation.

ACKNOWLEDGEMENTS

We would like to thank all our colleagues who took part in the work presented in this review. The services and computer time of the Computing Center of the University of Bonn have been essential to the present study. The financial support granted to this work by the Deutsche Forschungsgemeinschaft is gratefully acknowledged. We thank Prof. T.C. Killian and Prof. C.A. Gottlieb for making their results available to us prior to publication.

References

1. G. Richards, Nature **278** 507 (1979)
2. W.A. Goddard III, Science **227** 917 (1985)
3. H.F. Schaefer III, Science **231** 1110 (1986)
4. C.W. Bauschlicher, S.R. Langhoff, P.R. Taylor, Adv.Chem.Phys. **77** 103 (1990)
5. C.W. Bauschlicher, S.R. Langhoff, Science **254** 394 (1991)
6. C.C.J. Roothaan, Rev. Modern Phys. **23** 69 (1951); **32**, 179 (1960)
7. J.A. Pople, R.K.Nesbet, J.Chem.Phys. **22** 571 (1954)
8. S. Huzinaga, Phys.Rev. **120** 866 (1960); **122** 131 (1961)
9. S.F. Boys, Proc. Roy.Soc. (London) **A200** 542 (1950)
10. S. Huzinaga, J. Chem. Phys., **42** 1293 (1965)
11. T.H. Dunning, J. Chem. Phys., **53** 2823 (1970)
12. I. Shavitt, Methods of Electronic Structure Theory, H.F. Schaefer III (Ed.), Plenum Press, New York, 1977 p. 189
13. M.Born, J.R. Oppenheimer, Ann. Physik **84** 457 (1927)
14. B.A. Hess, C.M. Marian, S.D. Peyerimhoff *Ab-initio calculation of spin-orbit effects in molecules including electron correlation* to appear in: Modern Electronic Structure Theory, D. R. Yarkony (Ed.), World Scientific Publ. Company.
15. D.R. Hartree, Proc. Cambridge Phil. Soc. **24** 89 (1928); Proc.Roy.Soc. (London) **A141** 282 (1933)
16. V. Fock, Z. Phys. **61** 126 (1930)
17. R. Shepard, Adv. Chem. Phys. **67** 63 (1987)
18. B.O. Roos, P.R. Taylor, P.E.M. Siegbahn, Chem.Phys. **48** 158 (1980); P.E.M. Siegbahn, J. Almlöf, A. Heilberg, B.O. Roos, J.Chem.Phys **74** 2384 (1981)
19. S. Wilson *Electron Correlation in Molecules*, Clarendon Press, Oxford, 1984; R.J. Bartlett, Ann. Rev. Phys. Chem. **32** 359 (1981)
20. R.J. Buenker, S.D. Peyerimhoff, Theor.Chim.Acta **35** 33 (1974); **39** 217 (1975); R.J. Buenker, S.D. Peyerimhoff, W. Butscher Mol.Phys. **35** 771 (1978); R.J. Buenker, R.A. Philips, J.Molec.Struct. **123** 291 (1985); R.J. Buenker, Int. J. Quant. Chem. **29** 435 (1986)
21. S.R. Langhoff, E.R. Davidson, Int.J.quant.Chem. **8** 61 (1974)
22. P.O. Löwdin, J.Chem.Phys, **19** 1396 (1951)
23. P.O. Löwdin, in *Pertubation Theory and its Applications in Quantum Mechanics*, ed. C.H.Wilcox, Wiley, New York (1966)
24. Z. Gershgorn, I. Shavitt, Intern. J. Quantum Chem. **2** 751 (1968)
25. L.E.Nitzsche, E.R.Davidson, J.Chem.Phys. **68** 3103 (1977)
26. E.R.Davidson, L.E. McMurchie, S.I.Day, J.Chem.Phys. **74** 5491 (1981)
27. D.C.Rawlings, E.R.Davidson, Chem.Phys.Lett. **98** 424 (1983)
28. G.A. Segal and K. Wolf, Chem.Phys, **56** 321 (1981)
29. B. Engels, Theor. Chim. Acta, **86** 429 (1993)
30. B. Engels, Chem. Phys. Lett. **179** 398 (1991); B. Engels, J. Chem. Phys., **100** in press; H.U. Suter, B. Engels, J. Chem. Phys., **100** in press
31. C.Eckart, Phys. Rev. **47** 552 (1935)
32. E.B. Wilson, J.B. Howard, J.Chem.Phys. **4** 260 (1936); E.B. Wilson, J.C.Decius, P.C. Cross, Molecular Vibrations, McGraw-Hill, New York (1955)
33. J.K.G. Watson, Mol.Phys., **15** 479 (1968); **19**, 465 (1972)
34. A.R. Hoy, I.M. Mills, G. Strey, Mol.Phys. **24** 1265 (1972)
35. J. Tennyson, B.T. Sutcliffe, J. Mol. Spectr. **101** 71 (1983)
36. H.Diel, S. Flügge, A. Völkel, A. Weiguny, Z. Phys. **162** 1 (1961)
37. K.F. Freed, J.R. Lombardi, J. Chem. Phys. **45** 591 (1966)
38. M.Perić, M. Mladenović, S.D. Peyerimhoff, R.J. Buenker, Chem. Phys. **82**, 317 (1983)
39. G.D. Carney, L.L. Sprendel, C.W. Kern, Adv. Chem. Phys. **37** 305 (1978)
40. S. Carter, N.C. Handy, Mol.Phys. **47** 1445 (1982)
41. S. Carter, N.C. Handy, B.T. Sutcliffe, Mol. Phys. **49** 745 (1983)
42. B.T. Sutcliffe, Y. Tennyson, Mol. Phys. **58** 1053 (1986)
43. N.C. Handy, Mol. Phys. **61** 207 (1987)
44. S. Carter, N.C. Handy, Mol.Phys. **53** 1033 (1984)

45. B.T. Sutcliffe, Computational Techniques in Quantum Chemistry and Molecular Physics, Ed. G.H.F. Dierksen, B.T. Sutcliffe, A. Veilard, Reidel Dordrecht-Holland, (1975) p. 1; Quantum Dynamics of Molecules, Ed. R.G. Wooley, Plenum Press, New York and London (1980), p. 1; Current Aspects of Quantum Chemistry (Studies in Physical and Theoretical Chemistry, Vol 21), Ed.: R.Carbo (1982) p. 99
46. W.R. Thorson, I. Nakagawa, J. Chem. Phys., **33** 994 (1960)
47. T. Barrow, R.N. Dixon, G.Duxbury, Mol. Phys. **27** 1217 (1974)
48. J. T. Hougen, P.R. Bunker, J.W.C. Johns, J. Mol. Spectr. **34**, 136 (1970)
49. P.R. Bunker, J.M.R. Stone, J. Mol. Spectr. **41** 310 (1972)
50. A.R. Hoy , P.R. Bunker, J. Mol. Spectr. **52** 439 (1974)
51. P.R. Bunker, B.M. Landsberg, J. Mol. Spectr. **67** 374 (1977)
52. P.R. Bunker, Molecular Symmetry and Spectroscopy, Academic Press, New York (1979)
53. A.Sayvetz, J. Chem. Phys **7** 383 (1939)
54. G. Herzberg, E. Teller, Z. Phys. Chem. **B21** 410 (1933)
55. R. Renner, Z. Phys **92** 172 (1934)
56. D.A. Ramsay, J. Chem. Phys., **27** 971 (1957); K. Dressler, D.A. Ramsay, Phil. Trans.R.Soc., **A251** 553 (1959)
57. J.A. Pople, H.C. Longuet-Higgins, Mol. Phys. **1** 372 (1958)
58. CH. Jungen, A.J. Merer, Molecular Spectroscopy: Modern Research, Vol 2, ed. K.N. Rao, Academic Press, NewYork (1976)
59. CH. Jungen, A.J. Merer, Mol.Phys. **40** 1 (1980)
60. R. N. Dixon, Mol. Phys. **9** 357 (1965)
61. H. Köppel, W. Domcke, L.S. Cederbaum, J. Chem. Phys. **74** 2945 (1981)
62. G. Duxbury, Molecular Spectroscopy, Vol. 3, Billing & and Sons, Guildford and London (1975) p. 497
63. J. M. Brown, F.Jorgenson, Adv. Chem. Phys. **52**, Ed.: I. Prigogine and S.A. Rice, John Wiley & Sons (1983) p. 117
64. H. Köppel, W. Domcke, L.S. Cederbaum, Adv. Chem. Phys. **72**, Ed. I. Prigogine and S.A. Rice, John Wiley & Sons (1984) p.59
65. M. Perić, S.D. Peyerimhoff, R.J. Buenker, Int. Rev. Phys., Chem. **4** 85 (1985)
66. S. Carter, N.C. Handy, Mol. Phys. **52** 1367 (1984)
67. S. Carter, N.C. Handy, P. Rosmus, G.Chambaud, Mol.Phys. **71** 605 (1990)
68. M. Brommer, P. Rosmus, S. Carter, N.C. Handy, Mol. Phys. **77** 549 (1992)
69. M. Brommer, B. Weis, B. Follweg, P. Rosmus, S. Carter, N.C. Handy, H.-J. Werner, P.J. Knowles, J. Chem. Phys. **98** 5222 (1993)
70. C. Petrongolo, J. Chem. Phys. **89** 1297 (1988)
71. M. Perić, S.D. Peyerimhoff, Mol. Phys. **78** 855 (1993)
72. M. Perić, S.D. Peyerimhoff, R.J. Buenker, Z. Phys. **24** 177 (1992)
73. W. Kolos, Adv. in Quant. Chem., Ed.: P.-O. Löwdin, Academic Press, New York-London (1970) p. 99
74. G. Hirsch, P.J. Bruna, S.D Peyerimhoff, R.J. Buenker, Chem. Phys. **45** 335 (1980); R.J. Buenker, G. Hirsch, S.D. Peyerimhoff, P.J. Bruna, J. Römelt, M. Bettendorf, C. Petrongolo, Current Aspects of Quantum Chemistry, Studies in Physical and Theoretical Chemistry, Vol 21, Ed.: R. Carbo, Elsevier, Amsterdam (1981) p. 81
75. R.J. Buenker, M. Perić, S. D. Peyerimhoff, R. Marian, Mol. Phys. **43** 987 (1981)
76. M. Perić, S.D. Peyerimhoff, R.J. Buenker, Mol. Phys. **49** 379 (1983)
77. J. Radić-Perić, M. Perić, Z. Naturforsch. **429** 103 (1987)
78. M. Perić, R.J. Buenker, S.D. Peyerimhoff, Mol.Phys. **71** 673 (1990)
79. M. Perić, S.D. Peyerimhoff, R.J. Buenker, Mol.Phys. **71** 693 (1990)
80. M. Perić, R.J. Buenker, S.D. Peyerimhoff, Mol.Phys. **59** 1283 (1986)
81. CH. Jungen, K.-E. Hallin, A.J. Merer, Mol. Phys. **40** 25 (1980)
82. G. Duxburry, R.N. Dixon, Mol. Phys. **43** 255 (1981)
83. M. Perić, M. Mladenović, J. Fejzo, C.M. Marian, P.J. Bruna, Chem. Phys. Lett **88** 547 (1982)
84. M. Perić, R. Runau, J. Römelt, S.D. Peyerimhoff, R.J. Buenker, J. Mol. Spectr. **78** 309 (1979)
85. H.A. Bethe, E.E. Salpeter, Quantum Mechanics of One- and Two-Electron Atoms,

Springer (1957); S.R. Langhoff, C.V. Kern, Modern Theoretical Chemistry, Plenum Press (1977) p. 381
86. W. Weltner, Jr.., Magnetic Atoms and Molecules, Scientific and Academic eds., New York van Nostrand Reinhold (1983)
87. R. McWeeny, J. Chem. Phys. **42** 1717 (1965)
88. A. Carrington, D.H. Levy, T.A. Miller, Adv. Chem. Phys. **18** 149 (1970); I. Engelbrecht, J. Hinze, Adv. Chem. Phys. **44** 1 (1980)
89. R.A. Frosch, H.M. Foley, Pys. Rev **88** 1337 (1952)
90. M. Perić, B. Engels, S.D. Peyerimhoff, J. Mol. Spectr. **150** 70 (1991)
91. M. Perić, S.D. Peyerimhoff, R.J. Buenker, Can. J. Chem. **59** 1318 (1981); M. Perić, M. Krmar, Bull Soc. Chim. Beograd **47** 43 (1982)
92. G. Herzberg, J.W.C. Johns, Proc.R.Soc. A **298** 142 (1967)
93. W. Reuter, S.D. Peyerimhoff, Chem. Phys **160** 11 (1992)
94. B. Engels, M. Perić, J. Chem. Phys. **97** 7629 (1992)
95. S.-K. Shih, S.D. Peyerimhoff, R.J. Buenker, M. Perić, Chem. Phys. Lett., **55** 206 (1978)
96. G. Herzberg, Electronic Spectra of Polyatomic Molecules, Princeton, van Nostrand, (1966)
97. G. Herzberg, J.W.C. Johns, Proc. R. Soc. A **295** 107 (1966)
98. M. Perić, S.D. Peyerimhoff, R. J. Buenker, Chem. Phys. Lett. **105** 44 (1984)
99. R.N. Dixon, G. Duxbury, D.A. Ramsay, Proc. R. Soc. A **296** 137 (1967)
100. S. D. Peyerimhoff, R.J. Buenker, Chem. Phys. **42** 167 (1979)
101. M. Perić, R.J. Buenker, S.D. Peyerimhoff, Astrophys. Lett. **24** 69 (1984)
102. M. Kroll, J. Chem. Phys. **63** 319 (1975)
103. D. A. Ramsay; provisional experimental value; privat communication (1981)
104. W. Reuter, M. Perić, S.D. Peyerimhoff, Mol. Phys. **74** 569 (1991)
105. H. Lew, Can. J. Phys. **54** 2028 (1976); H. Lew, R. Grolean, Can. J. Phys. **65** 739 (1987)
106. B.M. Dinelly, M.W. Crofton, T. Oka, J. Mol. Spectr. **127** 1 (1988)
107. H. Lew, I. Heiber, J. Chem. Phys. **58** 1246 (1973)
108. J. E. Reutt, L.S. Wang, Y.T. Lee, D.A. Shirley, J. Chem. Phys. **85** 6928 (1986)
109. B. Nestmann, M. Perić Chem. Phys. **89** 257 (1984)
110. P.J. Bruna, personal communication
111. M. Perić, R.J. Buenker, S.D. Peyerimhoff, Can. J. Chem. **57** 2491 (1979); M. Perić, Chem. Phys. Lett. **76** 573 (1980)
112. J.M. Berthou, B. Pascat, H. Guenebaut, D.A. Ramsay, Can. J. Phys. **50** 2265 (1972)
113. P.F. Zittel, W.C. Lineberger, J. Chem. Phys. **65** 1236 (1976)
114. H. Guenebaut, B. Pascat, J.M. Berthou, J. Chim. Phys. **62** 867 (1965)
115. R.N. Dixon, G. Duxbury, Chem. Phys. Lett., **1** 330 (1967)
116. P.J. Bruna, G. Hirsch, M. Perić S.D. Peyerimhoff, R.J. Buenker, Mol. Phys. **40** 521 (1980); M. Perić, M. Krmar, Bull Soc. Chim. Beograd **45** 531 (1980)
117. G. Duxbury, M. Horani, J. Rostas, Proc. R. Soc. A, **331** 109 (1972)
118. D.W. Turner, C. Baker, A.D. Baker, C.R. Brundle, Molecular Photoelectron Spectroscopy, Wiley-Interscience, (1970)
119. A.W. Potts, W.C. Price, Proc. R. Soc. A., **326** 181 (1972)
120. J. Delwiche, P. Natalis, J.E. Collin, Int. J. Mass. Spectom.Ion Phys., **5** 443 (1970)
121. D.C. Frost, A. Katrib, C.A. McDowell, R.A.N. McLean, Int. J. Mass. Spectom.Ion Phys., **7** 485 (1971)
122. S. Durmaz, G.H. King, R.J. Suffolk, Chem. Phys. Lett. **13** 304 (1972)
123. P.J. Bruna, C.M. Marian, Chem. Phys. **37** 425 (1979)
124. G. Herzberg, Q. Rev. Chem.Soc. **25** 201 (1971)
125. M. Perić, S.D. Peyerimhoff, R. J. Buenker, Can. J. Chem. **61** 2500 (1983)
126. C. M. Woodman, J. Molec. Spectr. **33** 311 (1970)
127. M.R. Jacox, D.E. Milligan, J. Chem. Phys. **46** 184 (1967)
128. D.M. Lindsay, J.L. Gole, J.R. Lombardi, Chem. Phys. **37** 333 (1979)
129. J. Radić-Perić, J. Römelt, S.D. Peyerimhoff, R.J. Buenker, Chem. Phys. Lett. **50** 344 (1970)
130. L. Gausset, G. Herzberg, A. Lagerquist, B. Rosen, Astrophys. J. **142** 45 (1965)
131. W. Weltner Jr., D. McLeod, Jr., J. Chem. Phys. **40** 1305 (1964); **45** 3096 (1966)

132. M. Perić, J. Radić–Perić, Chem. Phys. Lett. **67** 138 (1979)
133. M. Perić, B.A. Hess, R.J. Buenker, Mol. Phys. **58** 1001 (1986)
134. R.N. Dixon, Proc. R. Soc. A **252** 165 (1960)
135. D.E. Milligan, M.E. Jacox, J.Chem. Phys. **47** 5157 (1967)
136. A. Carrington, A.R. Fabris, B.J. Howard, N.J.D. Lucas, Mol. Phys. **20** 961 (1971)
137. M. Perić, K. Bhanuprakash, R.J. Buenker, Mol. Phys. **65** 403 (1988)
138. J.W.C. Johns, Can. J. Phys. **39** 1738 (1961)
139. D.K. Russel, M. Kroll, R.A. Beaudet, D.A. Dows, Chem. Phys. Lett. **20** 153 (1973); D.K. Russel, M.Kroll, R.A. Beaudet, J. Chem. Phys. **66** 1999 (1977)
140. K.G. Weyer, R.A. Beaudet, R. Straubinger, H. Walther, Chem. Phys. **47** 171 (1980)
141. B.S. Snowden, PhD Thesis, Vanderbilt University (1963)
142. V. Saraswathy, J.J. Diamond, G.A. Segal, J. Phys. Chem. **87** 718 (1983)
143. G. Herzberg, Molecular Spectra and Molecular Structure I: Spectra of Diatomic Molecules, Van Nostrand Reinhold Company (1950)
144. B. Weis, S. Carter, P. Rosmus, H.–J. Werner, P.J. Knowles, J. Chem. Phys. **91** 2818 (1989)
145. R.A. Back, C. Willis, D.A. Ramsay, Can. J. Chem. **52** 1006 (1974)
146. M. Perić, R. J. Buenker, S.D. Peyerimhoff, Can. J. Chem. **55** 1533 (1977)
147. R.A. Back, C. Willis, D.A. Ramsay, Can. J. Chem. **56** 1575 (1978)
148. J.R. Thomas, B.J. DeLeeuw, G.Vacek, T.D. Crawford, Y. Yamaguchi, H.F. Schaefer III, J. Chem. Phys., **99** 403 (1993)
149. P. Botschwina, M. Horn, J. Flügge, S. Seeger, J. Chem. Soc. Faraday Trans. **89** 2219 (1993)
150. M. Staikova, M. Perić, B. Engels, J. of Mol. Spectr. **163** 221 1994
151. M. Perić, S. D. Peyerimhoff, R.J. Buenker, Mol. Phys. **55** 1129 (1985)
152. M. Perić, R.J. Buenker, S. D. Peyerimhoff, Mol. Phys. **62** 1323 (1987)
153. M. Perić, R.J. Buenker, S. D. Peyerimhoff, Mol. Phys. **64** 843 (1988)
154. H. Thümmel, M. Perić, S. D. Peyerimhoff, R.J. Buenker, Z. Phys. D **71** 637 (1990)
155. M. Perić, S. D. Peyerimhoff, R.J. Buenker, J. of Mol. Spectr. **148** 180 (1991)
156. M. Perić, W. Reuter, S. D. Peyerimhoff, J. of Mol. Spectr. **148** 201 (1991)
157. M. Perić, B. Engels, S. D. Peyerimhoff, J. of Mol. Spectr. **150** 56 (1991)
158. B. Engels, M. Perić, W. Reuter, S. D. Peyerimhoff, F. Grein, J. Chem. Phys. **96** 4526 (1992)
159. M. Perić, B. Engels, J. Chem. Phys. **97** 4996 (1992)
160. H. Lorenzen–Schmidt, M. Perić, S. D. Peyerimhoff, J. Chem. Phys. **98** 525 (1993)
161. M. Perić, S.D. Peyerimhoff, J. Chem. Phys. **98** 3587 (1993)
162. M. Perić, S. D. Peyerimhoff, J. Mol. Struct., **297** 347 (1993)
163. M. Staikova, B. Engels, M. Perić, S.D. Peyerimhoff, Mol. Phys. **80** 1485 (1993)
164. G.C. Lie, E. Clementi, J. Chem. Phys. **60** 1275 (1974)
165. T.H. Dunning Jr., P.J. Hay, Modern Theoretical Chemistry, Vol 2, Ed. H.F. Schaefer III, Plenum, New York (1977)
166. X.–M. Song, T.A. Cool, J. Chem. Phys. **96** 8664 (1992)
167. P.J.H. Thossen, P.M. Goodwin, T.A. Cool, J. Chem. Phys. **84** 5334 (1988)
168. P.J.H. Thossen, T.A. Cool, D.A. Webb, E.R. Grant, J. Chem. Phys. **88** 617 (1988)
169. J.S. Bernstein, X.–M. Song, T.A. Cool, Chem. Phys. Lett. **145** 188 (1988)
170. F.T. Smith, Phys. Rev. **179** 111 (1969)
171. A. Macias, A. Riera, J. Phys. B. **11** 489 (1978)
172. C.A. Mead, D.G. Truhlar, J. Chem. Phys., **77**, 6090 (1982)
173. H.–J. Werner, W. Meyer, J. Chem. Phys. **74** 5802 (1981)
174. C. Petrongolo, G. Hirsch, R.J. Buenker, Mol. Phys. **70** 825 (1990)
175. G. Hirsch, R.J. Buenker, C. Petrongolo, Mol. Phys. **70** 835 (1990)
176. G. Hirsch, R. J. Buenker, C. Petrongolo, Mol. Phys. **73** 1085 (1991)
177. R. De Vivie–Riedle, M.C. van Hemert, S.D. Peyerimhoff, J. Chem. Phys. **92** 3613 (1989)
178. P.G. Carrick, A.J. Merer, R.F. Curl, J. Chem. Phys. **78** 3652 (1983)
179. W.–B. Yan, J.L. Hall, J.W. Stephens, M.L. Richnow, R.F. Curl, J. Chem. Phys. **86** 657 (1986)
180. M. Vervolet, M. Herman, Chem. Phys. Lett. **144** 48 (1988)
181. H. Kanamori, K. Seki, E. Hirota, J. Chem. Phys. **87** 73 (1987)

182. H. Kanamori, E. Hirota, J. Chem. Phys. **89** 3962 (1988)
183. K. Kawaguchi, T. Amano, E. Hirota, J. Mol. Spectr. **131** 58 (1988)
184. D.E. Milligan, M.E. Jacox, L. Abouaf–Marguin, J. Chem. Phys. **46** 4562 (1967)
185. M. E. Jacox, Chem. Phys. **7** 424 (1975)
186. R.A. Shepherd, W.R.M. Graham, J. Chem. Phys. **86** 2600 (1987)
187. M.E. Jacox, W.B. Olsen, J. Chem. Phys. **86** 3134 (1987)
188. W.–B. Yan, C.B. Dane, D. Zeitz, J.L. Hall, R.F. Curl, J. Mol. Spectr. **123** 486 (1987)
189. J. W. Stephens, W.–B. Yan, M. L. Richnow, H. Solka, R.F. Curl J. Mol. Struct. **190** 41 (1988)
190. W.P. Kraemer, B.O. Ross, P.R. Bunker, P. Jensen, J. Mol. Spectr. **120** 236 (1986)
191. J.N. Murell, S. Carter, L.O. Halonen, J. Mol. Spectr. **93** 307 (1982)
192. C. Blindauer, M. Perić, U. Schurath, J. Mol. Spectr. **158** 177 (1993)
193. S.D. Peyerimhoff, R.J. Buenker, Can. J. Chem. **57** 3182 (1979)
194. M. Perić, C. M. Marian, unpublished results (1985)
195. B. Roos, P. Siegbahn, Theor. Chim. Acta **8** 61 (1974)
196. CH. Jungen, K.–E. J. Hallin, A.J. Merer, Mol. Phys. **40** 65 (1980)
197. V. Barone, A. Grand, C. Minichina, R. Subra, J. Chem. Phys. **99** 6787 (1993)
198. K. Funken, B. Engels, S.D. Peyerimhoff, F. Grein, Chem. Phys. Lett. **172** 180 (1990)
199. K. Funken, Diplom Thesis, Bonn, (1988)
200. T. Sears, Comput. Phys. Reports **2** 1 (1984)
201. F.B. van Duijneveldt, Techn. Report RJ 945, IBM Research Laboratory, San Jose (1971)
202. B. Engels, S.D. Peyerimhoff, Mol. Phys. **67** 583 (1989)
203. M. Frisch, J.A. Pople, J.S. Binkley, J. Chem. Phys. **80** 3265 (1984)
204. AIP document no. PAPS JCPSA–97–4996–7
205. G.W. Hills, J. M. Cook, R.F. Curl, F.K. Tittel, J. Chem. Phys. **65** 823 (1976)
206. J. M. Cook, G.W. Hills, R.F. Curl, J. Chem. Phys. **67** 1450 (1977)
207. T.C. Steimle, J.M. Brown, R.F. Curl, J. Chem. Phys. **73** 2552 (1980)
208. J.M. Cook, G.W. Hills, J. Mol. Spectr. **94** 456 (1982)
209. G.W. Hills, C.R. Brazier, J.M. Brown, J.M. Cook, R.F. Curl, J. Chem. Phys. **76** 240 (1982)
210. J. M. Cook, G.W. Hills, J. Chem. Phys. **78** 2144 (1983)
211. M. Kanada, S. Yamamoto, S. Saito, J. Chem. Phys. **94** 3423 (1991)
212. T.H. Dunning, J. Chem. Phys. **90** 1007 (1989)
213. I. Carmichael, J. Chem. Phys. **91** 1072 (1989)
214. B. Fernandez, P. Jørgensen, J. Byberg, J. Olsen, T. Helgaker, H. J. A. Jensen, J. Chem. Phys. **97** 3412 (1992)
215. L. B. Knight Jr., M. Viniski, P. Miller, C.A. Arrington, D. Feller, J. Chem. Phys. **91** 4468 (1989)
216. W.R.M. Graham, K.I. Dismuke, W. Weltner, Jr., J. Chem. Phys. **63** 2264 (1975)
217. K.D. Tucker, M.L. Kutner, P. Thaddeus, Astrophys. J. **193** L115 (1974)
218. K.D. Tucker, M.L. Kutner, Astrophys. J. **222** 859 (1978)
219. R. Y. Saykally, L. Veseth, K.M. Evenson, J. Chem. Phys. **80** 2247 (1984)
220. M. Bogey, C. Demuynck, J.L. Destombes, Astron. Astrophys. **144** L15 (1985)
221. J.M. Vrtilek, C.A. Gottlieb, W.D. Langer, P. Thaddeus, R.W. Wilson, Astrophys. J. **296** L35 (1985)
222. C.A. Gottlieb, E.W. Gottlieb, P. Thaddeus, Astophys. J. **264** 740 (1983)
223. D.R. Woodward, J.C. Pearson, C.A. Gottlieb M. Guélin, P. Thaddeus, Astron. Astrophys. **186** L14 (1987)
224. M. Bogey, C. Demuynck, J. Destombes, Mol. Phys. **66** 955 (1989)
225. E.T. Cochran, E.J. Adrian, V.A. Bowers, J. Chem. Phys. **40** 213 (1964)
226. T.C. Killian, C.A. Gottlieb, private communication (1994)

Theoretical Studies of the Electronic Spectra of Organic Molecules

Björn O. Roos, Markus Fülscher, and Per-Åke Malmqvist
Department of Theoretical Chemistry, Chemical Centre, P.O.B. 124, S-221 00 Lund, Sweden

and

Manuela Merchán and Luis Serrano-Andrés
Departamento de Química Física, Universidad de Valencia, Dr. Moliner 50, Burjassot, E-46100 Valencia, Spain

June-1994

Abstract The complete active space (CAS) SCF method in conjunction with multiconfigurational second-order perturbation theory (CASPT2) has been used to study the electronic spectra of a large number of molecules. The wave functions and the transition properties are computed at the CASSCF level, while dynamic correlation contributions to the excitation energies are obtained through the perturbation treatment. The methods yield energies, which are accurate to at least 0.2 eV, except in a few cases, where the CASSCF reference function does not characterize the electronic state with sufficient accuracy. The applications comprise: the polyenes from ethene to octatetraene (*cis*- and *trans*-forms); a number of cyclic pentadienes; norbornadiene; benzene, phenol, phosphabenzene, and the azabenzenes; free base porphin; and the nucleic acid base monomers cytosine, uracil, thymine, and guanine. Finally, the photochemistry of the molecules aminobenzonitrile (ABN) and dimethylaminobenzonitrile (DMABN) has also been studied, in particular the double fluorescence that occurs in DMABN. Taken together these studies comprise large amounts of new spectroscopic data of high accuracy, which either confirm existing assignments of experimental data or lead to new predictions and qualitative as well as quantitative understanding of a large number of electronic spectra. Most studies are restricted to ground state geometries (vertical energies), but in a few cases (octatetraene, ABN, and DMABN) also excited state geometries have been determined, thus yielding 0-0 transition energies and emission spectroscopic data.

S. R. Langhoff (Ed.), Quantum Mechanical Electronic Structure Calculations with Chemical Accuracy, 357–438.

Contents

1 Introduction

The calculation of electronic excitation energies and intensities of conjugated organic molecules has been a frustrating challenge for quantum chemistry since the first *ab initio* studies of the benzene molecule [1, 2]. Errors of more than 2 eV in computed excitation energies were obtained in these early attempts. The frustration is due to the contrast between these *ab initio* results and the relative success of the semi-empirical π-electron theories developed in the fifties (for a review of the Pariser-Parr-Pople model see the articles in vol. 37:4 of *Int. J. Quantum Chem.* (1990)). Already the early very approximate calculations on the benzene molecule by Goeppert-Mayer and Sklar [3] gave a better description of the electronic spectrum of the benzene molecule than the first *ab initio* studies. It was not until very extensive MRCI calculations were performed [4] that results of comparative accuracy could be obtained, but even this study resulted in excitation energies, which differed from experiment by up to 0.5 eV.

It was quite clear that the reason for the large errors was the difficulties for the *ab initio* methods to include the dynamic correlation effects. Benzene and molecules of similar size contain many valence electrons, which must be included in the correlation treatment. In a conventional configuration-interaction (CI) approach the size of the CI expansion can easily become very large and the demands on the basis set may also be exhaustive. Of special importance is the $\sigma \rightarrow \pi^*$correlation effects, which manifest themselves as a dynamic polarization of the σ framework. It is possible to model this polarization by an effective Hamiltonian approach, where the σ electrons are represented as a polarizable medium. It was shown that such a treatment leads to a reduction of the Coulomb repulsion terms in the Hamiltonian much along the spirits of the PPP model [5]. Such a model thus explains the physics behind a large fraction of the correlation contribution to the excitation energies. It is, for example, clear that electronic states with a dominantly ionic character will be more affected than covalent states. The $\sigma \rightarrow \pi^*$correlation effects are thus very state dependent and omission of them leads to erratic predictions for the relative energies of different excited states. This is a major reason why methods like CI singles (CIS) should not be used even though they may seem appealingly simple [6]. It has been claimed in a recent application to the electronic spectrum of butadiene that dynamic electron correlation effects are included in a CIS treatment [7]. It is correct that a certain type of single excitations (with respect to the ground state) are doubly excited (de-excitation coupled to excitation) with respect to the excited state and therefore describe electron correlation. They give, however, only a small fraction of the correlation energy and the treatment is unbalanced since the corresponding effect occurring in the ground state is not included (a pure Hartree-Fock ground state is used). The CIS method is also by its definition unable to treat electronic states, which contain a large fraction of doubly excited configurations. As will be illustrated below, this is a very common situation in many electronic spectra. Agreement with experiment obtained with the CIS technique is therefore fortuitous, at least for states where electron correlation is important (valence excited states).

The multireference (MR) CI method was recently used to study the electronic spectrum of the pyrimidine molecule [8]. It was shown in this application that very large CI expansions were needed in order to converge the results for the excitation energies. Such convergence could actually not be achieved without extending the length of the CI expansion beyond the limits set by the MRCI program (about 3 million configuration functions (CFs)). The remaining error in the computed excitation energies was in some cases 0.4 eV. Although this error might be acceptable, the test showed that the MRCI method is not a practical tool for studies of electronic spectra of larger organic molecules.

This review will describe an alternative approach, where the treatment of the electron correlation effects is divided up into two parts. First, static correlation effects are included by performing complete active space (CAS) SCF calculations for the states of interest, either separately for each state or, more commonly, as average calculations for a set of states of a given symmetry (for a review of the CASSCF method see [9]). The active orbitals are chosen such that all CFs with weights appreciably different from zero are included in the CAS wave function for all states treated in a given calculation. In conjugated molecules the active orbital space comprises in general the π valence orbitals, the lone-pair orbitals, and (depending on the case) a set of Rydberg orbitals. This level of theory will include the near-degeneracy effects which are very common in excited states. Remaining (dynamic) correlation effects are included in a second step, where the CASSCF wave function is used as the reference in a second-order perturbation treatment, performed in the full CI space generated by the molecular orbitals [10] - [12]. Such a procedure has been shown to give an adequate description of the dynamic correlation effects, and has the virtue that it can be applied with rather extended basis sets so that larger molecules can be treated. The approach has been applied in calculating a number of properties of molecules in their ground and excited states [13] - [17]. However, the most exciting applications have been in electronic spectroscopy. The method has been used in a large number of studies of electronic spectra of a variety of organic molecules ranging in size from ethene to porphin [18] - [28], and also to some transition metal atoms and complexes [29] - [32]. Here we shall review the applications in organic chemistry. The paper is organized as follows: In the next section we discuss the methods and how they are used in applications in spectroscopy. We shall then discuss in turn the electronic spectra of polyenes (from ethene to octatetraene), of benzene and related ring systems, of systems with two interacting double bonds, of the porphin molecule, the double fluorescence occurring in the excited states of aminobenzonitriles, and finally the spectra of the nucleotide base monomers and related systems.

2 The CASSCF/CASPT2 approach to spectroscopy

To compute excitation energies and transition properties of organic conjugated molecules using *ab initio* methods is a non-trivial task for several reasons:

- The molecules are usually large, which sets severe limitations on the basis set. It is therefore important to use methods, which grow as slowly as possible with the size of the basis set. The n^4 dependence in the integral calculation or the n^5 dependence in an integral transformation is usually not the crucial step. It is instead the treatment of electron correlation effects that will decide the basis set limit of a given method.

- The electronic structure of excited states is often of a complex nature with a strong coupling of different electronic configurations, which mixes not only single excitations, but frequently also involves double excitations. It is then not possible to base the treatment on simple single configuration based methods. A multiconfigurational treatment is needed.
- Different excited states have different character and a general method must be able to treat all of them at the same level of accuracy. Valence excited states are different from Rydberg states. Among the valence excited states we find those which are well characterized as single excitations form the ground state, while others have a more complex nature. Correlation effects vary significantly with the nature of the states. Thus dynamic polarization effects dominate for states with strong ionic components in the wave function, while states of a more covalent character have correlation energies that are similar to those of the ground state.
- Quantitative results can only be obtained when the correlation treatment encompasses all valence electrons in the system. This is due to the strong coupling between the π-electron and the polarizable σ-electron framework. Therefore methods have to be used, which allows such a treatment for systems with many electrons. Multireference (MR) CI methods have severe limitations in this respect, since the size of the CI space grows quickly with the number of correlated electrons.
- A complete study of the electronic spectrum in the UV region, say up to about 8 eV (60000 cm^{-1}), will include a large number of states. When the symmetry is low all the states have to be treated in the same calculation. One severe difficulty is then to be able to locate all states of interest in the first step of the calculation, where a reference function for the correlation treatment is determined. The correlation energy can vary significantly for the different states, thereby reversing their order, and states with especially large correlation contributions may have very high reference energies. Inspection of MRCI results reported in the literature often shows that important states are missing, simply because they were not included among the reference states.

Below we shall discuss in more detail these different aspects of the computational problems of excited state calculations and show how they are solved in the present approach. All calculations that are described in the present work have been performed with the MOLCAS quantum chemistry software [33], and the reader is referred to the user's guide for details about the possibilities offered by this set of programs. MOLCAS can be obtained from the University of Lund by contacting one of the authors (MPF or BOR).

2.1 Basis sets

It is clear from the above listing of the problems to be solved that quantitative results can only be obtained when the basis set is large enough to be able to treat the different types of excited states on a comparable level of accuracy. This means that the basis set must be able to describe equally well valence and Rydberg states, and also flexible enough to cover the states of an ionic nature at the same time as the less diffuse covalent states. The basis set must also be extensive enough to describe the major differential correlation effects. In essence this implies the use of at least one set of $3d$-type functions for each first row atom.

A test of the basis set effects on the computed excitation energies and transition moments was performed for the pyrazine molecule [34]. The study comprised the six $\pi \rightarrow \pi^*$ and two $n \rightarrow \pi^*$ electronic states. No Rydberg states were included, since they are known to be much less basis set dependent as soon as good quality Rydberg type orbitals are added. Basis sets of the ANO type were used [35]. These ANOs are based on a rather extensive primitive set (hereafter referred to as ANO-L), which for hydrogen has the size $(8s4p3d)$ and for the atoms Li–Ne $(14s9p4d3f)$. Later applications have used more modest primitive sets for the construction of the ANOs (hereafter referred to as ANO-S): H$(7s3p)$, Li–Ne$(10s6p3d)$, Na–Ar$(13s10p4d)$, Sc–Zn$(17s12p9d4f)$, and Ga–Kr$(17s15p9d)$ [36]. A number of different ANO basis set sizes were used in the pyrazine study, the smallest being of double-zeta quality in the valence shell, $(3s2p/2s)$ (only one s-type ANO is used for the $1s$-orbitals in C and N), and the largest an extended set with two $3d$-type functions on carbon and nitrogen and two $2p$-type functions on hydrogen.

The results of the study are presented in Table I. Inspection of the data shows that basis sets without $3d$-type functions on the first-row atoms will lead to $\pi \rightarrow \pi^*$ excitation energies that are up to 0.8 eV too large. The $n \rightarrow \pi^*$ excitation energies are much less sensitive. The maximum difference between the $(4s3p1d/2s)$ basis set and the large $(5s4p2d/3s2p)$ set is on the other hand only 0.17 eV. The influence of a third p-type and a fourth s-type function on carbon and nitrogen on the excitation energies is small. The same is true for the $2p$-type functions on hydrogen. We notice that the large basis set calculation deviates from experiment by between 0.0 and 0.2 eV, an error which is then inherent to the method. We can conclude from these results that calculations without $3d$-type functions are of limited value. The smallest recommended basis set is $(3s2p1d)$ for first row atoms and $(2s)$ for hydrogen. However, preferably $(4s3p1d)$ should be used. This basis set, together with H$(2s)$, is used in most of the applications to be discussed below. With a practical limit of about 250 functions (planar molecules, no further symmetry) this limits the size of the molecules that can be treated to about 14 atoms, not counting the hydrogens. Higher symmetry will increase the maximum size. The largest molecule without symmetry treated until now is dimethylaminobenzonitrile, $N(CH_3)_2C_6H_4(CN)$ with 11 first row atoms and 10 hydrogens. The basis set size is 174.

The electronic spectrum will contain valence as well as Rydberg excited states. It is important to be able to describe both at a balanced level. This is so even if the main interest is in the valence excited states. Excited states with artificial mixing of Rydberg and valence states are easily obtained if an inadequate basis set is used, which does not contain a good representation of the Rydberg orbitals. One straightforward way to represent these orbitals in the basis set is to add diffuse

TABLE I
Excitation energies (CASPT2) for valence excited states in pyrazine as a function of the basis set[a].

	(32/2)	(42/2)	(43/2)	(421/2)	(431/2)	(431/21)	(432/32)	(542/32)	exp.[b]
$\pi \rightarrow \pi^*$ states									
1^1B_{2u}	5.10	5.09	5.05	4.79	4.76	4.76	4.76	4.75	4.8
1^1B_{1u}	7.46	7.44	7.39	6.75	6.74	6.72	6.69	6.70	6.5
2^1B_{1u}	8.22	8.17	8.41	7.57	7.69	7.68	7.53	7.57	7.7
2^1B_{2u}	8.23	8.21	8.50	7.77	7.87	7.85	7.74	7.70	"
2^1A_g	8.90	8.89	8.85	8.27	8.22	8.17	8.22	8.19	-
1^1B_{3g}	8.89	8.88	8.81	8.39	8.36	8.34	8.31	8.26	-
$n \rightarrow \pi^*$ states									
1^1B_{3u}	3.88	3.91	3.93	3.85	3.81	3.79	3.85	3.63	3.8
1^1A_u	4.61	4.65	4.67	4.63	4.60	4.57	4.63	4.52	-

[a]Energies in eV. Basis set of ANO type based on the primitive set C,N($14s9p4d$), H($8s4p$). The contracted basis set label has the meaning: (421/21) = C,N($4s2p1d$), H($2s1p$), etc.

[b]For detailed references to experiment see [34].

TABLE II

Exponents used for the Rydberg orbitals for molecules and cations

s-type	p-type	d-type
Neutral molecules		
.02462393	.04233528	.06054020
.01125334	.01925421	.02744569
.00585838	.00998821	.01420440
.00334597	.00568936	.00807659
.00204842	.00347568	.00492719
.00132364	.00224206	.00317481
.00089310	.00151064	.00213712
.00062431	.00105475	.00149102
Monocharged cations		
s-type	p-type	d-type
.09849573	.16934112	.24216081
.04501337	.07701682	.10978277
.02343352	.03995284	.05681759
.01338390	.02275744	.03230637
.00819369	.01390272	.01970875
.00529457	.00896824	.01269924
.00357238	.00604256	.00854849
.00249725	.00421901	.00596406

functions at the atomic centra. However, such a procedure has several disadvantages. A large number of new basis set functions are added, which increases the computational cost. How well they describe the Rydberg orbitals will depend on the geometry of the molecule and the orbital exponents used. The procedure is not well defined. It has been found in practical applications that such a procedure also leads to artificial mixing of Rydberg and valence excited states. In the CASPT2 method the occurrence of a large number of diffuse functions gives rise to intruder state problems, which are difficult to handle. An alternative method is to add diffuse functions centred at one point in the molecule (the centre of charge is often used). This method was used successfully in the study of the cyclic pentadienes [25]. Here we shall describe a more well defined variant of the same approach [37].

The method is based on the universal Gaussian basis sets devised by Kaufman *et al.* for representing Rydberg and continuum wave functions [38] and contain the following computational steps:

- Perform a CASSCF calculation on the system with one electron removed. The centre of charge for this system is used as the centre for the Rydberg basis functions. When Rydberg series converging to more than one ionization limit are to be studied, the calculation should be repeated for each of the ionic states. The average location of the charge centra is then used as the Rydberg orbital centre. This calculation can be performed with a minimal basis set. It may be added that the results are not very sensitive to the precise location of the centre.

- The minimal basis set is now extended with a set of primitive functions with low exponents generated as described by Kaufman *et al.* The universal exponents to be used for neutral molecules and mono-charged cations are given in Table II. The same calculations are repeated with the extended basis set.
- The lowest virtual orbitals of each angular momentum component are selected from the resulting sets of MOs. The MB part of the functions is set to zero. A fictitious density matrix corresponding to having one electron in each of these orbitals is constructed and is averaged for each of the ionic states included. The components of the *p*- and *d*-type functions are also averaged.
- The resulting density matrix is diagonalized and the corresponding natural orbitals are used as basis functions for the Rydberg states. Normally 1–2 functions of *s*-, *p*- and *d*-type are added to the ANO basis set.

The number of basis functions added to the basis set is 9n, where n is the number of Rydberg sets used. With n=1 (the normal choice for 3*s*,3*p*, and 3*d* Rydberg states) only a small extension of the original basis set is made. The fact that each Rydberg orbital is constructed as a linear combination of a sizable set of primitives (8) leads in general only to a minor increase in the time used in the integral calculation. In a number of applications to be discussed below it is shown that this choice of Rydberg orbitals leads to a consistent and accurate description of the Rydberg excited states and the interaction between them and the valence excited states.

2.2 CASSCF AND THE SELECTION OF ACTIVE ORBITALS

The calculation consists of two steps. First CASSCF wave functions are determined for each of the excited states. This wave function is in the second step used as the reference function for the second-order perturbation treatment. The one-electron basis consists of the molecular orbitals generated by the CASSCF calculation. The choice of the active orbitals for the CASSCF wave function is the crucial step in the whole calculation. The maximum number of orbitals that can be used is 12–15 depending on the symmetry of the molecule and the number of active electrons. This number sets the limit to how many excited states can be treated.

The active molecular orbitals (AMOs) must encompass all orbitals, which have an occupation number appreciably different from zero or two in any of the electronic states of a given symmetry, which is to be included in the calculation. Thus, the number of AMOs will depend on to what energy in the spectrum the calculation is to cover. It is possible to use a different set of AMOs for states of different symmetry. Apart from describing the excitation process, the AMOs must also cover the near-degeneracy effects in the ground state and the different excited states. This means, for example, that the active space should include all valence π-orbitals in conjugated systems. However, for larger molecules (more than eight π-centres) it is usually possible to leave the most stable π-orbitals inactive, implying that they will remain doubly occupied in all excited states studied in the calculation. Correlation contributions involving these orbitals will then be treated only at the CASPT2 level of the calculation.

As an example of the problems one may encounter during the selection of the AMOs, consider the norbornadiene molecule [22]. Suppose we decide to excite out of the two ethylenic π-type orbitals only. This, of course, eliminates the possibility of including $\sigma \rightarrow \pi^*, \sigma^*$, or R excitations in the study. Should such excitations appear within the studied energy range, they will show up as intruder states in the CASPT2 treatment. As a result it may become necessary to add new orbitals to the active space. Before considering such possibilities we start with an AMO space comprising the four quasi π-orbitals of symmetries a_1, b_1, b_2, and a_2 (norbornadiene has C_{2v} symmetry). We denote this active space (1111). It is large enough for the treatment of the two lowest triplet states and the lowest singlet state (1^1A_2), which corresponds to the excitation $\pi(b_1) \rightarrow \pi^*(b_2)$. However, already the second singlet state in norbornadiene is of Rydberg type ($\pi(b_1) \rightarrow 3s$). Adding the $3s$,$3p$,$3d$ Rydberg orbitals to the AMOs, increases the size of the active space to (5332), that is, 13 orbitals. The excited states that can now be included in the calculation are the $\pi \rightarrow \pi^*$ states and two sets of Rydberg excitations arising from excitations out of the occupied π-orbitals of a_1 and b_1 symmetry, respectively. The two sets of Rydberg states can be expected to be rather close in energy, due to the relatively weak interaction between the double bonds.

To perform CASSCF/CASPT2 calculations with four active electrons in 13 AMOs for a system with C_{2v} symmetry is quite possible, although somewhat expensive. However, as it turned out in the actual application [22], it became necessary to extend the active space in order to avoid intruder states of the $\pi \rightarrow \sigma^*$ type. The AMO space then becomes too large. One solution is to use different sets of AMOs for the different excitation types. The procedure is perhaps best illustrated by an explicit example: Consider the Rydberg orbitals of b_1 symmetry ($3p_x$ and $3d_{xz}$). An active space consisting of them plus the valence orbitals yields a CAS CI space, which includes the Rydberg states $\pi(b_1) \rightarrow 3p,3d$ (A_1) and $\pi(a_1) \rightarrow 3p,3d$ (B_1). In practice the active space used was (1411), due to the occurrence of an intruder state. States of other symmetries and corresponding to other excitation types can be treated in a similar way. For norbornadiene, four different active spaces were used for the different symmetries and excitation types (see [22] for details and also below). This more complicated solution should be avoided unless it is necessary, that is, unless the size of the active space is too large to be handled without partitioning. It is important to emphasize the non-trivial nature of the selection of the active space. It requires some experience and some prior knowledge about the electronic structure of the molecule to be studied. It may also be necessary to iterate the procedure, if the CASPT2 program complains about large coefficients in the first-order wave function.

Having selected the active space, the CASSCF calculations can now be performed. It remains to be decided how many states should be included for each symmetry. The minimum number is determined on physical grounds. For example, with the active space (1411) in norbornadiene we aim at computing two Rydberg states of A_1 and two of B_1 symmetry. These two calculations are best done as state averaged calculations. Then, only two CASSCF calculations have to be performed instead of four. The state averaging ensures that the wave functions are orthogonal to each other and non-interacting. CASSCF calculations on individual excited states also often have convergence problems, since several states of the same symmetry may occur close in energy. Several tests have been made, which shows that the approximation introduced by the state averaging is very good and has only minor effects on the computed energies. The fact that the orbitals are not completely optimal for each state is also to a high degree corrected for by the CASPT2 calculation, since the first-order CI space includes singly excited configurations.

Once the CASSCF calculation has been performed for a given set of states it is important to inspect the result to ensure that the desired states are included. This is often not the case when the number of wanted wave functions equals the number of states included in the state averaging. Instead new unexpected states may appear. It is then necessary to extend the number of roots, such that all states of interest are included among them. The new roots may be higher Rydberg states, doubly excited states, etc. Some valence excited states appear very high up in the CASSCF spectrum, since their dynamic correlation energy is large. It should be emphasized that it is only when a good qualitative description of the wave functions are obtained at the CASSCF level of approximation that one can expect the second-order treatment of the remaining correlation error to be effective.

2.3 Multiconfigurational second-order perturbation theory

The CASPT2 method will not be described in detail here. The reader is instead referred to the original papers [10, 11] and a recent review article [12]. The method computes the first-order wave function and the second-order energy in the full space of the configurations generated by the MO basis. The zeroth-order Hamiltonian is constructed from a Fock type one-electron operator that reduces to the Møller-Plesset form in the closed-shell case. The contributions from inactive electron pairs to the correlation energy is thus computed at such a level of approximation. Two formulations of the zeroth-order Hamiltonian will be used in the applications to be discussed below. The first is based on only the diagonal elements of the Fock matrix. A direct diagonalization of the first-order equations is then possible. The method is labeled PT2D. It has the drawback that it is not invariant to orbital rotations, which can be a serious problem, especially in calculation of potential curves. An invariant method is obtained when also the non-diagonal elements of the Fock matrix are included in the zeroth-order Hamiltonian. The first-order equation then has to be solved using iterative methods, which is somewhat more time-consuming. The reward is, however, a more consistent solution. This approach will be labelled PT2F. It has in a number of applications been shown to be more accurate than PT2D and is the recommended method. If not otherwise stated, this is the method used in all applications.

The CASPT2 equations are formulated exclusively in terms of one-, two-, and three-body density matrices and are therefore independent of the size of the CASSCF reference function. The limiting factor is not the number of correlated electrons, but the number of active orbitals, which determines the size of the density matrices. In practice, not more than 12 orbitals can be used in the case of no symmetry partitioning. The number may be extended to 14–15 for highly symmetric molecules.

A good measure of the reliability of the CASPT2 result is given by the weight, ω, of the CASSCF reference function in the (normalized) zero plus first-order wave function. This weight will depend on the number of correlated electron pairs and an approximate estimate is obtained from the formula:

$$\omega = (1 + \alpha)^{-N/2}$$

where N is the number of electrons and α is a constant in the range 0.01–0.02. This equation reflects the fact that CASPT2 is a size-extensive approach. Consequently ω will go to zero when the number of electrons rises towards infinity. If ω comes out considerably smaller than the above estimate there is reason to check the calculation in more detail. Second-order perturbation theory is only a valid approximation in cases where the perturbation is small. When the coefficient of one or more terms in the first-order wave function becomes large (in practice larger than 0.05–0.10 in magnitude) the reason is most likely the occurrence of small denominators in the perturbation equations. Configurations with zeroth-order energies close to the reference energy give rise to large coefficients in the first-order wave function. We shall call such terms *intruder states*, even if the name is then used somewhat inappropriately. When intruder states appear in the first-order wave function, the second-order treatment is no longer valid. A calculation where this happens has to be reconsidered. The best cure of the problem is to extend the active space, such that the intruder state is included within the CAS CI space. It is then non-interacting with the reference function and will not appear in the first-order wave function at all. However, it is not always possible to extend the active space further. The CASPT2 treatment is then not possible and alternative methods have to be considered. In special cases it may be possible to simply disregard the intruder state. Such cases are characterized by a very small interaction between the intruder state and the reference function. In spite of a small denominator, the contribution to the second-order energy remains small. This often happens when higher Rydberg states occur nearly degenerate (in the zeroth-order Hamiltonian) to a given excited state. This is not a too uncommon situation in calculations using extended basis sets. If the Rydberg orbital can be localized, the best solution is to simply delete it from the MO basis. Alternatively, it may be included in the active space. Both approaches have been used with very similar results.

2.4 Transition moments and oscillator strengths — the CASSI approach

In order to understand the electronic spectrum of a molecule, one needs to compute not only the excitation energies, but also the transition moments and the corresponding intensities. No theory has been developed to perform such calculations at the same level of accuracy as is used for the energies. However, experience has shown that the transition densities are not very sensitive to the dynamic correlation effects, as long as the large terms in the full CI expansion of the wave function are included within the reference space — the CASSCF function. To compute the transition densities at this level of theory is therefore an acceptable approximation. However, even this is not a trivial task when the molecular orbitals for the two CASSCF wave functions involved in the transition are not orthonormal, which happens when the two wave functions have been individually optimized.

A method has been developed, which performs such a calculation by means of a transformation of the orbital bases to a bi-orthonormal form with a subsequent transformation of the CASSCF wave functions [39, 40]. The method works efficiently also with very long and complex CAS CI spaces. It has the limitation that calculations for all involved states have to be performed with the same number of inactive and active orbitals (per symmetry). The name CASSI (CAS state interaction) has been given to the method, since it can be used to compute the first- and second-order transition density matrices, and consequently also the Hamiltonian matrix in a basis consisting of CASSCF wave functions. This technique is useful when a set of independently optimized CASSCF solutions have been obtained. The CASSI method can then be used to check their orthogonality and their interaction. The interaction matrix can be diagonalized, yielding a new set of solutions which are mutually orthogonal and non-interacting. This feature of the CASSI method has, for example, been used to study charge transfer in aqueous Fe^{2+}–Fe^{3+} complexes [41]. The original CASSCF wave functions were the localized solutions and CASSI was used to mix them and obtain symmetry adapted wave functions and potential curves from which the charge transfer rate could be computed.

Here, the CASSI method will be used to compute the transition densities and the corresponding oscillator strengths. The latter needs also information about the energy separations. These will be taken from the CASPT2 results, since they are often strongly affected by dynamic correlation effects (in contrast to the transition densities). This mixed approach has in a number of applications proven to give results with good accuracy.

3 Applications

3.1 Introduction

In this chapter a number of applications of the CASSCF/CASPT2 approach in spectroscopy will be discussed. The emphasis will be on the analysis of the different spectra and the properties of the excited states more than on the technical details of the calculations. They have all been performed using the method that was discussed in the preceding chapter. Some of the early applications have used Rydberg basis functions placed on the atomic centra. For butadiene, the calculations have later been repeated with the recommended set of Rydberg functions placed in the charge centroid on the cation, resulting in improved energies and assignments. It should be clearly stated that many of these studies have been non-trivial to carry through. Often the process has had to be reiterated with an extended active space and/or an increased number of roots in the CASSCF treatment. The final calculation has all desired states correctly described at the CASSCF level and does not give any intruder states in the CASPT2 wave function. Some of the calculations are also time-consuming, since they have been performed with extended basis sets and the number of active orbitals close to the practical limit. However, it should be mentioned that almost all calculations have been made on workstations of the type IBM RS/6000.

Most studies will treat both valence excited states and Rydberg states. However, for some of the larger molecules it has been necessary to delete the Rydberg orbitals from the MO basis set in order to be able to locate the most important valence states, which at the CASSCF level of theory appear high up in the spectrum. In these cases the less interesting Rydberg spectrum has not been studied. The valence excited states can often be characterized as either covalent or ionic. The wave function of the former states is dominated by covalent valence bond structures. The dynamic correlation energy is usually of the same order as that of the ground state. However, multiply excited configurations often appear in these states with large coefficients, which makes the electronic structure complex. Typical examples are the 1^1E_{2g} state in benzene and the 2^1A_g state in butadiene. Other states are dominated by ionic valence bond structures. They are usually well described by one or a few singly excited configurations. Typical examples are the V-state in ethene, the 1^1E_{1u} state in benzene, and the 1^1B_u state in butadiene. Dynamic correlation effects are large in this type of valence excited states, and are dominated by the dynamic polarization effects in the σ framework. These states therefore appear in the CASSCF state manifold with much too high energies, which sometimes makes it difficult to locate them. It may also lead to an erroneous mixing with Rydberg states, which happen to be close in energy. Such situations will be described below for ethene and butadiene.

The first set of applications to be discussed are the linear polyenes. This set of calculations comprises the molecules ethene to octatetraene. The emphasis has been on the location and assignment of the first two valence excited states, an issue that has been extensively discussed in the literature, both experimentally and theoretically. For octatetraene the calculations also include an optimization of excited state geometries, and from there a determination of the emission spectrum. These studies have led to a consistent and accurate theoretical interpretation of the electronic spectrum of the shorter polyenes, which may be used as a basis in coming experimental and theoretical discussions of the spectroscopies of this important class of compounds.

The next set of applications are concerned with molecules with two interacting double bonds. The molecules included are the cyclic pentadienes and norbornadiene. The spectroscopies of all these molecules are closely related to each other and can be discussed within a common framework. There are also great similarities to the spectra of the linear polyenes, in particular with respect to the lowest valence excited states. Again, the theoretical study is able to give a detailed description of the different spectra. A number of new assignments have been made. It may be stated that the spectroscopy for the pentadienes has now (for the first time) been fully described and understood.

A number of aromatic systems have been studied. Here the results obtained for the molecules benzene, the azabenzenes, phenol, and phosphabenzene will be discussed. These studies have resolved a number of problems discussed in the experimental literature. Among them is the location of the $n \rightarrow \pi^*$ in relation to the $\pi \rightarrow \pi^*$ states in the heterocyclic ring systems.

The study of the four lowest transitions in the porphin molecule has a somewhat different character than the other investigations reported here. The molecule is large and these preliminary calculations have therefore been performed with a DZ ANO basis set without 3*d*-type functions. This will, of course, make the results somewhat uncertain. Also, porphin has 26 π electrons in 24 orbitals, an impossibly large active space for a CASSCF/CASPT2 calculation. The system thus offers a possibility to test an alternative way of selecting the active space, which is based on a preceding singles and doubles CI calculation. The resulting natural orbital occupation numbers are used to systematically increase the active space until convergence in the CASPT2 energies is obtained. A maximum of eleven orbitals was needed to obtain converged excitation energies for the bands in the spectrum below 3.5 eV (four bands). This method has been used also in other applications where an *a priori* selection of the AMOs was found difficult.

In one section we discuss the so-called TICT (twisted intramolecular charge transfer) problem. This application has a different character from the others, since it is concerned with the dynamics of the excited states. The study involves the molecules aminobenzonitrile (ABN) and dimethylaminobenzonitrile (DMABN). The latter molecule exhibits a dual fluorescence phenomenon, which is absent in ABN. The study investigates this phenomenon by computing the properties of the excited states as a function of the rotation and inversion of the amino group. The TICT process is fully supported by the theoretical study.

Finally we discuss the electronic spectra of the nucleotide base monomers: uracil, thymine, cytosine, adenine, and guanine. These calculations only involve the 4–6 lowest valence excited $\pi \rightarrow \pi^*$ states and the two lowest $n \rightarrow \pi^*$ states. Nevertheless, it has been possible to use the results to explain the major features in the spectra of these molecules. A number of experimental problems have been resolved and there is agreement between computed and experimental transition energies and moments. Of particular interest is the direction of the transition moments. The experimental determination of this property is ambiguous, since two different directions are possible from the same set of experimental data. The theoretical results are accurate enough to allow a conclusive selection of the correct value.

The CASSCF/CASPT2 approach has until now been applied to about 50 different molecular systems. The results are consistently of high accuracy both for vertical transition energies and 0-0 or vertical emission bands. The largest errors have occurred for the V state in ethene (0.4 eV) and the corresponding state in *trans*-butadiene (0.3 eV). All other results are of higher accuracy, at least in cases where an unbiased comparison to experiments in the gas phase can be made. It is, of course, possible to use the present approach also to study solvent shifts on the transition energies using any of the reaction field Hamiltonians available in the literature. Such calculations are planned for the base monomers, where all experimental data come from solution or crystal spectra.

3.2 The electronic spectra of linear polyenes

3.2.1 Introduction

The spectroscopy of linear polyenes has been intensively studied by many scientists since the original measurement of their absorption spectra by Hauser and Kuhn [42]. They discovered that the lower-energy absorption band decreased its energy with the addition of more −C=C− units to the polyene chain. These experiments confirmed that light is absorbed by the molecule as a whole and formed the basis for many theories about the conjugation across double bonds [43]. The importance of the systems is related to the role of their photoisomerization in the process of vision and the biological production of energy [44] and, since the discovery of the high conductivity of doped polyacetylenes [45], to the properties of the conducting polymers [46]-[48]. Key problems in the spectroscopy of polyenes have been the ground state structure of the different isomers and the order and nature of the two low-lying valence singlet excited states. Since Hudson and Kohler [49, 50] presented spectroscopic evidence that a low-lying ^1A_g state was the lowest singlet excited state in diphenyloctatetraene, much of the discussions have focused on problems related to the nature of the lowest excited singlet state. Transition to this state is one-electron forbidden and the excitation energy is difficult to establish from experiment, masked as it is by the strong optically allowed 1^1B_u band. The evidence of emission from the 2^1A_g state in octatetraene [51, 52] and longer polyenes enhanced the need for a covering theoretical study. Extensive reviews of the spectroscopy of polyenes during the last three decades can be found elsewhere [53]-[55].

The understanding of the excited states of the short polyenes has remained a challenge for quantum chemistry since the first theoretical studies were performed [56, 57]. Twenty years ago [50, 58], the 2^1A_g state was shown to require a theoretical description including at least doubly excited configurations. The state has a two-photon allowed character (the failure of simpler molecular orbital models, based on single excitation CI, to describe multiconfigurational states is still illustrated in the literature [7]). In the present paper we are going to focus our interest on the vertical valence and Rydberg low-lying excited states of the shortest linear polyenes from ethene to octatetraene in different isomers. In the latter molecule, a study of also the adiabatic excited states aids in understanding the emission spectrum of larger polyenes and the properties of the peculiar 2^1A_g state. A more comprehensive description of the results can be found in the original papers [24]-[28].

The π spectrum of linear polyenes involve from two to eight π electrons and valence π orbitals in going from ethene to octatetraene. A large number of valence $\pi \rightarrow \pi^*$ singly excited states can arise from this structure. Using butadiene as a model, the ground state π electronic configuration $(1a_u)^2(1b_g)^2(2a_u)^0(2b_g)^0$ leads to four singly excited singlet (and triplet) valence states. The HOMO $\rightarrow$ LUMO (1^1B_u in the *trans-* and 1^1B_2 in the *cis*-isomer) transition is dipole allowed and the most intense in the spectrum. The next valence state of the same symmetry, HOMO−1 $\rightarrow$ LUMO+1, is high in energy. The two quasi-degenerate excitations HOMO−1 $\rightarrow$ LUMO and HOMO $\rightarrow$ LUMO+1 mix to give the $^1A_g^-$ and $^1A_g^+$ states [59]. Transitions to the A_g states are dipole forbidden in the *trans* isomer, but not in the *cis* isomer, where the oscillator strengths can be compared. The minus state has the lower energy and a low intensity, while the plus state is pushed up in energy and carries most of the intensity. This simple scheme is complicated by the presence of higher excitations. In the lowest energy part of the spectrum, the doubly excited configuration (HOMO,HOMO) $\rightarrow$ (LUMO,LUMO) mixes into the $2^1A_g^-$ state. The transition will be two-photon allowed but with low intensity, and is difficult to locate under the strong band of the 1^1B_u state. Two low-lying valence singlet and triplet states are therefore expected from butadiene to octatetraene. A number of higher valence states are possible in the larger molecules. The following study also includes the Rydberg series HOMO $\rightarrow 3s, 3p, 3d$.

3.2.2 The ethene molecule

Although ethene is not strictly a polyene, it is the building block of the longer chain systems. Its spectroscopy is complicated by the strong coupling of the π-electrons to the σ-electron framework upon excitation and also by the fact that the weakening of the double bond makes the excited state nonrigid. One singly excited singlet and one triplet valence state is obtained from the $b_{3u} \rightarrow b_{2g}$ ($\pi \rightarrow \pi^*$) excitation. The electronic spectrum of the molecule has been extensively studied. It is dominated by the intense and broad band corresponding to the $\pi \rightarrow \pi^*$ valence excitation to the 1^1B_{1u} (V) state. The vertical excitation energy has been estimated to be about 8.0 eV. The equilibrium conformation of both the singlet V and the triplet 1^3B_{1u} (T) state have 90° twisted structures. The CASPT2 result for the vertical transition to the 1^1B_{1u} state, 8.40 eV, is 0.4 eV above the experimentally estimated vertical value. A recent extensive MRCI study [60] gave the value 7.94 eV, which should be rather accurate. The deviation is related to the diffuse character of the CASSCF wave function for this state ($\langle z^2 \rangle$ 44.1 au^2 instead of 16.8 au^2 in the MRCI study [60]), and is the largest deviation found for any of the excitation energies reported in the present review. In order to monitor the importance of using more contracted orbitals in the reference function, a study of the excitation energies and $\langle r^2 \rangle$ expectation values as a function of an external perturbation has been performed [24]. The perturbation is chosen such that the Rydberg states are pushed up in energy and therefore interact less with the valence excited state. The conclusion from this study was that the error in the excitation energy of the 1^1B_{1u} state of ethene is not an artifact of the CASPT2 approach, but is due to the erroneous treatment of the valence-Rydberg mixing in the CASSCF reference function. This situation does not arise in the calculation of the 1^3B_{1u} state where the computed energy is 4.39 eV, in agreement with the experimental value, 4.36 eV [61, 62]. The $\langle z^2 \rangle$ value for the triplet state is close to the ground state value, illustrating the absence of any Rydberg mixing in this case. Table III presents computed and experimental excitation energies and oscillator strengths for the excited states in ethene. The states considered were, in addition to the ground state, the T and V states plus the singlet and triplet Rydberg states of $3s$, $3p$, and $3d$ character. The excitation energies and the singlet-triplet splittings are in agreement with the experimental values, except for the V state. A more detailed discussion of the results can be found in the original paper [24]. The results are in general agreement with earlier studies, except for the 1^3B_{1u} state, which was

TABLE III

Calculated and experimental excitation energies and oscillator strengths in ethene.

	Excitation energy (eV)					Oscillator strength
			Other results			
State	PT2F	Exp	CIS[a]	MRCI[b]	CI1[c]	
Singlet states						
1^1B_{1u} (Valence)	8.40	≈8.0[d]	7.78	7.96	7.96	0.160
1^1B_{3u} ($1b_{3u} \rightarrow 3s$)	7.17	7.11[e–h]	7.10	7.13	7.26	0.067
1^1B_{1g} ($1b_{3u} \rightarrow 3p\sigma$)	7.85	7.80[i,j]	7.68	7.86	7.93	forbidden
1^1B_{2g} ($1b_{3u} \rightarrow 3p\sigma$)	7.95	7.90[j]	7.83	7.89	8.01	forbidden
2^1A_g ($1b_{3u} \rightarrow 3p\pi$)	8.40	8.28[j]	8.10	8.21	8.36	forbidden
2^1B_{3u} ($1b_{3u} \rightarrow 3d\sigma$)	8.66	8.62[e–g,k]	8.71	8.73	8.80	0.0009
1^1A_u ($1b_{3u} \rightarrow 3d\pi$)	8.94		8.83	9.04		forbidden
3^1B_{3u} ($1b_{3u} \rightarrow 3d\delta$)	9.03	8.90[e–g,l]	8.92	9.31		0.0006
1^1B_{2u} ($1b_{3u} \rightarrow 3d\delta$)	9.18	9.05[e,k]	8.88	8.99		0.077
2^1B_{1u} ($1b_{3u} \rightarrow 3d\pi$)	9.31	9.33[m]			8.99	0.078
Triplet states						
1^3B_{1u} (Valence)	4.39	4.36[m]	3.54	4.35		
1^3B_{3u} ($1b_{3u} \rightarrow 3s$)	7.05	6.98[f]	6.88			
1^3B_{1g} ($1b_{3u} \rightarrow 3p\sigma$)	7.80	7.79[f]	7.60			
1^3B_{2g} ($1b_{3u} \rightarrow 3p\sigma$)	7.90		7.72			
2^3A_g ($1b_{3u} \rightarrow 3p\pi$)	8.26	8.15[f]	7.75			
2^3B_{3u} ($1b_{3u} \rightarrow 3d\sigma$)	8.57	8.57[f]	8.63			
1^3A_u ($1b_{3u} \rightarrow 3d\pi$)	8.94					
3^3B_{3u} ($1b_{3u} \rightarrow 3d\delta$)	8.97					
2^3B_{1u} ($1b_{3u} \rightarrow 3d\pi$)	9.07					
1^3B_{2u} ($1b_{3u} \rightarrow 3d\delta$)	9.09					

[a]CIS results from ref. [6]
[b]MRCI results from ref. [261]
[c]CI1 results from ref. [262]
[d]Estimated vertical excitation energy from earlier theoretical work [60, 261, 262]
[e]Ref.[263]. [f]Ref. [264]. [g]Ref. [265]. [h]Ref. [266]. [i]Ref. [267]. [j]Ref. [268]. [k]Ref. [269]. [l]Refs. [270]. [m]Ref. [61, 62].

obtained 0.8 eV too low by CIS [6].

The experimental ground state geometry was used in the calculations. The ANO-L basis set, contracted to $4s3p2d$ for carbon and $3s2p$ for hydrogen was used. To this set was added 2 s-type, 2 p–type, and 1 d-type primitive Gaussian functions with smaller exponents centered on the carbon atoms [24].

3.2.3 Trans- *and* cis-*1,3-butadiene*

The electronic structure of the excited states of butadiene has been discussed extensively in the literature both experimentally and theoretically [24],[53]-[55]. Much of the theoretical work is concerned with the geometry of the excited states and the mechanistic features of the photochemistry [63], but the present interest relates only to the vertical spectrum of the molecule. Two occupied π ($1a_u$ and $1b_g$) and two unoccupied π^* ($2a_u$ and $2b_g$) orbitals constitute the valence π-orbitals of the *trans* isomer. Three low-lying valence states appear up to 8.0 eV, one 1^1B_u and two ^1A_g states. Only the transition to the former state is optically allowed. The absorption spectrum of butadiene has as a major feature a band starting at 5.73 eV with a maximum at 5.92 eV [55]. The high energy tail of the band contains contributions from the $1b_g \rightarrow 3s3p$ Rydberg series. The allowed 1^1B_u state has a somewhat less diffuse character than that of the equivalent state in ethene, and consequently a smaller error is expected in the computed energy. Table IV presents the experimental and theoretical values for the excitation energies and oscillator strengths in *trans* and *cis*-1,3-butadiene. The results presented for the *trans* form are new [64]. They have been obtained using the prescription for the Rydberg orbitals described in the methods section above. The earlier study [24] was made with a more primitive choice of Rydberg basis functions. With the new basis set, more Rydberg states could be localized and it was also possible to locate the second valence excited state (3^1A_g). This state has, however, not been computed for the *cis* conformer, since it lies too high in the CASSCF spectrum. The Rydberg basis set was in this case slightly different. A experimental ground state geometry was used in the calculations on the *trans* isomer. A pointwise CASPT2 optimization of the geometry of the *cis* isomer was carried out. The ANO-L basis set was used contracted to $4s3p1d$ functions for the carbon and $2s1p$ functions for hydrogen.

All methods used to compute the excitation energies in butadiene yield the largest error for the 1^1B_u valence state. The error in the CASPT2 value, 6.24 eV, is 0.32 eV. Only the valence shell effective Hamiltonian (VSEH) study by Graham and Freed [65] obtains a lower result, 6.14 eV, using third-order perturbations. The corresponding state in the *cis* isomer is computed at 5.58 eV with an estimated experimental value of 5.49 eV [66]. Here, the interaction with the Rydberg state is smaller and consequently the CASPT2 can better handle the dynamical correlation contribution. The calculations on this state in butadiene together with the V state in ethene illustrates one of the principal limitations of the CASPT2 approach: when the reference state is not well represented by the CASSCF wave function the second-order approach will not be able to handle the remaining correlation effects in an efficient way. Higher-order terms will have to be included, which have a modifying effect on the reference function. Nevertheless, the resulting error is not very large, the largest being the 0.4 eV deviation obtained for the V state in ethene. The error in the reference function is due to an erroneous interaction with a nearby Rydberg state. The interacting matrix element is small and the interaction therefore becomes efficient only in a very small energy range. The problem arises, since the valence and Rydberg states have different errors at the CASSCF level of theory and therefore are misplaced with respect to each other. Accidentally they may become close in energy and a strong interaction results, which gives the valence state a too diffuse character and, correspondingly, a too compact Rydberg state. The CASPT2 will give a correlation energy that is too small for the valence state and too large for the Rydberg state. This is clearly illustrated by the energy of the ^1B_u states in *trans* butadiene (cf. Table IV).

TABLE IV

Calculated and experimental excitation energies (eV) and oscillator strengths in *trans* and *cis*-1,3-butadiene[a].

State	Excitation energy, *trans*						Osc. str.	Energy	Osc. str.
			Other results						
	PT2F	Exp	CIS[b]	SAC-CI[c]	MRCI[d]	VSEH[e]	*trans*	*cis* PT2F	*cis*
Singlet states									
1^1B_u (Valence)	6.24	5.92[f–h]	6.21	6.39	6.48	6.14	0.661	5.58[i]	0.218
2^1A_g (Valence)	6.24	?	?	7.00	6.53	6.19	forb.	6.04	0.008
1^1B_g ($1b_g \rightarrow 3s$)	6.21	6.21[g,h]	6.11				forb.	6.05	forb.
1^1A_u ($1b_g \rightarrow 3p\sigma$)	6.52	-	6.45				0.003	6.48	forb.
2^1B_u ($1b_g \rightarrow 3p\pi$)	6.57	6.66[f,h]	6.99	7.05		7.00	0.109	6.86	0.087
2^1A_u ($1b_g \rightarrow 3p\sigma$)	6.66	6.80[f,h]	6.61				0.039	6.67	0.040
3^1A_g (Valence)	6.97	7.0-7.3[f,g]					forb.		
4^1A_g ($1b_g \rightarrow 3d\pi$)	7.28	7.28[f]	7.19	7.33			forb.	7.43	0.001
2^1B_g ($1b_g \rightarrow 3d\delta$)	7.39	7.36[j]	7.22				forb.	7.32	0.0001
3^1B_g ($1b_g \rightarrow 3d\delta$)	7.42	7.36[j]	7.25				forb.	7.34	forb.
5^1A_g ($1b_g \rightarrow 3d\pi$)	7.48	7.48[g,h]	7.45			7.49	forb.	7.43	0.003
4^1B_g ($1b_g \rightarrow 3d\sigma$)	7.51	7.48[g,h]	7.39				forb.	7.44	forb.
Triplet states									
1^3B_u (Valence)	3.15	3.22[f]	2.62	3.48	3.18	3.23		2.81	
1^3A_g (Valence)	4.80	4.91[f]	4.33	5.15	4.83	4.90		4.74	
1^3B_g ($1b_g \rightarrow 3s$)	6.17	6.1[j]	6.00					5.99	
1^3A_u ($1b_g \rightarrow 3p\sigma$)	6.50							6.64	

[a]The labels for the *cis* molecule should be changed into the C_{2v} symmetry [28].
[b]CIS results from ref. [7].
[c]SAC-CI results from ref. [271].
[d]MRCI results from ref. [272].
[e]VSEH results from ref. [65].
[f]Refs. [273] – [275]. [g]Refs. [276, 277].
[i]The experimental transition is estimated at 5.49 eV [66].
[j]Ref. [7].

One of the major issues concerning the electronic spectrum of the polyenes is the relative location of the two lowest valence states. In butadiene, the $2^1A_g \leftarrow X$ band origin was recently estimated by resonance Raman scattering experiments to lie 0.2–0.3 eV below the $1^1B_u \leftarrow X$ band origin [67]-[69]. The 2^1A_g state is described by a multiconfigurational wave function, mainly composed of the doubly excited configuration $(1b_g)^2 \rightarrow (2a_u)^2$ and the one-electron promotions $1b_g \rightarrow 2b_g$ and $1a_u \rightarrow 2a_u$. The vertical excitation has not been identified from experiment, and the theoretical results place it slightly above the 1^1B_u vertical state. Our computed CASPT2 result, 6.24 eV, is in agreement with the VSEH value 6.19 eV [65]. The 0-0 transition energies have not been computed for butadiene, but a corresponding study on octatetraene shows that the difference for this and the vertical energy is much larger for the 2^1A_g state (see below). This state offers a striking example of the inability of simple theoretical models based on single excitations to deal with important excited states. A recently published study of butadiene using the CIS method [7] did not find any evidence of a valence ^1A_g excited state up to 9.13 eV, 3 eV above experimental and other theoretical evidence. Also, a singlet-triplet splitting of 4.8 eV was obtained, while the experimental value is 2.7 eV.

Also the third valence excited singlet state, 3^1A_g, has been located in the spectrum of the *trans*-isomer. The major terms in the wave function are the plus combination of the HOMO−1→LUMO and HOMO→LUMO+1 configurations with some contributions from doubly excited configurations. Transition to this state is forbidden in *trans*-butadiene, while the corresponding state in the pentadienes is strongly allowed. The character of the same state in *cis*-butadiene is therefore an interesting question, which we shall try to resolve in a coming study. The computed excitation energy for 3^1A_g is 6.97 eV. Doering and McDiarmid reported in 1980 a feature found between 7.0 and 7.3 eV in their electron impact spectrum to be valence in character and assigned it to the 2^1A_g state [70]. When this state was later identified at considerably lower energy the assignment of the band around 7 eV was changed to a 3*p* Rydberg state [71]. With the new theoretical evidence at hand, the original assignment appears to be correct, except that it is the third state of ^1A_g symmetry, not the second. This reassignment will affect the assignments made for the 3*p* Rydberg states slightly.

Table IV shows that the computed Rydberg and triplet states of the molecule are in agreement with experiment and also with most previous theoretical results. In our earlier study [24] the 2^1B_u state was found 0.3 eV below the experimental value (assigned to the feature at 7 eV as described above). With the new assignment, the error has decreased. It should, however, be noted that most other theoretical and experimental work do assign this state to the band at 7 eV [65, 71]. The results for the $1b_g \rightarrow 3d$ Rydberg series shown in Table IV are presented here for the first time.

Very little is known about the spectrum of *cis*-butadiene. The energies predicted here can be expected to be of an accuracy even better than that obtained for the *trans* isomer, since the Rydberg mixing into the valence states is smaller in this case. They thus constitute a well founded prediction of the spectrum. The third valence excited state of ^1A_g symmetry remains to be located, however.

TABLE V

Calculated and experimental excitation energies (eV) and oscillator strengths in *trans* and *cis*-1,3,5-hexatriene[a].

	Excitation energy, *trans*				Osc.str. *trans*		Energy *cis*	Osc. str. *cis*
			Other results					
State	PT2F	Exp	GVB[b]	QCI[c]	TW	QCI[c]	PT2F	*cis*
Singlet states								
1^1B_u (Valence)	5.05	4.95[d]	6.56	5.14	0.7894	1.24	5.00	0.617
2^1A_g (Valence)	5.16	5.21[e]	5.87	5.74	forbidden		5.04	0.0008
1^1A_u ($2a_u \rightarrow 3s$)	5.64	5.67[f]	5.97		0.0015		5.69	0.0010
1^1B_g ($2a_u \rightarrow 3p\sigma$)	5.76	5.85[f]	6.00	5.74	forbidden		5.91	0.0152
2^1B_g ($2a_u \rightarrow 3p\sigma$)	6.05	6.06[d,f]	6.20	5.99	forbidden		6.11	forb.
3^1A_g ($2a_u \rightarrow 3p\pi$)	6.09	6.2[d,f,g]	6.26	6.51	forbidden		6.14	0.0107
3^1B_u ($2a_u \rightarrow 3d\pi$)	6.11	6.2[d,f,g]	6.27	6.10	0.1706	0.035	6.29	0.0703
2^1B_u (Valence)	6.24	6.2[d,f,g]			0.1114		5.70	0.017
4^1B_u ($2a_u \rightarrow 3d\pi$)	6.40	6.40[d,f]	6.68		0.1917		6.75	0.0001
2^1A_u ($2a_u \rightarrow 3d\delta$)	6.53	6.53[f]	6.81		0.0015		6.64	forb.
3^1A_u ($2a_u \rightarrow 3d\sigma$)	6.61	6.53[f]	6.72		0.0218		6.82	0.0025
4^1A_u ($2a_u \rightarrow 3d\delta$)	6.77	6.74[d,f]	7.07		0.0022		6.70	forb.
Triplet states								
1^3B_u (Valence)	2.54	2.61[d]	2.71	2.84			2.57	
1^3A_g (Valence)	4.10	4.11[d]	4.32				3.94	
1^3A_u ($2a_u \rightarrow 3s$)	5.63						5.67	
1^3B_g ($2a_u \rightarrow 3p\sigma$)	5.77						6.06	

[a]Present results for the *trans* molecule.
The labels for the *cis* molecule should be changed into the C_{2v} symmetry [28].
[b]GVB-CI results from ref. [279, 280].
[c]QCI-CI6 results from ref. [281, 282].
[d]Ref. [77].
[e]Two-photon spectrum from ref. [81].
[f]Refs. [283, 284]. [g]Refs. [79, 285].

3.2.4 Trans- *and* cis-*1,3,5-hexatriene*

The hexatriene molecule has received less attention than the other polyenes, although recent studies have been devoted to describing the spectroscopy of its low-lying states [72]-[76]. Table V shows the computed and experimental vertical excitation energies and oscillator strengths for the two stable isomers. The results obtained for the *trans* isomer will be discussed in more detail: The most intense band of the spectrum matches with the 1^1B_u valence state, which is computed at 5.05 eV, near the experimental value of 4.95 eV [77]. The diffuse character of the state decreases on increasing the size of the molecule. The equivalent state of the *cis* conformer, 1^1B_2, is computed at 5.00 eV. The experimental investigations [77]-[80] report the state at 4.92 eV, only slightly below the *trans* state, which is in agreement with the theoretical prediction. On the other hand, recent accurate resonance Raman [76] and multiphoton ionization [73, 74] studies have located the origin of the band below the 1^1B_2 state. A two-photon absorption spectrum [81] shows a maximum at 5.21 eV for the *trans* isomer, which perfectly matches with the vertical excitation to the two-photon allowed 2^1A_g state, computed at the CASPT2 level at 5.16 eV. Table V also presents a new valence state at 6.24 eV and gives assignments for the Rydberg $3s3p3d$ series.

Experimental ground state geometries have been used in the calculations for both isomers. The calculations were performed with the ANO-L basis set contracted $4s2p1d$ for carbon and $2s1p$ for hydrogen. To this set was added $2s2p2d$ primitive Gaussian functions with smaller exponents [25].

3.2.5 All-trans-*1,3,5,7-octatetraene*

Octatetraene is the shortest polyene that exhibits emission and for which the transition to the 2^1A_g state can be clearly seen in one- and two-photon absorption spectra [54, 55]. The most intense peak of the absorption spectrum appears at 4.41 eV [80],[82]-[84] and it shows a perfect mirror emission in the gas phase [82]. The lowest singlet 0-0 transition has been, however, identified at 3.59 eV as corresponding to the 2^1A_g state [82],[85]-[87]. Although previous evidence suggested that the fluorescence came exclusively from the 1^1B_u state in the gas phase [84, 83], recent experiments seem to definitively also establish emission from the lowest 2^1A_g state [51, 52]. Our theoretical study of the absorption and emission spectra of octatetraene confirms the existence of dual fluorescence, which has been observed in this system.

Tables VI and VII summarize the computed and experimental data of the octatetraene spectroscopy. The study [26] starts with the CASSCF (ANO basis set) optimization of the ground and two low-lying singlet states equilibrium geometry. It was found that the experimental crystal X-ray geometry [88] underestimated the double bonds by around 0.02 Å . An important increase of the double bonds and a corresponding decrease of the single CC bonds appears in the 2^1A_g state, in agreement with an experimental suggestion [87]. A complete description of the optimization procedures and geometries is presented in the original paper [26]. The lowest computed vertical singlet state is the 2^1A_g state at 4.38 eV, which is symmetry forbidden. The most intense 1^1B_u valence state appears slightly above, at 4.42 eV, perfectly matching the experimental data. No value for the vertical

TABLE VI

Calculated and experimental vertical excitation energies (eV) and oscillator strengths in *All-trans*-1,3,5,7-octatetraene.

State	Excitation energy					Osc.str.
			Other results			
	PT2F	Exptl	MC[a]	QCI[b]	EH[c]	
Singlet states						
2^1A_g (Valence)	4.38		5.64	5.21	4.88	forb.
1^1B_u (Valence)	4.42	4.41[d]		4.76		1.8316
1^1B_g ($2b_g \rightarrow 3s$)	5.32					forb.
1^1A_u ($2b_g \rightarrow 3p\sigma$)	5.48					$3\ 10^{-6}$
2^1A_u ($2b_g \rightarrow 3p\sigma$)	5.68					0.0054
2^1B_u ($2b_g \rightarrow 3p\pi$)	5.70	5.69[e]		5.65		0.0213
3^1B_u (Valence)	5.83	5.88[f]				0.0100
3^1A_g ($2b_g \rightarrow 3d\pi$)	6.10	6.04[f]				forb.
4^1A_g (Valence)	6.56					forb.
5^1A_g (Valence)	7.14					forb.
4^1B_u (Valence)	8.44					0.0019
Triplet states						
1^3B_u (Valence)	2.17	2.10[g]		2.45	2.07	
1^3A_g (Valence)	3.39	3.55[g]				
1^3B_g ($2b_g \rightarrow 3s$)	5.31					
1^3A_u ($2b_g \rightarrow 3p\sigma$)	5.45					

[a]MCSCF/STO-3G results from ref. [286].
[b]QCI results form ref. [287].
[c]Effective Hamiltonian results from ref. [288].
[d]Refs. [51, 80],[82]-[84],[289].
[e]Ref. [82]. [f]Refs. [82, 84]. [g]Ref. [84].

2^1A_g state is reported, and the estimation of 3.97 eV [54] seems not to be very accurate. Further, assignments have been made for the triplet states and higher features (cf. Table VI and original paper). The 0-0 transitions to the low-lying states can be computed as the difference from the ground state to the equilibrium geometries of the excited states. Table VII shows the CASPT2 values 3.61 and 4.35 eV for the 2^1A_g and 1^1B_u states respectively, confirming the experimental results, which yield 3.59 and 4.41 eV (cf. Table VII), respectively. Furthermore, the emission maxima can be computed from the equilibrium geometry of the excited state vertically to the ground state. The computed values 2.95 and 4.14 eV match the experimental gas-phase values at ≈3.1 and 4.20 eV (cf. Table VII). To summarize, we have confirmed the presence of the 2^1A_g state of octatetraene below the optically active 1^1B_u state even in vertical absorption. The weak fluorescence observed in the emission spectrum of the molecule starting around 3.5 eV is confirmed here to originate from the 2^1A_g state. The absence of fluorescence in earlier gas-phase studies has probably been due to problems with the resolution of

TABLE VII

Calculated and experimental energies (eV) for the low-lying singlet states in *All-trans*-1,3,5,7-octatetraene.

State	Emission maxima			0-0 transition			Vertical absorption		
	TW[a]	Exp	Previous	TW[a]	Exp	Previous	TW[a]	Exp	Previous
2^1A_g	2.95	≈3.1[b]	3.23[c]	3.61	3.59[d]	3.91[c],4.15[e] 3.89[f]	4.38		4.88[c],5.21[e] 5.64[f]
1^1B_u	4.14	4.20[g]		4.35	4.41[h]	4.56[e]	4.42	4.41[h]	4.76[e]

[a]Present CASPT2F results.
[b]Gas-phase fluorescence, refs [51, 52].
[c]Effective Hamiltonian results from ref. [288].
[d]Extrapolated from solution and gas-phase values from refs. [52, 82, 86].
[e]QCI results from ref. [287].
[f]MCSCF/STO-3G results from ref. [286].
[g]Optical spectroscopy in solution, ref. [82].
[h]Optical, electron energy loss, and multiphoton spectra. Refs. [51, 80],[82]-[84],[289].

the spectrum.

The calculations were performed with the ANO-L basis set contracted $4s3p1d$ for carbon and $2s1p$ for hydrogen. To this set was added optimized $1s1p1d$-type diffuse primitive Gaussian functions [26].

3.2.6 Summary and conclusions

The theoretical study of the excited states of linear polyenes has given a deeper insight into the complex electronic structure of these systems. The obtained excitation energies and oscillator strengths are in agreement with experimental data. The dipole-allowed 1^1B_u state has been shown to be the lowest vertical state in all the studied polyenes except octatetraene, where the 2^1A_g state is below, even at the vertical absorption geometry. This suggests that in longer polyenes 2^1A_g is also the lowest excited state. The 1^1B_u state is computed to be too diffuse in character in the shorter polyenes till hexatriene due to the inability of the CASSCF method to deal with the valence-Rydberg mixing in such states. Anyway, the maximum deviation from experiment has been 0.4 eV in ethene, decreasing to 0.2 and 0.1 eV in butadiene and hexatriene, respectively. The position of the 2^1A_g states in the shorter polyenes can be related to the well-known absence of fluorescence in these systems [54, 55]. The study of the adiabatic low-lying excited singlet states in octatetraene has made it possible for the first time to fully describe the main features of the absorption and emission spectra, including the 0-0 transitions and the emission maxima. The deviation from experimental data is less than 0.1 eV and the most recent findings in the fluorescence of the octatetraene molecule have been confirmed. Further extensive studies on the adiabatic states of the shorter polyenes will be necessary in order to fully understand the photophysics of these species.

3.3 THE INTERACTING DOUBLE BONDS

The quantitative (and also qualitative) interpretation of the electronic spectra of penta-heterocyclic compounds has remained a challenge for both experiment and theory since the first calculations were performed and the first UV spectrum was measured. Cyclopentadiene is the prime example of a ring shaped molecule with a conjugated π-electron system. On the one hand, this molecule can be related to the simplest aromatic heterocycles as pyrrole, furan, and thiophene. On the other hand, the structure of cyclopentadiene resembles that of polyenes, in particular, *cis*-butadiene. In the series *cis*-butadiene, cyclopentadiene, norbornadiene, the latter is the most complicated molecule. Bicyclo[2.2.1]hepta-2,5-diene, norbornadiene, is a formally unconjugated, strained, nonplanar diene, in which both indirect conjugation and π,σ interaction must be considered. In this chapter we shall present, analyze, and compare the results obtained for cyclopentadiene (CP), pyrrole, furan [25], thiophene [27], and norbornadiene (NB) [22]. The resulting transition energies and oscillator strengths made conclusive assignments possible both for valence and Rydberg states, while they differed substantially from previous theoretical *ab initio* studies.

For the qualitative description of the transitions it is helpful to consider first the interaction of two independent ethylenic moieties. The interaction of the degenerate bonding, π_b, and the antibonding, π_a, MOs of the two ethylenic units results in the following four MOs: $\pi_1 = \pi_b + \pi_b$, $\pi_2 = \pi_b - \pi_b$, $\pi_3 = \pi_a + \pi_a$, and $\pi_4 = \pi_a - \pi_a$, leading to the orbital ordering $b_1(\pi_1)$, $a_2(\pi_2)$, $b_1(\pi_3)$, and $a_2(\pi_4)$ in the C_{2v} point group of the heterocyclic systems (and also in *cis*-butadiene). In addition, the CH_2 unit in cyclopentadiene and the heteroatom in the remaining compounds contribute with a valence-type π-orbital of b_1 symmetry of higher ionization energy. Norbornadiene, the model system for dominant through-space type interaction of unconjugated double bonds [89, 90] shows a normal orbital ordering, $a_1(\pi_1)$, $b_1(\pi_2)$, $b_2(\pi_3)$, $a_2(\pi_4)$, the same as is found in conjugated dienes, although the symmetries of the different π orbitals differ from them, because of the different orientation of the C_2 axis. From this structure we expected to find the following valence excited singlet states. The one-electron promotions $\pi_2 \rightarrow \pi_3$ and $\pi_1 \rightarrow \pi_4$ give rise to two states of B_2 symmetry in CP and related five-membered ring systems (A_2 symmetry in NB). The one-electron promotions $\pi_1 \rightarrow \pi_3$ and $\pi_2 \rightarrow \pi_4$ result in a pair of states of A_1 symmetry in the heterocyclic systems. The corresponding electronic configurations will mix to form A_1^- and A_1^+ states. The minus state has the lower energy and a low intensity, while A_1^+ will be pushed up and will carry most of the intensity (B_2^- and B_2^+ states in NB). There is an additional low-lying state of A_1 symmetry formed by two-electron promotions, mainly $(\pi_2)^2 \rightarrow (\pi_3)^2$. In CP and related cyclic systems, this third state of A_1 symmetry can interact with the other two A_1 states and further alter their energies. No such interaction is possible in NB. In addition, triplet excited valence B_2 and A_1 states are expected in the five-atom rings (A_2 and B_2 states in NB). Rydberg states are expected to be interleaved among the valence excited states. Because of the closeness of the energies of the π_1 and π_2 orbitals in pyrrole, thiophene, and even more pronounced in NB, Rydberg states arising from excitations out of both these orbitals are predicted to be close in energy. As a result, the spectra have a complex structure, which makes the assignments of the different bands particularly difficult.

The results are summarized in Tables VIII-XII. For more detailed information the reader is referred to the original references [22, 25, 27]. Below we shall compare the theoretical results with the experimental data and previous *ab initio* results for each molecule separately. A discussion is next included where comparisons are made among the electronic spectra of the different systems and general trends are rationalized.

3.3.1 *Cyclopentadiene*

The UV spectrum of CP shows three major features around 5.3, 6.2, and 8.0 eV. From the analysis of the optical absorption and energy loss spectra made by McDiarmid *et al.* [91]-[93] and the electron impact study of Frueholz *et al.* [94], a detailed description of the excited states in the molecule is obtained.

The lowest energy singlet-singlet transition in CP gives rise to an absorption band with an intensity maximum around 5.26 eV [91, 94, 95]. All the experimental data are consistent with a singlet-singlet symmetry-allowed transition and an assignment to the 1^1B_2 state. The CASPT2 value obtained for the excitation energy is 5.27 eV. The computed intensity is 0.15, which places it as the second most in-

TABLE VIII

Calculated and experimental excitation energies and oscillator strengths in cyclopentadiene.

	Excitation energy (eV)				
			Other results		
State	PT2F	Exptl	SAC-CI[a]	RPA[b]	CIS[c]
Singlet states					
1^1B_2 (Valence)	5.27	5.26[d–f]	6.48	5.54	5.77
1^1A_2 ($1a_2 \rightarrow 3s$)	5.65	5.68[d,f,g]	5.76	5.95	
1^1B_1 ($1a_2 \rightarrow 3p\sigma$)	6.24	6.25[d,f,g]	6.33	6.44	
2^1B_2 ($1a_2 \rightarrow 3p\pi$)	6.25	6.31[d,f,g]	6.42	6.53	
2^1A_2 ($1a_2 \rightarrow 3p\sigma$)	6.30	6.26[h]	6.38	6.59	
2^1A_1 (Valence)	6.31	6.2[i]	7.81	8.01	
3^1A_2 ($1a_2 \rightarrow 3d\delta$)	6.85			7.06	
3^1B_2 ($1a_2 \rightarrow 3d\pi$)	6.86	6.80[d]		7.25	
2^1B_1 ($1a_2 \rightarrow 3d\delta$)	6.87			7.09	
4^1A_2 ($1a_2 \rightarrow 3d\sigma$)	6.91			7.13	
3^1A_1 ($1a_2 \rightarrow 3d\pi$)	6.93	7.05[d]	7.12	7.22	
4^1A_1 (Valence)	7.89	7.9,8.0[d,e]		8.80	
3^1B_1 ($2b_1 \rightarrow 3s$)	7.95	8.03[d]	8.17	8.26	
Triplet states					
1^3B_2 (Valence)	3.15	3.10[d,f]	3.52		2.77
1^3A_1 (Valence)	4.90	>4.7[d]	5.50		4.48
1^3A_2 ($1a_2 \rightarrow 3s$)	5.63		5.82		
1^3B_1 ($1a_2 \rightarrow 3p\sigma$)	6.25		6.36		

	Oscillator strength		
		Previous	
State	This work	SAC-CI[a]	RPA[b]
Singlet states			
1^1B_2 (Valence)	0.1475	0.159	0.115
1^1A_2 ($1a_2 \rightarrow 3s$)		forbidden	
1^1B_2 (Valence)	0.0306	0.031	0.057
1^1B_1 ($1a_2 \rightarrow 3p\sigma$)	0.0248	0.0044	0.024
2^1B_2 ($1a_2 \rightarrow 3p\pi$)		forbidden	
2^1A_2 ($1a_2 \rightarrow 3p\sigma$)	0.0003	0.0193	0.4235
2^1A_1 (Valence)		forbidden	
3^1A_2 ($1a_2 \rightarrow 3d\delta$)	0.0101		0.0006
3^1B_2 ($1a_2 \rightarrow 3d\pi$)	0.0003		0.0009
2^1B_1 ($1a_2 \rightarrow 3d\delta$)		forbidden	
4^1A_2 ($1a_2 \rightarrow 3d\sigma$)	0.0208	0.00001	0.0013
3^1A_1 ($1a_2 \rightarrow 3d\pi$)	0.4415		0.2478
4^1A_1 (Valence)	0.0252	0.0202	0.0005

[a]SAC-CI results from ref. [96]
[b]SCF-RPA results from ref. [97]
[c]CIS results from ref. [98]
[d]Electron impact spectroscopy from ref. [94]
[e]Optical absorption and electron energy-loss spectroscopy from ref. [91]
[f]UV absorption data from ref. [95]
[g]UV absorption data from ref. [92]
[h]Two-photon resonant multiphoton ionization spectrum from ref. [93]
[i]Estimation based on experimental data, ref. [91]

tense transition in the spectrum, in agreement with the experimental findings and with the results of Nakatsuji *et al.* [96]. Their SAC-CI study gives an energy of 6.48 eV for the 1^1B_2 state. As the main source of such a large deviation from experiment, the inadequate treatment of electron correlation, especially the dynamic $\sigma\pi$ polarization effects was pointed out [25]. The 1^1B_2 state is dominated by the single excitation $1a_2 \rightarrow 3b_1$. It is therefore not surprising that the RPA energy of Galasso [97], 5.54 eV, is in reasonable agreement with experiment. A larger deviation is however obtained with the CIS procedure, 5.77 eV, the transition energy reported by Zgierski and Zerbetto [98].

The next valence excited state is the 2^1A_1 state, which appears at 6.31 eV. This is the ${}^1A_1^-$ state. It is a mixture of the configurations $2b_1 \rightarrow 3b_1$ and $1a_2 \rightarrow 2a_2$, but includes a large fraction (about 43%) of the doubly-excited configuration $(1a_2)^2 \rightarrow (3b_1)^2$. The computed intensity is very low. This state has not been found in the spectrum, but McDiarmid [91] has estimated the excitation energy to be about 6.2 eV, in agreement with the present result. Due to the multiconfigurational character of this state, including a large fraction of doubly-excited configurations, the theoretical difficulties encountered in previous *ab initio* studies [96, 97], placing this transition at higher energies, are not surprising. On the other hand, a recent semiempirical estimate gives the energy of the 2^1A_1 state significantly lower [98].

The third valence excited state represents the most intense feature of the system. The calculations place the 4^1A_1 state at 7.89 eV in agreement with experiment [91, 94]. This is the ${}^1A_1^+$ state with an intensity of 0.44. The SAC-CI study [96] does not report this state, and the RPA calculations [97] place it at 8.80 eV with an intensity of 0.25.

The n=3 Rydberg series of CP has received special attention by Frueholz *et al.* [94] and McDiarmid *et al.* [92, 93]. Our assignment of the 1^1A_2 state at 5.65 eV is in agreement with the experimental findings [92, 94, 95] and earlier theoretical predictions [96, 97]. McDiarmid *et al.* [92, 93] have performed a careful analysis of the $3p$ Rydberg states and give the energies 6.25(B_1), 6.26(A_2), and 6.31(B_2) eV. The computed results are 6.24, 6.30, and 6.25 eV. Even though the order of the A_2 and B_2 states has shifted, the agreement with experiment must be considered as satisfactory. The first five members of the $R(3d)$ series have been calculated to lie in the energy range 6.85–6.93 eV. These states have not been directly observed in the spectrum, but Frueholz *et al.* [94] report two unassigned peaks at 6.80 and 7.05 eV, which could be related to the two most intense of the five transitions. The highest state studied, 3^1B_1, was the first component of the $2b_1$ Rydberg series. It is found to be in the region dominated by the intense valence ${}^1A_1^+$ state.

Calculations have been performed for the lowest triplet state in each of the four symmetry species. Two valence excited states occur at 3.15 (1^3B_2) and 4.90 eV (1^3A_1). The lowest energy transition observed in the CP spectrum clearly corresponds to the valence excited triplet state, 1^3B_2, with an intensity maximum occurring at 3.10 eV [94, 95]. The SAC-CI treatment [96] overestimates the excitation energy by about 0.4 eV and the CIS procedure [98] underestimates it by 0.3 eV. Our results predict about the same excitation energy for the 1^3A_1 state in CP (4.90 eV) and *cis*-butadiene (4.73 eV).

3.3.2 Pyrrole

The spectrum of pyrrole has been known for a long time [99]-[102] and recent measurements employing modern techniques have given a very detailed picture [103]-[108]. The main features are two broad bands located at about 6.0 and 7.5 eV. Superimposed on this structure are two sets of Rydberg series arising from excitations out of the highest ($1a_2$) and next highest ($2b_1$) occupied π-orbitals. In the experimental literature they are labeled R and R', respectively. In spite of numerous calculations employing a variety of computational methods, no complete assignment of the bands in the spectrum was available in the literature (see ref. [25] for a historical background).

The results show that the band around 6.0 eV is composed of two valence excited states plus a number of Rydberg states, thus giving a complex structure (cf. Table IX). According to the calculation, the 2^1B_2 state is nearly degenerate with the 2^1A_1 valence excited state. The 2^1A_1 state is not discussed at all in the experimental literature. The calculations locate the valence excited 2^1B_2 state at 6.00 eV with an intensity of 0.125. This is in perfect agreement with the findings reported by Bavia *et al.* [105], on the basis of both a vapor and a crystal spectrum, and the evidence found in the EI spectrum of Flicker *et al.* [107, 108]. Earlier theoretical studies have placed the $^1B_2$ valence excited state at energies above 7.0 eV [96, 109, 110].

The most intense feature in the spectrum is the maximum around 7.5 eV. It is natural to assume that the intensity underlying the many Rydberg series is due to a valence excitation. However, there was no *ab initio* evidence in support of this assumption. The most intense peak reported by Flicker *et al.* [107, 108] occurs at 7.54 eV. Similar results are given by Bavia *et al.* [105] from their vapor UV spectrum. Our result confirms the observations and locates an intense transition (6^1A_1) at 7.46 eV, with an oscillator strength of 0.33.

The calculated excitation energy for the first singlet-singlet transition is 5.08 eV, in agreement with the analysis of the UV spectrum of Mullen and Orloff [104] and the EI spectrum of Flicker *et al.* [107, 108]. It was however originally assumed [104] to be of the $\pi \rightarrow \pi^*$ type. Earlier theoretical studies [96, 109, 110] have resulted in excitation energies for the 1^1A_2 state in the range 5.20–5.25 eV. The nature of the first spin-allowed transition in pyrrole is thus conclusively established.

It is generally agreed that the three transitions corresponding to the $1a_2 \rightarrow 3p$ excitations are located in the band with the intense peak around 6.0 eV [105]. Our calculated energies for the $3p$ excitations are 5.78, 5.83, and 5.85 eV corresponding to the 1^1B_2, 2^1A_2, and 1^1B_1 states, respectively. Experimentally, several assignments have been attempted [103]-[105]. The energy difference between the two dipole-allowed transitions is computed to be 0.07 eV, in agreement with the results of Bavia *et al.* [105]. The calculations show that the two states have similar intensities. Other theoretical studies have positioned the $1a_2 \rightarrow 3p$ bands in the energy region 5.8–6.3 eV (cf. Table IX for details).

The first member of the second Rydberg series (R'), 2^1B_1 state, has a computed transition energy of 5.97 eV and an oscillator strength of only 0.006. Other theoretical studies give 5.96 eV [110] and 6.13 eV [96, 109]. Experimentally, the situation is rather confused [103, 105, 107, 108]. We concluded [25] that this Ry-

TABLE IX

Calculated and experimental excitation energies and oscillator strengths in pyrrole.

State	Excitation energy (eV)			Oscillator strength	
	PT2F	Exptl	Other results	This work	Previous[d]
Singlet states					
1^1A_2 ($1a_2 \rightarrow 3s$)	5.08	5.22[e,f]	5.22[b],5.25[c],5.20[d]	forbidden	
1^1B_2 ($1a_2 \rightarrow 3p\pi$)	5.78	5.86[g]	6.02[a],6.33[b],5.80[c],5.97[d]	0.0399	0.0421
2^1A_2 ($1a_2 \rightarrow 3p\sigma$)	5.83		6.03[c],5.95[d]	forbidden	
1^1B_1 ($1a_2 \rightarrow 3p\sigma$)	5.85	5.88[e,f,h,j]	6.05[b],5.63[c],5.85[d]	0.0239	0.0322
2^1A_1 (Valence)	5.92		6.51[a],6.73[b],6.53[c],7.20[d]	0.0195	0.0214
2^1B_1 ($2b_1 \rightarrow 3s$)	5.97		6.13[b],5.96[c],6.13[d]	0.0006	0.0016
2^1B_2 (Valence)	6.00	5.98[e,g,k]	7.07[a],7.70[b],7.52[d]	0.1253	0.0874
3^1B_1 ($1a_2 \rightarrow 3d\delta$)	6.40	6.43[g]	6.33[c]	0.0117	
3^1A_2 ($1a_2 \rightarrow 3d\delta$)	6.42		6.62[c]	forbidden	
4^1A_2 ($1a_2 \rightarrow 3d\sigma$)	6.51		6.88[c]	forbidden	
3^1B_2 ($1a_2 \rightarrow 3d\pi$)	6.53	6.50[e,i]	6.97[c],7.59[d]	0.0011	0.0930
3^1A_1 ($1a_2 \rightarrow 3d\pi$)	6.54	6.50[e,i]	7.12[c],6.68[d]	0.0010	0.0002
4^1B_1 ($2b_1 \rightarrow 3p\sigma$)	6.62	6.78[e,g,j]	6.70[c],6.70[d]	0.0161	0.0222
4^1A_1 ($2b_1 \rightarrow 3p\pi$)	6.65	6.78[e,g,j]	6.91[c],6.89[d]	0.0127	0.0024
5^1A_2 ($2b_1 \rightarrow 3p\sigma$)	6.77		7.07[c],6.85[d]	forbidden	
6^1A_2 ($2b_1 \rightarrow 3d\delta$)	7.31			forbidden	
5^1B_1 ($2b_1 \rightarrow 3d\delta$)	7.32	7.26[e]		0.0001	
5^1A_1 ($2b_1 \rightarrow 3d\pi$)	7.36	7.43[e]	8.18[c]	0.0047	
6^1B_1 ($2b_1 \rightarrow 3d\sigma$)	7.39	7.43[e]		0.0025	
4^1B_2 ($2b_1 \rightarrow 3d\pi$)	7.43	7.43[e]	8.10[c]	0.0005	
6^1A_1 (Valence)	7.46	7.54[e]		0.3261	
5^1B_2 ($1a_2 \rightarrow 4d\pi$)	7.72	≈7.4[g]		0.0043	
7^1A_1 ($1a_2 \rightarrow 4d\pi$)	7.88			0.0079	
Triplet states					
1^3B_2 (Valence)	4.27	4.21[e,i]	4.27[a],4.42[b],4.65[d]		
1^3A_2 ($1a_2 \rightarrow 3s$)	5.04		5.25[b],5.17[d]		
1^3A_1 (Valence)	5.16	5.10[i]	5.45[a],5.71[b],5.84[d]		
1^3B_1 ($1a_2 \rightarrow 3p\sigma$)	5.82		5.99[b],5.82[d]		

[a]SCF plus CI in the π orbital space, ref. [290]
[b]Selected-MRCI results, ref. [109]
[c]SDQ method from ref. [110]
[d]SAC-CI resuls from ref. [96]
[e]Electron impact spectroscopy from ref. [107, 108]
[f]UV absorption data from ref. [104]
[g]UV absorption data from ref. [105]
[h]UV absorption data from ref. [99, 100, 102]
[i]Electron impact spectroscopy from ref. [106]
[j]UV absorption data from ref. [101]
[k]UV absorption data from ref. [100]

dberg series has not been observed. The five components of the $R(3d)$ transitions have been computed to lie in the energy range 6.40–6.54 eV, with one of them (3^1B_1 at 6.40 eV) having an appreciable intensity. That is thus the state which should be most easily detectable experimentally [105]. A maximum around 6.5 eV in the EI spectrum of Van Veen [106] can be related to the 3^1B_2 and 3^1A_1 states.

The three origins of the $R'(3p)$ Rydberg series have been calculated to appear between 6.62 and 6.77 eV. As new assignments, the features around 6.78 eV, which occur in both the UV and EI spectra with larger intensities than those around 6.5 eV [99, 105, 107, 108] are proposed as the origins of the $R'(3p)$ series. As shown in Table IX, the series $R'(3d)$ are calculated to have origins in the energy region 7.31–7.43 eV. No experiment has identified these series in the spectrum in a conclusive manner.

The lowest triplet state is a valence excited state of $^3B_2$ symmetry. The calculated excitation energy, 4.27 eV, agrees with the EI value of 4.21 eV [106]-[108]. The second triplet valence excited state (1^3A_1) occurs at 5.16 eV, which has been identified in the EI spectrum of Van Veen [106] to appear at 5.10 eV. Earlier theoretical studies have, however, placed this state at considerably higher energies [96, 109, 110].

3.3.3 Furan

The spectrum of furan has been well documented experimentally [106]-[108],[111]-[116], although not so extensively as CP and pyrrole. The assignments of the strong optically-allowed Rydberg transitions have been revised several times. The latest assignments of Derrick *et al.* [116] give support to the Rydberg series $1a_2 \rightarrow np$, $1a_2 \rightarrow nd$, and $2b_1 \rightarrow ns$, with origins at 6.48, 7.28, and 7.38 eV, respectively. Electron-impact measurements have given additional information, in particular concerning dipole- and spin-forbidden transitions [106]-[108].

The 1^1B_2 state is computed to lie 6.04 eV above the ground state. It is clearly related to the second most intense feature in the energy-loss spectrum of Flicker *et al.* [107, 108]. Previous theoretical investigations [96, 117] have obtained considerably larger transition energies and have related this state to the maximum around 7.82 eV [107, 108], which we assigned [25] to the $^1A_1^+$ state. The second valence state is 2^1A_1 ($^1A_1^-$) at 6.16 eV. As for the previous systems the intensity is low. The third singlet valence excited state is $4^1A_1(^1A_1^+)$, at 7.74 eV. It corresponds to the most intense feature in the spectrum with a maximum at 7.82 eV [106]-[108],[113].

The first excited singlet state in furan is the origin of the $R(3s)$ Rydberg series. The computed transition energy 5.92 eV is in good agreement with the interpretation of Robin [112]. The components of the $3p$ and $3d$ Rydberg states have been computed to lie in the energy range 6.46–6.59 eV and 7.00–7.31 eV, respectively, in agreement with the available experimental evidence (cf. Table X for details). Somewhat larger deviation was however found for the transition energy of the $3s$ component of the R' Rydberg series respect to the experimental assignments [116].

As shown in Table X, the first two triplet states are of valence type, in agreement with the EI data [107, 108].

TABLE X

Calculated and experimental excitation energies and oscillator strengths in furan.

	Excitation energy (eV)				Oscillator strength	
			Other results			Previous
State	PT2F	Exptl	MRCI[a]	SAC-CI[b]	This work	SAC-CI[b]
Singlet states						
1^1A_2 $(1a_2 \rightarrow 3s)$	5.92	5.94[e]	5.94	6.27	forbidden	
1^1B_2 (Valence)	6.04	6.06[c,d]	7.58	7.50	0.1543	0.0309
2^1A_1 (Valence)	6.16		≈6.9	7.32	0.0015	0.0044
1^1B_1 $(1a_2 \rightarrow 3p\sigma)$	6.46	6.48[d,f]	6.52	6.74	0.0309	0.0426
2^1B_2 $(1a_2 \rightarrow 3p\pi)$	6.48	6.48[d,f]	6.52	6.80	0.0471	0.0891
2^1A_2 $(1a_2 \rightarrow 3p\sigma)$	6.59	6.61[c,d]		6.98	forbidden	
3^1A_2 $(1a_2 \rightarrow 3d\delta)$	7.00				forbidden	
3^1B_2 $(1a_2 \rightarrow 3d\pi)$	7.13				0.0074	
2^1B_1 $(1a_2 \rightarrow 3d\delta)$	7.15				0.0000	
4^1A_2 $(1a_2 \rightarrow 3d\sigma)$	7.22				forbidden	
3^1A_1 $(1a_2 \rightarrow 3d\pi)$	7.31	7.28[d,f]	7.46	7.59	0.0003	0.0000
3^1B_1 $(2b_1 \rightarrow 3s)$	7.21	7.38[d,f]	7.27	7.62	0.0192	0.0309
4^1A_1 (Valence)	7.74	7.82[c,d]			0.4159	
Triplet states						
1^3B_2 (Valence)	3.99	4.02[d,g]	4.08	4.40		
1^3A_1 (Valence)	5.15	5.22[d]	5.4-5.8	5.75		
1^3A_2 $(1a_2 \rightarrow 3s)$	5.86		5.82	6.20		
1^3B_1 $(1a_2 \rightarrow 3p\sigma)$	6.42		6.45	6.68		

[a]MRCI results from ref. [117]
[b]SAC-CI results from ref. [96]
[c]UV absorption data from refs. [111, 113, 114]
[d]Electron impact spectroscopy from ref. [107, 108]
[e]Assignment based on UV absorption data from ref. [112]
[f]Assignment based on UV absorption data from ref. [116]
[g]Electron impact spectroscopy from ref. [106]

3.3.4 Thiophene

The spectrum of thiophene is similar in appearance to the cyclopentadiene spectrum, more than to pyrrole and furan, as might have been expected [102, 118]. However, the nature of the transitions is different.

The first valence 2^1A_1 $\pi \rightarrow \pi^*$ state is computed to be at 5.33 eV, in agreement with optical spectroscopy [100, 102, 111],[119]-[126]. The second valence-excited singlet state, 1^1B_2, is predicted at 5.72 eV. The excitation energy for this second singlet state has also been reported from studies of optical spectroscopy [100, 102, 111, 119, 120]. The 2^1A_1 state is slightly more intense. Two higher valence $\pi \rightarrow \pi^*$ states were also located: the 3^1A_1 state at 6.69 eV and the 4^1B_2 state at 7.32 eV, the latter corresponding to the most intense transition in the spectrum up to 7.8 eV. It represents a unique feature of the thiophene spectrum. It is interesting to

TABLE XI

Calculated and experimental excitation energies and oscillator strengths in thiophene.

State	Excitation energy (eV)			Oscillator strength	
			Other results		Previous
	PT2F	Exptl	CI[i]	This work	CI[i]
Singlet states					
2^1A_1 (Valence)	5.33	5.33[a,b],5.39[c–e],5.48[f]	6.67	0.0885	0.068
1^1B_2 (Valence)	5.72	5.64[a,d,e,g],5.77[b]	7.83	0.0703	0.321
1^1A_2 $(1a_2 \rightarrow 3s)$	5.93	5.93[f],5.99[c,d]	7.12	forbidden	
1^1B_1 $(3b_1 \rightarrow 3s)$	6.23		7.41	0.0002	0.002
2^1B_1 $(1a_2 \rightarrow 3p\sigma)$	6.30		7.57	0.0303	0.020
2^1A_2 $(3b_1 \rightarrow 3p\sigma)$	6.35		7.77	forbidden	
2^1B_2 $(1a_2 \rightarrow 3p\pi)$	6.56	6.60[h]	7.73	0.0296	0.047
3^1A_2 $(1a_2 \rightarrow 3p\pi)$	6.58			forbidden	
3^1A_1 (Valence)	6.69	6.6[a,e,g],6.69[c]	9.05	0.1850	0.172
4^1A_1 $(3b_1 \rightarrow 3p\pi)$	6.76			0.0150	
3^1B_1 $(3b_1 \rightarrow 3p\pi)$	6.83			0.0200	
4^1A_2 $(1a_2 \rightarrow 3d\sigma)$	6.97			forbidden	
5^1A_2 $(1a_2 \rightarrow 3d\delta)$	7.08			forbidden	
5^1A_1 $(1a_2 \rightarrow 3d\pi)$	7.23	7.33[h]		0.0006	
4^1B_1 $(1a_2 \rightarrow 3d\delta)$	7.24			0.0010	
3^1B_2 $(1a_2 \rightarrow 3d\pi)$	7.28			0.0005	
4^1B_2 (Valence)	7.32	7.1[f],7.2[g]		0.3923	
5^1B_1 $(3b_1 \rightarrow 3d\sigma)$	7.37			0.0011	
5^1B_2 $(3b_1 \rightarrow 3d\pi)$	7.53			0.0022	
6^1A_1 $(3b_1 \rightarrow 3d\pi)$	7.57			0.0001	
6^1A_2 $(3b_1 \rightarrow 3d\delta)$	7.64			forbidden	
6^1B_1 $(3b_1 \rightarrow 3d\delta)$	7.67			0.00006	
7^1B_1 $(n \rightarrow \pi^*$,Val)	7.77			0.0333	
n^1A_1 $(n \rightarrow 3s)$	9.10			0.0479	
n^1A_2 $(n \rightarrow \pi^*$,Val)	9.69			forbidden	
n^1A_2 $(n \rightarrow 3p\pi)$	10.27			forbidden	
Triplet states					
1^3B_2 (Valence)	3.75	3.7[g],3.75[f]	5.12		
1^3A_1 (Valence)	4.50	4.6[g],4.62[f]	5.73		
1^3A_2 $(1a_2 \rightarrow 3s)$	5.88		7.08		
1^3B_1 $(1a_2 \rightarrow 3p\sigma)$	5.90		7.36		
2^3B_1 $(3b_1 \rightarrow 3s)$	6.18		7.54		

[a]Optical spectrum, refs. [111, 119, 120]
[b]Optical spectrum, ref. [121]
[c]Optical spectrum, ref. [122]. The measured oscillator strength for the 2^1A_1 transition is 0.1.
[d]Optical spectrum, ref. [123] – [126]
[e]Optical, ref. [102], MCD data ref. [291]
[f]Electron impact spectrum, ref. [107, 108]
[g]Electron impact spectrum, ref. [106]
[h]Analysis based on IPs, ref. [292]
[i]CI calculations, ref. [293]

note that the weight of the doubly excited configurations in the wave function of the 3^1A_1 state is nearly 30%, which gives an explanation for its low intensity compared to the equivalent state of the previous heterocyclic systems.

As shown in Table XI and discussed in detail elsewhere [27], the computed spectrum, which includes four valence excited singlet states and a large number of Rydberg states, including the $1a_2 \rightarrow$n=3 and $3b_1 \rightarrow$n=3 series, shows an overall agreement with the available experimental data.

3.3.5 Norbornadiene

Both a theoretical and an experimental investigation of the excited electronic states of norbornadiene has been recently performed [22]. The experimental investigation is a combined optical absorption and polarization-selected (2+1) resonant-enhanced multiphoton ionization (REMPI) spectroscopic study of the spectrum of static and jet-cooled samples of norbornadiene. The theoretical study uses the same techniques as presented above. The results from the combined theoretical and experimental analysis of the main electronic transitions are listed in Table XII. In what follows we give a brief outline of the results related to the valence singlet excited states.

The lowest singlet excited state is of A_2 symmetry. That the lowest singlet-singlet transition is optically forbidden was clearly established by Doering and McDiarmid [127]. The computed excitation energy, 5.28 eV, is in agreement with the observed value of 5.25 eV [127]. The next valence excited state is the 1^1B_2, at 6.20 eV, arising from the minus combination of the configurations $(a_1 \rightarrow b_2)$ and $(b_1 \rightarrow a_2)$, the $^1B_2^-$ state. The computed intensity is relatively low, 0.0081. The transitions in this region are assigned as a structured Rydberg transition and an underlying valence transition. The valence transition has its maximum around 6 eV. Considering the uncertainty in the experimental determination of the peak location and the expected error bars of the calculation, the agreement between theory and experiment is satisfactory. Previous theoretical calculations place this transition at higher energies [97, 128], the RPA calculation [97] with the largest intensity. Thus, neither of the previous limited *ab initio* methods seem capable of giving an adequate characterization of the $^1B_2^-$ valence state of NB. The third valence excited state is the most intense feature in the system. The calculation places the 2^1B_1 state at 6.48 eV with an intensity of 0.34 (this is the $^1B_2^+$ state) in agreement with experimental evidence [22].

The 5^1A_2 and 6^1A_1 valence states are calculated to be close in energy, with computed values of 7.36 and 7.49 eV, respectively. The results support the assignment of the second valence state of $^1A_2$ symmetry around 7.5 eV deduced from the electron-impact investigation [127]. It also supports the previous conclusion that the second $^1B_2$ transition lies lower in energy than the second $^1A_2$ transition [127]. The computed intensity for the 6^1A_1 state is very low, consistent with its doubly excited nature and the experimental inability to observe this transition. The RPA calculation [97] places the second $^1A_2$ valence-excited state at 9.05 eV, much too high in energy, and does not report the doubly excited valence state. The CIS calculation [128] places a $^1A_2$ state at 7.32 eV and an $^1A_1$ sate at 6.98 eV. However, as the characters of the states are not identified, and since Rydberg states of the

TABLE XII

Calculated and experimental excitation energies and oscillator strengths in norbornadiene.

State	Excitation energy (eV)				Oscillator strength	
			Other results			Previous
State	PT2F	Exp	RPA[a]	CIS[b]	This work	RPA[a]
Singlet states						
1^1A_2 (Valence)	5.28	$5.23^{c,d}$, 5.25^{e}	5.43	6.00	forbidden	
1^1B_1 ($b_1 \rightarrow 3s$)	5.73	5.85^{f}	6.18		0.0112	0.0181
2^1A_2 ($b_1 \rightarrow 3p_y$)	6.03		6.77		forbidden	
1^1B_2 (Valence)	6.20	5.89^{g}, $5.95^{c,e}$, 5.92^{d}, 6.0^{f}	6.75	6.67	0.0081	0.1536
2^1A_1 ($b_1 \rightarrow 3p_x$)	6.33	6.38^{e}, 6.35^{d}, 6.40^{f}	6.49		0.0105	0.0295
2^1B_1 ($b_1 \rightarrow 3p_z$)	6.33		6.81		$9\ 10^{-6}$	0.0295
3^1A_2 ($b_1 \rightarrow 3d_{yz}$)	6.34		7.40		forbidden	
2^1B_2 (Valence)	6.48	$6.65^{d,e}$, 6.7^{f}	8.57	7.49	0.3434	0.1314
3^1B_2 ($b_1 \rightarrow 3d_{xy}$)	6.50	6.6^{f}	7.10		0.0033	0.0473
3^1A_1 ($a_1 \rightarrow 3s$)	6.67	6.85^{e}, 6.84^{d}, 6.91^{f}	7.37		0.0481	0.0209
3^1B_1 ($a_1 \rightarrow 3p_x$)	6.75		7.82		0.0011	0.0001
4^1B_2 ($a_1 \rightarrow 3p_y$)	6.77		7.73		0.0102	0.0074
4^1B_1 ($b_1 \rightarrow 3d_{x^2-y^2}$)	6.79		7.20		0.0058	0.0166
5^1B_1 ($b_1 \rightarrow 3d_{z^2}$)	6.92		7.36		0.0088	0.0030
4^1A_1 ($b_1 \rightarrow 3d_{xz}$)	6.94	$7.2^{d,e}$	7.21		0.0016	0.0002
5^1B_2 ($a_1 \rightarrow 3d_{yz}$)	7.27				0.0216	
5^1A_1 ($a_1 \rightarrow 3p_z$)	7.29		7.73		0.0032	0.0019
4^1A_2 ($a_1 \rightarrow 3d_{xy}$)	7.29				forbidden	
5^1A_2 (Valence)	7.36	7.50^{e}	9.05	7.32	forbidden	
6^1B_1 ($a_1 \rightarrow 3d_{xz}$)	7.39	7.65^{e}			0.0213	
6^1A_1 (Valence)	7.49			6.98	0.0002	
7^1A_1 ($a_1 \rightarrow 3d_{x^2-y^2}$)	7.74	7.9^{e}			0.0176	
8^1A_1 ($a_1 \rightarrow 3d_{z^2}$)	7.86	8.0^{e}			0.0212	
Triplet states						
1^3A_2 (Valence)	3.42	3.4^{c}, 3.47^{d}				
1^3B_2 (Valence)	3.80	$3.9^{c,d}$				

[a] RPA results from ref. [97]
[b] CIS results from ref. [128]
[c] Electron impact data ref. [294]
[d] Electron impact data ref. [295]
[e] Electron impact data ref. [127]
[f] Optical and REMPI data [21]
[g] UV absorption data ref. [296, 297]

same symmetries also occur at similar energies, it is not possible to correlate these states with those computed by us or obtained in the RPA calculation [97].

A comparison of the accuracy of the present results with those obtained for cyclopentadiene, a molecule of similar size lacking indirect conjugation and $\sigma - \pi$

interaction, assess the ability of the CASPT2 method to handle those effects. Further considerations of the correlations from the obtained results are considered in the next section.

3.3.6 Discussion

Comparison of the spectra of the five-membered ring systems reveals that the expected similarity is only general. CP (and *cis*-butadiene) show a lowest and intense 1^1B_2 state around 5.3–5.6 eV, and a weak 2^1A_1 state around 6.0–6.3 eV. In pyrrole and furan the two states are almost degenerate, while the 2^1A_1 state in thiophene is clearly below the 1^1B_2 state. The trend in the intensities for the 2^1A_1 state also shows that thiophene is different from the other compounds. The oscillator strength is much larger and it is only in this molecule that the state has been clearly identified. In thiophene, and also in pyrrole, the $b_1 \rightarrow b_1$ single excitation dominates the wave function, which can be related to the orbital energy difference, which is smallest for these molecules [27]. The doubly excited configurations are not so important for the description of the 2^1A_1 state. Instead they will appear in the next state of $^1A_1$ symmetry, which then is moved to lower energy with a lower intensity. Thus, the strongest transition in CP, pyrrole, and furan corresponds to the $^1A_1^+$ state, but it corresponds to the second $^1B_2$ state in thiophene. The appearance of the second $^1B_2$ state in thiophene at lower energy than in the other systems can be related to the smallest energy difference between the π_1 and π_4 orbitals.

On the other hand, taking as a measure of the aromaticity in heterocyclic dienes the energy splitting between the highest occupied b_1 and a_2 orbitals and the corresponding splitting of the virtual pair, pyrrole and thiophene, having the smaller splittings, can be regarded as more aromatic than the other two. The electronic spectrum of CP and furan show greater similarity to that of conjugated dienes, while pyrrole and thiophene are different and may be described as quasi-aromatic.

As regards cyclopentadiene and norbornadiene it is worth noting that since the HOMO and HOMO−1 of CP are more than twice as far apart as they are in NB [22], comparably promoted states will be further apart in the former molecule than in the latter (in the absence of interactions with states external to the one-electron model). The spectra of NB and CP should be similar in character but the latter more widely spaced. In any case, the spectra can be interpreted in terms of the same basic type of excitations.

The lowest singlet state is mainly represented by the HOMO→LUMO excitation, which is dipole-allowed in CP and forbidden in NB, and for which the excitation energies are of comparable magnitude. The second valence transition corresponds to a (HOMO−1→LUMO) − (HOMO→LUMO+1) excitation and is therefore weak. The most intense transition is to the analogous plus state. It is found at higher energy in CP than in NB, consistent with the larger energy difference between the corresponding orbitals involved in the transition in CP. However, the minus state of CP and NB are found at similar energies. It is understood since in CP the minus, plus, and doubly excited states all are of the same symmetry and can interact. The contribution of doubly excited configurations in the A_1^- state of CP is significant (about 40%). This interaction causes a reduction in energy of the lowest A_1^- state from that predicted by a simple orbital model. Concomitant with this reduction in energy of the A_1^- state in CP, the third valence excited A_1 state is pushed up to above 8.5 eV. The doubly excited ${}^1A_1$ state of NB does not interact with any other valence-excited states. It is calculated to appear at a moderate energy with a very weak intensity, in accordance with its "pure" doubly excited character. The origin of this state in the interacting double bond model is a symmetric combination of one triplet excitation on each of the ethene moieties (TT). The low-lying excited states in NB of B_2 and A_2 symmetry originate from the V-state in ethene (NV). The TT energy of two non-interacting ethene molecules is 8.7 eV, while the NV state energy is about 8.0 eV. It is therefore not surprising that the three states appear with similar energies in NB.

Finally, it is important to realize that in order to arrive at a qualitative understanding of these systems high-level calculations are required. The present method first treats the important near degeneracies using the multiconfigurational CASSCF method. The CASSCF wave function is subsequently used as the reference function in a multiconfigurational perturbational treatment, the CASPT2 method. The latter accounts for dynamic correlation effects, which, since they contribute significantly to excited state energies, are necessary for a quantitative theoretical study of the spectra. From this quantitative determination, a meaningful qualitative rationalization and relationship of the spectra in five-membered ring systems and norbornadiene could be achieved.

3.4 Absorption spectra of benzene and various derivatives

The benzene molecule and derivatives thereof are key compounds in chemistry. Absorption and fluorescence spectroscopy are often used to identify these species and to extract information about their surroundings. Many of these systems are easily accessible and are therefore preferred systems for experimental investigations. As a consequence, benzene and several derivatives are characterized in fine detail. From the point of view of theory, the reason for the vivid interest in these molecules lies in the difficulties to describe their excited states, and they have been studied with a variety of approaches since the early days of quantum chemistry.

In this section we report calculations on the electronic spectra of benzene, a number of azabenzenes, phosphabenzene, and phenol. All these spectra expose similar features. Because of that, we shall first discuss the benzene molecule in detail. Second, for technical reasons, we focus on the di-azabenzenes. Pyridine, phosphabenzene, and phenol will be considered as a separate group. The different states in substituted benzenes are often referred to by using the same D_{6h} symmetry labels as is used in benzene itself. We shall adopt the same labelling, but enclose them within apostrophes to avoid misunderstanding.

3.4.1 The benzene molecule

One of the first *ab initio* configuration-interaction (CI) calculations on the spectrum of benzene was published by Buenker, Whitten and Petke [1]. These were full CI calculations within a CI-space spanned by 6 π-electrons and 6 π-orbitals. Except for the 1^1B_{2u} state, their results differed considerably from experimental values and early semi-empirical calculations. During the following years a number of more and more elaborate CI-calculations were presented [2, 129, 130] which, on the one hand improved on the basis sets, and, on the other hand increased the number of π-orbitals included. Although the results obtained were becoming better and better, the predicted excitation energies for the 1^1B_{1u} and 1^1E_{1u} states were still more than 1.0 eV too large. In an effort to include dynamic correlation effects Matos *et al.* [4] complemented CASSCF calculations with subsequent externally-contracted CI and CEPA calculations. These authors, for the first time, were able to predict the line position of the 1^1B_{1u} and 1^1E_{1u} states using *ab initio* techniques with an accuracy of 0.7 and 0.3 eV, respectively. Modern theoretical studies based on the MRDCI [131] and SAC-CI [132] approach achieved an accuracy of 0.5 eV or better. Only recently, we were able to predict the valence excited $\pi \rightarrow \pi^*$ states of benzene [20] with an accuracy of about 0.2 eV. In these calculations dynamic electron correlation effects are accounted for using the CASSCF/CASPT2 approach. They have been further refined since then.

In Table XIII results are presented from the most recent calculations [133] on the electronic spectrum of the benzene molecule. These data represent the most accurate *ab initio* results that have been obtained until now. The calculations were performed with ANO-L type basis sets of the following structure: C[$4s3p1d$], H[$2s1p$] and R[$1s1p1d$]. Here, R stands for Rydberg orbitals placed in the centre of the molecule. The experimental line positions of the valence excited singlet states are well known and reproduced by the calculations with a deviation of at most 0.1 eV. The experimental information is more uncertain for the $1^1B_{1g}(3d)$ and $1^1B_{1g}(3d)$ singlet Rydberg states. These lines were not observed directly, but extrapolated backwards from higher members of the series. That this extrapolation worked is shown by the agreement with the computed values, which differ from the empirical estimates by between 0.10 and 0.12 eV. The computed results for the excitation energy of the $2^1E_{1u}(3p)$ singlet Rydberg state differs more from the experimental value, 0.25 eV (assuming that the assignment is correct). We have no explanation for the somewhat larger deviation in this case. It is, however, not impossible that it is due to an erroneous interaction (at the CASSCF level) with the corresponding valence excited state (a situation similar to that obtained in the ethene and butadiene molecules).

It is instructive to regard the present calculations as benchmarks and to analyze the results in more detail. One way to classify the valence excited states of benzene is to divide them into two groups, one which is described by wave functions dominated by a single configuration, and a second which is exposed to strong multiconfigurational mixing. Both situations are frequently observed for excited state wave functions of a molecule. As an illustration we present in Table XIV the principal configurations and contributions from singly, doubly and triply excited configurations to selected singlet excited states of the benzene molecule. The ground state and the Rydberg states constitute the first group. As can be seen from the table, the 1^1B_{1u} and 1^1E_{1u} states also belong to this class. In contrast, the 2^1E_{2g} state clearly falls into the second class. It also becomes evident, that a balanced description of the most important correlation effects can barely be achieved using a single configuration reference method. The problem is further

TABLE XIII

Singlet excitation energies and oscillator strengths for the benzene molecule.

State	Excitation energy (eV)		Oscillator strength	
	computed	expt	computed	expt
($\pi-\pi^*$, $\pi-\sigma^*$) singlets				
1^1B_{2u}	4.84	4.90[a]	forbidden	
1^1B_{1u}	6.30	6.20[a,c,j]	"	
$1^1E_{1g}(3s)$	6.38	6.33[i]	"	
$1^1A_{2u}(3p)$	6.86	6.93[e,j]	0.07	
$1^1A_{1u}(3p)$	6.99		forbidden	
$1^1E_{2u}(3p)$	6.92	6.95[j]	"	
1^1E_{1u}	7.02	6.94[a,c]	0.82	1.05[b]
$2^1E_{1u}(3p)$	7.16	7.41[c]	0.0058	
$1^1B_{2g}(3d)$	7.57	7.46[g,h]	forbidden	
$1^1B_{1g}(3d)$	7.58	7.46[g,h]	"	
$2^1E_{1g}(3d)$	7.57	7.54[f]	"	
$3^1E_{1g}(3d)$	7.57		"	
$2^1A_{1g}(3d)$	7.74	7.80[e,f]	"	
$1^1E_{2g}(3d)$	7.76	7.81[e,f]	"	
$1^1A_{2g}(3d)$	7.81		"	
2^1E_{2g}	7.89	7.80[d]	"	

[a] Reference [134], [b] Reference [132], [c] Reference [135]
[d] Reference [136], [e] Reference [298], [f] Reference [299]
[g] Reference [300], [h] Reference [300], [i] Reference [301]
[j] Reference [302]

aggravated if the symmetry of the molecule is lowered by chemical substitution and therefore several states will belong to the same irreducible representation, permitting additional interactions.

Another way to group the states is by looking at the differential dynamic correlation contribution to the excitation energy as it is done in Table XV. We observe that the $\sigma\sigma$ contribution is virtually identical for all states. On the other hand, the $\sigma\pi$ and $\pi\pi$ contributions vary strongly. Hence, there is a group of states for which the excitation energy is hardly changed as compared to the CASSCF result. Others expose, however, a large differential dynamic correlation contribution, which may lead to a reordering of states. The 1^1E_{1u} and $2^1E_{1u}(3p)$ states in benzene are typical, well known representative of the second group. From the table it is also seen that the differential correlation contributions to the singlet excited valence states always leads to a lowering of the excitation energies (no lucky cancellation of terms!). Therefore, it is important for any method dealing with the dynamic correlation effects to include all electrons and reference functions.

To summarize we note the following properties of the valence singlet excited states of benzene:

– The 1^1B_{2u} state is dipole forbidden. It is seen in the spectra as a weak

TABLE XIV

Principal configurations and contributions from singly, doubly and triply excited configuration (in percent) in the CASSCF wave functions for the valence excited states of benzene.[a]

state	principal configuration	weight	single	doubles	triples
singlet excited valence states					
1^1A_{1g}	$(1a_{2u})^2(1e_{1g})^4$	88.1	88.1	8.0	-
1^1B_{2u}	$(1a_{2u})^2(1e_{1g})^3(1e_{2u})^1$	80.2	81.9	10.5	2.1
1^1B_{1u}	$(1a_{2u})^2(1e_{1g})^3(1e_{2u})^1$	91.5	91.5	1.2	-
1^1E_{1u}	$(1a_{2u})^2(1e_{1g})^3(1e_{2u})^1$	82.1	86.3	2.9	1.5
2^1E_{2g}	$(1a_{2u})^2(1e_{1g})^3(1b_{2g})^1$	14.0	48.2	40.0	1.4
	$(1a_{2u})^1(1e_{1g})^4(1e_{2u})^1$	34.2			
	$(1a_{2u})^2(1e_{1g})^2(1e_{2u})^2$	20.5			
singlet excited Rydberg states					
1^1E_{1g}	$(1a_{2u})^2(1e_{1g})^3(3s)^1$	86.3	86.3	-	-
2^1E_{1u}	$(1a_{2u})^2(1e_{1g})^3(3p_\pi)^1$	86.0	86.0	-	-
1^1A_{2u}	$(1a_{2u})^2(1e_{1g})^3(3p_\sigma)^1$	88.6	88.6	-	-
1^1E_{2u}	$(1a_{2u})^2(1e_{1g})^3(3p_\sigma)^1$	88.2	88.2	-	-
1^1A_{1u}	$(1a_{2u})^2(1e_{1g})^3(3p_\sigma)^1$	88.4	88.4	-	-

Only CI-coefficients larger than 0.1 were used to sum the contributions.

TABLE XV

Differential second-order correlation energy partitioned into σ-σ, σ-π, and π-π contributions with respect to the ground state of benzene. Energies are given in units of Hartree.

state	σ-σ	σ-π	π-π
singlet excited valence states			
1^1A_{1g}	0.0000	0.0000	0.0000
1^1B_{2u}	+0.0020	+0.0022	-0.0062
1^1B_{1u}	+0.0038	+0.0217	+0.0115
1^1E_{1u}	+0.0037	+0.0409	+0.0113
2^1E_{2g}	+0.0020	+0.0041	-0.0043
singlet excited Rydberg states			
1^1E_{1g}	+0.0058	-0.0131	+0.0027
2^1A_{2u}	+0.0062	-0.0144	+0.0010
2^1E_{2u}	+0.0062	-0.0138	+0.0012
2^1A_{1u}	+0.0062	-0.0132	+0.0008
2^1E_{1u}	+0.0058	-0.0291	-0.0087

band and gets the intensity by vibrational coupling. It is a covalent state. The calculated and observed vertical excitation energies are 4.84 and 4.90 eV [134], respectively.

- The 1^1B_{1u} is an ionic state, that is, the wave function exposes large coefficients for configuration state functions that correspond to ionic structures. The dynamic correlation contribution to the state energy is considerably larger than for the ground state. The predicted position, 6.28 eV, differs not more than 0.08 eV from the observed location, 6.20 eV [134, 135].

- Transitions to the 1^1E_{1u} state are dipole allowed with a band maximum at 6.94 eV [134, 135]. The experimental determined oscillator strength ranges from 0.6–1.1. This state is very ionic and is computed to be 7.09 eV higher in energy than the ground state.

- The 2^1E_{2g} is the highest valence excited stated included in the present study. It has only recently been observed and is located at 7.80 eV [136]. It is a covalent state which is characterized by a large contribution from double and higher excitations (more than 40% of the weight of the wave function stems from this class of configurations).

3.5 Pyrimidine, pyrazine, pyridazine and *s*-triazine

The azabenzenes are isoelectronic with benzene. Substitution of a C-H group by an nitrogen atom leads to a lowering of the symmetry of the molecule and to a splitting of the degenerate orbitals correlating with the e_{1g} and e_{2u} orbitals in benzene. As a consequence transitions to the "1^1B_{2u}" and "1^1B_{1u}" valence excited states become dipole allowed except in *s*-triazine. Therefore, one may expect at least a broadening of the "E_{1u}"-band, and, in addition, a number of weak $n \rightarrow \pi^*$ transitions.

A survey of experimental and theoretical studies on the electronic structure of the azabenzenes has been given by Innes *et al.* [137]. In addition, a comprehensive collection of vacuum UV absorption spectra and theoretical data has been presented by Bolovinos *et al.* [138]. Recently, a series of studies have also been published by Walker *et al.* [139]-[142], which includes vacuum UV absorption spectra, electron energy loss spectra, extensive compilations of theoretical data and MRDCI calculations. Besides the MRDCI calculations by Walker *et al.* only a few other *ab initio* studies on the electronic spectra of the present molecules have been reported. These investigations include various CI-techniques based on single determinant, GVB, and CASSCF reference functions.

The study of the azabenzenes were the first one (except benzene), which was performed with the CASSCF/CASPT2 method [18]. At this time the importance of a good representation of the Rydberg orbitals also in studies of valence excited states was not clear to us and corresponding basis functions were not included in the AO basis set. As a consequence the computed excitation energies are not as accurate as those obtained in later applications and in some cases intruder states of a quasi-Rydberg type caused problems. For the pyridine molecule we have performed new calculations with better basis sets (see below) and the differences in the results illustrate the importance of a good representation of the Rydberg orbitals.

All computations were done at the ground state geometry determined by microwave spectroscopy (c.f. Table 5 in Ref [137]). ANO-L basis sets of the following structure: C,N[$4s3p2d$] and H[$3s2p$] was used. To compute the $\pi \rightarrow \pi^*$ excited singlet states the active space was constructed by distributing the 6 π electrons among 12 π-orbitals. The 12-active-orbital rule could not be applied to the pyridazine molecule, due to the presence of Rydberg-type intruder states. Instead we used eight π-orbitals and four σ-orbitals. The AMOs used to determine the $n \rightarrow \pi^*$ singly excited states included two σ-orbitals for each lone pair. It was found sufficient to use only six π-orbitals. In two cases, pyridine and pyridazine, it was necessary to use somewhat larger active spaces to avoid intruder states of $\sigma \rightarrow \sigma^*$ and $\sigma \rightarrow \pi^*$ type. All excitation energies are computed using the diagonal Fock operator only and to overcome the intruder state problem we deleted the corresponding orbitals from the molecular orbital basis. The computed excitation energies and oscillator strength are collected in Table XVI including also experimental values.

In the following we shall briefly discuss each molecule individually. In general, it seems to be difficult to give an experimental measure of the energy separation of the states corresponding to a degenerate state in benzene. Hence, to compare with observed data we use the average energies.

Pyridazine (1,2-diazabenzene). The present study includes the three lowest vertically excited $\pi \rightarrow \pi^*$ singlet transitions and four $n \rightarrow \pi^*$ singlet excited states[141]. The difference between the calculated and the observed band positions range from 0.1 to 0.2 eV. Unfortunately, the reference weight of the "E_{1u}" states is somewhat small, which is due to Rydberg-type functions not included in the active space. The deficiency is much more severe for the "E_{2g}" states and therefore the results are not included. The present results agree with an earlier *ab initio* calculation reported by Walker and Palmer [141].

Pyrimidine (1,3-diazabenzene). The vacuum UV spectra shows three strong bands assigned to $\pi \rightarrow \pi^*$ excitations with maxima at 5.1, 6.7 and 7.6 eV. The lowest $n \rightarrow \pi^*$ 0-0 transition is observed at 3.85 eV [138]. The calculations reproduce the observed data with an error less than 0.2 eV. The present calculations are in qualitative agreement with earlier CI calculations [140], but are overall more accurate. It has been possible to compare the present approach with MRCI calculations [8], which have been designed to account for as much dynamic correlation effects as possible. The reference function was based on a CASSCF calculation similar to those performed here. The CI wave function comprised all singly and doubly excited states with respect to the reference functions with the added condition that there was at most one hole in the σ-orbitals. Thus dynamic polarization of the σ-orbitals was included but no $\sigma\sigma$ correlation terms. Such an approach can be justified keeping in mind that these terms were shown to be almost constant in the case of benzene.

These MRCI calculations give an illustrative example of the difficulties that such an approach has when it is applied in studies of excited states of systems where a larger number of electrons has to be correlated. In spite of pushing the

TABLE XVI

Singlet excitation energies and oscillator strengths of pyrimidine, pyrazine, pyridazine and s-triazine. Experimental values are given in parentheses[a].

state	pyridazine		pyrimidine	
	excitation energy	oscillator strength	excitation energy	oscillator strength
π–π^* singlet excited states				
1^1B_{2u}	4.86(4.9$_b$-5.0$_c$)	0.009(0.02$_b$)	4.93(5.12)	0.001(0.028)
1^1B_{1u}	6.61(6.2$_b$-6.5$_c$)	0.003(0.10$_b$)	6.72(6.7)	0.05(0.094)
1^1E_{1u}	7.39,7.50(7.1$_b$-7.3$_c$)	0.75,0.50	7.32,7.57(7.57)	0.58,0.79(0.72)
2^1E_{2g}			7.82,8.31	0.06,0.14
n–π^* singlet excited states				
	3.48(3.4$_b$)	0.01(0.006$_b$)	3.81(3.85)	0.02(0.005)
	3.66	forbidden	4.12	forbidden

state	pyrazine		s-triazine	
	excitation energy	oscillator strength	excitation energy	oscillator strength
π–π^* singlet excited states				
1^1B_{2u}	4.77(4.81)	0.07(0.062)	5.33(5.70)	forbidden
1^1B_{1u}	6.68(6.51)	0.08(0.10)	6.77(6.86)	forbidden
1^1E_{1u}	7.57,7.75(7.67)	0.76,0.66(0.72)	8.16(7.76)	0.61(0.73)
2^1E_{2g}	8.16,8.27	forbidden	8.03	0.21
n–π^* singlet excited states				
	3.58(3.83)	0.01(0.006)	3.81	forbidden
	4.37	forbidden	4.00(4.59)	0.015(0.013)

[a] Unless stated explicitly, the experimental data are taken from ref [138]
[b] Reference [137]
[c] Reference [141]

program to its limit, it was not possible to converge the MRCI calculations with respect to the choice of the reference space (which was selected from a CASSCF wave function). Thus the results remain somewhat ambiguous. The agreement with experiment was also not better than about 0.4 eV, which is twice the error obtained in the CASSCF/CASPT2 study. Even though a somewhat larger basis set was used here, the major difference is most likely due to the incompleteness of the MRCI calculations compared to the CASPT2 treatment, where a full CASSCF reference function could be used.

Pyrazine (1,4-diazabenzene). The lowest $\pi \rightarrow \pi^*$ excited state of pyrazine with an observed band maximum at 4.81 eV [138] is predicted to appear at 4.77 eV. The second and third singlet excited valence states have also been assigned, 6.51 and 7.67 eV [138]. The computed transition energies, 6.68 for the "$^1B_{1u}$" state and 7.57, 7.75 eV for the "$^1E_{1u}$" pair of states, are in agreement with experiment. In the energy range 8.0–9.0 eV the spectrum is rich in structures attributed to

Rydberg states and appear to hide the bands corresponding to the "$^1E_{2g}$" states.

Three of the lowest energy $n \rightarrow \pi^*$ transitions have been identified in experiment [138]. The computed excitation energies are consistent with the available data, and in addition place a fourth $n \rightarrow \pi^*$ excited state at 4.37 eV.

There have been a few reported *ab initio* calculations of the excited state spectrum of pyrazine. The most recent results are based on MRDCI calculations [141]. Earlier, Hackmeyer and Whitten [143] employed CI techniques using a single determinant closed shell reference function whereas Wadt *et al.* [144] approached the spectrum using CI methods based on a GVB reference function. These results are comparable to the present CASSCF calculations. In contrast, the theoretical excitation energies reported by Walker and Palmer [141] are in accord with the present PT2D calculations.

s-Triazine. The first three $\pi \rightarrow \pi^*$ absorption lines, which have been measured by vacuum UV spectroscopy, are located at 5.70, 6.86 and 7.76 eV [138]. In contrast to the di-azabenzenes, the computed excitation energies are somewhat more inaccurate. On the one hand, the "$^1B_{2u}$" state energy appears to be to low. On the other hand, the excitation energy for transition to the "$^1E_{1u}$" state is overestimated. Also, there is a large difference in the reference weight for these states. Since our programs are limited to point groups with twofold symmetry operations, the calculations were carried out in C_{2v} symmetry. Although the orbitals were restricted manually to have D_{3h} symmetry at the CASSCF level of approximation, symmetry breaking is observed in the CASPT2 calculation. The final resolution of this problem will have to await more extensive calculations.

3.5.1 Pyridine, phosphabenzene and phenol

The photophysics of pyridine has been investigated almost as extensively as benzene. In contrast, very little is known, experimentally or theoretically, about phenol and phosphabenzene. Therefore the application of the CASSCF/CASPT2 method to these compounds is primarily of predictive interest. To enable comparisons on an equal footing we repeated our earlier calculations on the pyridine molecule and report here our latest results.

The calculations on pyridine, phenol and phosphabenzene closely follow the general outline, that is to say, experimental ground state geometries [137, 145, 146] were used together with ANO-L type basis sets contracted to TZP quality in the valence shell. In addition the basis sets include contracted Rydberg ANO-type functions, one set for each angular momentum, which have been constructed as described in section 2. The active spaces were selected dependent on the type and symmetry of the excitations to be studied. The interested reader is referred to the original publications for further details.

In the following we shall only focus on the $\pi \rightarrow \pi^*$ and $n \rightarrow \pi^*$ singlet excited states for which the results are collected in Table XVII. However, we computed also the excitation energies for the members of the lowest Rydberg series and a number of triplet excited states for which the results are given elsewhere.

Pyridine (azabenzene). The vacuum UV spectrum of pyridine presented by Bolovinos *et al.* [138] shows three $\pi \rightarrow \pi^*$ absorption band systems located at 4.99, 6.38 and 7.22 eV. Recently, these findings were fully confirmed by Walker *et*

TABLE XVII

Singlet excitation energies and oscillator strengths of pyridine, phosphabenzene and phenol. Experimental values are given in brackets.

State	pyridine[a]		phophabenzene[b]		phenol[c]	
	excitation energy	oscillator strength	excitation energy	oscillator strength	excitation energy	oscillator strength
π–π^* singlets						
1^1B_{2u}	4.84(4.99)	0.018(0.029)	4.10(4.2)	0.002	4.53(4.51)[c]	0.007(0.020)
1^1B_{1u}	6.42(6.38)	0.005(0.085)	5.32(5.2)	0.17	5.80(5.77)[d]	0.005(0.132)
1^1E_{1u}	7.23,7.48(7.22)	0.82,0.64(0.90)	5.85,6.30	0.63,0.72	6.50,6.56(6.66)[d]	0.681,0.779[c]
2^1E_{2g}	7.96,7.95	0.03,0.158	6.67,7.21	0.004,0.001	7.14,7.72	0.108,0.013
n–π^* singlets						
	4.91(4.59)	0.009(0.003)	4.44(4.7)	0.06		
	5.17(5.43)	forbidden	4.63			

[a] The experimental data on pyridine are taken from Ref. [138]

[b] The experimental data on phophabenzene are taken from Ref. [152]

[c] The experimental data on phenol are taken from Ref. [153, 154]
The combined oscillator strength for the 1^1E_{1u} state is 1.1.

[d] Substituent shifts from benzene, cf. Ref. [156]

al [139]. The computed transition energy for the two lowest $\pi \rightarrow \pi^*$ singlet excited states, 4.84 and 6.42 eV, are in excellent agreement with the experimental data. The calculated excitation energies of the nearly degenerate states corresponding to the $^1E_{1u}$ state in benzene are 7.23 and 7.48 eV, respectively. The lower transition energy falls right on the top of the experimental band maximum and the second is 0.25 eV higher in energy.

The theoretical literature on the absorption spectrum of pyridine includes several reports of *ab initio* calculations. In general, the MRDCI calculations by Walker and Palmer [139] and the SAC-CI results presented by Kitao and Nakatsuji [147] are in qualitative agreement with experiment. In contrast, Foresman *et al.* [6] failed to reproduce the experimental state ordering and the computed excitation energies are spread over a wide range. It is also worth noting that the present results for the high-energy states differ somewhat from those obtained in our previous calculations [18], which were carried out together with the di-azabenzenes. The energy splitting of the nearly degenerate states was predicted to be larger, 0.65 eV. Preference is, however, to be given to the present results for the following reasons: The predicted splitting appears to be small enough to explain the experimental difficulties in separating the lines. Moreover, the calculations reported here are performed using superior basis sets with respect to the description of Rydberg states, which might have contaminated our earlier calculations as indicated by the weight, ω, of the reference states. They are almost identical for the present wave functions but differed substantially in our previous attempts. The second pair of nearly degenerate states are predicted to be located at 7.95 and 7.96 eV. These energies are $\approx$0.2 eV larger than those computed earlier.

The assignment of the lowest $n \rightarrow \pi^*$ band is, unfortunately, not straight forward. Goodman [148], Villa *et al.* [149] and Doering and Moore [150] report excitation energies of 4.6, 4.3 and 4.3 eV, respectively. The values are 0.3–0.6 eV lower than the computed result. Villa *et al.*, however, argued that this may be an adiabatic transition. To check this possibility we allowed the geometry to relax while insisting on C_{2v} symmetry. To reduce the computational costs the geometry optimizations were performed used Dunning's [151] basis sets. The bond angle at the nitrogen atom opens almost 10^0 for the $^1B_1$ state relative to the ground state and the C-N bond lengthens 0.056 Å. The computed adiabatic excitation energy, 4.37 eV, is 0.58 eV smaller than the vertical excitation energy and is in agreement with the experiment.

Phosphabenzene. UV/VIS and MCD solution spectra were published for phosphabenzene by Waluk *et al.* [152]. These spectra cover the range 4.0–5.5 eV. Within that region three transitions can be identified and are located at 4.2, 4.7 and 5.2 eV. Our results can be brought into agreement with these data if it is assumed that the lowest band in the MCD spectrum is due to transitions to the 1^1B_2 state. As a consequence, the second and third bands are then assigned to the 1^1B_1 and 2^1A_1 states, respectively. Such a correspondence would reduce the error of the singlet excited $\pi \rightarrow \pi^*$ valence states, in accordance with experience, to 0.1 eV whereas the $n \rightarrow \pi^*$ valence states would have a somewhat larger error, 0.26 eV. Using the PPP method combined with parameters derived from experimental ionization potentials and electron affinities as well as INDO/S calculations, Waluk *et al.* predicted the state ordering. To the authors knowledge there exist no other calculations on the absorption spectrum of phosphabenzene.

Phenol. Kimura and Nagakura [153] observed the absorption spectrum of phenol using vacuum UV techniques. The bands are located at 4.59 and 5.82 eV for the "$^1B_{1u}$" and "$^1B_{2u}$" states, respectively, and their vibrational fine structure is also known [154, 155]. Furthermore, they decomposed the so-called ^1B band — corresponding to $^1E_{1u}$ in benzene — into two bands, at 6.70 and 6.93 eV. However, this decomposition is very unconvincing. Using instead the well-known benzene spectrum together with generally accepted experimental substitution shifts (see, for example, Petruska [156]) gives 4.54, 5.77, and 6.66 eV. The "$^1B_{1u}$" transition energy is in fact very precisely known (4.507 eV) from modern two-photon spectroscopy [154].

The position of the low energy $\pi \rightarrow \pi^*$ bands are reproduced with an accuracy of 0.05 eV. The slightly split "$^1E_{1u}$" state is 0.13 eV lower than the 6.66 eV obtained by combining the benzene $^1E_{1u}$ energy with the substituent shift tabulated by Petruska [156]. This result is in agreement with the spectrum published by Kimura *et al.*, but is in contrast to their interpretation. The weak features observed at the high energy flank of the broad, slightly asymmetric peak may then be interpreted as the $^1E_{2g}$ state with an approximate maximum at about 7.2 eV. With these assumptions the error between the computed and observed spectrum is approximately 0.15 eV. The small splitting of the "$^1E_{1u}$" state in phenol is consistent with the larger body of spectroscopic data on the azabenzenes, where it has also been impossible, so far, to resolve the components of this band.

3.5.2 Summary

In this section we have presented theoretical transition energies for the vertical $\pi \rightarrow \pi^*$ and $n \rightarrow \pi^*$ singlet excited states of benzene and a number of derivatives. The present calculations on the benzene molecule are accurate to 0.1 eV and it is difficult to see how more advanced theory and calculations could improve upon these results.

The spectra of the azabenzenes are remarkably similar to that of the benzene molecule. In particular, we note that the lowest excited state, 1^1B_{2u}, is observed within a rather narrow energy band of $\approx$0.1 eV, except for *s*-triazine. Transition to higher states are shifted to the blue, and the states correlating with the degenerate states in benzene are split by $\approx$0.2 eV. In the low energy region of the spectrum, additional $n \rightarrow \pi^*$ states are seen. The $\pi \rightarrow \pi^*$ vertical excitation energies of the azabenzenes have been reproduced with an accuracy of 0.20 eV or better. The error is somewhat larger, 0.3 eV, for the $n \rightarrow \pi^*$ transitions in all cases where comparisons to unambiguous experimental data can be made. The only exception is *s*-triazine. Unfortunately, a direct comparison of the computed oscillator strengths with experiment cannot be made, since the experimental values also include vibrationally induced intensities. The situation is, however, less severe for the strong oscillators and the theoretical values are there in agreement with experiment.

By comparing the computed $\pi \rightarrow \pi^*$ transition energies we find that the spectra of phenol and phosphabenzene are shifted on the average by 0.6 and 1.07 eV, respectively, to lower energies relative to pyridine. On the other hand, the relative state separations are approximately identical for the three molecules. The line splitting of the states corresponding to the degenerate E-states in benzene is somewhat larger for phosphabenzene while it is very small for phenol. It is also interesting to note that the dynamic correlation contributions to the lower excited states, especially the very ionic "E_u"-states, hardly changes.

The theoretical and observed $\pi \rightarrow \pi^*$ transition energies for the pyridine molecule differ by at most 0.2 eV. Apparently, the error is somewhat larger for the $n \rightarrow \pi^*$ transitions. It was shown, however, that assuming the measured transition energy to correspond to an adiabatic transition, leads to much better agreement. There is only a limited number of reported experimental investigations on the electronic spectra of phenol and, in particular, phosphabenzene. The available data, however, support the assumption that the accuracy of the present results is comparable to that obtained for the pyridine molecule, provided that Kimura and Nagakura were mistaken in their decomposition of the "$^1E_{1u}$" band.

3.6 The porphin molecule

The inherent bottleneck of the CASPT2 method is the number of active orbitals and active electrons. For many systems, even when the number of π electrons might be exceedingly large, this limitation can be overcome in practice by a careful consideration of the differential correlation effects and still the CASPT2 method can be successfully applied. In order to analyze the performance of the methodology for a large system, where the π-valence electron correlation cannot be fully accounted for at the CASSCF level, the CASPT2 method was applied to determine the main features of the spectrum of free base porphin (FBP) shown in figure 1, a system involving 24 π-valence MOs and 26 π-valence electrons. As illustrated below and described in detail elsewhere [157], analysis of the electron correlation effects for the corresponding excited states of FBP makes feasible a proper theoretical determination of its spectrum. Porphyrins and their analogs participate in a wide variety of fundamental biological processes. Free base porphin can be considered the basic unit from which porphyrins derive and has consequently received considerable attention.

The electronic structure of FBP has been the subject of several *ab initio* studies [6],[158]-[170]. Due to the size of the molecule, most of the previous *ab initio* calculations carried out to interpret the electronic spectrum of FBP either were performed using a minimal basis set, correlating only the π electrons, or employing improved basis sets, but considering a limited number of σ and π MOs at a given incomplete correlated level. The truncation achieved in the one- and many-electron basis sets introduced an unknown degree of uncertainty in the computed excitation energies. In our study all σ and π electrons (except the $1s$ core electrons) are explicitly correlated within the space spanned by the one-electron basis set, with no further assumptions. As far as we are aware, it represents the first fully correlated *ab initio* study of the electronic spectrum of FBP. It makes possible an analysis of the validity of previous approximations and yields a full interpretation of the Q and B bands. Preliminary to the characterization of the excited states, the issue of the relative stability between the two possible ground-state geometries was addressed

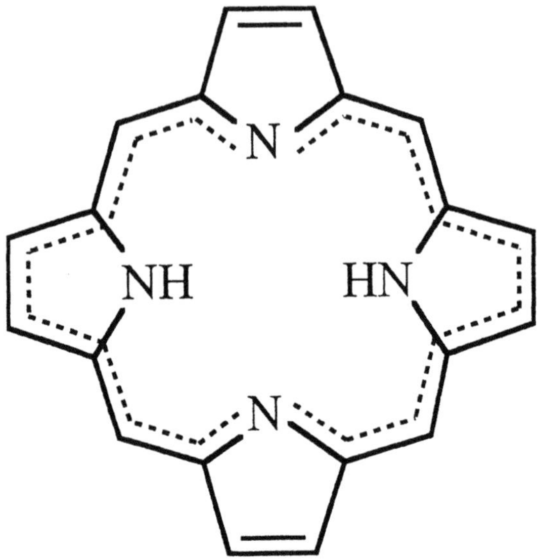

Fig. 1. Free base porphin structure.

[170]. The CASPT2 results yielded a D_{2h} ground state, which is about 1 eV more stable than the C_{2v} structure, in agreement with a recent geometry optimization performed at the MP2 level [169].

The electronic spectrum of FBP is characterized by three regions [163],[171]-[176]. The so-called Q band in the visible region is relatively weak. The intense Soret or B region occurs in the near-UV. The higher UV bands appearing in the third region are broad and diffuse. Experimentally the following features are known: (i) the Q band shows more structure than the B band, (ii) the lowest component of it has a weaker intensity and is polarized parallel to the inner hydrogen-hydrogen axis, and (iii) the B band presents parallel polarization on its low energy side and perpendicular polarization on the high energy side of the band.

The following qualitative picture is supported by the available experimental data. The spacing between the HOMO$-$1 and the lower π orbitals is significant. On the other hand, the two lowest unoccupied MOs have orbital energies close to the two highest occupied MOs [170]. The largest near-degeneracy effects ought therefore to occur within this four-orbital space. This is the basis for the four-orbital model originally suggested by Gouterman *et al.* [177]-[179]. These four valence-type π-orbitals are in order of increasing energy: $5b_{3u}$, $2a_u$, $4b_{2g}$, and $4b_{1g}$, the first two being doubly occupied in the closed-shell RHF wave function. From this structure, we expect to find four low-lying optically-allowed valence singly excited singlet states. The excitations $5b_{3u} \rightarrow 4b_{2g}$ and $2a_u \rightarrow 4b_{1g}$ give rise to two states of B_{1u} symmetry (parallel polarization), while the two excitations

TABLE XVIII

Calculated excitation energies (in eV) and oscillator strengths of the vertical singlet excited states of free base porphin correlating only the π electrons (πPT2F results) and including also the contributions of the σ electrons at second-order level. The CASSCF(4/4) wave function was used as reference function. The STO-3G basis set was employed throughout.

	Excitation energy (eV)									Osc.str.		
	CAS							Prev			Prev	
State	(4/4)[a]	πPT2F	PT2F	Exptl				b	c	TW	Exp[d]	b
1^1B_{1u}	3.99	1.85	1.20	1.98[e]	2.01[g]	2.02[h]	1.98[i]	2.27	2.77	$<10^{-4}$	0.02	0.02
1^1B_{2u}	4.56	3.48	2.74	2.42[e]	2.38[g]	2.39[h]	2.33[i]	3.42	3.15	0.01	0.07	0.04
2^1B_{1u}	5.76	4.69	3.74				3.14[j]	4.72	5.04	1.92		1.87
				3.33[f]	3.13[g]	3.17[h]					1.15[k]	
2^1B_{2u}	5.87	4.93	3.98				3.17[j]	5.29	5.71	1.94		2.55

[a] Four active orbitals (HOMO, HOMO-1, LUMO, and LUMO+1) and four active electrons.

[b] MRSDπCI results from ref. [163]

[c] CIS results from ref. [6]

[d] Ref. [163]

[e] Q(0-0) data from the optical spectra in gas phase at 393°C, ref. [174]

[f] Band maximum of the vapour optical spectra at 322°C, ref. [174]

[g] Optical spectra in benzene at 25°C, ref. [163, 174]

[h] Optical spectra in ethanol at -180°C, ref. [174]

[i] Site selected optical spectra in anthracene at low temperature, ref. [175]

[j] Absorption data in ethanol solid glass at -180°C, ref. [171]

[k] This value corresponds to the sum of oscillator strengths of two states, ref. [163]

$5b_{3u} \rightarrow 4b_{1g}$ and $2a_u \rightarrow 4b_{2g}$ result in two states of B_{2u} symmetry (perpendicular polarization). The corresponding electronic configurations of the same symmetry will mix, since they are quasi-degenerate, to form "minus" and "plus" states. The B_{1u}^- and B_{2u}^- states have the lower energy and low intensity, while the B_{1u}^+ and B_{2u}^+ states will be pushed up and carry most of the intensity. The relative location between the two minus (and two plus) states is more difficult to predict from these simple considerations and one has to rely on actual calculations.

We focus first on changes in the computed vertical transitions when electron correlation effects involving the π electrons and including both π and σ electron correlation contributions are accounted for. Employing the STO-3G basis set [180], the CASSCF(4/4) wave function (four active orbitals, HOMO, HOMO−1, LUMO, and LUMO+1, four active electrons) was used as reference function in the CASPT2 treatment. Two sets of calculations were performed: correlating only the π electrons, π-PT2F(4/4) results (σ MOs were kept frozen at second order), and including also the contributions of the σ electrons at the second-order level. The whole virtual space was employed in both cases. The results are listed in Table XVIII.

The optical spectra [174] of FBP show three major features around 2.0, 2.4, and 3.3 eV. The CASSCF(4/4) results deviate from these experimental data by more than 2 eV. As could be expected, the differential correlation effects involving all π electrons are important. Therefore, a significant improvement is noted at the π-PT2F(4/4) level. The agreement with experiment is particularly reasonable for the computed excitation energy of the 1^1B_{1u} state, 1.85 eV. The π-PT2F(4/4) results are consistent with the previous *ab initio* findings at a comparable level of theory. The MRSDπCI study of Nagashima *et al.* [163], employing a minimal basis set, gave excitation energies for the 1^1B_{2u} and 2^1B_{1u} states very close to the results obtained at the π-PT2F(4/4) level. The computed transition energies for the 1^1B_{1u} and 2^1B_{2u} states at the MRSDπCI level are however around 0.4 eV larger than the π-PT2F(4/4) results.

When electron correlation contributions of the σ electrons are also taken into account, PT2F(4/4) results, the nice agreement with experiment found for the 1^1B_{1u} transition energy is lost while further improvement in the description of the remaining states is achieved. The differential correlation contributions involving both σ and π electrons behave in such a way that the transition energy for the first singlet excited state becomes underestimated. An overestimation for the other three states still persists compared to experiment. Nevertheless, considering the limited one-electron basis set employed in these initial calibration calculations, the PT2F(4/4) result can be regarded as satisfactory; deviation from the experimental data is less than 0.8 eV. The computed oscillator strengths are also in agreement with experimental evidence [163]. As shown in Table XVIII using also the the STO-3G basis set, larger deviations were obtained with the CIS procedure (the same STO-3G basis set was used) [6]. The neglect of higher (doubly substituted) configurations in the CIS expansion makes a sizeable contribution to the differential correlation energy between the ground and the excited state, placing the excited states too high.

We therefore conclude from this part of the study that both σ and π electron

TABLE XIX

Calculated excitation energies (in eV) and oscillator strengths (within parentheses) of the vertical singlet excited states of free base porphin using the ANO-type C,N[3s2p]/H[1s] basis set.

State	PT2F(4/4)	PT2F(O/E)[a]	Exp[b]	Other results	
				CIS[c]	MRSD$\sigma\pi$CI[d]
1^1B_{1u}	1.33(0.008)	1.70(0.001)	1.98-2.02(0.02)	2.56	1.97
1^1B_{2u}	2.05(0.024)	2.26(0.016)	2.33-2.42(0.07)	2.71	2.40
2^1B_{1u}	2.99(1.676)	2.91(1.662)	3.14-3.33(1.15)	4.66	3.41
2^1B_{2u}	3.08(1.589)	3.04(1.542)		4.91	3.24

[a]Best estimate. O and E denote the number of active orbitals and electrons, respectively. $1^1B_{1u}(11/10)$; $1^1B_{2u}(7/6)$; $2^1B_{1u}(6/6)$; $2^1B_{2u}(6/6)$.
[b]See Table 1 for proper references to the experimental data.
[c]CIS results from ref. [6]
[d]MRSD$\sigma\pi$CI results from ref. [167]

correlation contributions are strongly differential. They have to be included in the theoretical treatment to give a reliable description of the electronic spectrum of FBP. For this purpose, larger one-electron basis sets have to be employed. In order to get more insight into the problem, the full π valence correlation was analyzed by using the restricted active space (RAS) SCF method [181, 182] together with the ANO-type C,N[3s2p]/H[1s] basis set [35]. This calculation had all the π-orbitals active, but allowed only single and double substitutions from the SCF ground state configuration. The results made it possible to scrutinize the validity of the CASSCF(4/4) wave function as a reference function in the CASPT2 treatment and to develop criteria to enlarge in a systematic way the number of active orbitals. The main results obtained in the final CASPT2 calculations are summarized in Table XIX.

The π-RASSCF results ought to give a good estimate of the full π-valence differential correlation energies, yielding results close to the full π-valence CASSCF treatment. The computed transition energies at this level of theory are too high, 2.60, 3.61, 4.84, and 4.89 eV for the 1^1B_{1u}, 1^1B_{2u}, 2^1B_{1u}, and 2^1B_{2u} states, respectively. Consistent with the previous experience using a minimal basis set, π electron correlation is not enough to accurately describe the electronic spectrum of FBP. However, analysis of the π-RASSCF wave functions gives further insight into the electronic structure of the excited states. The percentage of configurations within the four-orbital model in the corresponding π-RASSCF wave function lies in the range 72–77%. The lowest singlet state, 1^1B_{1u}, has the lowest percentage. It gives the clue to why in this case configurations involving orbitals outside of the four-orbital model are particularly important. This is also apparent in the occupation numbers of the corresponding RASSCF natural orbitals (NOs) [157]. It is therefore not unexpected that for the lowest singlet excited state a larger deviation from the experiment occurs when the CASSCF(4/4) wave function is used as reference function for the CASPT2 treatment (cf. Table XIX). That deviation is due to a non-negligible contribution from a number of additional orbitals in the description of the near-degeneracy effects in the lowest excited state.

Analysis of the magnitude of the occupation numbers of the π-RASSCF NOs made it possible to expand the active space in a systematic way towards a converged solution (in the sense of a full π-CAS reference function). It is possible to perform this expansion in a systematic way and stop when no further change of the CASPT2 energies occur. These 'converged' values for the transition energies are listed in Table XIX. Below we discuss these results and compare them with experimental data and previous *ab initio* results.

The improvement for the lowest singlet-singlet transition upon enlarging the one-electron basis is not significant at the PT2F(4/4) level (from 1.20 to 1.33 eV). As mentioned above, contributions outside the four-orbital model are particularly relevant for this transition. Convergence studies increasing the active space lead to 1.70 eV, as the best estimate for the vertical transition energy, which is in reasonable agreement with the available 0-0 optical absorption data [163, 174, 175]. Further enlargements of the one-electron basis set at PT2F(4/4) level affects the computed first singlet-singlet transition energy by less than 0.1 eV. We therefore conclude that the remaining deviation from experiment (about 0.3 eV), assuming that the 0-0 and vertical transition energies are similar, is mainly due to the still limited reference function. Notice, that in the CASSCF wave function, 10 electrons and 11 orbitals are active, which represent about half of the full π space.

Using a split-valence basis set the CIS result places the 1^1B_{1u} state at 2.56 eV [6]. Inclusion of diffuse functions do not significantly affect this value, which is in agreement with our own experience and consistent with the valence character of the states under consideration. The MRSD$\sigma\pi$CI study of Yamamoto *et al.* [167] locates it at 1.97 eV. In view of the preceding analysis showing the theoretical difficulties encountered, in particular for the appropriate description of the first singlet-singlet transition, the MRSD$\sigma\pi$CI result is somewhat surprising. In those CI calculations the lowest 48 σ orbitals were frozen. We only freeze the 1s MOs (lowest 24 σ MOs). In addition, a truncation of 60% of the σ correlating orbitals was used [167]. The whole virtual space is employed here. It was already noted, using the minimal basis set, that for the lowest singlet excited state both $\sigma\sigma$ and $\sigma\pi$ correlation make a sizable contribution to the excitation energy. Using the STO-3G basis set, the PT2F results is 1.20 eV (only 1s MOs frozen and employing the whole virtual space) (cf. Table XVIII). If the virtual σ MOs are deleted the transition energy increases to 1.58 eV. Thus, the effect of discarding the virtual σ MOs is to increase the excitation energy by 0.40 eV. If now the occupied σ are frozen, i.e., only π electrons are correlated using only the π virtual space, the corresponding PT2F value increases to 1.61 eV. If in addition, a few deep π MOs are not correlated, the excitation energy would further increase. We already know the result obtained by correlating only four electrons within four orbitals, the CASSCF(4/4) result: 3.99 eV. Thus, in the range from 1.2 to 4.0 eV there is a large variety of possible computed values, depending on how many electrons are correlated and how many σ and π MOs are employed. Such truncation might fortuitously lead to good agreement with experiment. It seems then reasonable to suppose that the excellent result obtained by Yamamoto *et al.* [167] can be related to a fortunate cancellation of errors in the limited treatment performed.

For the remaining singlet-singlet transitions, there is agreement between the theoretical data and experiment. Indeed, the PT2F(4/4) results already give a reasonable description for these states. It particularly holds for the B band (2^1B_{1u} and 2^1B_{2u} states), for which the computed transition energies are not so sensitive to the improvement of the reference function. The CIS results reported by Foresman *et al.* [6] give rise to transition energies which deviate from experiment with up to 1.6 eV. On the other hand, the MRSD$\sigma\pi$CI results [167] show a reverse ordering for the states associated to the B band.

To summarize, not withstanding that the reference function involves about half of the π-valence orbitals and electrons, the calculation accurately reproduces the excited state manifolds of FBP as to state energies, symmetries, and transition types. The computed excitation energies are found within 0.3 eV of the experimental energies and the oscillator strengths are in good relative agreement with their experimental values. The splitting of the Q bands is calculated to be 0.56 eV, about 0.1 larger than the experimental gas-phase value [174]. In agreement with polarization studies of tetraphenylporphin [172], the lower Q band shows parallel polarization and the higher Q band perpendicular polarization. The intensity of the former is weaker than the latter, in agreement with the experimental evidence [163]. Low temperature studies of the B band in ethanol glass resolved a splitting of 0.03 eV between equally intense transitions [171], to be compared to the computed splitting of 0.13 eV. The lowest state of the B band shows parallel polarization and the higher perpendicular polarization, in agreement with experiment [172]. It is found that electron correlation effects of the σ electrons, in addition to those of the π correlation contributions, have to be taken into account in order to give a balanced description of the excited states. Theoretical treatments based on only π-electron correlation, or limited models including the correlation of the σ electrons only partially, are not reliable.

3.7 Internal Charge Transfer and Dual Fluorescence in Aminobenzonitriles

3.7.1 Introduction

The photophysics of many donor-acceptor systems has been extensively studied since the discovery by Lippert *et al.* [183, 184] of the dual fluorescence exhibited by the p-dimethylamino-benzonitrile (DMABN) molecule (shown along with ABN in figure 2) in polar solvents. The observed higher-energy band corresponds to the expected aniline-like emission [185] taking place from a state (LE, locally excited state) with low polarity [186, 187]. The lower-energy band was characterized as originating from a strongly polar state with evidence of strong charge transfer [188]-[191] (the CT, charge transfer state). Models to explain the presence of such an emission range from an inversion of the low-lying singlet states by solvent effects [183] to solvent-induced couplings of the states, where the N inversion acts

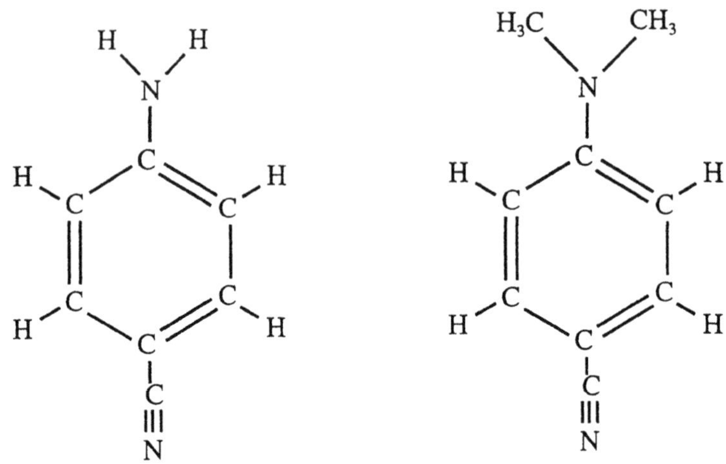

Fig. 2. p-Aminobenzonitrile (ABN) and P-Dimethylaminobenzonitrile (DMABN) structures.

as a promoting mode [192]. Strong arguments against a proposed proton transfer mechanism [193] or excimer [194] and exciplex [195, 196] formation with the solvent have been put forward [189, 190, 197, 198]. The most extended model to explain the dual fluorescence was suggested by Grabowski *et al.* [188, 199] and supports a "twisted intermolecular charge transfer" (TICT) state responsible for the lower-energy fluorescence band. The TICT model [186, 187] explains the dual fluorescence as occurring by means of a horizontal radiationless transition *i.e.*, from two conformations of the same molecule differing in the dihedral angle comprising the amino group and the phenyl ring. A large number of experimental facts support the TICT model. Compounds where the amino group rotation is inhibited by molecular bridging do not show the lower-energy CT emission [186],[200]-[202], while other systems with the planar conformation made unavailable by steric hindrance only show the CT band [186, 203, 204]. The internal charge transfer (ICT) reaction has been characterized in a large number of donor-acceptor systems. The photophysics and the general features of the phenomenon have been extensively reviewed [186, 187],[205]-[207] and recent works have been devoted to the dynamics of the ICT process [190]-[192],[208]-[213].

The theoretical studies on the key molecule (DMABN) are limited due to the size of the system and the complexity of the hypersurfaces. Semiempirical [214]-[221] and *ab initio* [216, 222] calculations have sketched the qualitative outlines of the problem. Most of them are in agreement with the main features of the TICT model. The DMABN and ABN (p-aminobenzonitrile) molecules have been recently studied [223] by using the CASSCF and CASPT2 theories. The ABN system was selected because it does not show any evidence of dual fluorescence even in strongly polar solvents [190]-[192]. The study begins with the optimization of the ground state of both molecules at the CASSCF level. The two low-lying singlet and triplet excited states were computed, together with the ground state, at the optimized ground state geometry and at the geometries resulting from the twisting of the amino group by 30, 60 and 90°. The calculations were repeated for two wagging angles, 0 and 21°, of the amino group in order to check the influence of the wagging motion along the twisting path. In addition, calculations on the three singlet states at a non-twisted geometry along the wagging coordinate, allow analysis of a suggested [192, 208] wagging mechanism for the ICT reaction. Basis sets of the ANO-S type contracted to $3s2p1d$ for N and C and $2s$ for H are used [36, 224]. The geometry optimizations were made with a DZP basis set [225]. The active space comprises the ten electrons and the nine orbitals of the π system of the benzonitrile plane, including the lone-pair electrons and orbital on the aniline-type nitrogen.

3.7.2 Nature of the low-lying excited states

In a simple model it can be assumed that the relative energies of the excited states are related to orbital SCF energy differences. The analysis of the orbitals shows that the most striking change in the orbital energies on going from the non-twisted to the 90° twisted conformation is the destabilization of the lone-pair N (lpN) orbital, while the HOMO (π_{ring}) and LUMO (π^*_{ring}) orbitals stabilize. It may be concluded that the most important change due to the twisting is on states with contributions of the lpN $\rightarrow \pi^*_{ring}$ excitation, *i.e.*, states of A symmetry. The structure of the CASSCF wave function for the low-lying singlet excited states shows that the lower state calculated at the ground state geometry (1^1B or S_1 in C_2 and C_1 symmetries, respectively) is a multiconfigurational state with dominant contributions from the HOMO $\rightarrow$ LUMO+1 and HOMO$-1 \rightarrow$ LUMO single excitations, and also with important contributions from doubly excited configurations. The second singlet excited state (2^1A or S_2 state) is dominated by a singly excited configuration: HOMO $\rightarrow$ LUMO. At the non-twisted geometries, the HOMO and LUMO orbitals are delocalized over the entire π system (benzonitrile moiety), and have important contributions of the lpN orbital, mainly at the totally planar (non-twisted and non-wagged) conformation. At the twisted geometries, the destabilization of the lpN orbital changes the nature of the 2^1A state, which now becomes a pure singly excited state described by the lpN $\rightarrow$ LUMO configuration. The state obtains a large dipole moment by the charge transfer from the localized lone-pair orbital to the benzonitrile moiety. The process is more important in DMABN than in ABN. It can be understood because the "pure" lpN $\rightarrow$ LUMO state at the non-twisted geometry is expected to have higher energy in ABN. As the size of the substituent chain on the nitrogen increases, the energy of the lpN $\rightarrow$ LUMO state is expected to decrease due to the increased polarizability of the ligands. Therefore, from a qualitative analysis, it can be concluded that the avoided crossing between the ^1A HOMO $\rightarrow$ LUMO and the ^1A lpN $\rightarrow$ LUMO states will determine the shape of the potential surfaces involved in the ICT process. When the energy gap between both states decreases at the non-twisted geometry, the crossing of the potential energy surfaces takes place at smaller twisting angles

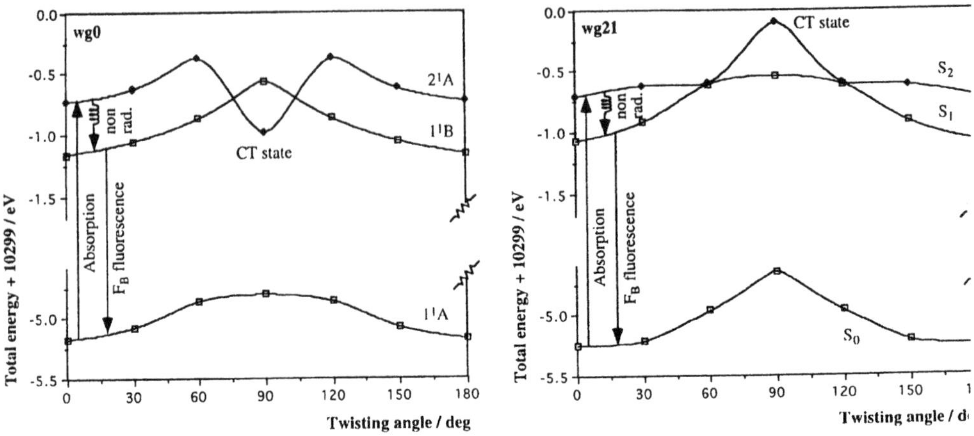

Fig. 3. Ground and low-lying singlet excited states potential curves in ABN along the twisting coordinate. The amino group is wagged by 0° (wg0) or 21° (wg21).

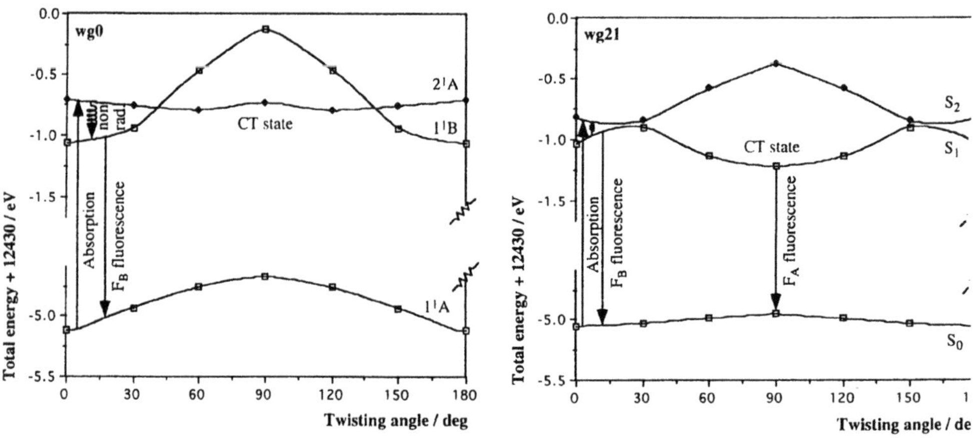

Fig. 4. Ground and low-lying singlet excited states potential curves in DMABN along the twisting coordinate. The dimethylamino group is wagged by 0° (wg0) or 21° (wg21).

(DMABN) leading to lower activation barriers for the process.

Figures 3 and 4 show the potential energy curves of the ground and low-lying singlet excited states of ABN and DMABN along the twisting path, both using a wagging angle of 0 (wg0) and 21° (wg21). A completely different behaviour is observed in the two molecules. In ABN all the states destabilize on going from the non-twisted to the twisted geometry, except the 2^1A state with a 0° wagging angle after 60° of twisting. The shape of the curves is related to the energy gaps among the states. The avoided crossing between the 2^1A and ^1A lpN → LUMO states occurs after 60° of twisting. Tables XX and XXI reflect the change in the nature of the computed 2^1A state, which after the strongly avoided crossing, becomes a charge transfer state with a large dipole moment. The isomerization barrier (around 0.36 eV) decreases somewhat at the 21° wagged structure, but the nature of the states does not change. The avoided crossing can be understood by inspection (at the twisted geometry) of the HOMO → LUMO state, which appears as the third state of ^1A symmetry with an energy of 5.81 eV and a dipole moment of 3.54 D.

In DMABN the singlet states exhibit a completely different behaviour. While the ground and 1^1B states also destabilize along the twisting coordinate, the 2^1A state stabilizes even near the non-twisted geometry. The stabilization can be related to a small energy gap, and therefore a stronger interaction, between the HOMO → LUMO and the lpN → LUMO states as was discussed above. The latter state strongly stabilizes along the twisting coordinate due to the localization of the lone-pair orbital, and the avoided crossing takes place in DMABN near a non-twisted geometry. Furthermore, the small gap between the 1^1B and 2^1A states also determine the shape of the potential curves. The states are closer in DMABN than in ABN. It means that (cf. Figure 4, wg21) they interact at smaller twisting angles. As the potential energy surface on the 21° wagged molecule has no symmetry (C_1 framework), the 1^1B and 2^1A states interact. They should be described as S_1 and S_2 and their nature does not keep the characteristics of the corresponding C_2 states. As Tables XX and XXI and Figure 4 (wg21) describe, the S_1 state in DMABN has 1^1B nature (low polarity and oscillator strengths) for less than 30° twisting angle. With a larger twist, the state progressively changes its nature to be the CT state observed at the perpendicular geometry (large polarity, charge transfer, and small oscillator strength due to the lack of overlap between the lpN orbital and the π system). The S_2 state has, near the ground state non-twisted geometries (absorption region) the nature of the vertical 2^1A state. Its large oscillator strength guarantees that most of the energy of the absorption will be carried by the S_2 state. After the 30° of twisting the S_2 state becomes the low polarity 1^1B state. Table XXI shows the effective charges computed for the different parts of the molecule. They illustrate the charge transfer process taking place in the CT state. The transfer of the charge is to some extent included in the S_2 state at the non-twisted geometries, +0.3e, and it strongly increases with the twisting to be +0.7e at the perpendicular geometry.

A final set of calculations analyzes the role of the inversion mode on the ICT process. It has been proposed [191, 192, 208] to be responsible for the CT state through pyramidalization of the amino nitrogen. We have computed the energies, oscillator strengths, dipole moments, and effective charges of the lowest singlet states of both ABN and DMABN at the non-twisted conformation along the wagging coordinate, *i.e.*, with several out-of-plane angles (from 0 to 45°) for the amino group [223]. To summarize briefly these results, no evidence of the strongly polar CT state responsible of the lower-energy fluorescence band could be found in this study. A dipole moment of 15–16 D has been measured for the CT emissive

TABLE XX

Computed energies (ΔE)[a], oscillator strengths[b] (f), and dipole moments[c] (μ) of the electronic states of ABN and DMABN at several geometries along the twisting coordinate (θ)[d].

	$\theta=0°$			$\theta=30°$			$\theta=60°$			$\theta=90°$		
State	ΔE	f	μ	ΔE	f	μ	ΔE	f	μ	ΔE	f	μ
	p-Aminobenzonitrile (ABN). 0° wagging angle.											
1^1A	–	-	6.89	–	-	6.58	–	-	5.83	–	-	5.5
1^1B	4.01	0.004	6.70	4.02	0.081	6.87	3.99	0.018	5.19	4.24	0.002	5.2
2^1A	4.44	0.361	12.42	4.45	0.324	12.83	4.49	0.099	13.01	3.82	0.000	15.7
1^3A	3.37	-	6.19	3.29	-	6.72	3.13	-	5.73	3.25	-	5.6
1^3B	3.88	-	7.74	4.04	-	7.26	4.20	-	5.35	4.11	-	5.8
	p-Aminobenzonitrile (ABN). 21° wagging angle.											
1^1A	–	-	6.30	–	-	6.23	–	-	5.27	–	-	4.8
1^1B	4.17	0.002	6.06	4.29	0.001	6.03	4.37	0.002	5.04	4.10	0.007	4.6
2^1A	4.54	0.358	12.08	4.59	0.331	12.00	4.35	0.207	12.81	4.54	0.000	15.6
1^3A	3.92	-	6.17	3.41	-	5.99	3.19	-	5.39	3.12	-	5.0
1^3B	3.93	-	6.89	4.16	-	6.47	4.23	-	5.43	3.96	-	5.2
	p-Dimethylaminobenzonitrile (DMABN). 0° wagging angle.											
1^1A	–	-	7.36	–	-	6.84	–	-	6.04	–	-	5.7
1^1B	4.05	0.010	7.58	3.99	0.004	6.99	4.30	0.001	6.13	4.55	0.003	5.7
2^1A	4.41	0.416	13.79	4.17	0.200	13.58	3.97	0.033	14.63	3.94	0.000	15.6
1^3A	3.66	-	6.84	3.58	-	6.90	3.64	-	6.00	3.65	-	5.5
1^3B	3.69	-	10.39	3.63	-	9.27	4.20	-	6.41	4.40	-	5.9
	p-Dimethylaminobenzonitrile (DMABN). 21° wagging angle											
1^1A	–	-	6.40	–	-	6.37	–	-	5.47	–	-	5.1
1^1B	4.02	0.003	6.33	4.18	0.004	6.35	4.40	0.003	5.40	4.58	0.006	4.8
2^1A	4.23	0.411	13.27	4.13	0.308	13.23	3.85	0.112	14.58	3.74	0.000	15.[illegible]
1^3A	3.50	-	6.33	3.63	-	6.25	3.85	-	5.44	3.92	-	5.1
1^3B	3.80	-	7.61	4.16	-	6.93	4.53	-	5.64	4.56	-	5.5

[a]Energies at the CASPT2 level (eV).
[b]Oscillator strengths computed with CASSCF transition moments and CASPT2 energies.
[c]Dipole moments at the CASSCF level (D).
[d]See Figures 3 and 4.

TABLE XXI

Effective charge distribution[a] in the ground (GS) and lowest electronic singlet and triplet excited states of ABN and DMABN at planar (θ=0°) and twisted (θ=90°) conformations.

Gr.	GS	1^1B	2^1A	1^3A	1^3B
		ABN. Non-twisted. Wag. 0°			
NH_2	-0.018	0.029	0.228	-0.023	0.088
Bz	0.196	0.151	0.006	0.209	0.043
CN	-0.178	-0.180	-0.234	-0.186	-0.131
		ABN. Non-twisted. Wag. 21°			
NH_2	-0.023	0.004	0.199	-0.012	0.045
Bz	0.199	0.176	0.048	0.184	0.107
CN	-0.175	-0.180	-0.247	-0.172	-0.152
		ABN. Twisted. Wag. 0°			
NH_2	-0.111	-0.123	0.692	-0.166	-0.150
Bz	0.277	0.305	-0.488	0.327	0.328
CN	-0.166	-0.182	-0.204	-0.161	-0.178
		ABN. Twisted. Wag. 21°			
NH_2	-0.207	-0.215	0.670	-0.151	-0.149
Bz	0.368	0.389	-0.473	0.311	0.318
CN	-0.161	-0.174	-0.197	-0.160	-0.169
		DMABN. Non-twisted. Wag. 0°			
NMe_2	-0.022	0.050	0.309	-0.019	0.203
Bz	0.207	0.129	-0.065	0.202	-0.042
CN	-0.185	-0.179	-0.244	-0.183	-0.161
		DMABN. Non-twisted. Wag. 21°			
NMe_2	-0.036	-0.007	0.256	-0.030	0.070
Bz	0.197	0.164	0.002	0.195	0.063
CN	-0.161	-0.157	-0.258	-0.165	-0.133
		DMABN. Twisted. Wag. 0°			
NMe_2	-0.087	-0.097	0.582	-0.098	-0.092
Bz	0.263	0.278	-0.346	0.272	0.287
CN	-0.176	-0.181	-0.236	-0.175	-0.195
		DMABN. Twisted. Wag. 21°			
NMe_2	-0.167	-0.177	0.660	-0.096	-0.093
Bz	0.317	0.339	-0.435	0.246	0.257
CN	-0.150	-0.162	-0.225	-0.150	-0.164

[a]CASSCF (10 electrons-9 orbitals) calculations.

state [186, 191]. Our calculations report for DMABN a dipole moment of 13.58 D for the 2^1A state at the ground state geometry (non-twisted and 10° wagged conformation) [223], while the maximum polarity for the state reached at the non-twisted conformation along the wagging coordinate is 13.79 D. Accordingly, the effective charges do not increase to larger values than +0.3e at the non-twisted geometries, while the results obtained at the twisted geometries are consistent with a state with strong charge transfer character (effective charges of +0.7e). Therefore, together with the observed unsensitivity of the polarity of the 2^1A state to the wagging mode as shown in Table XX, we have to conclude that, at least in DMABN, the twisting mechanism better describes the charge transfer than a pure wagging process.

As Table XX shows, the two-lowest triplet excited states do not give rise to any charge transfer process along the twisting coordinate, in agreement with experimental results [186, 187]. It has also been found that phosphorescence emission often occurs at energies above the TICT fluorescence [186]. The triplet TICT state corresponding to the singlet TICT state has been computed in DMABN at 4.03 eV, 0.3 eV above the singlet state and slightly, 0.1 eV, above the 1^3A state.

3.7.3 Mechanism for the Internal Charge Transfer (ICT) reaction

Table XXII summarizes the experimental and computed results on the absorption and lower-energy emission of ABN and DMABN. The suggested photophysical process leading to the dual fluorescence starts with the absorption of the energy at the ground state geometry. The most intense state is the S_2 (2^1A) state, computed in DMABN at 4.38 eV and an oscillator strength of 0.51. The S_1 (1^1B) state is computed at 4.19 eV with low intensity, 0.005. The S_2 state carries therefore most of the energy involved in the process. The polarization measurements made in different solvents [183, 188, 187] confirms two bands in the absorption spectrum, a weak low-lying B symmetry band, and a main A symmetry feature. From the absorption to the S_2 state, a non-radiative deactivation should be expected to the S_1 state, followed by a double process: the fluorescence from the LE state (S_1 state at small twisting angles) and the adiabatic photoreaction, via intramolecular twisting, to reach the CT state responsible of the lower-energy fluorescence. The character of the S_1 state progressively changes along the twisting from the initial LE state to the final CT state. Tables XX and XXI and Figure 4 (wg21) show the computed values for the properties of the low-lying singlet excited states of DMABN along the process. The shape of the S_1 potential energy curve is an approximation of the actual hypersurface, since the geometries of the excited states have not been fully optimized. The activation barrier expected for the ICT reaction from LE to CT states is, however, observed. Assuming the minimum of the 1^1B state (S_1 at small twisting angles) as the minimum of the LE emissive state, a barrier of 20 kJ/mol is obtained. The activation barrier of the ICT process has been measured to increase from 5 to 20 KJ/mol with decreasing solvent polarity [186, 191, 226]. The trend is related to the strong sensitivity of the CT fluorescence band to the polarity of the environment due to its large dipole moment, around 15–16 D [186, 191]. The band decreases its energy more than 1.5 eV on going from non-polar to polar solvents [191, 226].

TABLE XXII

Summary of experimental and computed excitation energies, oscillator strengths (f), and dipole moments (μ,D) of ABN and DMABN[a].

	Energies(eV)		f	μ (D)	
State	Exp[d]	Theo	Theo	Exp[e]	Theo
para-Aminobenzonitrile (ABN)[b,c]					
1^1A	-	-	-	6.6	6.30
1^1B	>4.0	4.17	0.002	-	6.06
2^1A	4.74	4.54(3.82)	0.357(0.0)	-	12.00(15.75)
para-Dimethylaminobenzonitrile (DMABN)[b,c]					
1^1A	-	-	-	6.6	7.24
1^1B	>4.0	4.19	0.005	10[g]	7.44
2^1A	4.42(3.22[f])	4.38(3.74)	0.510(0.0)	14(15)	13.58(15.45)

[a]Theoretical energies at the CASPT2 level, μ at the CASSCF level, and f computed using the CASSI approach and CASPT2 excitation energies.
[b]The values given first are for absorption. The theoretical geometries used are the ground state minimal energy structures: ABN, 0° twisted and 21° wagged and DMABN, 0° twisted and 10° wagged.
[c]The values in parentheses are for emission (F_A band). Minimal energy structures used for the CT state: ABN, 90° twisted and 0° wagged and DMABN, 90° twisted and 21° wagged.
[d]Values in n-Heptane. Band maxima. Ref. [192].
[e]Refs. [186] and [191].
[f]Values in the non-polar solvent cyclohexane. Ref. [191]. Around 3.8-3.9 eV has been measured in alkanes [226]. The gas-phase value, if were possible, should be expected at higher energy.
[g]See text for a discussion of this value.

Our computed dipole moment of 15.45 D and effective charge on N, +0.7e, are in agreement with this experimental result. The decrease of the activation barrier and the resulting occurrence of dual fluorescence in polar solvents are explained by the stabilization of the CT state. DMABN does not exhibit dual fluorescence in gas-phase, undoubtedly due to a too high activation barrier. The 1^1B state appears with low polarity, although our computed value is not strictly comparable with the measured dipole moment of 10 D [191] for the LE state, possibly since we have not computed the LE state energy minimum.

The ABN molecule does not exhibit dual fluorescence even in polar solvents [190]-[192]. The shape of the potential energy curves shown in Figure 3 suggest that the destabilization of the 2^1A state along the twisting path leads to a too high activation barrier for the system to proceed from the LE to the CT state.

3.7.4 *Summary and conclusions*

Three main electronic states are involved in shaping the low-lying excited state potential curves in ABN and DMABN. First, the energy gap between the ^1A HOMO $\rightarrow$ LUMO and ^1A lpN $\rightarrow$ LUMO states (larger in ABN) determines the behaviour of the 2^1A state along the twisting and causes the avoided crossing at larger (ABN) or smaller (DMABN) twisting angles (cf. Figures 1 and 2). On the other hand, the energy gap between the S_2 (2^1A) and S_1 (1^1B) states at the non-twisted geometry leads to higher (ABN) or lower (DMABN) energy barriers on the S_1 surface where the process from the LE to the CT state takes place. To summarize, it seems that a requirement for the occurrence of dual fluorescence and ICT in aminobenzonitriles and related molecules is a small energy gap between the interacting states, as has been suggested [192, 208], and that the isomerization leading to a twisted conformation of the amino group seems to better describe the charge transfer state responsible of the long-wavelength fluorescence appearing in solution.

3.8 THE NUCLEIC ACID BASE MONOMERS

3.8.1 *Introduction*

Detailed knowledge of the electronic structure of nucleic acid base monomers is important for the understanding of the static and dynamic properties of the genetic material. An example of practical use of these data, specifically the transition moment directions, is the determination of the G-C content of DNA based on the CD spectra. High-resolution vapor-phase spectra are not available for the constituents, and the measured spectra are dominated by broad, overlapping bands. Owing to this lack of resolution, individual band components cannot be unambiguously resolved. Because of that we have attempted for some time to gain insight into the spectral and structural features of the DNA/RNA base monomers.

The discussion of the electronic spectra of nucleic acid base monomers has to a large extent concentrated on the number of transitions and their polarization direction, e.g., there is a long-standing controversy concerning the existence and the origin of a band at 5.2 eV in cytosine. This discussion could possibly have been avoided with a common consistent interpretation of all data on the pyrimidine nucleotides at hand. To present such information is the main goal of this overview. The task is complicated by the fact that, up till now, no conclusive evidence has been reported for an $n \rightarrow \pi^*$ transition.

Over the years a wealth of information about the electronic spectra and transition moments of the constituents of the nucleic acids has been gathered. Both theoretical and experimental work has been carried out and was reviewed by Callis [227] in 1983. Most of the theoretical work has been based on semi-empirical calculations [228]-[239]. Within the last decade several *ab initio* quantum chemical studies were presented, which employed MCSCF, MRCI or RPA methods and different quality basis sets [240]-[244]. In spite of all this effort, the information at hand still remains incomplete and controversial. In this section we will present our latest results obtained using the CASSCF/CASPT2 approach.

The present calculations were performed with the averaged experimental heavy atom distances and angles reported by Taylor and Kennard. The N-H and C-H

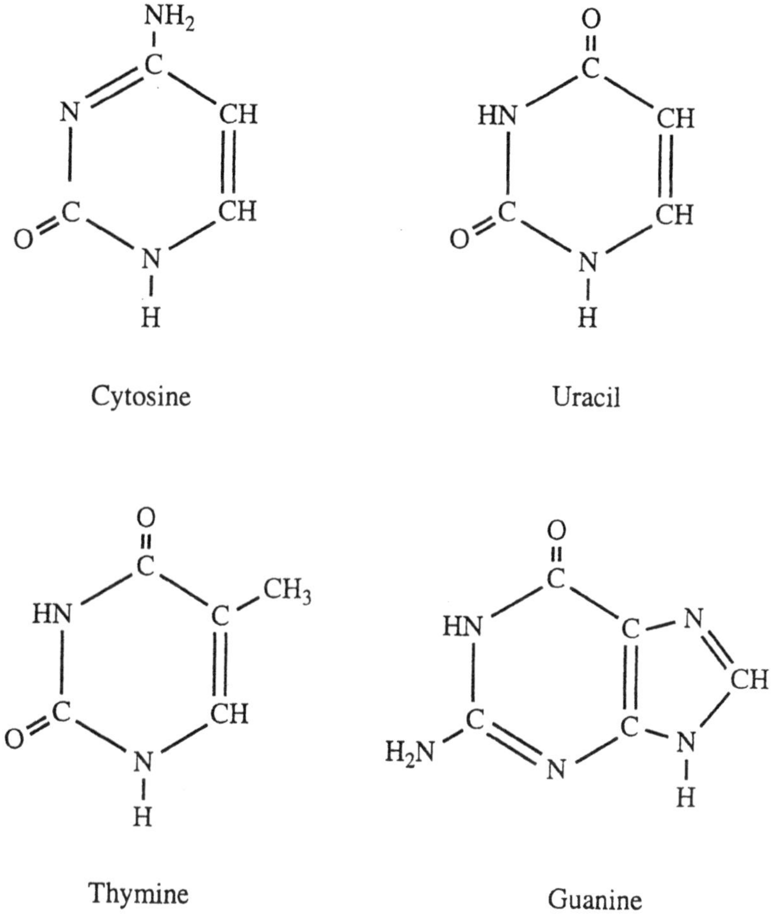

Fig. 5. Nucleic acid base monomer structures.

distances were assumed to be 1.01 and 1.08 Å, respectively, and we selected the tautomeric forms most abundant in aqueous solutions. Thymine is assumed to have the same geometry as uracil and the methyl group is attached to C_5 with a bond length of 1.53 Å. The structures for cytosine, uracil, thymine, and guanine are shown in figure 5. The ANO-L basis set, contracted C,N,O $4s3p1d$ and H $2s$ was used throughout together with one set of Rydberg functions optimized separately for each molecule. Because of the size of the nucleic acid molecules the following compromises were made: Preliminary CASSCF calculations correlating the 8 π-electrons highest in energy were used to obtain the Rydberg orbitals. The Rydberg orbitals were then deleted from the MO-space. To compute the

TABLE XXIII

Excitation energies, oscillator strengths (f,Θ), and dipole moments of singlet excited states of cytosine[a].

state	ΔE(CASSCF)	ΔE(CASPT2)	f	Θ	μ	experiment		
						ΔE	f	Θ
π–π^* transitions								
$2A'$	5.18	4.39	.06	61^0	1.86	4.6[b],4.6[c],4.6[d],4.6[e]	0.11[b],0.06[c],0.14[e]	$(6^0,54^0)$[e]
$3A'$	6.31	5.36	.11	-2^0	2.77	5.5[b],5.4[c],5.6[d],5.8[e]	0.07[c],0.13[e]	$(76^0,-17^0)$[e]
$4A'$	7.30	6.16	.86	-40^0	2.44	6.3[b],6.1[c],6.4[d],6.3[e]	0.68[b],0.115[c],0.36[e]	$(86^0,-27^0)$[e]
$5A'$	7.82	6.74	.15	15^0	3.68	6.7[c]	.07[c]	
σ–π^* transitions								
$1A''$	5.13	5.00	.005	–		5.0[d],5.3[e]	0.03[e]	

[a] Ground state properties: E(CASSCF)=-392.792572, μ=3.04, E(CASPT2)=-393.904916. The excitation energies and dipole moments are given in units of eV and au, respectively.

[b] Aquous solution spectra, Ref [303]

[c] VUV spectra of films absorbed onto a crystalline surface, Ref [252]

[d] Circular dichoism, Ref [248]

[e] Polarized absorption spectra of cytosine monohydrate crystals [249]

final $\pi \rightarrow \pi^*$ excited singlet state wave functions, the active space included 10 π-electrons and the dynamic correlation contributions were estimated using the PT2D approximation, only. For more details the interested reader is referred to the original publications [19, 245, 246].

In the following we first shall present and discuss our results on the pyrimidine nucleotides. The results are given in tabular form. They include the CASSCF and CASPT2 excitation energies in eV, the oscillator strength and the transition moment directions in the dipole approximation. Polarization angles are given according to the conventions defined by Hug and Tinoco[230]. In addition, the tables include the dipole moments (in atomic units). Our calculations on guanine are then reported in a separate paragraph, followed by a short summary. The corresponding calculations on adenine are not yet finished and will be reported later [246].

3.8.2 Cytosine, uracil and thymine

The three systems, cytosine, uracil and thymine, are isoelectronic and may be regarded as derivatives of pyrimidine. By comparison with the azabenzenes we might expect three to four $\pi \rightarrow \pi^*$ absorption bands at energies lower than 7 eV. For cytosine we may also expect a low energy $n \rightarrow \pi^*$ transition from the ring nitrogen located at energies somewhat larger than 4 eV. The promotion energy for the carbonyl oxygen electrons are more difficult to predict.

It is, however, also important to note that the pyrimidine nucleotides are 10 π-electron systems including eight centers and therefore it seems to be a better choice to select uracil as the parent compound. Thymine and cytosine can then be understood as weakly perturbed uracil systems and we might expect the absorption spectra to expose large similarities.

The number of electronic $\pi \rightarrow \pi^*$ transitions in the experimentally determined spectra is consistent with the expectations. There is general agreement concerning the first $\pi \rightarrow \pi^*$ absorption band. It is located for all three molecules at 4.5–4.6 eV. Common to cytosine, uracil and thymine is also two further $\pi \rightarrow \pi^*$ absorption maxima at 6.1–6.3 eV and 6.7–7.0 eV, respectively. Owing to the lack of resolution, the data concerning the gap between the first and third band is more scarce. For cytosine an additional band is seen at $\approx$5.4 eV by almost all spectroscopic methods, and at somewhat higher energies, $\approx$5.8 eV, for uracil and thymine.

To the authors knowledge three research groups suggested the existence of $n \rightarrow \pi^*$ transitions in the low energy region: Miles *et al.* [247, 248] and Źaloudek *et al.* [249] for cytosine at 5.0–5.3 eV, and Novoros and Clark [250] for uracil at 4.7 eV. However, Miles *et al.* [247, 248] found the Cotton effects not to be sensitive enough to unequivocally prove the existence. Źaloudek *et al.* observed a small component oriented perpendicular to the molecular plane, but preferred to assign it to a $\pi \rightarrow \pi^*$ transition. A selection of experimental data and references are included in the Tables XXIII-XXV, which also includes the results of the CASSCF/CASPT2 calculations.

Transition to the lowest $\pi \rightarrow \pi^*$ excited singlet state are computed to take place at 4.4 eV for cytosine, 4.8 eV for thymine and 5.0 eV for uracil. The calculations also predict an energy gap of $\approx$ 1 eV between the first and second singlet excited

TABLE XXIV

Excitation energies, oscillator strengths (f,Θ), and dipole moments of singlet excited states of uracil.[a]

state	ΔE(CASSCF)	ΔE(CASPT2)	f	Θ	μ	experiment		
						ΔE	f	Θ
π–π^* transitions								
$2A'$	6.91	4.97	.16	-2^0	2.36	$4.8^b,4.7^c,4.8^d,4.5^e$	$0.125^c,0.20^e$	$(7^0,-11^0)^d,(-9^0,16^0)^e$
$3A'$	7.04	5.79	.15	-24^0	1.13	$6.1\,^b,5.8^c,5.8^e$	0.114^c	$(-53^0,59^0)^e$
$4A'$	8.38	6.57	.19	62^0	2.73	6.26^c	0.165^c	
$5A'$	8.51	6.92	.76	-32^0	1.47	$6.8^b,7.0^f$		
σ–π^* transitions								
$1A''$	4.66	4.51	<.01	–	1.06	$4.9^c,4.7^e$		

[a] Ground state properties: E(CASSCF)=-412.675263, μ=1.72, E(CASPT2)=-413.839404. The excitation energies and dipole moments are given in units of eV and au, respectively. Transition moments and oscillator strengths are computed from a RASSI calculation including 12 states.

[b] Absorption spectra of aqueous solutions, Ref [304]

[c] Circular dichroism, Ref [305]

[d] Linear dichroism, Ref [258]

[e] Polarized absorption spectra of 1-methyluracil, Ref [250]

[f] Sublimed film spectra, Ref [251]

state. The excitation energies are 5.4 eV for cytosine and 5.8 eV for uracil and thymine. The third excited $\pi \rightarrow \pi^*$ singlet state is predicted to be located at 6.1 eV for cytosine and thymine. The computed energy for uracil is somewhat higher, but it is believed that this value is in error by ≈0.3 eV. In this case, the excitation energy appears to be very sensitive to the choice of the active space and is somewhat perturbed by an intruder state. The highest states included in the calculation are located at 6.7, 6.9 and 7.1 eV. The results also show that the $n \rightarrow \pi^*$ transitions lowest in energy appear at energies close to the first $\pi \rightarrow \pi^*$ excitation.

The computed $\pi \rightarrow \pi^*$ excitation energies are in accord with, e.g., the sublimed film data reported by Yamada and Fukutome [251] and Raksányi *et al.* [252]. They are also in agreement with the data extracted from linear dichroism measurements and the crystal absorption spectra using polarized light (for references refer to the tables). The alignment is somewhat more contradictory with CD and MCD data. These methods detect an electronic transition at energies ranging from 4.9 to 5.2 eV. However, if it is assumed to be an excitation of lone pair electrons — in principle, CD and MCD methods are capable of detecting these weak oscillators — most of the conflicts can be resolved if somewhat larger error bars, up to 0.5 eV, are accepted. Such large deviations can be accepted here since the measurements are typically done for the nucleosides and the spectra are affected by contributions from the sugar moieties.

Two methods have been applied to determine the polarization direction of the individual absorption bands: linear dichroism of the molecules dissolved in stretched polyvinyl films and polarization studies on single crystals. The most comprehensive experimental information on cytosine is given by Źaloudek *et al.* [249]. By combining polarized reflection spectra and stretched film dichroism data Źaloudek *et al.* arrived at the following polarization angles: $+6^0$, $+76^0$ and either -27^0 or $+86^0$ for the three lowest $\pi \rightarrow \pi^*$ bands, respectively. Our results are in conflict with this choice of angles, but agree well with the alternative directions reported by these authors. By comparison with a number of additional experimental data we estimate error bars of $\pm 20^0$ for the computed polarization directions.

Steward and Davidson [253] have studied the spectra of 1-methylthymine. Eaton and Lewis [254] as well as Novros and Clark [250] investigated the spectra of 1-methyluracil. These authors find the first transition to be polarized almost along the N1-C4 axis. They also note that the polarization is changed by $\approx 20^0$ compared to thymine. Assuming that the uncertainty in the calculated polarization angles is of the same order of magnitude as seen in the case of cytosine, then the predicted values are in good agreement with experiment. The data at hand, however, do not allow firm conclusions to be made about the sign of the polarization angle.

The effect of substituting a carbonyl group in uracil by an amino group is larger than we expected. The third excited state exhibits the strongest oscillator strength, whereas f is largest for the fourth transition in uracil and thymine. However, the associated polarization directions are almost constant, $\approx$-40^0 and the dipole moments are reduced relative to the ground state. This suggests that the fourth and fifth state of uracil have changed order in cytosine. There is also a marked difference in the dynamic correlation contributions. The excitation energy

TABLE XXV

Excitation energies, oscillator strengths (f,Θ), and dipole moments of singlet excited states of thymine[a].

state	ΔE(CASSCF)	ΔE(CASPT2)	f	Θ	μ	experiment		
						ΔE	f	Θ
$\pi-\pi^*$ transitions								
$2A'$	6.79	4.78	.17	18^0	2.59	4.5[b],4.6[c],4.7[d],4.5[e]	0.154[c]	$(51^0,-31^0)$[d],$(14^0,-19^0)$[e]
$3A'$	7.18	5.81	.17	-19^0	0.61	6.0[b],5.9[c],6.0[e]	0.206[c]	
$4A'$	8.30	6.12	.15	64^0	3.02	6.6[b],6.4[c]	0.162[c]	
$5A'$	8.58	7.10	.85	-31^0	1.19	7.0[f]		
$\sigma-\pi^*$ transitions								
$1A''$	4.77	4.67	<.01	–	1.11	4.9[c]		

Theory

[a] Ground state properties: E(CASSCF)=-451.721422, μ=1.97, E(CASPT2)=-453.0344769 The excitation energies and dipole moments are given in units of eV and au, respectively.

[b] Absorption spectra of aqueous solutions, Ref [306]

[c] circular dichroism, Ref [305]

[d] linear dichroism, Ref [258]

[e] polarized absorption spectra of 1-methylthymine, Ref [307]

[f] sublimed film spectra, Ref [251]

of the first excited state of cytosine is reasonably well described at the CASSCF level of approximation and differential correlation effects are not larger than 1.2 eV for any of the transitions considered here. In contrast, the CASSCF method overestimates the lowest $\pi \rightarrow \pi^*$ excitation energy of uracil by 2 eV.

It is also of interest to compare earlier theoretical studies with the present results. There have been a number of reported calculations of the electronic spectra of uracil, thymine and cytosine. The majority of these studies applied semi-empirical methods. In addition, Jensen *et al.* [241] carried out RPA calculations on cytosine using the Huzinaga/Dunning [151] basis sets. Recently, Petke *et al.* [244] reported very similar RPA calculations on cytosine and uracil but these authors also applied MRCI techniques. Matos and Roos [242] calculated the low-lying $\pi \rightarrow \pi^*$ state of cytosine using the CASSCF method. In common with the previous calculations they employed the Gaussian primitives determined by Huzinaga, but in order to achieve an optimal description of valence correlation effects they determined ANO-type contraction coefficients. Recently, Fülscher *et al.* [240] explored the spectra of the nucleic acid base monomers using the CASSCF/CASSI method and ANO-type basis sets derived from Duijneveldt's [255] primitive sets.

The present CASSCF calculations are in close agreement with those reported by Matos and Roos. It also has been shown by Fülscher and Roos [19] that the CASSCF/CASSI calculations performed by the same authors are hampered by basis set deficiencies causing, in particular, interaction with spurious states of Rydberg character. If the perturbing orbitals are removed from the MO basis sets the CASSCF results become virtually identical with the present data. The basic difference between the RPA and MRCI calculations and the CASSCF results is that the former predict a much larger number of excited states with transition energies lower than 7.0 eV. A careful analysis of the excited state properties reported by Petke *et al.* reveals that these calculations suffer from the same incompleteness of basis sets. Finally, Fülscher and Roos concluded that Petke's results calculated at the MRCI level of theory are consistent with the CASSCF prediction if all states with large second electric moment and small oscillator strength are discarded. Comparison to the RPA calculations are more difficult since this method does not allow the computation of state properties. Moreover, it is common knowledge that the RPA method yields good results for states that are dominated by a single excitation relative to the closed-shell single determinant reference state, but fails in most other instances and poorly accounts for differential correlation effects.

3.8.3 Guanine

Absorption spectra of guanine exhibit band maxima at approximately 4.5, 4.9 and 6.4 eV. CD and MCD spectra indicate that the 6.4 eV band is composite with a second transition being positioned at about 5.9 eV. Callis *et al.* [256] and Clark [257] have investigated the spectrum of single crystal 9-methyl guanine and guanine hydrochloride using reflectance techniques and obtained similar excitation energies. The polarization vectors of the first two transitions are approximately perpendicular to each other and the second transition is oriented nearly along the long molecular axis, -75^0. The polarization of the first band is generally accepted as being $+4^0$ in accord with results from linear dichroism studies [258, 259]. Un-

TABLE XXVI

Excitation energies, oscillator strengths (f,Θ), and dipole moments of singlet excited states of guanine[a].

state	ΔE(CASSCF)	ΔE(CASPT2)	f	Θ	μ	experiment		
						ΔE	f	Θ
$\pi-\pi^*$ transitions								
$2A'$	5.72	4.47	.20	-64^0	1.07	4.4[b],4.3[c],4.5[d,e],4.5[f]	.16[f]	$(4^0,-61^0)$[b],$(9^0,93^0)$[c],$(-4^0,35^0)$[f]
$3A'$	6.74	5.30	.09	52^0	2.72	5.0[b],5.0[c],5.0[d,e],4.9[f]	.25[f]	$(-88^0,31^0)$[b],$(95^0,3^0)$[c],(-75^0)[f]
$4A'$	7.18	5.63	.05	-90^0	3.10	6.1[d,e],6.1[f]	.41[f]	$(-71^0,-79^0)$[f]
$5A'$	8.45	6.83	.26	0^0	3.20	6.7[d,e],6.6[f]	.48[f]	$(-9^0,41^0)$[f]
$\sigma-\pi^*$ transitions								
$1A''$	4.83	4.81	<.01	-	1.80			

[a] ground state properties: E(CASSCF)=-539.607364,μ=2.67,E(CASPT2)=-541.227163
The excitation energies and dipole moments are given in units of eV and au, respectively.
[b] Linear dichroism of urcil dissolved in stretched poly-vinyl alcohol films, Ref [258]
[c] Linear dichroism of urcil dissolved in streched poly-vinyl alcohol films, Ref [259]
[d] CD spectrum of guanine 5'monophosphate, Ref [307]
[e] MCD spectrum of guanine 5'monophosphate, Ref [308]
[f] Solution absorbtion and polarized reflection spectra of 9-ethylguanine, Ref [257]

like the situation found for the pyrimidine systems, the experimental evidence is unusually clear. The calculated position (cf. Table XXVI) of the lowest energy band, 4.5 eV, is virtually identical to the experimental value. The transition energies of the second, third and fourth $\pi \rightarrow \pi^*$ singlet excited states are reproduced with an accuracy of $\approx$0.4 eV. The available experimental spectra are obtained for various molecular environments and a detailed comparison between the calculated and experimental oscillator strengths is not possible. The computed polarization directions disagree with the interpretation of single crystal experiments. The present theoretical results are, however, consistent with the second set of transition directions obtained by Matsuoka and Nordén: -61^0 and $+31^0$, respectively.

Two *ab initio* studies of the guanine spectrum have been presented. In particular, Fülscher *et al.* [240] reported CASSCF/CASSI studies of the guanine absorption spectrum and Petke *et al.* [243] carried out RPA and MRCI calculations. The calculated polarization directions from these studies are reasonably consistent giving values ranging from -60^0 to -30^0 for the first band and 30^0–70^0 for the second band in accord with the present results. A variety of semi-empirical calculations have also been published and the polarization angles are generally in agreement with the *ab initio* results [228]-[234]. It is difficult to have a decided conception about the source of the deviations of the calculated and observed polarization directions, and is is most likely due to accumulated effects of environmental perturbations and experimental limitations.

3.8.4 Summary

The present calculations on excited states of the nucleic acid base monomers have shown that it is possible to attack problems of such complexity using modern *ab initio* methods and correlating all valence electrons. Errors in the calculated excitation energies are found on the average to be 0.2 eV for the pyrimidine nucleotides and 0.4 eV for guanine. The computed polarization angles are, typically, uncertain within $\pm 20^0$. It should be recalled that several tautomeric forms are known for the nucleotides and, in addition, they may have different conformations in different environments. With this in mind, it is concluded that the agreement of the calculated and observed data is satisfactory.

It also appears that the current calculations are capable of providing a single consistent picture for the pyrimidine nucleotides. Specifically, we predict four $\pi \rightarrow \pi^*$ singlet excited states with transition energies smaller than 7.0 eV. In addition, the $n \rightarrow \pi^*$ transition lowest in energy is found to be close in energy to the first $\pi \rightarrow \pi^*$ transition at energies near 4.5 eV. As a consequence, the low energy features observed by mostly CD spectroscopy at $\approx$5.0 eV can be attributed to such a transition.

We close this chapter by quoting Professor Inga Fischer-Hjalmars [229]:

– *It may be presumptuous to suggest to experimentalists how to interpret their measurements. However, given the good agreement between calculations and observations as far as frequencies and intensities are concerned, it might be reasonable to pay some attention also to the calculated polarization directions, in particular for stronger transitions where most of the single excitations that contribute to the CI-sum will cooperate. For weak transitions, composed of differences between single excitations, the theoretical results might be more doubtful.*

4 Final remarks

A number of applications in molecular spectroscopy has been used to illustrate the behaviour of the CASSCF/CASPT2 approach for theoretical studies in this field. Altogether about 50 different molecular systems have to date been studied with the new method, ranging in size from ethene to porphine. The overall accuracy of the approach is surprisingly high. In most cases the excitation energies are within 0.2 eV of the experimental estimates. A few exceptions exist and they are explained by difficulties in defining the CASSCF reference function due to valence Rydberg mixing. It must be remembered, however, that the comparison with experiment is not straightforward. It is not possible to specify with high precision the value for the vertical excitation energy from the experimental spectrum, especially for symmetry forbidden transitions. The bands are often broad and diffuse and overlap each other. Also, for some systems gas-phase spectra are not available, and the effect of a solvent or a crystal environment has to be taken into account.

However, in spite of these difficulties theory is of great value in advancing the understanding of the experimental data. It should be strongly emphasized that its value depends entirely on the ability of the theory to give quantitatively accurate results. It is important that the states are obtained in correct order with reasonably correct character (Rydberg or valence) and intensities. If these conditions are not fulfilled, the theoretical data are of little value in aiding the interpretation of experiment. The same is true when the theory is used to study photochemical processes. The importance of predicting accurately the relative energies of different potential surfaces was nicely illustrated in the study of ABN and DMABN.

A method which is to be used in spectroscopy and photochemistry must also be general. It must be capable of treating all kinds of excited states with the same accuracy. The CASSCF/CASPT2 approach has this quality. The importance of this was nicely illustrated in the applications to the linear polyenes and the pentadienes. One of the major questions in the spectroscopy of these compounds is the relative location of the two lowest excited states, where one of them has a substantial doubly excited character. This situation is obviously very common, and methods that cannot handle such states with the same accuracy as singly excited states are therefore of limited interest in practical applications. It has been recently stated [260] that the CI singles (CIS) method [6] is "becoming widely used". This development is very unfortunate. The CIS method suffers from all the problems mentioned above: The electronic states are predicted in the wrong order; states not well described as singly excited are missing from the spectrum; etc. Attempts to correct these deficiencies using perturbation theory do not seem very successful.

The same criticism actually holds for all methods that are based on a single Hartree-Fock like reference configuration. The coupled-cluster techniques are of this type. Many excited states have a truly multiconfigurational structure. Trying to describe such situations starting from a single configuration must be very difficult. A converged solution will require at least triples in the cluster expansion, which will be very difficult to achieve for larger molecules.

We have in the present review reported on results obtained for small and medium sized molecules (the largest being porphin with 34 atoms). The potential for extending the applications to even larger systems is very high. One needs to rewrite the CASSCF and CASPT2 programs using direct schemes, which is quite feasible. The only real limitation of the method is the size of the active space. This limitation is, however, not linked directly to the size of the molecule. There are immediate plans to apply the approach in studies of excited states in amino acids and small peptides.

5 Acknowledgements

The research reported in this communication has been supported by a grant from the Swedish Natural Science Research Council (NFR), by IBM Sweden under a joint study contract, by the Cooperación Científico-Técnica (Ministerio de Asuntos Exteriores) of Spain, and by the DGICYT project PB91-0634.

References

1. R. J. Buenker, J. L. Whitten, and J. D. Petke. *J. Chem. Phys.*, **49**, 2261, (1968).
2. S. D. Peyerimhoff and R. J. Buenker. *Theor. Chim. Acta*, **19**, 1, (1970).
3. M. Goeppert-Mayer and A. L. Sklar. *J. Chem. Phys.*, **6**, 645, (1938).
4. J. M. O. Matos, B. O. Roos, and P.-Å. Malmqvist. *J. Chem. Phys.*, **86**, 1458, (1987).
5. P.-Å. Malmqvist and B. O. Roos. *Theor. Chim. Acta*, **83**, 191, (1992).
6. J. B. Foresman, M. Head-Gordon, J. A. Pople, and M. J. Frisch. *J. Phys. Chem.*, **96**, 135, (1992).
7. K. B. Wiberg, C. M. Hadad, G. B. Ellison, and J. B. Foresman. *J. Phys. Chem.*, **97**, 13586, (1993).
8. P.-Å. Malmqvist, B. O. Roos, M. P. Fülscher, and A. Rendell. *Chem. Phys.*, **162**, 359, (1992).
9. B. O. Roos. *Adv. Chem. Phys.*, **69**, 399, (1987).
10. K. Andersson, P.-Å. Malmqvist, B. O. Roos, A. J. Sadlej, and K. Wolinski. *J. Phys. Chem.*, **94**, 5483, (1990).
11. K. Andersson, P.-Å. Malmqvist, and B. O. Roos. *J. Chem. Phys.*, **96**, 1218, (1992).
12. K. Andersson and B. O. Roos. Multiconfigurational second-order perturbation theory. In R. Yarkony, editor, *Modern electron structure theory*, volume 1. World Scientific Publishing, New York (1994).
13. K. Andersson and B. O. Roos. *Int. J. Quantum Chem.*, **45**, 591, (1993).
14. R. González-Luque, M. Merchán, and B. O. Roos. *Chem. Phys.*, **171**, 107, (1993).
15. R. González-Luque, M. Merchán, P. Borowski, and B. O. Roos. *Theor. Chim. Acta*, **86**, 467, (1993).
16. R. Pou-Amérigo, M. Merchán, I. Nebot-Gil, and B. O. Roos. *J. Chem. Phys.*, submitted (1994).
17. K. Pierloot and B. O. Roos. *Inorg. Chem.*, **31**, 5353, (1992).
18. M. P. Fülscher, K. Andersson, and B. O. Roos. *J. Phys. Chem.*, **96**, 9204, (1992).
19. M. P. Fülscher and B. O. Roos. *to be published*, (1994).
20. B. O. Roos, K. Andersson, and M. P. Fülscher. *Chem. Phys. Letters*, **192**, 5, (1992).
21. B. O. Roos, L. Serrano-Andrés, and M. Merchán. *Pure & Appl. Chem.*, **65**, 1693, (1993).
22. B. O. Roos, M. Merchán, R. McDiarmid, and X. Xing. *J. Am. Chem. Soc.*, in press, (1994).
23. M. Rubio, M. Merchán, E. Ortí, and B. O. Roos. *Chem. Phys.*, **179**, 395, (1994).
24. L. Serrano-Andrés, M. Merchán, I. Nebot-Gil, R. Lindh, and B. O. Roos. *J. Chem. Phys.*, **98**, 3151, (1993).
25. L. Serrano-Andrés, M. Merchán, I. Nebot-Gil, B. O. Roos, and M. P. Fülscher. *J. Am. Chem. Soc.*, **115**, 6184, (1993).
26. L. Serrano-Andrés, R. Lindh, B. O. Roos, and M. Merchán. *J. Phys. Chem.*, **97**, 9360, (1993).
27. L. Serrano-Andrés, M. Merchán, M. P. Fülscher, and B. O. Roos. *Chem. Phys. Letters*, **211**, 125, (1993).
28. L. Serrano-Andrés, B. O. Roos, and M. Merchán. *Theor. Chim. Acta*, **87**, 387, (1994).
29. K. Andersson and B. O. Roos. *Chem. Phys. Letters*, **191**, 507, (1992).
30. K. Pierloot, E. Van Praet, L. G. Vanquickenborne, and B. O. Roos. *J. Phys. Chem.*, **97**, 12220, (1993).
31. K. Pierloot, E. Tsokos, and B. O. Roos. *Chem. Phys. Letters*, **214**, 583, (1993).
32. B. J. Persson, B. O. Roos, and K. Pierloot. *J. Chem. Phys.*, **xx**, xxxx, (1994).
33. K. Andersson, M. P. Fülscher, R. Lindh, P.-Å. Malmqvist, J. Olsen, B. O. Roos, A. J. Sadlej, and P.-O. Widmark. *MOLCAS Version 2 User's Guide*. Dept. of Theor. Chem., Chem. Center, Univ. of Lund, Lund, (1991).
34. M. P. Fülscher and B. O. Roos. *Theor. Chim. Acta*, **87**, 403, (1994).
35. P.-O. Widmark, P.-Å. Malmqvist, and B. O. Roos. *Theor. Chim. Acta*, **77**, 291, (1990).
36. K. Pierloot B. Dumez, P.-O. Widmark, and B. O. Roos. Medium size ANO basis sets for the atoms H-Kr. To be published (1994).
37. M. P. Fülscher, J. Lorentzon, P.-Å. Malmqvist, and B. O. Roos. A consistent method to construct basis sets for Rydberg orbitals. To be published, (1994).
38. K. Kaufmann, W. Baumeister, and M. Jungen. *J. Phys. B: At. Mol. Opt. Phys.*, **22**, 2223, (1989).
39. P.-Å. Malmqvist. *Int. J. Quantum Chem.*, **30**, 479, (1986).

40. P. Å. Malmqvist and B. O. Roos. *Chem. Phys. Letters*, **155**, 189, (1989).
41. G. Karlström and P.-Å. Malmqvist. *J. Chem. Phys.*, **96**, 6115, (1992).
42. K. W. Hauser and R. Kuhn. *Z. Phys. Chem. Alt.*, **B29**, 363, (1935).
43. S. Basu. In P. O. Löwdin, editor, *Adv. Quantum Chem.*, volume 11, page 33, Academic Press, New York (1978).
44. M. Ottolenghi. In J. N. Pitts, editor, *Adv. Photochemistry*, volume 12, page 97, Interscience, New York (1980).
45. C. K. Chiang, C. R. Fincher, Y. W. Park, A. J. Heeger, H. Shirakawa, E. J. Louis, S. C. Gau, and A. G. McDiarmid. *Phys. Rev. Lett.*, **39**, 1098, (1977).
46. M. Kertész. In P. O. Löwdin, editor, *Adv. Quantum Chem.*, volume 15, page 161, Academic Press, New York (1982).
47. K. Tanaka and T. Yamabe. In P. O. Löwdin, editor, *Adv. Quantum Chem.*, volume 17, page 251, Academic Press, New York (1985).
48. A. J. Heeger, S. Kivelson, J. R. Schriefer, and W. P. Su. *Rev. Mod. Phys.*, **60**, 781, (1988).
49. B. S. Hudson and B. E. Kohler. *Chem. Phys. Lett.*, **14**, 299, (1972).
50. B. S. Hudson and B. E. Kohler. *J. Chem. Phys.*, **59**, 4984, (1973).
51. W. G. Bouwman, A. C. Jones, D. Phillips, P. Thibodeau, Ch. Friel, and R. L. Christensen. *J. Phys. Chem.*, **94**, 7429, (1990).
52. H. Petek, A. J. Bell, Y. S. Choi, K. Yoshihara, B. A. Toungue, and R. L. Christensen. *J. Chem. Phys.*, **98**, 3777, (1993).
53. B. S. Hudson and B. E. Kohler. *Annu. Rev. Phys. Chem.*, **24**, 437, (1974).
54. B. S. Hudson, B. E. Kohler, and K. Schulten. In E. C. Lim, editor, *Excited States*, volume 6, page 1, Academic Press, New York (1982).
55. G. Orlandi, F. Zerbetto, and M. Z. Zgierski. *Chem. Rev.*, **91**, 867, (1991).
56. J. E. Lennard-Jones. *Proc. Roy. Soc.*, **A158**, 280, (1937).
57. N. S. Bayliss. *J. Chem. Phys.*, **16**, 287, (1948).
58. K. Schulten and M. Karplus. *Chem. Phys. Lett.*, **14**, 305, (1972).
59. R. Pariser. *J. Chem. Phys.*, **24**, 250, (1956).
60. R. Lindh and B. O. Roos. *Int. J. Quantum Chem.*, **35**, 813, (1989).
61. E. H. van Veen. *Chem. Phys. Lett.*, **41**, 540, (1976).
62. W. M. Flicker, O. A. Mosher, and A. Kuppermann. *Chem. Phys. Lett.*, **36**, 56, (1975).
63. M. Olivucci, I. N. Ragazos, F. Bernardi, and M. A. Robb. *J. Am. Chem. Soc.*, **115**, 3710, (1993).
64. L. Serrano-Andrés, M. Merchán, and B. O. Roos. *to be published*, (1994).
65. R. L. Graham and K. F. Freed. *J. Chem. Phys.*, **96**, 1304, (1992).
66. M. E. Squillacote, R. S. Sheridan, O. L. Chapman, and F. A. L. Anet. *J. Am. Chem. Soc.*, **101**, 3657, (1979).
67. R. R. Chadwick, D. P. Gerrity, and B. S. Hudson. *Chem. Phys. Lett.*, **115**, 24, (1985).
68. R. R. Chadwick, M. Z. Zgierski, and B. S. Hudson. *J. Phys. Chem.*, **95**, 7204, (1991).
69. D. L. Phillips, M. Z. Zgierski, and A. B. Myers. *J. Phys. Chem.*, **97**, 1800, (1993).
70. J. P. Doering and R. McDiarmid. *J. Chem. Phys.*, **73**, 3617, (1980).
71. R. McDiarmid. *Chem. Phys. Lett.*, **188**, 423, (1992).
72. R. J. Hemley, A. C. Lasaga, V. Vaida, and M. Karplus. *J. Phys. Chem.*, **92**, 945, (1988).
73. W. J. Buma, B. E. Kohler, and K. Song. *J. Chem. Phys.*, **92**, 4622, (1990).
74. W. J. Buma, B. E. Kohler, and K. Song. *J. Chem. Phys.*, **94**, 6367, (1991).
75. H. Petek, A. J. Bell, R. L Christensen, and K. Yoshihara. *J. Chem. Phys.*, **96**, 2412, (1992).
76. C. Westerfield and A. B. Myers. *Chem. Phys. Lett.*, **202**, 409, (1993).
77. W. M. Flicker, O. A. Mosher, and A. Kuppermann. *Chem. Phys. Lett.*, **45**, 492, (1977).
78. R. M. Gavin, S. Risemberg, and S. A. Rice. *J. Chem. Phys.*, **58**, 3160, (1973).
79. R. M. Gavin and S. A. Rice. *J. Chem. Phys.*, **60**, 3231, (1974).
80. D. G. Leopold, V. Vaida, and H. F. Granville. *J. Phys. Chem.*, **81**, 4210, (1984).
81. T. Fujii, A. Kamata, M. Shimizu, Y. Adachi, and S. Maeda. *Chem. Phys. Lett.*, **115**, 369, (1985).
82. R. M. Gavin, C. Weisman, J. K. McVey, and S. A. Rice. *J. Chem. Phys.*, **68**, 522, (1978).
83. L. A. Heimbrook, B. E. Kohler, and I. J. Levy. *J. Chem. Phys.*, **81**, 1592, (1984).
84. M. Allan, L. Neuhaus, and E. Haselbach. *Helv. Chim. Acta*, **67**, 1776, (1984).
85. M. F. Granville, G. R. Holton, B. E. Kohler, R. L. Christensen, and K. L. D'Amico. *J. Chem. Phys.*, **70**, 593, (1979).
86. K. L. D'Amico, C. Manos, and R. L. Christensen. *J. Am. Chem. Soc.*, **102**, 1777, (1980).
87. M. F. Granville, G. R. Holton, and B. E. Kohler. *J. Chem. Phys.*, **72**, 4671, (1980).
88. R. H. Baughman, B. E. Kohler, I. J. Levy, and C. Spangler. *Synth. Met.*, **11**, 37, (1985).

89. R. Hoffmann, E. Heilbronner, and R. Gleiter. *J. Am. Chem. Soc.*, **92**, 706, (1970).
90. R. Hoffmann. *Acc. Chem. Res.*, **4**, 1, (1971).
91. R. McDiarmid, A. Sabljic, and J.P. Doering. *J. Chem. Phys.*, **83**, 2147, (1985).
92. R. McDiarmid and A. Sabljic. *J. Phys. Chem.*, **95**, 6455, (1991).
93. A. Sabljic, R. McDiarmid, and A. Gedanken. *J. Phys. Chem.*, **96**, 2442, (1992).
94. R.P. Frueholz, W.M. Flicker, O.A. Mosher, and A. Kuppermann. *J. Chem. Phys.*, **70**, 2003, (1979).
95. D.F. Evans. *J. Chem. Soc.*, page 1735, (1960).
96. H. Nakatsuji, O. Kitao, and T. Yonezawa. *J. Chem. Phys.*, **83**, 723, (1985).
97. V. Galasso. *Chem. Phys.*, **153**, 13, (1991).
98. M.Z. Zgierski and F. Zerbetto. *J. Chem. Phys.*, **99**, 3721, (1993).
99. G. Scheibe and H. Grieneisen. *Z. Phys. Chem.*, **B25**, 52, (1934).
100. S. Menczel. *Z. Phys. Chem.*, **125**, 161, (1927).
101. L.W. Pickett, M.E. Corning, G.M. Wieder, D.A. Semenov, and J.M. Buckley. *J. Am. Chem. Soc.*, **75**, 1618, (1953).
102. G. Horváth and Á.I. Kiss. *Spectrochim. Acta*, **23A**, 921, (1967).
103. P.J. Derrick, L. Åsbrink, O. Edqvist, B.Ö. Jonsson, and E. Lindholm. *Int. J. Mass Spectrom. Ion Phys.*, **6**, 191, (1971).
104. P.A. Mullen and M.K. Orloff. *J. Chem. Phys.*, **51**, 2276, (1969).
105. M. Bavia, F. Bertinelli, C. Taliani, and C. Zauli. *Mol. Phys.*, **31**, 479, (1976).
106. E.H. Van Veen. *Chem. Phys. Letters*, **41**, 535, (1976).
107. W.M. Flicker, O.A. Mosher, and A. Kuppermann. *J. Chem. Phys.*, **64**, 1315, (1976).
108. W.M. Flicker, O.A. Mosher, and A. Kuppermann. *Chem. Phys. Letters*, **38**, 489, (1976).
109. W. Butscher and K.-H. Thunemann. *Chem. Phys. Letters*, **57**, 224, (1978).
110. D.C. Rawlings and E.R. Davidson. *Chem. Phys. Letters*, **98**, 424, (1983).
111. W.C. Price and A.D. Walsh. *Proc. R. Soc. London Ser. A*, **179**, 201, (1941).
112. M.B. Robin. *Higher Excited States in Polyatomic Molecules*, volume 2. Academic Press, New York, (1975).
113. L.W. Pickett. *J. Chem. Phys.*, **8**, 293, (1940).
114. L.W. Pickett, N.J. Hoeflich, and T.C. Liu. *J. Am. Chem. Soc.*, **73**, 4865, (1951).
115. K. Watanabe and T. Nakayama. *J. Chem. Phys.*, **29**, 48, (1958).
116. P.J. Derrick, L. Åsbrink, O. Edqvist, B.Ö. Jonsson, and E. Lindholm. *Int. J. Mass Spectrom. Ion Phys.*, **6**, 161, (1971).
117. K.-H. Thunemann, R.J. Buenker, and W. Butscher. *Chem. Phys.*, **47**, 313, (1980).
118. B. Ellis and P.J.F. Griffiths. *Spectrochim. Acta*, **21**, 1881, (1965).
119. A. Weissberger, E.C. Taylor, and S. Gronowitz, editors. *Thiophene and its derivatives*, volume 3 and 44 of *The Chemistry of Heterocyclic Compounds*. Interscience, New York, (1952/1992).
120. W.C. Price and A.D. Walsh. *Quart. Rev. London*, **2**, 73, (1943).
121. E. Jones and I.M. Moodie. *Tetrahedron*, **21**, 2413, (1965).
122. G. Di Lonardo, G. Galloni, A. Trombetti, and C. Zauli. *J. Chem. Soc. Faraday Trans. II*, **68**, 2009, (1972).
123. G. Varsányi, L. Nyulászy, T. Veszprémi, and T. Narisawa. *J. Chem. Soc. Perkin Trans. II*, page 761, (1982).
124. L. Nyulászy and T. Veszprémi. *Chem. Scripta*, **26**, 629, (1986).
125. L. Nyulászy and T. Veszprémi. *J. Mol. Struc.*, **140**, 253, (1986).
126. L. Nyulászy and T. Veszprémi. *Chem. Scripta*, **28**, 331, (1988).
127. J.P. Doering and R. McDiarmid. *J. Chem. Phys.*, **75**, 87, (1981).
128. M.Z. Zgierski and F. Zerbetto. *J. Chem. Phys.*, **98**, 14, (1993).
129. P. J. Hay and I. Shavitt. *Chem. Phys. Letters*, **22**, 33, (1973).
130. P. Rancurel, B. Huron, L. Praud, J. P. Malrieu, and G. Berthier. *J. Mol. Spectrosc.*, **60**, 259, (1976).
131. M. H. Palmer and I. C. Walker. *Chem. Phys.*, **133**, 113, (1989).
132. O. Kitao and H. Nakatsuji. *J. Chem. Phys.*, **87**, 1169, (1987).
133. J. Lorentzson, P.-Å Malmqvist, M. P. Fülscher, and B. O. Roos. *to be published*, (1994).
134. A. Hiraya and K. Shobatake. *J. Chem. Phys.*, **94**, 7700, (1991).
135. E. N. Lassettre, A. Skerbele, M. A. Dillon, and K. J. Ross. *J. Chem. Phys.*, **48**, 5066, (1967).
136. N. Nakashima, H. Inoue, M. Sumitani, and K. Yoshihara. *J. Chem. Phys.*, **73**, 5620, (1980).
137. K. K. Innes, I. G. Ross, and W. R. Moomaw. *J. Mol. Spectrosc.*, **132**, 492, (1988).

138. A. Bolovinos, P. Tsekeris, J. Pjillis, E. Pantos, and G.Andritsopoulus. *J. Mol. Spectrosc.*, **103**, 240, (1984).
139. I. C. Walker, M. H. Palmer, and A. Hopkirk. *Chem. Phys.*, **141**, 365, (1989).
140. M. H. Palmer, I. C. Walker, M. F. Guest, and A. Hopkirk. *Chem. Phys.*, **147**, 19, (1990).
141. I. C. Walker and M. H. Palmer. *Chem. Phys.*, **153**, 169, (1991).
142. M. H. Palmer and I. C. Walker. *Chem. Phys.*, **157**, 187, (1991).
143. H. Hackmeyer and J. L. Whitten. *J. Chem. Phys.*, **54**, 3739, (1971).
144. W. R. Wadt, W. A. Goddard, and T. H. Dunning. *J. Chem. Phys.*, **65**, 438, (1976).
145. G. Portalone, G. Schulz, A Domenicano, and I. Hargittai. *Chem. Phys. Letters*, **197**, 482, (1992).
146. T. C. Wong and L. S. Bartell. *J. Chem. Phys.*, **61**, 2840, (1974).
147. O. Kitao and H. Nakatsuji. *J. Chem. Phys.*, **88**, 4913, (1988).
148. L. Goodman. *J. Mol. Spectrosc.*, **6**, 109, (1961).
149. E. Villa, A. Amirav, and E. C. Lim. *J. Chem. Phys.*, **92**, 5393, (1988).
150. J. P. Doering and J. H. Moore *jr.*. *J. Chem. Phys.*, **56**, 2176, (1972).
151. T. H. Dunning jr. *J. Chem. Phys.*, **53**, 2823, (1970).
152. J. Waluk, H.-P. Klein, A. J. Ashe III, and J. Michl. *Organometallics*, **8**, 2804, (1989).
153. K. Kimura and S. Nagakura. *Mol. Phys.*, **9**, 117, (1965).
154. S. J. Martinez III, J. C. Alfano, and D. H. Levy. *J. Mol. Spectrosc.*, **152**, 80, (1992).
155. J. K. Rice and R. W. Anderson. *J. Am. Chem. Soc.*, **90**, 6793, (1986).
156. J. Petruska. *J. Chem. Phys.*, **34**, 1120, (1961).
157. M. Merchán, E. Ortí, and B.O Roos. To be published, (1994).
158. J. Almlöf. *Int. J. Quantum Chem.*, **8**, 915, (1974).
159. D. Spangler, G.M. Maggiora, L.L. Shipman, and R.E. Christoffersen. *J. Am. Chem. Soc.*, **99**, 7470, (1977).
160. J.D. Petke, G.M. Maggiora, L.L. Shipman, and R.E. Christoffersen. *J. Mol. Spectr.*, **71**, 64, (1978).
161. R.E. Christoffersen. *Int. J. Quantum Chem.*, **26**, 573, (1979).
162. D.C. Rawlings, E.R. Davidson, and M. Gouterman. *Int. J. Quantum Chem.*, **26**, 251, (1984).
163. U. Nagashima, T. Takada, and K. Ohno. *J. Chem. Phys.*, **85**, 4524, (1986).
164. A. Ghosh, J. Almlöf, and P.G. Gassman. *Chem. Phys. Letters*, **186**, 113, (1991).
165. P.G. Gassman, A. Ghosh, and J. Almlöf. *J. Am. Chem. Soc.*, **114**, 9990, (1992).
166. X.-Y. Li and M.Z. Zgierski. *J. Phys. Chem.*, **95**, 4268, (1991).
167. Y. Yamamoto, T. Noro, and K. Ohno. *Int. J. Quantum Chem.*, **42**, 1563, (1992).
168. A. Ghosh and J. Almlöf. *Chem. Phys. Letters*, **213**, 519, (1993).
169. J. Almlöf, T.H. Fischer, P.G. Gassman, A. Ghosh, and M. Häser. *J. Phys. Chem.*, **97**, 10964, (1993).
170. M. Merchán, E. Ortí, and B. O. Roos. *Chem. Phys. Letters*, in press (1994).
171. C. Rimington, S.F. Mason, and O. Kennard. *Spectrochim. Acta*, **12**, 65, (1958).
172. B.G. Anex and R.S. Umans. *J. Am. Chem. Soc.*, **86**, 5026, (1964).
173. L. Edwards, D.H. Dolphin, and M. Gouterman. *J. Mol. Spectr.*, **35**, 90, (1970).
174. L. Edwards, D.H. Dolphin, M. Gouterman, and A.D. Adler. *J. Mol. Spectr.*, **38**, 16, (1971).
175. B.F. Kim and J. Bohandy. *J. Mol. Spectr.*, **73**, 332, (1978).
176. J. Dalton, L.R. Milgrom, and S.M. Pemberton. *J.C.S. Perkin II*, page 370, (1980).
177. M. Gouterman. *J. Chem. Phys.*, **30**, 1139, (1959).
178. M. Gouterman, G. Wagniere, and L.C. Snyder. *J. Mol. Spectr.*, **11**, 108, (1963).
179. C. Weiss, H. Kobayashi, and M. Gouterman. *J. Mol. Spectr.*, **16**, 415, (1965).
180. W.J. Hehre, R.F. Stewart, and J.A. Pople. *J. Chem. Phys.*, **51**, 2657, (1969).
181. J. Olsen, B.O. Roos, P. Jørgensen, and H.J.Aa. Jensen. *J. Chem. Phys.*, **89**, 2185, (1988).
182. P.-Å. Malmqvist, A. Rendell, and B. O. Roos. *J. Phys. Chem.*, **94**, 5477, (1990).
183. E. Lippert, W. Lüder, F. Moll, W. Nägele, H. Boos, H. Prigge, and I. Seibold-Blankenstein. *Angew. Chem.*, **78**, 125, (1961).
184. E. Lippert, W. Lüder, and H. Boos. In A. Magnini, editor, *Advances in Molecular Spectroscopy*, page 443, Oxford, (1962). Pergamon Press.
185. C. J. Seliskar, O. S. Khalil, and S. P. McGlynn. In E. C. Lim, editor, *Excited states*, volume 1, page 231, New York, (1974). Academic Press.
186. Z. R. Grabowski, K. Rotkiewicz, A. Siemiarczuk, D. J. Cowley, and W. Baumann. *Nouv. J. Chim.*, **3**, 443, (1979).
187. E. Lippert, W. Rettig, V. Bonačić-Koutecký, F. Heisel, and J. A. Miehé. *Adv. Chem. Phys.*, **68**, 1, (1987).

188. K. Rotkiewicz, K. H. Grellman, and Z. R. Grabowski. *Chem. Phys. Letters*, **19**, 315, (1973).
189. C. Rullière, Z. R. Grabowski, and J. Dobkowski. *Chem. Phys. Letters*, **137**, 498, (1987).
190. U. Leinhos, W. Kühnle, and K. A. Zachariasse. *J. Phys. Chem.*, **95**, 2013, (1991).
191. W. Schuddenboom, S. A. Jonker, J. M. Warman, U. Leinhos, W. Kühnle, and K. A. Zachariasse. *J. Phys. Chem.*, **96**, 10809, (1992).
192. K. A. Zachariasse, Th. von der Haar, A. Hebecker, U. Leinhos, and W. Kühnle. *Pure & Appl. Chem.*, **65**, 1745, (1993).
193. E. M. Kosower and H. Dodiuk. *J. Am. Chem. Soc.*, **98**, 924, (1976).
194. O. S. Khalil, R. H. Hofeldt, and S. P. McGlynn. *Chem. Phys. Letters*, **17**, 479, (1972).
195. R. J. Visser and C. A. Varma. *J. Chem. Soc. Faraday Trans. II*, **76**, 453, (1980).
196. M. C. C. Delange, D. T. Leeson, K. A. B. Vankuijk, A. H. Huizer, and C. A. Varma. *Chem. Phys.*, **174**, 425, (1993).
197. T. Kobayashi, M. Futakami, and O. Kajimoto. *Chem. Phys. Letters*, **130**, 63, (1986).
198. O. Kajimoto, M. Futakami, T. Kobayashi, and K. Yamasaki. *J. Chem. Phys.*, **92**, 1347, (1988).
199. A. Siemiarczuk, Z. R. Grabowski, A. Krowczynski, M. Asher, and M. Ottolenghi. *Chem. Phys. Letters*, **51**, 315, (1977).
200. K. Rotkiewicz, Z. R. Grabowski, A. Krowczyński, and W. Kühnle. *J. Lumin.*, **12–13**, 877, (1976).
201. G. Wermuth and W. Rettig. *J. Phys. Chem.*, **88**, 2729, (1984).
202. W. Rettig and R. Gleiter. *J. Phys. Chem.*, **89**, 4676, (1985).
203. W. Rettig and F. Marchner. *Nouv. J. Chim.*, **7**, 425, (1983).
204. H. Bischof, W. Baumann, N. Detzer, and K. Rotkiewicz. *Chem. Phys. Letters*, **116**, 180, (1985).
205. W. Rettig. *Angew. Chem. Int. Ed. Engl.*, **25**, 971, (1986).
206. V. Bonacić-Koutecký, J. Koutecký, and J. Michl. *Angew. Chem. Int. Ed. Engl.*, **26**, 170, (1987).
207. K. Bhattacharyya and M. Chowdhury. *Chem. Rev.*, **93**, 507, (1993).
208. K. A. Zachariasse, Th. von der Haar, A. Hebecker, U. Leinhos, and W. Kühnle. In *XVIth International Conference on Photochemistry*, Vancouver, August 1993.
209. T. Okada, N. Mataga, W. Baumann, and A. Siemiarczuk. *J. Phys. Chem.*, **91**, 4490, (1987).
210. J. Hicks, M. Vandersall, Z. Babarogic, and K. B. Eisenthal. *Chem. Phys. Letters*, **116**, 18, (1985).
211. J. Hicks, M. Vandersall, E. V. Sitzmann, and K. B. Eisenthal. *Chem. Phys. Letters*, **135**, 413, (1987).
212. J. D. Simon and S.-G. Su. *J. Phys. Chem.*, **94**, 3656, (1990).
213. O. Kajimoto, T. Nayuki, and T. Kobayashi. *Chem. Phys. Letters*, **209**, 357, (1993).
214. O. S. Khalil, J. L. Meeks, and S. P. McGlynn. *Chem. Phys. Letters*, **39**, 457, (1976).
215. D. J. Cowley and A. M. Peoples. *J. Chem. Soc. Chem. Comm.*, 352, (1977).
216. W. Rettig and V. Bonacić-Koutecký. *Chem. Phys. Letters*, **62**, 115, (1979).
217. F. D. Lewis and B. Holman III. *J. Phys. Chem.*, **84**, 2326, (1980).
218. J. Lipinski, H. Chojnacki, Z. R. Grabowski, and K. Rotkiewicz. *Chem. Phys. Letters*, **70**, 449, (1980).
219. J. P. LaFemina, C. B. Duke, and A. Paton. *J. Chem. Phys.*, **87**, 2151, (1987).
220. C. Cazeau-Dubroca, S. Ait Lyazidi, P. Cambou, A. Periqua, Ph. Cazeau, and M. Pesquer. *J. Phys. Chem.*, **93**, 2347, (1989).
221. D. Majumdar, R. Sen, K. Bhattacharyya, and P. Bhattacharyya. *J. Phys. Chem.*, **95**, 4324, (1991).
222. S. Kato and Y. Amatatsu. *J. Chem. Phys.*, **92**, 7241, (1990).
223. L. Serrano-Andrés, M. Merchán, B. O. Roos, and R. Lindh. *J. Am. Chem. Soc.*, submitted (1994).
224. R. Pou-Amérigo basis set for H. *Private communication.*
225. T. H. Dunning and P. J. Hay. In *Modern Theoretical Chemistry*, page 1, Plenum Press, New York (1976).
226. K. A. Zachariasse, private communication.
227. P. R. Callis. *Ann. Rev. Phys. Chem.*, **34**, 329, (1983).
228. P. R. Callis. *Photochemistry and Photobiology*, **3**, 315, (1986).
229. I. Fischer-Hjalmars and A. Henriksson-Enflo. *Int. J. Quantum Chem.*, **37**, 517, (1990).
230. W. Hug and I. Tinoco jr. *J. Am. Chem. Soc.*, **95**, 2803, (1973).
231. Y. J. l'Haya H. Ito. *Bull. Chem. Soc. Jpn*, **49**, 3466, (1976).
232. V. I. Danilov, V. I. Pechenaya, and N. V. Zheltovsky. *Int. J. Quantum Chem.*, **17**, 307, (1980).

233. V. A. Kuprievich and O. Shramko. *Int. J. Quantum Chem.*, **16**, 833, (1979).
234. S. K. Srivastava and P. C. Mishra. *Int. J. Quantum Chem.*, **18**, 827, (1980).
235. S. K. Srivastava and P. C. Mishra. *Int. J. Quantum Chem.*, **16**, 1051, (1979).
236. H. Berthod, C. Giessner-Prettre, and A. Pullman. *Int. J. Quantum Chem.*, **1**, 123, (1967).
237. H. Fugita, A. Imamura, and C. Nagata. *Bull. Chem. Soc. Jpn*, **41**, 2017, (1968).
238. J. C. Packer, J. S. Avery, J. Ladik, and G. Biczo. *Int. J. Quantum Chem.*, **3**, 79, (1969).
239. M. L. Bailey. *Theor. Chim. Acta*, **16**, 309, (1970).
240. M. P. Fülscher, P.-Å. Malmqvist, and B. O. Roos. Ab initio quantum chemical calculations of excitation energies and transition moments for the nucleic acid base monomers. In J.R. Lakowicz, editor, *Time-resolved Laser Spectr. in Biochem. II*, page 1204. Proc. SPIE, (1990).
241. H. J. Aa. Jensen, H. Koch, P. Jorgensen, and J. Olsen. *Chem. Phys.*, **119**, 297, (1988).
242. J. M. Matos and B. O. Roos. *J. Am. Chem. Soc.*, **110**, 7664, (1988).
243. J. D. Petke, G. M. Maggiora, and R. E. Christoffersen. *J. Am. Chem. Soc.*, **112**, 5452, (1990).
244. J. D. Petke, G. M. Maggiora, and R. E. Christoffersen. *J. Phys. Chem.*, **96**, 6992, (1992).
245. J. Lorentzson, M. P. Fülscher, and B. O. Roos. *To be published*, (1994).
246. J. Lorentzson, M. P. Fülscher, and B. O. Roos. *To be published*, (1994).
247. D. W. Miles, M. J. Robins, R. K. Robins, M. W. Winkley, and H. Eyring. *J. Am. Chem. Soc.*, **91**, 825, (1969).
248. D. W. Miles, M. J. Robins, R. K. Robins, M. W. Winkley, and H. Eyring. *J. Am. Chem. Soc.*, **91**, 831, (1969).
249. F. Źaloudek, J. S. Novros, and L. B. Clark. *J. Am. Chem. Soc.*, **107**, 7344, (1985).
250. J. S Novros and L. B. Clark. *J. Phys. Chem.*, **90**, 5666, (1986).
251. T. Yamada and H. Fukutome. *Biopolymers*, **6**, 43, (1968).
252. K. Raksányi, I. Földváry, J. Fidy, and L. Kittler. *Biopolymers*, **17**, 887, (1978).
253. R. F. Stewart and N. Davidson. *J. Chem. Phys.*, **39**, 255, (1963).
254. W. A. Eaton and T. P. Lewis. *J. Chem. Phys.*, **53**, 2164, (1970).
255. F. B. van Duijneveldt. Gaussian basis sets for the atoms H–Ne for use in molecular calculations. Technical Report RJ945, IBM Research, (1970).
256. P. R. Callis, B. Fanconi, and W. T. Simpson. *J. Am. Chem. Soc.*, **93**, 6679, (1971).
257. L. B. Clark. *J. Am. Chem. Soc.*, **99**, 3934, (1977).
258. Y. Matsuoka and B. Nordén. *J. Phys. Chem.*, **86**, 1378, (1982).
259. A. F. Fucaloro and L. S. Forster. *J. Am. Chem. Soc.*, **93**, 6443, (1971).
260. M. Head-Gordon, R. J. Rico, M. Oumi, and T. Lee. *Chem. Phys. Letters*, **219**, 21, (1994).
261. C. Petrongolo, R. J. Buenker, and S. D. Peyerimhoff. *J. Chem. Phys.*, **76**, 3655, (1982).
262. L. E. McMurchie and E. R. Davidson. *J. Chem. Phys.*, **67**, 5613, (1977).
263. R. S. Mulliken. *J. Chem. Phys.*, **66**, 2448, (1977).
264. D. G. Wilden and J. Comer. *J. Phys. B*, **13**, 1009, (1980).
265. R. McDiarmid. *J. Phys. Chem.*, **84**, 64, (1980).
266. A. J. Merer and R. S. Mulliken. *Chem. Rev.*, **69**, 639, (1969).
267. A. Gedanken, N. A. Kuebler, and M. B. Robin. *J. Chem. Phys.*, **76**, 46, (1982).
268. B. A. Williams and T. A. Cool. *J. Chem. Phys.*, **94**, 6358, (1991).
269. K. E. Johnson, D. B. Johnston, and S. Libski. *J. Chem. Phys.*, **70**, 3844, (1979).
270. P. A. Snyder, P. N. Schatz, and E. W. Rowe. *Chem. Phys. Lett.*, **110**, 508, (1984).
271. O. Kitao and H. Nakatsuji. *Chem. Phys. Lett.*, **143**, 528, (1988).
272. L. Serrano-Andrés, J. Sánchez-Marín, and I. Nebot-Gil. *J. Chem. Phys.*, **98**, 7499, (1992).
273. O. A. Mosher, W. M. Flicker, and A. Kuppermann. *Chem. Phys. Lett.*, **19**, 332, (1973).
274. O. A. Mosher, W. M. Flicker, and A. Kuppermann. *J. Chem. Phys.*, **59**, 6502, (1973).
275. W. M. Flicker, O. A. Mosher, and A. Kuppermann. *Chem. Phys.*, **30**, 307, (1978).
276. R. McDiarmid. *Chem. Phys. Lett.*, **34**, 130, (1975).
277. R. McDiarmid. *J. Chem. Phys.*, **64**, 514, (1976).
278. J. P. Doering and R. McDiarmid. *J. Chem. Phys.*, **75**, 2477, (1981).
279. M. A. Nascimento and W. A. Goddard III. *Chem. Phys. Lett.*, **60**, 197, (1979).
280. M. A. Nascimento and W. A. Goddard III. *Chem. Phys.*, **53**, 265, (1980).
281. R. J. Cave and E. R. Davidson. *Chem. Phys. Lett.*, **148**, 190, (1988).
282. R. J. Cave and E. R. Davidson. *J. Phys. Chem.*, **92**, 614, (1988).
283. J. P. Doering, A. Sabljic, and R. McDiarmid. *J. Phys. Chem.*, **88**, 835, (1984).
284. A. Sabljic and R. McDiarmid. *J. Chem. Phys.*, **82**, 2559, (1985).
285. D. E. Post, W. M. Hetherington, and B. S. Hudson. *Chem. Phys. Lett.*, **35**, 259, (1975).
286. M. Aoyagi, I. Ohmine, and B. E. Kohler. *J. Phys. Chem.*, **94**, 3922, (1990).

287. R. J. Cave and E. R. Davidson. *J. Phys. Chem.*, **92**, 2173, (1988).
288. M. Said, D. Maynau, and J. P. Malrieu. *J. Am. Chem. Soc.*, **106**, 580, (1984).
289. L. A. Heimbrook, J. E. Kenny, B. E. Kohler, and G. W. Scott. *J. Chem. Phys.*, **75**, 4338, (1981).
290. K. Tanaka, T. Nomura, T. Noro, H. Tatewaki, T. Takada, H. Kashiwagi, F. Sasaki, and K. Ohno. *J. Chem. Phys.*, **67**, 5738, (1977).
291. R. Håkansson, B. Nordén, and E.N. Thulstrup. *Chem. Phys. Letters*, **50**, 305, (1977).
292. P.J. Derrick, L. Åsbrink, O. Edqvist, B.Ö. Jonsson, and E. Lindholm. *Int. J. Mass Spectrom. Ion Phys.*, **6**, 177, (1971).
293. G.L. Bendazzoli, F. Bertinelli, P. Palmieri A. Brillante, and C. Taliani. *J. Chem. Phys.*, **69**, 5077, (1978).
294. R.P. Frueholz, W.M. Flicker, O.A. Mosher, and A. Kuppermann. *J. Chem. Phys.*, **70**, 1986, (1979).
295. M. Allan. *J. Electron Spectr. Relat. Phenom.*, **48**, 219, (1989).
296. M.B. Robin and N.A. Kuebler. *J. Chem. Phys.*, **44**, 2664, (1966).
297. M.B. Robin and N.A. Kuebler. *J. Mol. Spectr.*, **33**, 274, (1970).
298. R. L. Whetten, K.-J. Fu, and E. R. Grant. *J. Chem. Phys.*, **79**, 2620, (1983).
299. R. L. Whetten, S. G. Grubb, C. E. Otis, A. C. Albrecht, and E. R. Grant. *J. Chem. Phys.*, **82**, 1115, (1985).
300. S. G. Grubb, R. L. Whetten, A. C. Albrecht, and E. R. Grant. *Chem. Phys. Letters*, **108**, 420, (1984).
301. P. M. Johnson. *J. Chem. Phys.*, **64**, 4143, (1976).
302. P. M. Johnson and G. M. Korenowski. *Chem. Phys. Letters*, **97**, 53, (1983).
303. H. Morita and S. Nagakura. *Theor. Chim. Acta*, **11**, 279, (1968).
304. L. B. Clark and I. Tinoco Jr. *J. Am. Chem. Soc.*, **87**, 11, (1965).
305. W. C. Brunner and M. F. Maestre. *Biopolymers*, **14**, 555, (1975).
306. R. A. Cox, D. Voet, W. B. Gratzer and P. Doty. *Biopolymers*, **1**, 193, (1963).
307. C. A. Sprecher and W. C. Johnson jr. *Biopolymers*, **16**, 2243, (1977).
308. J. C. Sutherland and K. Griffin. *Biopolymers*, **23**, 2715, (1984).

Index

I

J

K

L

M

N

O

P

W

X

Y

Z

Understanding Chemical Reactivity

1. Z. Slanina: *Contemporary Theory of Chemical Isomerism.* 1986 ISBN 90-277-1707-9
2. G. Náray-Szabó, P. R. Surján, J. G. Angyán: *Applied Quantum Chemistry.* 1987 ISBN 90-277-1901-2
3. V. I. Minkin, L. P. Olekhnovich and Yu. A. Zhdanov: *Molecular Design of Tautomeric Compounds.* 1988 ISBN 90-277-2478-4
4. E. S. Kryachko and E. V. Ludeña: *Energy Density Functional Theory of Many-Electron Systems.* 1990 ISBN 0-7923-0641-4
5. P. G. Mezey (ed.): *New Developments in Molecular Chirality.* 1991 ISBN 0-7923-1021-7
6. F. Ruette (ed.): *Quantum Chemistry Approaches to Chemisorption and Heterogeneous Catalysis.* 1992 ISBN 0-7923-1543-X
7. J. D. Simon (ed.): *Ultrafast Dynamics of Chemical Systems.* 1994 ISBN 0-7923-2489-7
8. R. Tycko (ed.): *Nuclear Magnetic Resonance Probes of Molecular Dynamics.* 1994 ISBN 0-7923-2795-0
9. D. Bonchev and O. Mekenyan (eds.): *Graph Theoretical Approaches to Chemical Reactivity.* 1994 ISBN 0-7923-2837-X
10. R. Kapral and K. Showalter (eds.): *Chemical Waves and Patterns.* 1995 ISBN 0-7923-2899-X
11. P. Talkner and P. Hänggi (eds.): *New Trends in Kramers' Reaction Rate Theory.* 1995 ISBN 0-7923-2940-6
12. D. Ellis (ed.): *Density Functional Theory of Molecules, Clusters, and Solids.* 1995 ISBN 0-7923-3083-8
13. S.R. Langhoff (ed.): *Quantum Mechanical Electronic Structure Calculations with Chemical Accuracy.* 1995 ISBN 0-7923-3264-4

Kluwer Academic Publishers – Dordrecht / Boston / London

GPSR Compliance
The European Union's (EU) General Product Safety Regulation (GPSR) is a set of rules that requires consumer products to be safe and our obligations to ensure this.

If you have any concerns about our products, you can contact us on

ProductSafety@springernature.com

In case Publisher is established outside the EU, the EU authorized representative is:

Springer Nature Customer Service Center GmbH
Europaplatz 3
69115 Heidelberg, Germany

www.ingramcontent.com/pod-product-compliance
Ingram Content Group UK Ltd.
Pitfield, Milton Keynes, MK11 3LW, UK
UKHW021504200726
13877UKWH00027B/482
9789401101943